Natural History Collections

More information about this series at http://www.springer.com/series/15333

Lothar A. Beck
Editor

Zoological Collections of Germany

The Animal Kingdom in its Amazing Plenty at Museums and Universities

Springer

Editor
Lothar A. Beck
Department of Biology, Zoological Collection
Philipps-University of Marburg
Marburg, Germany

Series Editors
Lothar A. Beck
Department of Biology, Zoological Collection
Philipps-University of Marburg
Marburg, Germany

Hans-Dieter Sues
Department of Paleobiology
National Museum of Natural History
Washington, DC
USA

ISSN 2510-1862 ISSN 2510-1870 (electronic)
Natural History Collections
ISBN 978-3-319-44319-5 ISBN 978-3-319-44321-8 (eBook)
DOI 10.1007/978-3-319-44321-8

Library of Congress Control Number: 2017939361

Printed on acid-free paper

This Springer imprint is published by Springer Nature
The registered company is Springer International Publishing AG
The registered company address is: Gewerbestrasse 11, 6330 Cham, Switzerland

Contents

List of Figures

Part I
General Aspects

Chapter 1
In Past and Recent Times: On the Significance of Zoological Collections in Germany—An Introduction

Lothar A. Beck

Abstract A short historical view on the development of zoological collections in Germany from the cabinets of wonder in the past to the recent scientific museums and modern collections point to the great importance of these collections for research and education. Zoological collections have been and will be basic especially in the fields of biodiversity, ecology and evolution. They also are a reflection of advancement of science theory and methodology, national and international networks as well as of the biographies of important zoologists. That's what this compilation of most of the German zoological collections is aimed at accordance with recent recommendations of the science council and associations of museums.

Keywords History of science • Taxonomical research • Biodiversity • Digitalisation • Environmental education

1.1 Past Times

It was during the time of Carl von Linné and Georges Cuvier—before the great evolutionary biologists Charles Darwin and Alfred Wallace were born and before Jean Baptiste Lamarck wrote his "Philosophie Zoologique"—that the first zoological collections were established. These early collections still had a long way to go before they could gain their current-day significance in the research on biodiversity, ecology and, global climate change and their importance for evolutionary biology, taxonomy, systematics, and environmental education.

For more than 250 years, zoological collections in Germany have played an important scientific role in universities and in natural history museums. Following the zeitgeist in Europe at the time, "Scientific Museums" were founded in many German cities, inspired by the "cabinets of curiosities" and "cabinets of wonder" of

L.A. Beck (✉)
Department of Biology, Animal Evolution, Systematics and Didactics of Biology, Zoological Collection of the Philipps-University Marburg, Karl-von-Frisch-Str. 8, 35032 Marburg, Germany
e-mail: beck@staff.uni-marburg.de

L.A. Beck (ed.), *Zoological Collections of Germany*, Natural History Collections,
DOI 10.1007/978-3-319-44321-8_1

the sixteenth, seventeenth and eighteenth centuries, which had been established mostly by the nobility, well-travelled merchants, or scholars. Along with objects that gave some scientific insight (models, experiments), they included various plant specimens and, first and foremost, animal specimens, arranged according to mere aesthetic aspects (Davenne and Fleurent 2012; Mauriès 2002).

After the Linnaean system of binary nomenclature had gained recognition in the eighteenth century, collecting activities concentrated on illustrating the diversity of species in our world while trying to acquire specimens of particularly unusual objects (such as a kiwi, a kakapo, and a duck-billed platypus, which are still part of some of the older collections), collecting, sorting, and conserving became the principal task of a collection.

During the nineteenth and twentieth centuries, the number of land, freshwater, and marine faunistic objects increased considerably, a development that can be attributed to the general practice of placing reference specimens for the redefining of animal species in publicly accessible institutes. These extremely numerous objects (in Germany alone more than 100 million, despite wartime losses) became the foundation of many zoological collections.

This book's articles show the highs and lows in the public's appreciation of the developing philosophies of science and the compartmentalising disciplines of zoology and their respective protagonists. The articles also discuss the influence World Wars I and II (cf. Griffin 1946) and the division of Germany (1948–1990) had on the collections. In the historical paragraphs, many collectors, donors, and scholars are mentioned. The data in these articles are a treasure trove for biographical research and the advancement of science theory in a national and international context.

During the second half of the twentieth century, the significance of museums and collections waned while new fields of biological research (e.g. animal physiology, neurobiology, cytology, and molecular biology and genetics) emerged. In fact, some collections were dissolved and their objects sold, many disappeared into basements, attics, or back rooms, as space was needed for modern scientific methods. Cutbacks in personnel also played a part. Despite improved identification skills, universities downsized research and courses in systematics, ecology, distribution, and benefit of organisms. All this was stigmatised as outdated. Descriptive biology and research on the conservation of animal species and their function in the ecosystem found some refuge in nature conservation, but, for the most part, it was relegated to regionality and was not very theory oriented.

1.2 Recent Times

In the 1980s, museum associations had already drawn attention to the constant loss of personnel and resources they had to suffer and also to their dwindling significance (Meyer 1983; Deutscher Museumsbund 1984).

Since the conference in Rio (UNCED 1992) and prompted by factors like methodological progress, scientific need, and insight, it had become obvious that biodiversity benefits humankind (Wilson 1992) and that a reversion to natural resources would inevitably require research in zoological museums and collections. These collections are far more than just "cabinets of wonder". They are guardians of scientific evidence and the findings of generations of scientists. They are places where objects are being collected, documented, interpreted, and presented to the public (Kress 2014, Nature Editorial 2014). The objects are studied, and the evidence, e.g. of taxonomical research (type material), is stored, frequently in the form of DNA bar codes.

Research on biodiversity would hardly be possible without material from collections: modern research focuses on the origin of biodiversity (genetics, phylogenetics, taxonomy; cf. Raupach and Knebelsberger 2015), the structure of regional and global biodiversity, ecological features and characteristics of species and their integration into the ecosystem, quantitative and qualitative distribution patterns, and, finally, the sustainable use of biodiversity by humans. All this can be facilitated by digitalisation of the objects, as the emergence of national and international data banks shows, but there are financial and practical limits.

Animals are three dimensional; they are anatomical objects and can be examined by electron microscopy, laser scans, X-ray, tomography, etc. They have a qualitative tissue structure, contain useful DNA, and are a physicochemical treasure trove.

Research requires authentic and originally conserved material. Therefore, the objects should not be disposed of after they have been digitalised. Who can foretell the next-generation methods, some of which might depend on conserved object material? The widespread cross-linking to medical science, forensics, palaeontology, botany, geology, or climate research all requires the original object. Serendipity is a well-known phenomenon in research.

Furthermore, museums and zoological collections are stepping up their efforts and activities in the field of teaching and publicity. This important function had its beginnings in the nineteenth century, mainly at the universities, as discussed in many articles in this book. Objects are an integral part of many courses, especially in the field of biodiversity and environmental education. Teacher training plays a vital role, as millions of school children are among the visitors to museums and collections.

Publicity has now received a considerable boost in many establishments, owing to exhibitions, which are educationally well presented and, at the same time, inspiring, beautiful, and true to life. They aim at the intensification of emotional ties to nature, but also at "infotainment", a trend that makes larger museums seem like tourist attractions.

Associations of museums and university collections, data portals, and recommendations of the science council (Wissenschaftsrat 2011) have supported these new trends in Germany. Suitable programmes sponsored by grants from DFG, BMBF, and VolkswagenStiftung are already in effect. This book is concerned with both the above-mentioned trends and the recommendations of the science council. Where possible, up-to-date information of most of the zoological collections in Germany has been compiled and placed at national and global disposal.

Many museums and universities have published descriptions of their own collections, most of them in German; they also issue their own publication series, keep detailed home pages, and are present in social media. A widening of the current spectrum and an updating of information in English seemed only sensible.

References

Davenne C, Fleurent C (2012) Cabinets of wonder. Abrams, New York

Deutscher Museumsbund (1984) Memorandum Universitätsmuseen. Museumskunde 1984(49): 107–108

Griffin FJ (1946) Present state of some German museums. Nature 157:631–633

Kress WJ (2014) Valuing collections. Science 346(6215)

Mauriès P (2002) Cabinets of curiosities. Thames & Hudson, London

Meyer KO (1983) Zur Lage der zoologischen Universitätssammlungen. Museumskunde 1983(48): 113–119

Nature Editorial (2014) Save the museums. Nature 515:311–312

Raupach MJ, Knebelsberger T (2015) Von wegen "antiquiert": Taxonomie im Wandel. Biol unserer Zeit (BIUZ) 5(45):315–321

UNCED United Nations Conference on Environment and Development (1992) United Nations Convention on Biological Diversity

Wilson EO (1992) The diversity of life. Harvard University Press, Cambridge, MA

Wissenschaftsrat (2011) Empfehlungen zu wissenschaftlichen Sammlungen als Forschungsinfrastrukturen. Berlin 28 Jan 2011, Drs. 10464-11

Chapter 2
Germany's Zoological Collections: An International and Personal View on an Important Historical and Contemporary Scientific Resource

Hinrich Kaiser, George R. Zug, and Aaron M. Bauer

Abstract Germany's zoological collections are among the oldest, most diverse, and historically most interesting natural history collections. From the perspective of three herpetologists with particular interest in taxonomy, we recount how invaluable visits to German collections have been to our own research efforts. Not only have we been able to use our findings in multiple publications, we also list examples of how we discovered new species and rediscovered specimens long thought lost, and we see in these experiences reasons to believe that many treasures in Germany's herpetological collections remain to be discovered. We also provide a brief historical overview of how Central Europe's and Germany's political structures in the seventeenth and eighteenth centuries led to a more diverse pattern for the formation of natural history collections than elsewhere, and we comment on the effects that the Second World War had on the well-being and continuity of collections. At the beginning of the twenty-first century, German zoological collections are a nexus for zoological research and a globally important scientific treasure.

H. Kaiser (✉)
Department of Biology, Victor Valley College, 18422 Bear Valley Road, Victorville, CA 92395, USA

Department of Vertebrate Zoology, National Museum of Natural History, Smithsonian Institution, Washington, DC 20013, USA
e-mail: hinrich.kaiser@vvc.edu

G.R. Zug
Department of Vertebrate Zoology, National Museum of Natural History, Smithsonian Institution, Washington, DC 20013, USA

A.M. Bauer
Department of Biology, Villanova University, 800 Lancaster Avenue, Villanova, PA 19085, USA

L.A. Beck (ed.), *Zoological Collections of Germany*, Natural History Collections,
DOI 10.1007/978-3-319-44321-8_2

Keywords Cabinets of curiosity • Historical collections • Berlin • Bloch • von Borcke • Waldenburg • Senckenberg • Mertens • Bonn • Böhme • Taxonomy • *Carlia nigrauris* • *Lygosoma quadrupes* • *Boiga irregularis* • *Boa constrictor* • Reptiles

The three of us are systematists with geographically broad research interests in amphibians and reptiles, and because of this specific expertise, our knowledge of German zoological collections has a herpetological bias that informs this presentation. As in many other zoological disciplines, our research often requires the examination of reference specimens in historical zoological collections, and this has naturally involved visits to the nexus of European museum collections: Germany. We do not use the term "nexus" lightly: Germany is where the confluence of history has produced multiple zoological collections of significance that have endured despite the ravages of war and the decades-long influence of the Iron Curtain. However, while simply the age of a scientific establishment may be considered a testament to its success and longevity, it is through its place in the continuum of scientific research across the years where its importance is best documented.

German zoological collections, which generally originated as cabinets of curiosity that served as showpieces on the estates of aristocrats, wealthy merchants, physicians, or pharmacists—the wealthiest individuals in their day, with an interest in tangible oddities from nature's stores—have had a perennial presence in the scientific panoply and have stood out in the scientific literature since at least the early 1800s. Our purpose in these brief comments is to provide a review from three "outsiders," who have different backgrounds but share the experience of working with and in German zoological collections, with a focus on herpetology. HK is German-American who never studied in Germany but has been visiting German zoological collections and colleagues as part of his research. AMB collected data for his dissertation at the Zoologisches Forschungsmuseum Alexander Koenig in Bonn, has worked extensively on the history and type specimens of the Museum für Naturkunde in Berlin, and has visited most of the large and small herpetological collections in Germany. GRZ is curator emeritus at the US National Museum of Natural History, one of America's most venerable zoological collections, and has six decades of experience in the curation of museum specimens; his research and travels have included several German collections. What follows is part historical review and personal history, to showcase the importance of German zoological collections to zoological research today.

One of the reasons why German zoological collections stand out in a global perspective is their diversity. There exist not only several large, "living," and growing collections of importance, but there is also a number of smaller historical collections. Unlike the development of major collections in imperial Britain or France, for example, collection building in Germany took a different trajectory due to the country's politically fragmented nature until unification in the late nineteenth century. As a consequence, instead of forming a single, dominant museum alongside relatively inconsequential provincial collections, Germany experienced the growth of numerous substantial collections, representing the diversity of kingdoms, principalities, and free cities.

Whereas early German collections in some cities grew in importance as a result of their location in a political power base, there are other interesting collections of historical status. Among the largest and oldest herpetological (and zoological) collections are those in Berlin, Frankfurt, and München (Munich in English), all of which are rich in type material. These collections were already important by the early nineteenth century and they eventually employed some of Germany's best and most prolific herpetologists. As the capital of a unified Germany, Berlin in particular benefited from a flood of material that arrived from the German colonial territories in the time between unification in the 1870s and the First World War. In addition, Waldenburg, a collection seeded by the seventeenth-century curiosity cabinet of the Linck family, is home to what has been considered the world's oldest surviving collection of fluid-preserved herpetological specimens (Fig. 2.1; Bauer and Wahlgren 2013; Böhme 2014). The collection in Halle, from the same time, had its origin as one of the earliest examples of a teaching collection in an instructional complex at the local orphanage.

The Second World War was a critical period in the history of German natural history collections (Crumly 1984). While the large collections survived, often through the extraordinary efforts of curators and staff, many other collections were destroyed or so severely damaged that their roles as major research centers were permanently compromised. Examples of this include the museums of Dresden, Magdeburg, Wiesbaden, and Breslau (modern Wroclaw, Poland). With the partition of Germany in the Cold War era, the Berlin collection ended up in the German Democratic Republic, with no serious competition as a leading museum. However, whereas Dresden had been destroyed, the collection made a new beginning and developed once again into an important herpetological collection driven by motivated individuals, especially under the leadership of Fritz Jürgen Obst. These collections, however, were poorly known to researchers in the West, and they remained underfunded, understaffed, and underused for decades. In the West, the Senckenberg Museum, with a herpetology section given continuity and leadership by Robert Mertens, retained its prewar prestige. Other important collections, like München and Hamburg, also retained their significance. Although the university museum in Bonn was destroyed, the postwar era saw the meteoric rise of the Forschungsmuseum Alexander Koenig, previously a private, mostly ornithological collection (Böhme 2014). Under Wolfgang Böhme, the herpetological collection incorporated various other collections, including the historically rich Göttingen collection, as well as those from Kiel and Heidelberg, initiated aggressive student-driven collection growth, and has now grown to become one of the largest in Germany.

Fig. 2.1 Specimen of *Boa constrictor* (I2013A3) from the Linck Collection in the Naturalienkabinett Waldenburg. This specimen was figured in a plate by Scheuchzer in 1735 and is one of the oldest spirit-preserved reptiles in the world (see Bauer and Wahlgren 2013)

As a doctoral student, AMB came to Germany to collect data and examine types in a number of German collections, especially Bonn and Berlin. He was struck by a major difference between American and German collections. Having developed later in the nineteenth century than their German counterparts, American museums (with the exception of the Academy of Natural Sciences in Philadelphia), had mostly expanded within a context of modern, organized science, such that collections were often well studied by their collectors or by resident curators, within a short time of their arrival; as a consequence, collections grew in a relatively orderly fashion. As a result, most American collections are large by European standards, possess well-documented specimens, and provide few surprises. The murky past of older German collections, on the other hand, which often began as personal or royal collections and grew piecemeal through the early decades of the nineteenth century, continue to provide many opportunities for the rediscovery of lost types, historical specimens, and unrecognized new species.

The disruption of the Second World War imposed more significant disorder on the more orderly German collections of the twentieth century than in other parts of Europe. It was often assumed, and sometimes written (e.g., Taylor 1969; Frost 1985), that many, maybe even most, of the types in German herpetological collections had been destroyed. However, as AMB and HK discovered in Berlin, that number was almost negligible and not only hundreds of types, but even most of the foundational collection of Bloch, obtained by the museum at its inception in 1810, were still extant (Fig. 2.2; see also Bauer 1999). This sent AMB on a 30-year odyssey, exploring the oldest specimens in the collection, from skinks described by Bloch in 1776 (Bauer and Günther 2006) to geckos and lacertids collected by Pallas in his explorations of the Russian empire (Bauer and Günther 1991, 1995), to the African snakes and lizards sent to Lichtenstein by German apothecaries from the Cape region in the 1810s and 1820s (Bauer 2004). There are few collections worldwide with such a historic depth in terms of their collection.

GRZ has an entirely different perspective on German collections, rooted in a professional association with herpetological collections since 1958, beginning as a graduate student with time spent at the Florida Museum of Natural History, University of Florida (Gainesville), and the Museum of Zoology at the University of Michigan (Ann Arbor), before assuming a curatorship at the National Museum of Natural History (Washington, DC) in 1968. As a long advocate of visiting herpetological collections rather than borrowing, he has worked in the collections of most American museums at least briefly at some point in his career. His European collection sojourns began in 1975 with a visit to the Natural History Museum (London, United Kingdom), continuing with a trip to the Rijksmuseum van Natuurlijke Historie (now Naturalis; Leiden, the Netherlands) and the Zoological Collections at the University of Amsterdam. In the early 1990s, he visited German collections, including those in Frankfurt, Bonn, Berlin, and Munich, specifically to gather data for his various Pacific Islands lizard projects.

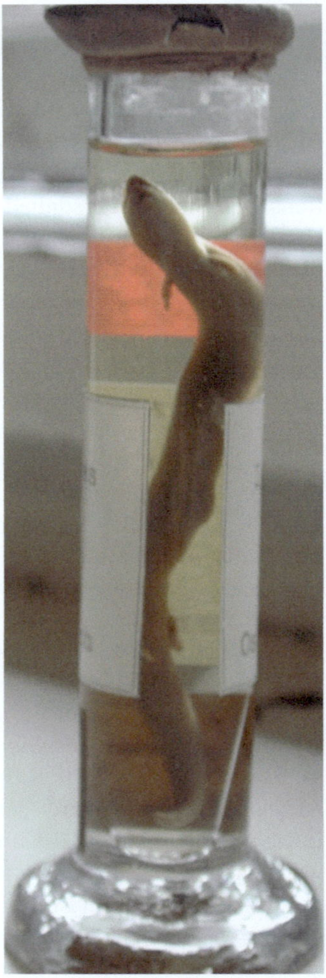

Fig. 2.2 Syntype (ZMB 1276) of *Lacerta serpens* (syn. *Lygosoma quadrupes*), an Asian skink. The species was described by Marcus Bloch in 1776 and was part of the foundation collection of the Berlin Museum. A large number of Bloch's specimens have survived the centuries (see Bauer 1999; Bauer and Günther 2006)

Compared to some of the work in American collections GRZ conducted, his experience with those in Germany was one accompanied by discovery and wonder. Working in research labs that carry the weight of history, tight places in historic buildings, on specimens that sometimes had lain unstudied for over a century, left a lasting, positive impression. It appeared that portions of the collections remained in a Sleeping Beauty slumber, not neglected but in stasis and awaiting renewed research activity. Remarkably, some of these collections even included guest researcher accommodations (as in Berlin and Bonn)! Beyond the age of collections, it also became apparent that many senior researchers from all over the globe began to discover the uniqueness and importance of German collections. One visit at the right time could result in lasting international collaborations.

To provide a glimpse of an extraordinary historic German herpetological collection, Berlin is a key example, and all three of us experienced it as it rose from the ashes since the fall of the Berlin Wall in 1989. During visits in the early 1990s, we observed a mature tree growing in collapsed parts of the building and a collection that had basically retained its nineteenth-century atmosphere. Specimens were kept in the finest, made-to-order glass-stoppered bottles, identified only by labels floating on the inside or glued to the outside—but not attached to the specimen. Many specimens were in jars they obviously did not belong to, a remnant of the time when specimens from broken bottles simply needed to be saved in a wartime hurry. The collection catalog was still handwritten, with continuity since the earliest entries in Wilhelm Peters's finely inked cursive. In the old cellars, hundreds of plaster-covered fossils remained untouched since their arrival, some perhaps a century or more earlier. And nighttime security was provided by guards with three Doberman pinschers, which made for adventurous evening encounters for late-working visiting researchers billeted in the basement guest quarters. Today, the Museum für Naturkunde is a modern facility, whose herpetological collection is kept in an astonishing glass-walled facility amidst the public displays of the museum, a giant testament to German collections of yesteryear and their relevance today.

HK's appreciation for German herpetological collections is not native but was brought on by the need to examine historical specimens from the Caribbean. This led him and AMB to an as-yet unpublished listing of several hundred "lost" amphibian types and generated in him a fondness for a different type of "fieldwork," removed from the outdoors yet essentially just as productive when it comes to making discoveries. Lately, our collective museum "fieldwork" has led to the discovery of several new reptile species, from as far afield as the outer reaches of the Indonesian archipelago (Fig. 2.3; see Zug 2010), some from areas that nowadays may no longer be able to support wildlife.

There is no doubt that the last decade has seen an encouraging renaissance when it comes to the upgrading and use of German herpetological collections. And, with Snow White now fully awake, the true value of the collections is only now becoming apparent. Not only has type material been located and cataloged, it happens with some frequency that we find the unexpected in some of the older collections. Recent work in Berlin, Frankfurt, Dresden, and Munich has led to the discovery of several new species, as well as the rediscovery of valuable type specimens that had been considered lost (e.g., Fig. 2.4). Even with a concerted effort by curators and researchers over the years, the size and variety of German herpetological collections are still being unveiled. The inescapable conclusion is that in some of the world's oldest collections, there is still much that can be discovered and needs to be evaluated. As we strive to gain a more complete knowledge of the globe's zoological diversity, preservation of these collections, and especially their modernization and use, is imperative not only for the thousands of specimens but especially for the protection of the biodiversity they represent.

Fig. 2.3 A recently described species of four-fingered skink, *Carlia nigrauris* Zug, 2010 from Tinjil Island, off the coast of Java, Indonesia. GRZ discovered specimens of this form in the Senckenberg collection, where they had been accessioned some four decades earlier. Robert Mertens, curator of herpetology at the museum at the time of their arrival, had recognized the uniqueness of these specimens in his 1957 report on the West Javan herpetofauna (Mertens 1957) but decided not to describe them owing to his limited knowledge of *Carlia* morphological variation. (**a**) Dorsal view of the body and (**b**) lateral view of the head of the holotype of *C. nigrauris* (SMF 53916)

Fig. 2.4 Holotype (ZMB 2583) of *Coluber irregularis* (= *Boiga irregularis*). This specimen was originally part of the collection of Willem Janssen and had been studied by the famous zoologist Blasius Merrem at Marburg. It eventually reached the Berlin Museum in 1818. Despite being a common, widespread species as well as an important invasive species, its presence in the Berlin collection was not recognized until 2013 (Bauer and Günther 2013) (Photo: © with Frank Tillack)

References

Bauer AM (1998) [1999] South Asian herpetological specimens of historical note in the Zoological Museum, Berlin. Hamadryad 23:133–149

Bauer AM (2004) Early German herpetological observations and explorations in southern Africa, with special reference to the Zoological Museum of Berlin. Bonn Zool Beitr 52:193–214

Bauer AM, Günther R (1991) An annotated type catalogue of the geckos (Reptilia: Gekkonidae) in the collection of the Zoological Museum, Berlin. Mitt Zool Mus Berlin 67:279–310

Bauer AM, Günther R (1995) An annotated type catalogue of the lacertid lizards in the Zoological Museum, Berlin (Reptilia: Squamata: Lacertidae). Mitt Zool Mus Berlin 71:37–62

Bauer AM, Günther R (2006) A syntype of *Lacerta serpens* Bloch, 1776 (Reptilia: Squamata: Scincidae): the oldest type specimen in the founding collection of the Zoological Museum of Berlin. Bibl Herpetol 6:17–25

Bauer AM, Günther R (2013) Origin and identity of the von Borcke collection of amphibians and reptiles in the Museum für Naturkunde in Berlin: a cache of Seba specimens? Zoosyst Evol 89:167–185

Bauer AM, Wahlgren R (2013) On the Linck collection and specimens of snakes figured by Johann Jakob Scheuchzer (1735)—the oldest fluid-preserved herpetological collection in the world? Bonn Zool Bull 62:220–252

Böhme W (2014) Herpetology in Bonn. Mertensiella 21. vi + 256 pp

Crumly CR (1984) Saving a legacy: natural history collections in Germany before and after World War II. Curator (NY) 27:205–219

Frost DR (ed) (1985) Amphibian species of the world. A taxonomic and geographical reference. Allen Press and the Association of Systematics Collections, Lawrence

Mertens R (1957) Amphibien und Reptilien aus dem äussersten Westen Javas und von benachbarten Eilanden. Truebia 24:83–105

Taylor EH (1969) Wiegmann and the herpetology of Mexico. In: Wiegmann AFA (ed) Herpetologia Mexicana [Facsimile reprint]. Society for the Study of Amphibians and Reptiles, Notre Dame, pp 3–6

Zug GR (2010) An outlying *Carlia* population from Java and comments on species groups within the genus *Carlia* (Reptilia: Squamata: Scincidae). Proc Calif Acad Sci, Ser 4(61):389–408

Chapter 3
Research Collections in Germany: Modern Trends in Methods of Sorting, Preserving, and Research

Ulrich Joger

Abstract Research methods applied in zoological collections are reviewed. Collection management and data processing are briefly introduced referring to the most integrative database systems. In the section on conservation techniques, the new task of preserving tissues for molecular research is emphasized.

The main focus of this overview is to provide some insight into modern imaging techniques like 3D and CT scanning, as well as into the expanding wealth of molecular techniques which have made museum collections extremely valuable sources for phylogenetics and population genetics, as well as for applications in monitoring biodiversity, such as DNA bar coding.

Modern techniques may be increasingly sophisticated and costly, yet there is no reason why they should not be applied even to smaller collections, as a research cooperation with universities and other large research institutions is always possible.

Keywords Natural history collections • Molecular • phylogenetic methods • Databases • Imaging

3.1 Zoological Collections in Germany

Contrary to botany, where living collections in botanical gardens play an important role (www.bgci.org) and to microbiology where there is a central collection of microorganisms and cell cultures [Deutsche Sammlung von Mikroorganismen und Zellkulturen (www.dsmz.de)], zoological collections are limited to museums and to universities. Due to the late unification of Germany (first in 1871, again in 1990) and to the federal structure of the republic, there is no dominating central natural

---

U. Joger (✉)
Staatliches Naturhistorisches Museum Braunschweig, Pockelsstraße 10, 38106 Braunschweig, Germany
e-mail: info.snhm@3landesmuseen.de; http://www.naturhistorisches-museum.de/

L.A. Beck (ed.), *Zoological Collections of Germany*, Natural History Collections,
DOI 10.1007/978-3-319-44321-8_3

Table 3.1 The German zoological collections, ordered by size and administrative constitution

	Leibniz institution	State owned	University	Communal/private
Large	X	X		
Large but composite		X		
Medium size		X	X	X
Small			X	X

history museum (as there is no dominating university, neither) but a wealth of medium-sized museums and a few large ones.[1]

Zoological collections in Germany are very diverse, both in legal constitution and in size. They can roughly be subdivided (not counting special cases[2]) into categories according to Table 3.1.

For the purpose of this overview, most small museums can be disregarded, as they usually do not harbor research collections and/or have no curators with scientific background. Some old university collections are an exception, as they may harbor important specimens, and the respective scientific maintenance may be fulfilled by university personnel such as professors (if interest in taxonomy or phylogeny is still present among them) or taxidermists (if such a position still exists). The problematical situation of undercurated university collections has recently been realized. In 2011, the German research council (Wissenschaftsrat) issued recommendations for research infrastructure. In this publication (Wissenschaftsrat 2011), scientific collections are treated on the same level of importance as university libraries, and a coordinating body for university collections is recommended.

In the aftermath of this important first step, several universities have nominated a spokesperson for collections. A Society for University Collections (Gesellschaft für Universitätssammlungen) was founded in 2012. It strengthens the position of individual collection managers and helps bringing attention of university deans to neglected university collections.

Medium-sized collections should not be underestimated as there are quite many of them in Germany. They may harbor old and valuable specimens. Some are centered around local paleontological highlights like the Jura Museum at Eichstätt, where famous *Archaeopteryx* fossils have been found, or the communal natural history museum of Chemnitz, built among Paleozoic tree trunks found in that city.

[1]For about 75 years (1870–1945), the Berlin Museum für Naturkunde was kind of a central German natural history museum and thus acquired the largest zoological collection of all German museums, but time was too short to dominate the others, and as it was located in Eastern Germany, it suffered from a delayed development during the time of the German Democratic Republic (1949–1990).

[2]Special cases are collections linked to an aquarium, e.g., the Löbbecke Museum and Aquazoo at Düsseldorf and the Deutsches Meeresmuseum with its Ozeaneum at Stralsund. Their live animal displays clearly dominate over their museum collections, providing very attractive sources of income.

Others have worldwide zoological collections. For instance, the Braunschweig Natural History Museum, founded in 1753, owns the largest ornithological collection in northern Germany, with about 60,000 preserved birds (mostly from the nineteenth century) and more than 10,000 bird eggs. About half of the world's bird species are represented. The zoological museum of the University of Kiel, founded in 1775, houses about 600,000 specimens or series of specimens, with a focus on marine fauna.

The composite museums are difficult to evaluate. They are large sized, but their natural history collections are usually valuated less than the collections of art or ethnography housed in the same institutions. This is reflected by the internal division of manpower in these institutions. Typically, their director is not a zoologist or paleontologist. They have only few curators in natural history. Examples of this kind are the Hessian state museums at Wiesbaden and Darmstadt, the Landesmuseum at Hannover, and the Übersee-Museum at Bremen. If only the natural history collections of these museums are evaluated, they must be regarded "medium size."

Medium-sized museums have recently been in the focus of a special program of the Volkswagen Foundation. Research grants have been issued for a couple of projects, aimed at strengthening research activities with this group of collections. The rationale for this program was provided by an inquiry among natural history museums executed by the Natural History group of the "Deutscher Museumsbund" (DMB). In this inquiry among small- and medium-sized natural history museums, a total of 37 million collection items (mainly zoological specimens) were counted. The real number may be much higher.

The large German natural history museums, with some of them being research institutions in the Leibniz community and others state museums, are linked together in the DNFS (German Nature Research Collections). They will present themselves in the forthcoming chapters. These important institutions are especially strong in lobbying for the sake of natural history collections and have already achieved a change of public attitude regarding nature museums. They are no longer regarded as old fashioned, backward stores of curious dead animals, but increasingly as valuable research institutes on issues like biodiversity loss, species preservation, and evolution. This is in line with a striking increase in the application of high technology techniques (see below).

3.2 Collection Management and Electronic Data Processing

3.2.1 Database Systems

Digitization of collection data has been an issue for decades, yet development and progress in the IT community has been more rapid generally than the implementation of collection management programs, at least in Germany. Compared with the USA, the degree of digitization in German natural history collections is still low.

Early pilot projects to develop collection data systems based on programs like Access, dBase or Filemaker have had little lasting success. Programs designed for museums were usually not fit for the special requirements of natural history collections. In this situation, two important German collections developed successfully operational database systems on their own: "SESAM" by the Senckenberg group and "Diversity Workbench" (DWB) by the Bavarian State Collections. Both are available for external users, but their general distribution is still limited. SESAM and the new system "AQUILA" is mainly used by Senckenberg institutions and by several collections of marine organisms, whereas DWB has its stronghold in botanical collections.

In the course of a metadatabase project of five German natural history museums supported by the German Research Foundation (DFG), we compared the capabilities of both SESAM and DWB with the free database system "Specify"—an American database program with already wide international application. Specify appeared superior to SESAM and DWB in the following features:

- Adaptability to specific user
- Import of existing data files
- User-friendliness (low in DWB)
- Data held on the user's own server (not possible in SESAM)

Together with its greater potential for the future, Specify was thus recommended for further use.

While—thanks to Linnaean systematics and to the International Commission on Zoological Nomenclature—there is no doubt which hierarchical system should be used for zoological databases, the main problems come with the practical application to gigantic numbers of specimens stored. Usually collection managers have no manpower available for the additional task to "feed" millions of existing data into a new database format. Thus, promising pilot projects ended with the temporary positions that had been financed by a grant and were not continued routinely. In the ongoing project mentioned above, the five German natural history museums explore means to combine their collections via a metadatabase which should be accessible in the Internet (see below).

3.2.2 Data Exchange and Metadata

One main feature of databases should be the general interchangeability of data by Internet. Ideally, users interested in a specific taxon should be able to trace easily all specimens of that taxon, stored in diverse collections in different countries around the world. For this end, a number of approaches have been implemented:

Global Biodiversity Information Facility (*GBIF*) is an international initiative for providing digital data on the worldwide biodiversity for free via the Internet. GBIF is meant not only to provide collection data but also field observation data on the respective species. Nevertheless, all eight GBIF "nodes" (coordinating centers of competence) in Germany are based in large collections (www.gbif.de). Currently (September 2015) of the animal group richest in species, insects, the GBIF metadatabase holds about one million entries. This is a tenfold increase since 2010, but still far from the real number of specimens held in German collections. Among vertebrates, the mammal collection of the Museum für Naturkunde Berlin is the only significant one included.

The Internet portal Biological Collection Access Services (*BioCASE*) is linked to GBIF via the Consortium for European Taxonomic Facilities (CETAF). It provides tools for assessing data on individual species (www.biocase.org).

SYNTHESIS is an EU-based initiative supporting both GBIF and BioCASE. Its principal aim is to improve the quality of and increase access to digital collections and data within natural history institutions.

The German-based *DNA Bank Network* (www.dnabank-network.org) and the *Global Genome Biodiversity Network* (http://data.ggbn.org/) are linked with GBIF and SYNTHESIS. They embrace large museums, botanical gardens, as well as microbial collections (DSMZ in Germany) in order to coordinate their efforts to establish genetic sample collections. The number of tissues, cultures, and DNA samples stored is still limited (less than 200,000 from 20,000 species).

Bar code of Life Data (*BOLD*) Systems Database is an online platform where currently more than 420,000 DNA bar codes (163,000 species) can be found (www.boldsystems.org). See below for DNA bar coding.

Summarizing the state of the art, there has been considerable progress to facilitate access to natural history data, but in daily work, data accessibility is still limited to a tiny subset of the data. Natural history collections in Germany should pay more attention to sharing data with each other and with the Internet community.

3.3 Conservation of Museum Specimens

3.3.1 Conservation of Whole Animals

Traditionally whole animals in zoological collections are preserved either dry or wet (in preservation liquid). The former method applies to taxidermally treated mammals and birds (rarely reptiles or fish), skeletons and bones, insects, mollusk shells, and other dried invertebrates. Chemical treatment against damage by insects has varied over the centuries: Inorganic chemicals like arsenic have been replaced by organic insecticides in the twentieth century. Nowadays, chemical-free treatment with gas (CO_2) is the state of the art. It must be repeated regularly.

Wet collections mainly preserve fish, amphibians, reptiles, and soft-bodied invertebrates in glass containers. Traditionally, formaldehyde was used, at least for initial preservation, but since that chemical is notorious for damaging DNA which is not the case with ethanol, the latter is now recommended (but see Poinar et al. 2010).

3.3.2 *Conservation of Tissues*

Tissue collections (blood samples or muscle tissue for DNA extraction) have been increasingly installed in modern natural history museums and universities. Tissues may be preserved in 96 % ethanol in the field and kept at room temperature for a time. They should be transferred to a standard freezer (−18 °C) as soon as possible. If blood samples are taken in the lab, ethanol is not necessary. Some labs use deep-freezing down to −80 °C, but lower temperatures (liquid nitrogen) are only needed if proteins are to be preserved.

3.4 Modern Research Methods

3.4.1 *Morphological Analysis*

3.4.1.1 Electron Microscopy

If collections have access to a scanning electron microscope (SEM), character states can be distinguished that are invisible with traditional methods. This is especially the case with cuticular or keratinized ornamentation (e.g., Joger 1984), mollusk shells, or egg shells (Schleich and Kästle 1988). Care must be taken if ethanol-preserved materials are used, as ethanol may soak soluble lipids from tissues, with the effect that outer layers of the skin may get lost.

3.4.1.2 3D Imaging and Laser Scanning

3D imaging has been greatly simplified in recent years. Cost-efficient scanners have been developed, and new software allows to integrate handheld laser scanning into perfectly three-dimensional images (Joger and Ritter 2012). Recently laser light has been replaced by visible light, and even color scanning is now possible.

Applications of laser scanning are manifold. It is especially useful for investigating fossil and recent bones. Bones of different individuals of a species may be virtually scaled in order to reconstruct a skeleton, even from fragments or individuals of different size. Another application of 3D imaging is scanning of insects which, of course, need a higher resolution than bones.

3.4.1.3 X-Rays and Tomography

X-ray machines have been used for decades to visualize internal structures of fossil or recent animals. Computer tomography (CT) is much better in resolution, and this technique produces three-dimensional images composed of many virtual "cuts" through a body without destroying it.

Synchrotron radiation X-ray tomography (SRXT) has become an expensive but highly desired nondestructive technique for the internal analysis of museum specimens, especially archeological and paleontological samples, artwork, and museum specimens, obtaining complete three-dimensional recordings at mm to submicrometer resolution. For example, previously unknown anatomical features of *Archaeopteryx* could be detected at the European Synchrotron Radiation Facility (Grenoble, France). Even a new species of pseudoscorpion from amber was described as *Pseudogarypus synchrotron* (Henderickx et al. 2012).

While synchrotrons are giant machines only found in a few institutions worldwide, microcomputed tomography (micro-CT) is increasingly applied to museum specimens. More and more large natural history museums and university collections invest in a micro-CT device.

Micro-CTs are different from mini-CTs. The latter operate with an X-ray detector and radiation source that is rotated around the examined object. The setup is analogous to clinical CTs.

In real micro-CTs, the examined object is rotated within the light path. Radiation source–object distance and object–detector distance can be freely varied, thereby allowing optimization of the magnification level (Schambach et al. 2010).

Applications of micro-CT in natural history are manifold: 3D visualizations of the interior structure of fossils, animal skulls, and insects have been done (Abel et al. 2012; Van de Kamp et al. 2011). For example, the inner ear of dinosaurs, crocodiles, and birds could be compared. Interference of micro-CT radiation with the preservation of DNA (which is needed for molecular genetic studies, see below) has been discussed but refuted (Paredes et al. 2012).

3.4.2 Molecular Genetic Analyses

3.4.2.1 Protein and DNA Preservation

In the early time of molecular phylogenetics, DNA sampling was difficult, and proteins were the main target molecules. These were either allozymes (isoenzymes) or blood proteins such as hemoglobin or albumin (Joger 1996). They could not be extracted from preserved museum specimens, and therefore fresh tissues or blood samples had to be frozen for research. Immunological methods were applied to infer genetic distances from protein similarities (Joger 1984).

Proteins lose their tertiary structure (and their primary structure partly) when in contact with ethanol. Therefore, methods based on enzymes or immunological distances cannot be applied to ethanol-preserved tissues. On the contrary to proteins, DNA is well preserved in ethanol and is only slowly degrading. It may be extracted from ethanol-preserved tissues with some effort, yet in small quantities. Given the enormous size of the genome, early efforts to analyze DNA extracted from tissues suffered from difficulties to identify homologous genes. This could be overcome by a DNA–DNA hybridization technique which used the ability of single DNA strands to bind to their homologous counterpart but not to other sites (Sibley and Ahlquist 1983). Charles Sibley and his group used interspecific DNA–DNA hybridization in countless experiments to reconstruct the phylogenetic tree of birds. However, the invention of polymerase chain reaction (PCR) in 1985 opened a much easier way to analyze DNA from partly degraded samples. This was a real revolution in molecular genetics, as well as in its application to museum samples.

3.4.2.2 Ancient DNA and the PCR Revolution

Before the invention of the polymerase chain reaction (PCR), DNA sequencing required laborious procedures, including bacterial cloning, and degraded DNA from museum samples could only be sequenced in exceptional cases. Mullis and Falloona (1987) used specific primers (short DNA strands which initiate replication of a defined section of a DNA molecule) and a polymerase from a thermophilic bacterium (*Thermus aquaticus*) to allow cyclic heating of a reaction mixture without stopping the amplification of the desired DNA strand. Heating is necessary to separate the double strand into single strands in order to repeat the process. Thus DNA could, for the first time, be amplified multiple times (in theory indefinitely, from a single molecule). In 1993, Mullis received the Nobel Prize for his invention.

PCR primers are able to find very rare intact target sequences in a large mass of degraded or junk DNA. Therefore, it became now possible to study "ancient" DNA–DNA from museum samples—and even subfossil DNA. This chance was immediately taken. Some important steps in ancient DNA research are shown in Table 3.2.

Historically, mitochondrial DNA was sequenced before nuclear DNA (e.g., Lenk et al. 1999). This is due to the fact that numerous mitochondria are present in every cell, and therefore, the probability to find an intact target sequence within the mitochondrial genome is much larger than to find a nuclear sequence. The large amounts of nuclear sequences attained since 2006 became only possible by "next-generation sequencing" methods (see below). In addition to reliable phylogenies based on large portions of the genome, also tracing ancient allele polymorphisms become possible (e.g., Lalueza-Fox et al. 2007).

As far as zoological museum specimens are concerned, they are now being more and more included in molecular phylogenetic or population genetic studies. Since Diamond's enthusiastic article in *nature*, "Old dead rats are valuable" (1990), museum specimens are also seen as sources of genetic information (Thomas et al. 1990; Cooper 1994; Ellegren 1994; Hauf et al. 1995, 1999). This holds for all organic museum specimens (bones, feathers, dried skins, insects, ethanol-preserved bodies, and frozen materials). A few examples:

– Martinkova and Searle (2006) sequenced mitochondrial genes out of 268 samples of hides of stoats (*Mustela erminea*), from 18 museum collections.

Table 3.2 Milestones in Palaeogenetics

1985	Invention of PCR by Kary Mullis
1987	First automated sequencers commercially available
1989	First DNA from a fossil human bone (3500 y.b.p)
1990	First population genetic study using museum samples (Thomas et al. 1990)
1994	First short mitochondrial DNA sequence of mammoth (Hagelberg 1994; Höss et al. 1994)
1997	First mitochondrial DNA sequence of *Homo neanderthalensis* (Krings et al.)
1999	First nuclear sequences of a mammoth (Greenwood et al. 1999)
2001	First complete mitochondrial DNA of an extinct species (Moa)
2006	More than 10 million base pairs of mammoth DNA sequenced (Poinar et al. 2006)
2010	Neanderthal genome sequenced (Green et al. 2010)

Amplification of 700 base pair fragments was still possible from 30 to 40 % of over 100-year-old specimens.

– Guschanski et al. (2013) sequenced complete mitochondrial genomes of 63 species of African primates, out of 92 museum specimens (bone, skulls, skins), and reconstructed their phylogenetic tree. Seventy-seven percent of the museum specimens yielded suitable DNA.
– Döppes et al. (2014) compared DNA from a mummy of an Alpine hare (*Lepus timidus*) with living hares, identifying a formerly unknown lineage.

3.4.2.3 DNA Bar Coding

Many research activities in natural history collections are part of the global challenge to unravel the "tree of life" and to unveil earth's biodiversity. "DNA bar coding" is an effort to standardize a simple molecular approach in order to compare as many organisms as possible with an identical genetic marker.

DNA bar coding uses a rather short but universal mitochondrial gene fragment (cytochrome c oxidase subunit I, COI or cox 1) which is present in all animals (Hebert and Gregory 2005). The idea is to establish a database of COI sequences from as many species as possible (Murphy et al. 2013). This database can then be used to identify sibling species, eggs or larvae, and fragmented bodies in stomach contents or shed skins (Kress and Erickson 2012). Unknown species would be recognized easily. Automatic bar coding instruments are being designed to enable every researcher to get their own bar codes. A Consortium for the Barcode of Life (CBOL) has been established.

Although this is mainly an application for field research, a number of natural history museums (in Germany: Bavarian State Collections, München, Stuttgart Natural History Museum, and Museum Alexander Koenig, Bonn; www.bolgermany.de) are involved.

A striking feature of DNA bar coding is its methodological ease. However, in the light of breathtaking technological progress (see below), the question is emerging whether bar coding may sooner or later become obsolete.

3.4.2.4 Next-Generation Sequencing

A number of sophisticated techniques for DNA amplification and sequencing have been developed recently (Shapiro and Hofreiter 2012), and there are more and more being developed. Combinations of these techniques are getting tested for diverse applications. The output of DNA sequencing has become orders of magnitude higher and faster. Next-generation sequencing is going to provide the base for a second revolution in molecular genetics (after the invention of PCR in 1985). Usually, they are used with a high-throughput automated sequencer, which is able to produce data in the range between one gigabase and one terabase in a single run. I can only mention a few of these techniques here:

Multiplex PCR is a DNA amplification technique, which simultaneously applies specific primers for several DNA fragments, thus saving time. It is especially useful in combination with high-throughput single nuclear polymorphisms (SNP) genotyping.[3]

Restriction-site associated DNA sequencing (RADseq): In this approach, genomic DNA is digested with restriction enzymes; obtained fragments are PCR amplified and sequenced; finally, single nuclear polymorphisms (SNPs) are identified and genotyped. We used RADseq to detect hybridization between species of snakes (Zinenko et al. 2016).

Illumina sequencing (sequencing by DNA synthesis): This technology utilizes a terminator-based method that detects single bases, while fluorescently labeled nucleotides are incorporated into a DNA template. All four bases are present simultaneously; they can be detected base by base using different emission wavelengths.

Pyro-sequencing is an alternative to conventional Sanger (chain terminating) sequencing. It combines a polymerase with a chemiluminescent enzyme (luciferase) and, in its latest automated version, can generate 400 Mb in a 10-h machine run.

In an international team of university groups and museums, we used the two-step multiplex PCR (Römpler et al. 2006), PTS tagging, and pyro-sequencing to obtain complete mitochondrial genomes from ten mammoth samples (Chang et al., manuscript in preparation). In brief, we divided 78 primer pairs that span the mammoth mitochondrial genome into two, nonoverlapping pools of 39 primer pairs. Products of the four multiplex PCR amplifications then served as templates for singleplex amplifications.

[3]Genotyping is the measurement of general genetic variation in the genome.

Other new methods like microarray-based whole genome sequencing, targeted sequencing, exome sequencing, or transcriptomics (RNA sequencing) are being used with increasing frequency for "museomics" (genetic sequencing from collection material).

Generally, there is no need for a museum to invest in an automated sequencer, as sequencing can be ordered commercially from specialized service companies, or it can be done in cooperation with a university laboratory. It is, however, advisable that museums install a laboratory for DNA extraction and, possibly, amplification, too.

References

Abel RL, Laurini C, Richter M (2012) A biologist's guide to 'virtual' micro-CT preparation. Palaeontol Electron 15:1–16

Cooper A (1994) DNA from museum specimens. In: Herrmann B, Hummel S (eds) Ancient DNA. Heidelberg, Berlin, pp 149–165

Diamond JM (1990) Old dead rats are valuable. Nature 347:334–335

Döppes D, Gill-Frerking H, Joger U, Rosendahl W, Stümpel N (2014) A mountain hare mummy from the Zillertaler Alps. In: Yearbook of mummy studies. München, vol 2, pp 23–29

Ellegren H (1994) Genomic DNA from museum bird feathers. In: Herrmann B, Hummel S (eds) Ancient DNA. Heidelberg, Berlin, pp 211–217

Green RE, Krause J et al (2010) A draft sequence of the Neandertal genome. Science 328:710–722

Greenwood AD, Lee F, Capelli C, Possnert G, Pääbo S (1999) Evolution of endogenous retrovirus-like elements of the Woolly Mammoth (Mammuthus primigenius) and its relatives. Mol Biol Evol 16:840–847

Guschanski K, Krause J, Sawyer S, Valente LM, Bailey S, Finstermeier K, Sabin R, Gilissen E, Sonet G, Nagy ZT, Lenglet G, Mayer F, Savolainen V (2013) Next-Generation Museomics disentangles one of the largest primate radiations. Syst Biol 62:539–554

Hagelberg E (1994) Mitochondrial DNA from ancient bones. In: Herrmann B, Hummel S (eds) Ancient DNA. Heidelberg, Berlin, pp 195–204

Hauf J, Baur A, Chalwatzis N, Joger U, Zimmermann FK, Lazarev P (1995) Selective amplification of a mammoth mitochondrial cytochrome b fragment using an elephant-specific primer. Curr Genet 27:486–487

Hauf J, Joger U, Zimmermann FK, Lazarev P, Vartanyan SL (1999) Protein and nucleic acid sequences of Woolly Mammoth cytochrome b and the phylogenetic position of *Mammuthus* within the Elephantidae. In: Haynes G, Klimowicz J, Reumer JWF (eds) Mammoths and the Mammoths Fauna: studies of an extinct ecosystem. Proceedings of the 1st international mammoth conference, St. Petersburg, Russia, Deinsea (Rotterdam), vol 6, pp 211–217

Hebert PDN, Gregory TR (2005) The promise of DNA barcoding for taxonomy. Syst Biol 54:852–859

Henderickx H, Tafforeau P, Soriano C (2012) Phase-contrast synchrotron microtomography reveals the morphology of a partially visible new *Pseudogarypus* in Baltic amber (Pseudoscorpiones: Pseudogarypidae). Palaeontol Electron 15:1–11

Höss M, Pääbo S, Vereshchagin NK (1994) Mammoth DNA sequences. Nature 370:333

Joger U (1984) Morphologische und biochemisch-immunologische Untersuchungen zur Systematik und Evolution der Gattung Tarentola. Zool Jb Anat 112:137–256

Joger U (1996) Molekularbiologische Methoden in der phylogenetischen Rekonstruktion. Zool Beiträge 37(1):77–131

Joger U, Ritter A (2012) Die Geburt des Braunschweiger *Spinophorosaurus*. Vermessung von Dinosaurierknochen mit 3D-Technologie. IQjournal (Braunschweig) 1/2012:12–13

Kress WJ, Erickson DL (eds) (2012) DNA barcodes: methods and protocols. Springer protocols methods in molecular biology, p 858

Lalueza-Fox C, Römpler H, Caramelli D, Stäubert C, Catalano G, Hughes D, Rohland N, Pilli E, Longo L, Condemi S, de la Rasilla M, Fortea J, Rosas A, Stoneking M, Schöneberg T, Bertranpetit J, Hofreiter M (2007) A melanocortin 1 receptor allele suggests varying pigmentation among Neanderthals. Science 318:5855

Lenk P, Joger U, Fritz U, Heidrich P, Wink M (1999) Phylogeographic patterns in the mitochondrial cytochrome b gene of the European pond turtle, *Emys orbicularis* (Linnaeus). Mol Ecol 8:1911–1912

Martinkova N, Searle JB (2006) Amplification success rate of DNA from museum skin collections: a case study of stoats from 18 museums. Mol Ecol 6:1014–1017

Mullis KB, Falloona FA (1987) Specific synthesis of DNA in vitro visa a polymerase catalized chain reaction. Methods Enzymol 155:335–350

Murphy RW, Crawford AJ, Bauer AM, Che J, Donnellan SC, Fritz U, Haddad CFB, Nagy ZT, Poyarkov NY, Vences M, Wang W-Z, Zhang Y-P (2013) Cold code: the global initiative to DNA barcode amphibians and nonavian reptiles. Mol Ecol Resour 13:161–167

Paredes UM, Prys-Jones R, Adams M, Groombridge J, Kundu S, Agapow P-M, Abel RL (2012) Micro-CT X-rays do not fragment DNA in preserved bird skins. J Zool Syst Evol Res 50:247–250

Poinar H, Schwartz C, Qi J, Shapiro B, Mac Phee RDE, Buigues B, Tikhonov A, Huson DH, Tosho LP, Auch A, Rampp M, Miller W, Schuster C (2006) Museomics to paleogenomics: large scale sequencing of mammoth DNA. Science 311:392–394

Poinar H, Okello J, Zurek J, Devault A, Kuch M, Okwi AL, Wewankambo N, Bimenya G, Poinar D (2010) A quantitative and qualitative comparison of total nucleic acid (DNA and RNA) yields from formalin-fixed paraffin-embedded tissues using ten different extraction methods. Anal Biochem 400:110–117

Römpler H, Rohland N, Lalueza-Fox C, Willerslev E, Kuznetsova T, Raberder G, Bertrandpetit J, Schöneberg P, Hofreiter M (2006) Nuclear gene indicates coat-color polymorphism in mammoths. Science 313:62–65

Schambach SJ, Bag S, Schilling L, Groden C, Brockmann MA (2010) Application of micro-CT in small animal imaging. Methods 50:2–13

Schleich HH, Kästle W (1988) Reptile egg-shells SEM Atlas. Fischer, Stuttgart

Shapiro B, Hofreiter M (eds) (2012) Ancient DNA: methods and protocols. Methods Mol Biol 840:155–170

Sibley C, Ahlquist J (1983) Phylogeny and classification of birds based on the data of DNA-DNA hybridization. Curr Ornithol 1:245–292

Thomas WK, Pääbo S, Villablanca FX, Wilson AC (1990) Spatial and temporal continuity of kangaroo rat populations shown by sequencing mitochondrial DNA from museum specimens. J Mol Evol 31:101–112

Van De Kamp T, Vagovic P, Baumbach T, Riedel A (2011) A biological screw in a beetle's leg. Science 333(6038):52. doi:10.1126/science.1204245, PMID 21719669

Wissenschaftsrat (2011) Empfehlungen zu Forschungsinfrastrukturen. Köln

Zinenko O, Sovic M, Joger U, Gibbs L (2016) Hybrid origin of European Vipers (*Vipera orlovi* and *V. magnifica*) from Caucasus determined using genomic scale DNA markers. BMC Evol Biol 16:76. doi: 10.1186/s12862-016-0647-7

Chapter 4
National and International Collection Networks

Cornelia Weber

Abstract In the last few years, the number of networks of scientific collections, which by definition include zoological collections, has increased enormously. The resulting increased visibility of these collections' holdings has made it possible to considerably improve not only awareness and appreciation of them but also their usability and use as scientific resources.

This article provides a brief overview of the most important zoological and natural history collection networks and focuses especially on university collections, which represent an important part of scientific infrastructures for research and teaching.

Keywords Scientific infrastructure • Research • Teaching • University collections

4.1 Introduction

In the last 20 years, the awareness and the perception of research museums and scientific collections in the scientific community have fundamentally changed, largely due to the increasing use of the Internet. At the same time, new technical developments have significantly facilitated the establishment of networks.

These developments made it possible to present publicly inaccessible or practically unknown collections on the Internet and thereby to the scientific community. Initially, this took the form of simple contact information or collection descriptions with representative illustrations. Later, web-based databases and repositories appeared and made interregional cooperation possible. These were followed by portals and information systems covering certain regions or kinds of collections. Noteworthy examples include ZEFOD, Zentralregister botanischer und zoologischer Forschungssammlungen in Deutschland,[1] the central register of botanical and zoological research collections in Germany; the Information System

[1] http://zefod.genres.de (Retrieved on 27 July 2015).

C. Weber (✉)
Humboldt-Universität zu Berlin, Unter den Linden 6, 10099 Berlin, Germany
e-mail: weber@wissenschaftliche-sammlungen.de; http://www.wissenschaftliche-sammlungen.de

L.A. Beck (ed.), *Zoological Collections of Germany*, Natural History Collections,
DOI 10.1007/978-3-319-44321-8_4

on Collections and Museums at Universities in Germany[2]; and GBIF—the Global Biodiversity Information Facility.[3] The visibility of the collections and their holdings has made it possible to not only considerably increase awareness of them but to improve their usability and use as scientific resources as well.

Documentation on the Internet shows that, in Germany, a large share of zoological collections is held at universities. The database of the Coordination Centre for Scientific University Collections in Germany alone lists 49 collections with zoological holdings, including three forest zoology collections.[4] The first university collections, like the zoological collection of the Friedrich-Alexander-Universität Erlangen-Nürnberg, were already established in the eighteenth century (Herrmann and Weidemann 2007). No one knows how many objects are actually being kept at universities, but we do know that among them are a large number of type specimens, underscoring the importance of academic collections. Even though many objects and even entire collections have been integrated into other holdings, including nonuniversity holdings, or have been transferred to organisations with a different organisational form, such as in the case of the Berlin Museum für Naturkunde (Museum of Natural History),[5] today academic collections continue to represent an important scientific infrastructure for research and teaching.[6] The realisation that academic collections must be retained as decentralised infrastructure, with respect for their diversity and specific local characteristics, has triggered an enormous variety of activities, especially in the last few years.[7] For this reason, one section of this article is dedicated to university collections.

[2]http://www.universitaetssammlungen.de, English version: http://www.universitaetssammlungen.de/?setLocale=en (Retrieved on 27 July 2015).

[3]http://www.gbif.org (Retrieved on 27 July 2015).

[4]http://portal.wissenschaftliche-sammlungen.de (Retrieved on 27 July 2015).

[5]Shortly before its 200th anniversary on 1 January 2009, the Museum für Naturkunde Berlin was transformed into a public foundation and admitted to the Leibniz Gemeinschaft. The museum is the successor to three museums established in 1810 when the Berlin University on Unter den Linden was founded: the Anatomical-Zootomical Museum, the Mineralogical Museum (from 1814) and the Zoological Museum.

[6]See the web portal Wissenschaftliche Sammlungen, http://portal.wissenschaftliche-sammlungen.de (Retrieved on 27 July 2015).

[7]See news on that website of the Coordination Centre for Scientific University Collections in Germany (in German): http://wissenschaftliche-sammlungen.de/de/nachrichten/aktuelles (Retrieved on 27 July 2015).

4.2 Networks of University Collections

At the turn of the twenty-first century, many different actors and stakeholders came together to cooperatively develop concepts and strategies and to expand the use of university collections. Initially, interest in these collections as academic cultural heritage predominated; later, more emphasis was put on scientific aspects.

The first network in Europe was Universeum—Academic Heritage and European Universities: Responsibility and Public Access.[8] Universeum was founded at the Martin Luther University Halle-Wittenberg in April 2000. Twelve of the oldest and most renowned European universities were involved in the project.[9] The network members "discussed and passed the Declaration of Halle which, for the very first time, defined the common aim to facilitate the access of a broader public to the—often hidden—treasures of the old European academic institutions" (Bremer 2001): "Universities must acknowledge their wide cultural roles. Academic collections and museums provide special opportunities for experiencing and participating in the life of the university. These collections serve as active resources for teaching and research as well as unique and irreplaceable historical records. In particular, the collections of the oldest European universities provide windows for the public on the role of the university in helping to define and interpret our cultural identity. By valuing and promoting this shared academic heritage, our institutions demonstrate a commitment to the continued use of these resources by a broad public".[10] Today, the network Universeum—European Academic Heritage Network has 120 members from 21 countries, and it is "concerned with academic heritage in its broad sense, tangible and intangible. It aims at the preservation, study, access and promotion of university collections, museums, archives, libraries, botanical gardens, astronomical observatories, etc".[11] The activities of its working groups are central to this effort, namely, Recent Heritage of Science (2012) and Databases and Portals (2015).[12]

While Universeum has concentrated on the European region, the International Committee for University Museums and Collections (UMAC),[13] a committee of the International Council of Museums (ICOM),[14] is active worldwide: "UMAC was founded in 2001 to preserve and provide access to a significant part of national and

---

[8]Now Universeum—European Academic Heritage Network, see: http://universeum.it (Retrieved on 27 July 2015).

[9]University of Amsterdam, Humboldt University of Berlin, Bologna University, University of Cambridge, University of Groningen, Martin Luther University of Halle-Wittenberg, University of Leipzig, Royal College of Surgeons of England, University of Oxford, University of Pavia, Uppsala University, Utrecht University.

[10]Declaration of Halle, 16 April 2000.

[11]http://universeum.it (Retrieved on 27 July 2015).

[12]See http://universeum.it/working_groups.html (Retrieved on 27 July 2015).

[13]http://icom.museum (Retrieved on 27 July 2015).

[14]http://umac.icom.museum (Retrieved on 27 July 2015).

international heritage with its many unique and unusual collections".[15] Like Universeum, the network meets once a year to exchange expertise in the field.

A nationwide movement addressing the concerns of collections has existed in Germany since 2010. It has its origins in the symposium "University Museums and Collections in Academic Practice. Tasks—Concepts—Perspectives" held that year at Humboldt-Universität zu Berlin,[16] which triggered developments that have had a measurable impact upon the work in and with university collections in recent years. The network represents several million objects that are currently spread throughout over 900 collections at 85 universities.[17] Collection representatives, custodians and coordinators are represented in the network, where they work together to develop strategies to ensure the continued existence of academic collections and to more actively use collections in research, teaching and education. This community meets once a year for a conference at which current developments, projects and questions important to collections are presented and discussed.[18]

The network is managed, maintained and supported by the Coordination Centre for Scientific University Collections in Germany,[19] which was established by the Federal Ministry of Education and Research in 2012 following a recommendation by the German Council of Science and Humanities on scientific collections as research infrastructures.[20] The goal is to further develop university collections as decentralised research infrastructures as well as the establishment of a sustainable interdisciplinary and cross-site network.

In addition, local networks currently exist at some traditional German universities with a large number of collections. These networks take the form of working groups and strive to strengthen professionalisation of the work with and in collections and to enable the permanent use of and access to them.

The Society for University Collections (Gesellschaft für Universitätssammlungen e.V.), founded in 2012, is dedicated to fostering and promoting university collections and represents the common interests of such collections to decision-makers in science and research policy, external sponsors and the general public.[21]

---

[15]See the brochure about UMAC at http://publicus.culture.hu-berlin.de/umac/pdf/UMACFlyer.pdf (Retrieved on 27 July 2015).

[16]Universitätsmuseen und –sammlungen im Hochschulalltag. Aufgaben – Konzepte – Perspektiven, http://universitaetsmuseen.hu-berlin.de (Retrieved on 27 July 2015).

[17]See key figures: http://portal.wissenschaftliche-sammlungen.de/kennzahlen/1/global (as of July 2015).

[18]http://wissenschaftliche-sammlungen.de/de/netzwerk/netzwerk-universitaetssammlungen (Retrieved on 27 July 2015).

[19]http://wissenschaftliche-sammlungen.de (Retrieved on 27 July 2015).

[20]The German Council of Science and Humanities, Recommendations on Scientific Collections as Research Infrastructures, Drs. 10464-11, Berlin 28 January 2011, available at http://www.wissenschaftsrat.de/download/archiv/10464-11.pdf (German) and http://www.wissenschaftsrat.de/download/archiv/10464-11-11_engl.pdf (English) (Retrieved on 27 July 2015).

[21]http://gesellschaft-universitaetssammlungen.de (Retrieved on 15 September 2016).

4.3 Network of Scientific Collections

The network Scientific Collections International (SciColl), which was launched and established between 2011 and 2013, goes far beyond the interests and concerns of natural science collections. SciColl is "a global consortium devoted to promoting the use and impact of object-based scientific collections across disciplines, including archaeology, biology, biomedicine, earth and space sciences, technology and others".[22] A central tenet of the network is "that the next generation of interdisciplinary research needs access to collections across traditional disciplinary boundaries"[23]: "we will facilitate a new generation of interdisciplinary research that relies on collections in different fields".[24]

4.4 Networks of Natural History Collections

There is a variety of networks of natural history collections. In addition, a great number of natural science organisations and societies have special working groups that address the topic of collections, for example, the German Ornithologists' Society's (Deutsche Ornithologen-Gesellschaft e.V.) working group on ornithological collections[25] and the curators' working group of the Society for Biological Systematics (Gesellschaft für Biologische Systematik—GfBS),[26] as well as digital networks like the DNA Bank Network, whose main goal is to improve access to well-documented genetic resources for biodiversity research,[27] resources which are also important for zoological collections. As it is not possible to cover all existing organisations in this short article, only the most relevant networks established to promote and support natural history collections will be listed.

Two networks at the national level should be mentioned: The German Natural History Research Collections Consortium (Deutsche Naturwissenschaftliche Forschungssammlungen—DFNS)[28] comprises more than 100 million objects in natural history collections. The association promotes cooperation and coordination between its member institutions.[29] The Humboldt Ring is an association of six natural history institutes and collections in Germany[30]: Its "primary strategic goal is

[22]http://scicoll.org/about.html (Retrieved on 27 July 2015).

[23]http://scicoll.org/research.html (Retrieved on 27 July 2015).

[24]http://scicoll.org/about.html (Retrieved on 27 July 2015).

[25]http://www.do-g.de/fachgruppen (Retrieved on 27 July 2015).

[26]http://www.gfbs-home.de (Retrieved on 27 July 2015).

[27]http://www.dnabank-network.org (Retrieved on 27 July 2015).

[28]Deutsche Naturwissenschaftliche Forschungssammlungen (DNFS), http://www.dnfs.de (Retrieved on 27 July 2015).

[29]http://www.dnfs.de/seite/mitglieder (Retrieved on 27 July 2015).

[30]http://humboldt-ring.de/index.php/de/mitglieder (Retrieved on 27 July 2015).

raising recognition of the national and international significance of the Humboldt Ring, while at the same time retaining federal autonomy [...] primary substantive goal is to promote and expand innovative, integrative research, including the necessary research infrastructure in biodiversity and Earth system research".[31]

At the European level, the collections are represented by the Consortium of European Taxonomic Facilities (CETAF), a consortium of scientific institutions—natural history and zoological museums, research institutes and botanical gardens—in Europe. The aim is "to promote training, research and understanding in systematic biology and palaeobiology, and facilitate access to information (collections) and the expertise of its member institutions across Europe".[32]

The Biological Collection Access Service for Europe (BioCASE) "is a transnational network of biological collections of all kinds. BioCASE enables widespread unified access to distributed and heterogeneous European collection and observational databases using open-source, system-independent software and open data standards and protocols".[33]

The Society for the Preservation of Natural History Collections (SPNHC) was founded in 1985 and "is an international society whose mission is to improve the preservation, conservation and management of natural history collections to ensure their continuing value to society".[34] Today the network has more than 600 members around the world.

The International Committee for Museums and Collections of Natural History (NATHIST), a committee of the International Council of Museums (ICOM), is also active worldwide. It is "concerned with the conservation of biological diversity in museum collections as well as in the natural environment, the scientific study of the world's natural heritage and the education of the wider public through museum displays, conferences, field trips, etc".[35] The interests and concerns of natural history collections are also additionally represented in the Natural History Collections Working Group of the ICOM Committee for Conservation (ICOM-CC). It "promotes the discussion and dissemination of information on ethical and technical issues concerning the acquisition, preservation and conservation of natural history objects and collections".[36]

[31]http://humboldt-ring.de/index.php/de/ziele (Retrieved on 27 July 2015).

[32]http://www.cetaf.org/about-us/what-cetaf (Retrieved on 27 July 2015).

[33]http://www.biocase.org (Retrieved on 27 July 2015).

[34]http://www.spnhc.org (Retrieved on 27 July 2015).

[35]http://icom.museum/the-committees/international-committees/international-committee/international-committee-for-museums-and-collections-of-natural-history (Retrieved on 27 July 2015).

[36]http://www.icom-cc.org/33/working-groups/natural-history-collections (Retrieved on 27 July 2015).

4.5 Networks Open to All Types of Collections

The largest network that is open to all types of collections is the International Council of Museums (ICOM), which now has 35,000 members, 118 national committees, 30 international committees, 5 regional alliances and 20 affiliated organisations.[37] Today it has a presence in 136 countries and territories. ICOM Europe is one of its five regional alliances.[38] Europe also hosts the Network of European Museum Organisations (NEMO).[39] It was founded "as an independent network of national museum organisations representing the museum community of the member states of the Council of Europe".[40]

The largest national committee of the International Council of Museums by membership is the German committee, with over 5000 members, making it the largest organisation of museums and museum specialists in Germany.[41]

The German Museums Association (DMB), founded in 1917, is the oldest association representing the interests of all museums.[42] The DMB's central working organs are its special interest groups and working groups. One of its special interest groups is concerned with natural science museums.

4.6 Closing Remarks

All of these networks, large and small, virtual and physical, not only contribute to increasing awareness and appreciation of collections but also to a considerable improvement in scientific infrastructures.

The work that they do could hardly be carried out by individuals or single collections; such work requires alliances of organisations that allow many different actors to contribute their expertise. Such networks are essential to the professionalisation of collections work. The more we are able to successfully intermesh existing infrastructures to create a sustainable, superregional network that cuts across disciplinary and geographical boundaries, the better we will be able to utilise the potential of scientific collections for research, teaching and education.

[37] http://icom.museum/icom-network (Retrieved on 27 July 2015).

[38] http://www.icom-europe.org (Retrieved on 27 July 2015).

[39] http://www.ne-mo.org (Retrieved on 27 July 2015).

[40] http://www.ne-mo.org/about-us.html (Retrieved on 27 July 2015).

[41] http://www.icom-deutschland.de (Retrieved on 27 July 2015).

[42] Deutscher Museumsbund, http://www.museumsbund.de (Retrieved on 27 July 2015).

References

Bremer T (2001) Foreword. In: Bremer T, Wegener P (Coordinators) Alligators and astrolabes. Treasures of University Collections in Europe, Halle

Herrmann K, Weidemann B (2007) Zoologische Sammlung. In: Andraschke U, Ruisinger MM (eds) Die Sammlungen der Universität Erlangen-Nürnberg, catalogue for the exhibition Ausgepackt. Die Sammlungen der Universität Erlangen-Nürnberg, Stadtmuseum Erlangen, Nuremberg, pp 213–222

Chapter 5
Legal and Ethical Challenges: From Collection Management to Access and Benefit-Sharing

Cornelia Löhne, Peter Giere, and Dirk Neumann

Abstract Building, using, and managing zoological research collections are complex and demanding tasks, not only from a scientific point of view. In fact, scientists and collection managers are also embedded in a multifaceted sphere of conventions, regulations, and legislation. An important international framework for the exploration and conservation of biodiversity is the United Nations' Convention on Biological Diversity (CBD). However, the CBD does not only focus on conservation and sustainable use of biodiversity. It also sets out basic principles for a fair and equitable sharing of benefits arising from its utilization. Those principles have been implemented in national laws and international agreements on access and benefit-sharing, such as the Nagoya Protocol. In many cases, however, those laws turn out to be critical impediments for the access to and exchange of biological material, research results, and other information within the scientific community. The article will provide an overview on the concept of access and benefit-sharing, the Nagoya Protocol and its implementation in Europe and Germany, as well as challenges and recommendations for collection management. It will also shortly address other regulations affecting the preparation and transportation of zoological samples, i.e., the Convention on International Trade in Endangered Species of Wild Fauna and Flora (CITES), European legislation on animal by-products, and international rules for the air shipment of dangerous goods.

C. Löhne (✉)
Zoologisches Forschungsmuseum Alexander Koenig – Leibniz-Institut für Biodiversität der Tiere, Adenauerallee 160, 53113 Bonn, Germany

Botanische Gärten, Rheinische Friedrich-Wilhems-Universität Bonn, Meckenheimer Allee 171, 53115 Bonn, Germany
e-mail: c.loehne@uni-bonn.de

P. Giere
Museum für Naturkunde – Leibniz-Institut für Evolutions- und Biodiversitätsforschung, Invalidenstr. 43, 10115 Berlin, Germany

D. Neumann
Zoologische Staatssammlung München, Münchhausenstraße 21, 81247 Munich, Germany

L.A. Beck (ed.), *Zoological Collections of Germany*, Natural History Collections,
DOI 10.1007/978-3-319-44321-8_5

Keywords Access and benefit-sharing • Nagoya protocol • Genetic resources • Collection management

5.1 Introduction: A Historical Perspective

Collecting, collating, and comparing specimens of animals, plants, and other organism are fundamental methods of taxonomy, the science of describing, naming, and classifying life on Earth. Carolus Linnaeus himself, the eighteenth-century pioneer of taxonomy, was a strong advocate of scientific collecting (Müller-Wille 2007). Linnaeus was engaged in a dense, international network of scientists with whom he intensively exchanged information and biological specimens, and he encouraged his students and colleagues to travel, to explore nature, and to collect specimens wherever possible (Sörlin 2000). During the Linnaean era, collections gained the importance as primary resources, archives, and laboratories for natural sciences they have today. The development of natural history collections was mainly driven by the scientific motivation to discover, describe, and classify organismal life (Pomian 1994); legal and ethical aspects were of minor importance. Ever since then, natural history collections play an important role not only in scientific research but also in collecting and exchanging collection material with other researchers and institutions. In the course of the twentieth century, however, other factors than the scientific interests of collectors, researchers, and institutions more and more influenced the way how scientific collection material was obtained and exchanged. Especially the increasing complexity of international environmental regulations and nature conservation treaties restricted the freedom of scientist and collectors.

The first important milestone was the inauguration of the "Convention on International Trade in Endangered Species of Wild Fauna and Flora" (CITES) in 1973, which was designed to monitor the transfer of protected species and to minimize the threat to species by international trade. In the 1960s it had been realized that the demand of products derived from threatened species, especially in Europe and the United States of America, resulted in an increased trade of such products. Species protected by CITES are listed in the three appendices of the treaty. CITES is implemented individually by national legislation of the parties of the convention. Any transborder transfer of protected species, no matter whether with commercial or noncommercial intent, has to comply with the respective laws of the country of origin and the destination country (see Textbox 1). However, the international community soon realized the potential impediments for research on such species and established a simplified process for transfers between registered scientific institutions.

Besides CITES as the most prominent example, collectors and collections also have to comply with national conservation laws, veterinary regulations, and the rules and conditions of carriers or carrier organizations (e.g., ICAO and IATA for air transport). More recently, new European legislation on the import of the so-called animal by-products and regulations on the transportation of "dangerous goods" has become relevant for zoological collections. See Textbox 1 for an overview on these regulatory frameworks and the challenges arising out of them. Renner et al. (2012) provide a comprehensive analysis of the respective stipulations that need to be considered when collecting or transferring material from in situ or ex situ sources.

Textbox 1: Overview on Regulations Affecting the Preparation and Transportation of Zoological Samples

CITES. The "Convention on International Trade in Endangered Species of Wild Fauna and Flora" is, at the European level, implemented by the Council Regulation (EC) No 338/97 and the Habitats Directive (Council Directive 92/43/EEC), supporting the protection of wild fauna and flora. Some practical problems arise from discordant lists of protected species included in the respective annexes, with the European legislation being more inclusive by adding a fourth annex (D). CITES, however, provides simplified measure for the exchange of biological material of protected species between noncommercial research institutions, registered at the CITES secretariat. Currently this register comprises almost 800 research and collection institutions worldwide. Shipments between such institutions must be labeled accordingly and are, thus, exempt from CITES custom checks.

Animal by-product legislation. EU Regulation No. 1069/2009 and its amendment (EU No. 142/2011) give detailed rules on the import of the so-called animal by-products (defined as entire bodies or parts of animals, products of animal origin, or other products obtained from animals, which are not intended for human consumption, including oocytes, embryos, and semen). Even though recent amendments allow derogation of veterinary inspection if samples agree with the concept of safe sourcing or safe treatment, legal import of such consignments is at the discretion of the national authority of the member state of destination, with (EU) No. 142/2011 setting out the minimum requirements. The minimum period for notification to veterinarian border inspection posts for animal by-products (e.g., preserved biological material returned from fieldwork in the checked luggage) is 12 h prior to arrival. Safe treatment includes, e.g., fixation in 4 % formalin, preservation in ethanol (min 96 %), drying (only for insects and spiders),

(continued)

and boiling or warm water maceration (in case of mammals) with subsequent hydrogen peroxide cleaning of bones.

Dangerous Goods Regulations. Postal carriers and airlines have strict rules for the transport of the so-called "dangerous goods" (especially flammable liquids or corrosive chemicals). All substances used for fixation and preservation of zoological collection material ("safe treatment" as described above) are restricted for transport, e.g., under the Universal Postal Union's Convention, the International Civil Aviation Organization (ICAO), or the corresponding Dangerous Goods Regulations of civil airlines under the International Air Transportation Association (IATA). Under ICAO/IATA Special Provision A180, however, biological material treated with either ethanol, isopropanol, or formalin solution is not regarded as dangerous goods, if specific packing and marking requirements are met. This has been adapted by the UPU recently in their Model Transport Framework Agreement under point 1.11, which recognizes the exception of certain dangerous goods provided for in the ICAO/IATA Dangerous Goods Regulations, and thus allows preserved specimens for transportation.

Renner et al. (2012) provide further details on the abovementioned regulations, including proper documentation of postal shipments, custom law, and clearance, and discuss observed practical problems.

5.2 Access and Benefit-Sharing: The Concept and Its Implementation

Another groundbreaking event in international policy was the adoption of the Convention on Biological Diversity (CBD) during the United Nation's Conference on Environment and Development in Rio de Janeiro in 1992. The CBD finally came into force in December 1993. It is not only remarkable because the conservation and sustainable use of biodiversity were for the first time acknowledged as an official objective of international policy. It also reflects an important change of paradigm: genetic resources (see definitions in Textbox 2) were no longer considered common heritage of mankind, but member states were given the sovereign rights over the genetic resources within their borders. It is open to countries to grant free or to restrict access to their genetic resources and to establish requirements for a fair and equitable sharing of benefits arising from the utilization of those resources. Thereby, the CBD introduced an economic aspect to biodiversity.

Textbox 2: Glossary on Access and Benefit-Sharing (ABS)

Access. The acquisition of genetic resources (GR) or associated traditional knowledge (TKaGR) from a providing country. This term has not been defined in the CBD or the NP and may, thus, be used differently by some countries or organizations. The European Union limits this term to the acquisition of GR or TKaGR from providing countries that are parties of the NP.

Benefits. Not defined in the NP, but may be (1) monetary when research and development leads to a commercial product (e.g., royalties, milestone payments, licensing fees) or (2) nonmonetary (e.g., technology transfer, enhancement of research skills, sharing research results, research partnerships, access to scientific information and research results).

Convention on Biological Diversity (CBD). A United Nations' treaty that came into force on 29 December 1993. It has three major objectives: (1) the conservation of biodiversity, (2) its sustainable use, and (3) the fair and equitable sharing of benefits arising from the utilization of genetic resources.

Genetic resource (GR). Any material of plant, animal, microbial or other origin that contains functional units of heredity and that is of actual or potential value (Definition according to CBD).

Mutually agreed terms (MAT). An agreement reached between the providers of genetic resources and users on the conditions of access and use and the benefits to be shared between both parties.

Nagoya Protocol (NP). A subsidiary agreement to the CBD that implements Article 15 (ABS) and Article 8j (traditional knowledge). The NP came into force on 12 October 2014 and its full title reads: "Nagoya Protocol on Access to Genetic Resources and the Fair and Equitable Sharing of Benefits Arising from their Utilization to the Convention on Biological Diversity."

Prior informed consent (PIC). The permission given by the competent national authority of a provider country to a user prior to accessing genetic resources, in line with an appropriate national legal and institutional framework. It is a legal document that states what the user can and cannot do with the material.

Traditional knowledge (TK). There is currently no generally accepted definition of TK. The interpretation of the World Intellectual Property Organization (http://www.wipo.int/tk/en/tk/) might, however, be helpful to understand the concept: "[TK] is knowledge, know-how, skills and practices that are developed, sustained and passed on from generation to generation within a community, often forming part of its cultural or spiritual identity." The Nagoya Protocol only covers TK associated with genetic resources (TKaGR), not TK as a separate element.

Utilization of genetic resources. To conduct research and development on the genetic and/or biochemical composition of genetic resources, including through the application of biotechnology as defined in Article 2 of the Convention (definition according to the Nagoya Protocol).

The general provisions on access and benefit-sharing (ABS) are laid down in Article 15 of the CBD. According to this, access to genetic resources shall be subject to a "prior informed consent" (PIC) of the providing country's governmental authority and shall be based on "mutually agreed terms" (MAT), which have to be negotiated between the providing country and the user. Thus, since 1993 the access to biological material in the wild or in collections is inevitably linked with ABS. Many countries later on established national access legislation, which scientists need to take into account when collecting biological samples in the field or acquiring genetic resources from such countries. In some countries, ABS regulations turned out to be quite obstructive even for basic research on biodiversity, mainly due to intransparent or overly strict access regulations and bureaucratic difficulties in obtaining the necessary permits (e.g., Jinnah and Jungcurt 2009; Martinez and Biber-Klemm 2010).

The ultimate objective for introducing an economic aspect was to generate revenues that help developing countries to conserve their biodiversity (Rosendal 2000), to prevent the so-called "biopiracy," and to control the exploitation of genetic resources. Providing countries aimed at having control over the flow of their genetic resources through the different instances of scientific use and commercialization, i.e., through the so-called genetic resources value chain. In that sense, basic research and scientific collections are considered to be relevant intermediary players along that value chain, even though their focus is clearly noncommercial (Brahy and Louafi 2007; Martinez and Biber-Klemm 2010). During the negotiation process toward an international regime on access and benefit-sharing, the scientific community was very actively involved and demanded that such a regime should provide for a continuous facilitated access to genetic resources for noncommercial research purposes (Schindel et al. 2009).[1] In spite of those efforts, however, the apprehension prevailed that the transition from noncommercial research to applied research and commercialization remains blurred in many instances and that exemptions for noncommercial research would create loopholes for the commercial exploitation of genetic resources (Buck and Hamilton 2011). In fact, traditional scholarly standards such as the exchange of biological samples and the publishing of research results, knowledge, and information on genetic resources may eventually facilitate the subsequent use of genetic resources and associated information by third parties (Laird et al. 2002). Therefore, national and international ABS laws and regulations usually also cover the noncommercial use of genetic resources by scientists and scientific collections.

This holds true also for the Nagoya Protocol (NP, Textbox 2), a supplementary agreement to the CBD that entered into force on 12 October 2014. The NP is the result of a long political debate about an international regime on access and benefit-sharing (Buck and Hamilton 2011). It specifies the provisions laid out in Article

[1]Based on Schindel et al. (2009), it was suggested to define noncommercial research as "research with the goal of adding knowledge to the public domain, without restrictions or proprietary ownership."

15 of the CBD, including the provisions on traditional knowledge (TK) associated with genetic resources (Article 8j, CBD), and calls upon member states to introduce legislative measures governing access to genetic resources and compliance of users. Since the coming into force of the Nagoya Protocol, scientists (and other users of genetic resources) have to consider ABS at two different levels: (1) when obtaining genetic resources, they have to abide by the providing country's ABS laws and comply with any mutually agreed terms and (2) when utilizing genetic resources, they have to fulfill the national compliance legislation in their home country.

Within the European Union, the relevant obligations of the NP are implemented by EU Regulation No. 511/2014.[2] According to this regulation, each EU member state has to designate a competent national authority which has to monitor and check users of genetic resources and associated traditional knowledge. In Germany, this will be the task of the Federal Agency for Nature Conservation (Bundesamt für Naturschutz), in collaboration with some other federal bodies (Bundesanstalt für Landwirtschaft und Ernährung, Robert-Koch-Institut, Patent- und Markenamt). Each user of genetic resources is obliged to undertake due diligence in order to obtain and utilize genetic resources in line with the provisions of the Nagoya Protocol and has to keep all relevant documentation for a minimum of 20 years after ending the utilization of a specific genetic resource. The competent national authority is entitled to undertake on-the-spot checks of users and to issue sanctions against illegal utilization of genetic resources (e.g., penalties and seizure of the resources). Furthermore, users of genetic resources or associated traditional knowledge have to report to the authority in the event of (1) receiving external funding for research projects involving genetic resources or (2) bringing a product based on genetic resources on the European market. Especially the first checkpoint is most relevant for biodiversity researchers and collections and will, without doubt, increase the bureaucratic burden in basic research considerably. Here, the European legislation unfortunately did not follow the provisions of the NP, which requests each member country to "create conditions to promote and encourage research which contributes to the conservation and sustainable use of biological diversity, particularly in developing countries, including through simplified measures on access for non-commercial research purposes" (Art. 8a, NP). The EU regulation does not govern access to genetic resources of EU member states. Instead, each member state may establish access laws individually, and in some European countries (e.g., France, Hungary, Spain), such legislation is already in place or underway.

---

[2]Full title: "REGULATION (EU) No 511/2014 of the European Parliament and of the Council, of 16 April 2014, on compliance measures for users from the Nagoya Protocol on Access to Genetic Resources and the Fair and Equitable Sharing of Benefits Arising from their Utilization in the Union." This regulation is applicable since the coming into force of the Nagoya Protocol, with the exception of Articles 4 (on user obligations), 7 (on monitoring of users), and 9 (checks) becoming applicable with one year delay on 12 October 2015.

5.3 Challenges Arising from the Nagoya Protocol and Its Implementation

Most of the challenges for biodiversity researchers and collections stem from the rather imprecise and insufficient terminology used in the CBD, the Nagoya Protocol, and their European and national implementations. When the term "genetic resources" was defined 25 years ago as "material containing functional units of heredity and with actual or potential value," it was targeted at living material suitable for analysis of DNA or other biochemical compounds, but not on preserved material as found in collections. Today, DNA can be extracted from almost any biological material, even from plant remains hundreds of years old or from subfossil bones. Thus, "genetic resources" is far more inclusive than originally intended, and ABS may affect research disciplines that are only indirectly linked to biology, such as earth sciences (genetic resources present in water or soil samples) or archaeology.

Even more challenging is the definition of "utilization" as "research and development on the genetic and/or biochemical composition of genetic resources" (see also Textbox 2). "Research and development" remain undefined and what is covered by "utilization" is unclear. Will there be a divide between classical morphological studies (which do not touch the "genetic or biochemical composition") and studies involving analysis of DNA or other molecules extracted from biological material? Does the act of downloading and analyzing sequence data from online databases imply "utilization of genetic resources" in the sence of the Nagoya Protocol? These questions are just two examples that illustrate the dimension of the problem. It is obvious that collection management is challenged with additional documenting and reporting requirements covering all relevant instances of utilization within the institution and any transfer of material. In view of these new challenges, the Consortium of European Taxonomic Facilities (CETAF), the major European network for natural history collections, developed a set of guidance documents including a code of conduct, a general use statement that should guide MAT and PIC negotiations, and a more detailed description of best practice for taxonomic collections (available from the organization's website, www.cetaf.org). Similar standards have been developed by other stakeholder groups, such as the Global Genome Biodiversity Network (GGBN, for DNA repositories), the World Federation for Culture Collections (WFCC, for collections of microorganisms), or the International Plant Exchange Network (IPEN, for botanic gardens). Those provide general guidance on compliance with legal and ethical challenges arising from the CBD and the Nagoya Protocol. The actual workload and responsibility, however, will be on the collection managers in each institution.

5.4 How to Handle ABS? Recommendations for Researchers and Collection Managers

Some general recommendations to collection managers and researchers can be drawn from the legal framework and its implications as discussed above. Those recommendations are laid out in detail in CETAF's Code of Conduct and Best Practice and shall be given here in a condensed form:

1. Institutions and researchers should *acquire* only such biological material that has been obtained in line with the Nagoya Protocol and the providing countries' legislation (no matter whether from in situ or ex situ sources). This affects the standard procedures for field trips, which need to take into account some time prior to collecting for obtaining PIC and MAT from the competent national authority (besides obtaining other permits, such as research or collection permits). The same applies to accepting incoming material from other countries that are parties to the Nagoya Protocol, either through loan or unsolicited shipping or brought in by students, colleagues, or guest scientists. Information on each country's access regulations and competent authorities can be found on an internet portal maintained by the CBD secretariat, the ABS Clearing House (https://absch.cbd.int).
2. Institutions need to *manage* collections and associated data in a way that the provider of the biological material can be traced at any instance in the collection and research workflow and that any related terms and conditions are easily accessible. All relevant information on access to genetic resources, especially whether documents such as PIC and MAT are needed (restricted access) or not (free access), as well as the documents themselves and information on utilization of the material (who, when, how, etc.), should to be stored permanently. For transfers to third parties (permanent or temporary), documents on the legal status of the respective genetic resources may need to be forwarded.
3. Researchers and institutions should be sure on the status of biological material and *use* it only in line with the terms and conditions under which it was acquired. Special consideration should be given to any restrictions regarding specific analytical methods, the publication of research results or information (e.g., DNA raw data), and the transfer to third parties. If the researcher or an institution intends to use the material in a way not covered by the original terms and conditions, the respective competent national authority should be contacted in order to seek new PIC and negotiate new MAT. Note that all original agreements might become void in case of infringements, deliberate or unintentional.
4. Any *benefits* derived from the utilization of genetic resources must be shared fairly with the providing country and the local cooperation partners as agreed in negotiated terms. In the noncommercial context of basic research and natural history collections, this is usually done by nonmonetary benefits such as the transfer of knowledge, capacity building, joint publications, etc. It is recommended to document such benefit-sharing (including that undertaken

during field work, such as training of local students) and to present it to authorities and the public in order to increase trust among providers of genetic resources.

5. Institutions are advised to develop *internal policies*, guidelines, and workflows that help to comply with ABS. Such policies need to address all relevant steps of acquisition, storage, utilization, and transfer of genetic resources and associated information. It is also necessary that institutions train their staff and inform scientific visitors about the principles of access and benefit-sharing and the legal and practical implementations.

5.5 Conclusions and Outlook

To be clear, the principle of access and benefit-sharing is a very meaningful and politically essential instrument that helps implementing the CBD. The three major objectives of the CBD have equal weight, and therefore the conservation of biodiversity goes along with its sustainable utilization and the sharing of benefit arising thereof. Institutions and researchers that want to contribute to the exploration and conservation of biodiversity have the moral obligation to act in line with the spirit of the CBD and the provisions of the Nagoya Protocol. Nevertheless, the obligations laid out in the NP and the respective European and national legislation add a considerable bureaucratic burden to already existing laws and regulations biodiversity researchers and collections have to comply with.

Existing standards and best practices developed by the scientific community (e.g., CETAF, GGBN, WFCC; see above) might help individual scientists and institutions to understand and abide by those regulations. A broad adoption and implementation of such standards by the scientific community will also help to increase transparency and trust among providing countries, which might in the long term lead to simplified procedures on access and utilization for noncommercial scientific purposes. At the present stage, however, a substantial alleviation of the situation, e.g., by a registering system for scientific institutions comparable to CITES, is not in sight. On the contrary, there seems to be a trend among providing countries to grant access to their genetic resources only with very strict MAT, including prohibition of transfer to third parties. Free exchange of materials, research results, and other information within the scientific community is, however, one of the fundamental principles of science since Linnaean times, as recently stressed by the CBD's Global Taxonomy Initiative. However, this might now become compromised. Therefore, scientists should build and engage in strong scientific networks, get involved with policy and decision-making and make their voice heard.

References

Brahy N, Louafi S (2007) The role of the research sector in ABS governance. Working Papers No. 9. Institut du développement durable et des relations internationals, Paris, 19 pp http://www.iddri.org/Publications/The-role-of-the-research-sector-in-ABS-governance

Buck M, Hamilton C (2011) The Nagoya Protocol on access to genetic resources and the fair and equitable sharing of benefits arising from their utilization to the convention on biological diversity. Rev Eur Commun Int Environ Law 20:47–61

Jinnah S, Jungcurt S (2009) Could access requirements stifle your research? Science 323:464–465

Laird S, Alexiades M, Bannister K, Posey D (2002) Publication of biodiversity research results and the flow of knowledge. In: Laird S (ed) Biodiversity and traditional knowledge – equitable partnership in practice. Earthscan, London, pp 77–101

Martinez SI, Biber-Klemm S (2010) Scientists – take action for access to biodiversity. Curr Opin Environ Sustain 2:1–7

Müller-Wille S (2007) Collection and collation: theory and practice of Linnaean botany. Stud Hist Philos Sci C 38(3):541–562

Pomian K (1994) Sammlungen – eine historische Typologie. In: Grote A (ed) Macrocosmos in microcosmo: Die Welt in der Stube; zur Geschichte des Sammelns 1450 bis 1800. Leske & Budrich, Opladen, pp 107–126

Renner SC, Neumann D, Burkart M, Feit U, Giere P, Gröger A, Paulsch A, Paulsch C, Sterz M, Vohland K (2012) Import and export of biological samples from tropical countries – considerations and guidelines for research teams. Organisms, Diversity and Evolution 12:81–98

Rosendal GK (2000) The convention on biological diversity and developing countries. Kluwer, Dordrecht

Schindel DE, Häuser CL, Miller SE, participants in an international workshop: Bavikatte K, Beck E, Burks C, Davies N, Desmeth P, du Plessis P, Garrity G, Geeta R, Haas F, Holm-Müller K, Huntley B, Kamau EC, Kim W, Lyal C, Marinoni L, Martinez S, Matsuura K, Ni KJ, Ong P, Schönwitz R, Wahiche JD (2009) Preserving international access to genetic resources for non-commercial biodiversity research – submission of views from participants in the International Workshop on the topic of "Access and Benefit-sharing in Non Commercial Biodiversity Research", Bonn, 17–19 Nov 2008. UNEP/CBD/WG-ABS/8/INF/6. https://www.cbd.int/doc/?meeting=ABSWG-08. Last visited 27 July 2015

Sörlin S (2000) Ordering the world for Europe: science as intelligence and information as seen from the northern periphery. In: MacLeod R (ed) Nature and empire: science and the colonial enterprise. University of Chicago Press, Chicago, pp 51–69

Chapter 6
Modern Exhibition Concepts

Annette Scheersoi

Abstract Most museum visitors do not know that extensive collections of specimens are stored "behind the scenes" and that these collections are used for current research in biology (e.g. on biodiversity or climate change). In recent years, many museums have developed new exhibition concepts in order to inform the public about this important aspect of museums. In this chapter, several approaches are presented that are used by German natural history museums in order to introduce a broad audience to the fascinating world of curators and taxonomists, to show them the value of the collections and to explain the research being done on these collections.

Keywords Exhibition concepts • Collection items • Scientific methods • Interest

In most exhibitions at natural history museums, visitors are only shown a small selection of specimens, sometimes less than 1 % of the entire collection (Eder 2006). Depending on the goal, these selected objects are exhibited to convey and support a certain biological topic (evolution, systematics, ecology, etc.). Most visitors do not suspect that extensive collections are being stored "behind the scenes" at the museum and are being used for current zoological research. In recent years, in order to raise public awareness about this aspect of museums, many German museums have developed new exhibition concepts. The goal is to introduce a broad audience to the fascinating world of curators and taxonomists, to show them the value of the collections and to explain the research being done on these collections in a contemporary way. Typical scientific methods like observing, comparing, sorting, describing and identifying are highlighted here. The exhibitions use various approaches to give visitors access to the specimens in the collection and encourage them to engage in-depth with what is being shown. Emotional aspects play an important role here, created either by the original objects themselves or by the way in which they are presented, in order to spark the visitors' attention and interest.

A. Scheersoi (✉)
Didaktik der Biowissenschaften, Nees-Institut, Rheinische Friedrich Wilhelms-Unversität Bonn, Meckenheimer Allee 170, 53115 Bonn, Germany
e-mail: a.scheersoi@uni-bonn.de

L.A. Beck (ed.), *Zoological Collections of Germany*, Natural History Collections,
DOI 10.1007/978-3-319-44321-8_6

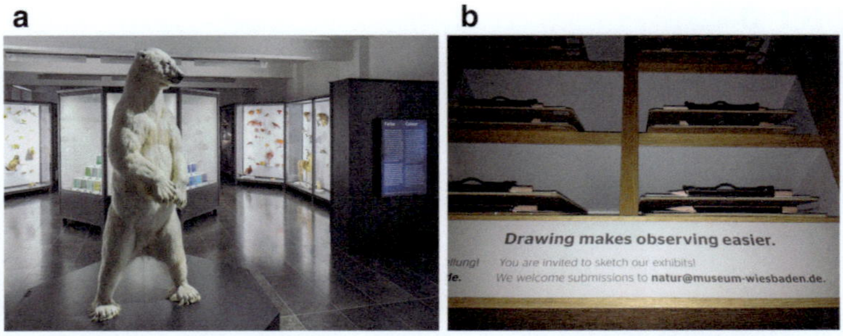

Fig. 6.1 (**a**, **b**) Wiesbaden Museum: A view of the "Color" room (Photo: © with Bernd Fickert, 2015) and drawing boards to promote precise observation (Photo: © with Annette Scheersoi)

6.1 Aesthetics

In the Wiesbaden Museum's new permanent exhibition "Aesthetics of Nature", objects from the natural history collection are not organized by biological themes, but are presented in four theme rooms (form, colour, movement and time) based on aesthetic aspects, so that the content can be accessed through the senses (Hoffmann 2014). A special lighting concept is also used to stage the objects. The goal is to build a bridge between science and art and to fascinate visitors in order to awaken their enthusiasm for nature. This presentation style is expected to attract new visitor groups, which can be especially important in a multipurpose facility. The exhibition intentionally avoids using hands-on media; instead, the focus is on observation, which is a central approach in biology. In order to encourage people to engage with the objects through precise observation, visitors are given stools and drawing boards so they can deepen and document their observations (Fig. 6.1).

6.2 Participation

Museum Koenig in Bonn is also counting on the effect of the original in its research exhibition "Unravelling the nature of diversity", presenting selected parts of its scientific collection that are organized in collection boxes and alcohol jars (Fig. 6.2). In order to show how research is done, the exhibit also presents typical research instruments (collection devices, preparation microscope, identification key, etc.).

In addition, there is an exhibit that lets even the youngest visitors experience the description of a new species ("Species nova") and encourages them to "publish" their findings afterward by hanging them up on a poster wall (next to research posters created by the scientists). This exhibit is an imaginary animal designed by the taxidermists, which visitors can study based on established criteria. They are asked to draw it and describe how it looks, its environment and what it eats.

a

b

Fig. 6.2 (**a**, **b**) Zoological Research Museum Koenig: Collection boxes and "Species nova" (Photos: © with Thomas Gerken, ZFMK)

Children are especially enthusiastic about this activity, while their parents can focus on the presentation of various research departments.

6.3 Stories and Personal Connections

The Natural History Museum of Mainz chose a narrative approach for its anniversary exhibit "Kept safely. Valuable objects tell their fascinating stories". Exemplary objects were chosen from the various collections (zoology, botany, palaeontology and mineralogy) and presented individually at different stations. For each item in the collection, a poster gives its name and where it was found, and tells a story about it (Fig. 6.3). Special research methods are also introduced *en passant*. There is separate textual information for younger visitors, indicated by the museum's mascot.

Visitors are also asked to participate actively and to solve a "stone puzzle": they are asked to make hypotheses and write down what they think an unidentified item in the collection might be. The most creative hypotheses are chosen by the museum employees and presented to the other visitors on a bulletin board as the "Idea of the Week".

The stories told by objects were also the focus of the exhibit "When skeletons talk" at the Natural History Museum in Basel. Here, visitors were able to look at original human skeletons to learn why the museum's collection is so important and which research methods modern natural sciences use to gain information by looking at bones. In addition to traditional hands-on elements, there were also audio installations on the individual fates of specific skeletons, thus creating emotional connections (Fig. 6.4).

6.4 A Look "Behind the Scenes"

In several museums, like Überseemuseum Bremen and Museum für Naturkunde Berlin, parts of the research collections have been integrated into the exhibition area, giving visitors a look at the researchers' work.

At the "Übermaxx" display rooms in Bremen, visitors can open drawers and pull out herbarium slipcases. Just like the collection areas that are not open to the public, the display rooms do not provide any didactic information. The collections' diversity mainly invites people to observe and compare the items. In particular, objects that cannot be used as exhibition pieces are made accessible to visitors in the display room (Becker 2006). That applies to skins and specimens preserved in alcohol, for instance, which are the best way to store samples for scientific work. The alcohol collection is also the main focus in Berlin. Around 276,000 jars are arrayed here in an artistic way; the collection is fascinating simply on the basis of its size and amazes people with its diversity (Fig. 6.5).

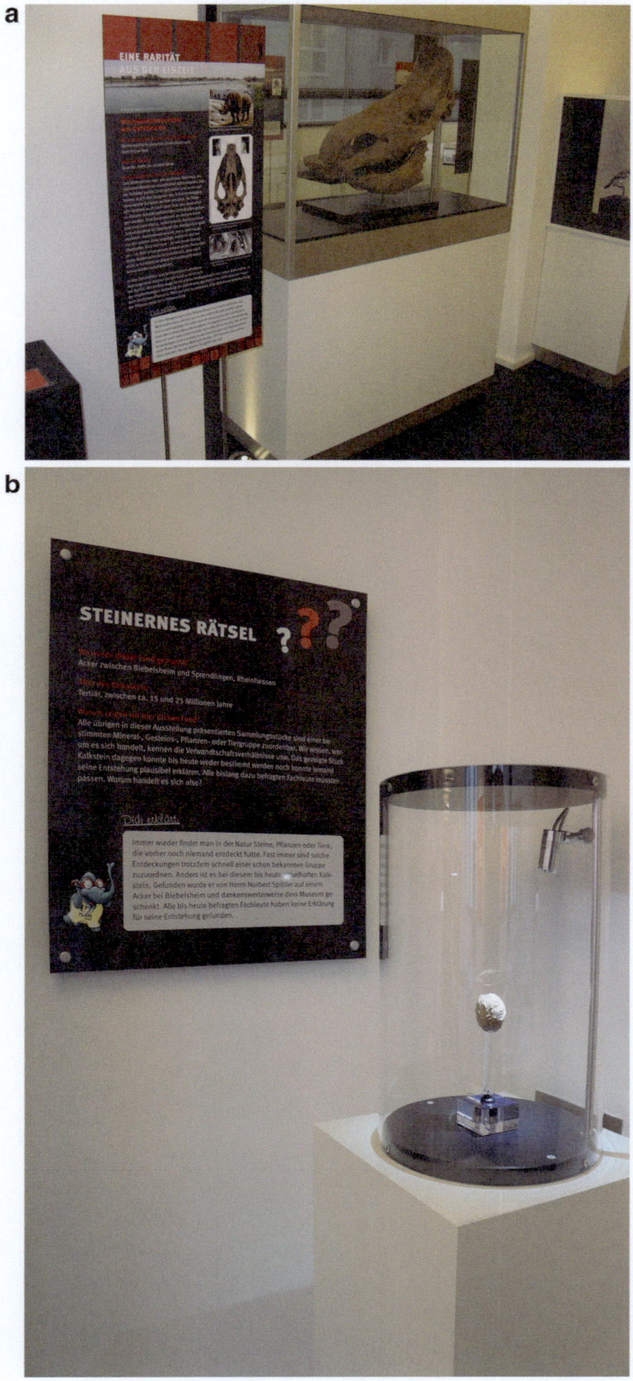

Fig. 6.3 (**a**, **b**) Natural History Museum Mainz: Collection item (Photo: © with Naturhistorisches Museum Mainz, Herbert Lutz) and the "stone puzzle" (Photo: © with Naturhistorisches Museum Mainz, Thomas Engel)

a

b

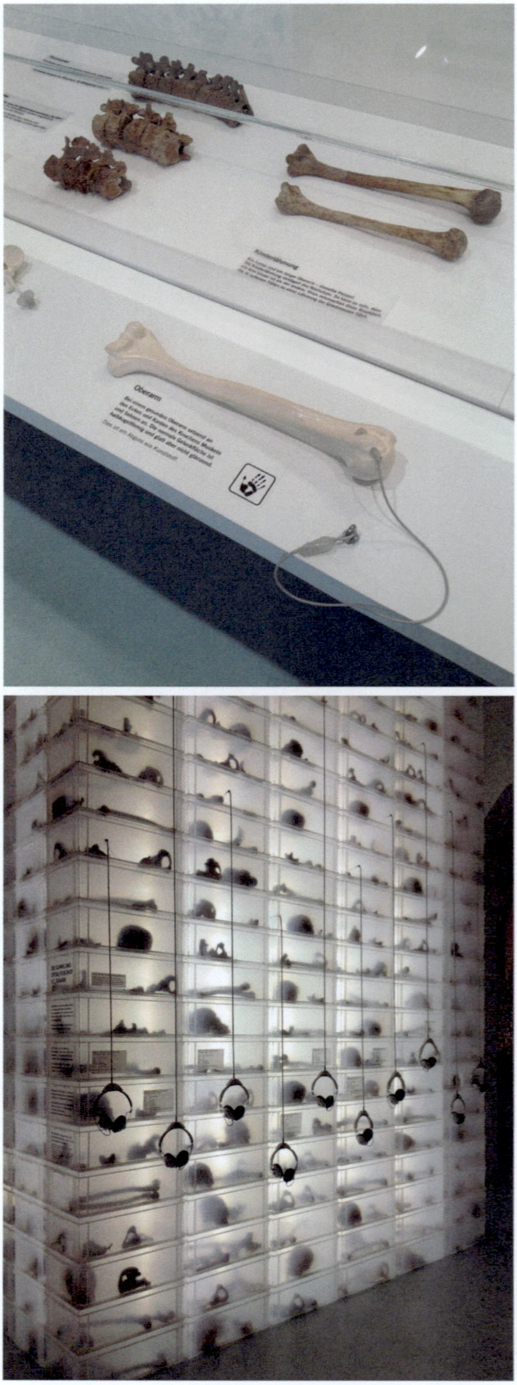

Fig. 6.4 Natural History Museum Basel: hands-on element and audio installation (Photos: © with Annette Scheersoi)

Fig. 6.5 (**a**, **b**) Display rooms: Übersee-Museum Bremen (Photo: © with Übersee-Museum, Matthias Haase) and Museum für Naturkunde Berlin (Photo: © with Museum für Naturkunde Berlin, Carola Radke)

6.5 The Relationship Between Human Beings and Nature

Another way to give a large audience access to the collection items and collections, and to demonstrate their importance, was realized at Ruhr Museum Essen. The "Terra cognita" exhibition focused on the collection items, but consciously added a cultural historical perspective, for instance, with poetry and prose excerpts to accompany the natural objects (Stottrop 2006). In addition, themed pull-out drawers contained original documents that demonstrated the nineteenth century nature lovers' passion for collection and research (for instance, through letters and painstakingly labelled specimen slides). In this way, the exhibition recognized the significance of these collection items not just in specialized terms but also as documents of the scientific discovery process and as part of cultural history. A "study room" with microscope access, a library and themed drawers interactively picks up on the subject of "collecting" (Fig. 6.6).

Another special exhibition at Ruhr Museum Essen, "Down and Under. The natural culture of the Ruhr region", was dedicated to the relationship between human beings and nature, presenting the collection items as witnesses to how nature has been treated and perceived.

Fig. 6.6 (**a**, **b**) Ruhr Museum: mammoth skeleton and "study room" in the "Terra cognita" exhibition; preserving jars, filled with soil samples polluted by heavy metal in the "Down and Under" exhibition (Photos: © with Ruhr Museum, Michael Rasche)

References

Becker P-R (2006) Neue Konzepte für Schausammlungen? Museumskunde 71(1):26–32

Eder J (2006) Sammlung, Forschung und Vermittlung – Spezifika naturwissenschaftlicher Museen. In: Neue Präsentationsformen in Naturkundemuseen. Museumskunde 71(1):10–13

Hoffmann D (2014) Ästhetik der Natur: Die neue Dauerausstellung im Museum Wiesbaden. Nat Museum 4:38–41

Stottrop U (2006) Das Potential naturkundlicher Sammlungen im Hinblick auf neue Präsentations- und Vermittlungsformen zum Thema Natur-Kultur. Museumskunde 71(1):33–41

Part II
The Collections

Chapter 7
ASCHAFFENBURG: The Collection of the Bavarian Royal National Academy of Forestry Aschaffenburg Is on Its Way to Becoming a Modern Museum of Natural History

Martin Hoepfner

Abstract The zoological collections of the Museum of Natural Science in the city of Aschaffenburg date back to the Bavarian Royal National Academy of Forestry, which from 1807 to 1910 existed under various names. The permanent exhibition today shows an extensive section of the domestic, but also the exotic bird and mammal fauna. Of special scientific value are the entomological collections, for example, the Singer Collection. With the establishment of Aschaffenburg's Museums Quarter, new opportunities have arisen for a new conception of these natural history collections.

Keywords Museum of Natural Science in the city of Aschaffenburg • Bavarian Royal National Academy of Forestry • Singer Collection • Aschaffenburg's Museums Quarter • Spessart • Bavarian Untermain • von Dalberg • News of the Museum of Natural Science of the city of Aschaffenburg (NaMA) and releases of the Museum of Natural Science of the city of Aschaffenburg (MiMA) • Society of Natural Sciences Aschaffenburg • Christian Schad Museum

---

Translation: Christine Taylor-Dorenkamp

M. Hoepfner (✉)
Museum of Natural Science Aschaffenburg, Schönborner Hof/West Wing, Wermbachstraße 15, 63739 Aschaffenburg, Germany

City Museums of Aschaffenburg, Schlossplatz 4, 63739 Aschaffenburg, Germany
e-mail: Martin.Hoepfner@museen-aschaffenburg.de; http://www.museen-aschaffenburg.de

L.A. Beck (ed.), *Zoological Collections of Germany*, Natural History Collections,
DOI 10.1007/978-3-319-44321-8_7

The Museum of Natural Science is located in the west wing of Schönborn Court (Museen der Stadt Aschaffenburg, Ines Otschik)

7.1 From a Forestry School to a Zoological Collection

Since 1970, the Natural Science Museum of the city of Aschaffenburg has been housed in the centrally located former palace of the Counts of Schönborn of 1681. As one of the seven exhibition sites of the city's museums, this house shows the geological-mineralogical and botanical and zoological collections with regional references to Spessart and the Bavarian Untermain. The zoological collection remains a focal point of the museum holdings that have accumulated for over 200 years.

However, it is precisely the zoological collection that owes its origin less to the intention of displaying animals in a publicly accessible natural history collection, rather to the parallel development of a remarkable educational institution.

In 1807, among other pioneering educational and cultural institutions, the forerunner of what would later become the Royal Academy of Forestry had been established as a simple, private forestry school. This school was approved and promoted by none other than Arch-Chancellor and Prince-Primate Karl Theodor Anton Maria von Dalberg, himself a native resident of Aschaffenburg. As early as 1808, Dalberg elevated the status of the forestry school to that of State Academy, where forestry students from all regions were trained in both practical and scientific aspects of forestry. Among other things named in the Royal Academy's 1820 renewal were the curricula concerning the forest and hunting mammals, the forest and hunting birds, and the "forest insectology," as well as the provision of two halls for establishing the library and the natural history collections (Papius 1820).

During an interruption of the Academy's operation from 1832 to 1844, in addition to all other collections, the collections of birds, bird eggs, mammals, and insects were housed in Würzburg. For the reopening of the Academy in 1844, all the collections were returned and transported back to Aschaffenburg by a river barge via the River Main (Scherg).

Apparently, it was early on when tendencies toward establishing a permanent collection began to develop as one could only infer based on an administrative order given by the incumbent ministry of that time, which admonishes that the high cost of the plethora of mammals and birds was "not to give room to spectator-curiosity, but to procure only that which is required for teaching" (Scherg).

Nevertheless, the zoological collection grew steadily, until the dissolution of the Academy in 1910, and, despite its original purpose, was known far beyond the borders of the city as a public display collection (von Herrlein 1857).

In 1859, for example, collections were publicized as such: "European birds: 550 species in some 1200 specimens . . ." and "Foreign Birds: 320 species in about 400 specimens" and "A collection of European beetles from 5000 species in 10,000 pieces" (Forestry Communications, Volumes 9–10).

The zoological collection today displays specimens not only of domestic but also of exotic birds (Museen der Stadt Aschaffenburg, Ines Otschik)

7.2 Achievements of the Aschaffenburg Society of Natural Sciences

Apart from the aforementioned interruption from 1832 to 1844, the Forestry Academy continued its operation under various names in Aschaffenburg, until its relocation to Munich in the year 1910.

At the initiative of individual professors and the Society of Natural Sciences that was founded in 1878, large parts of the natural history collections were kept in Aschaffenburg even after the dissolution of the Forestry Academy.

Founding members of this Society were professors of the Forestry Academy and other scientifically interested personalities of Aschaffenburg, who, in 1911, began presenting the museum collections in the former Academy's building. For many decades thenceforth, members of the Society of Natural Sciences meritoriously volunteered to supervise the municipal museum, enlarged the collections, and tied in with its own publication a series of the publications of the Forest Academy: News of the Museum of Natural Science of the city of Aschaffenburg (NaMA) and releases of the Museum of Natural Science of the city of Aschaffenburg (MiMA). To this day, the Society of Natural Sciences continues these series of publications and looks after the library that has grown through the extensive exchange of publications.

Between 1927 and 1966, Wilhelm Noll, who later became director of municipal operations, took over as the curator and volunteer manager of the museum. Thanks to him that much of the zoological collections could be evacuated during the Second World War and later retrieved. Although significant losses were recorded, in contrast to the completely destroyed collections in neighboring Hanau and Würzburg, the holdings in Aschaffenburg were largely preserved. The entomological collection was practically unscathed. This may well be regarded as a stroke of luck, because, just at this time, the entomological collections proved itself to be of particularly high scientific value within the museum. In the following decades, the zoological holdings grew vigorously once again, mainly through the collecting activities of individual Society members, as well as through purchases and donations. After the war, the glaring lack of space for the museum was only first recognized when the collections were exhibited and stored, beginning in 1970, in the renovated west wing of Schönborn Court. Since then, the permanent zoological exhibition has undergone only minor changes, and almost all innovations in the field of museum conception and design have passed it by (Schmittner 2007). In addition to the resulting impression given to museum visitors of the late 1960s museum presentation style, the nineteenth-century collection cabinets of the Forestry Academy strongly suggest the history of the collections' development.

Till this day visitors are able to relate to the history of the zoological collections (Museen der Stadt Aschaffenburg, Ines Otschik)

7.3 Outstanding Scientific Value of the Entomological Collections

In today's permanent collection, displayed on 250 m^2 of exhibition space, is an extensive cross section of the collections, in particular, not only domestic fauna but also the exotic bird and mammal fauna consisting of mostly historical specimens. Still other specimens can be found in the archive such as the herpetological collection and a collection of fishes with various real-skin preparations, collections of mollusks, and parts of the scientific legacy of Hans Stadler.

An outstanding importance is given to the zoological collection of the Museum of Natural Science due to its entomological collections, constituted by the collections of Singer, Elbert, Fröhlich, and Lepinski. These are the core of academically relevant collections for the Lower Franconia area, the Rhine-Main area, and somewhat beyond.

The Singer Collection includes, among other things, extensive collections of beetles and bugs not only of local relevance but also specimens from Central and Southern Europe and is thus the centerpiece of the total collection, bearing outstanding international importance. The accompanying documents, with their detailed collection data, are essential to the faunal surveys being conducted today (Burmeister 2003).

Part of the entomological collections is still today on display; the majority, however, is kept safely in storage.

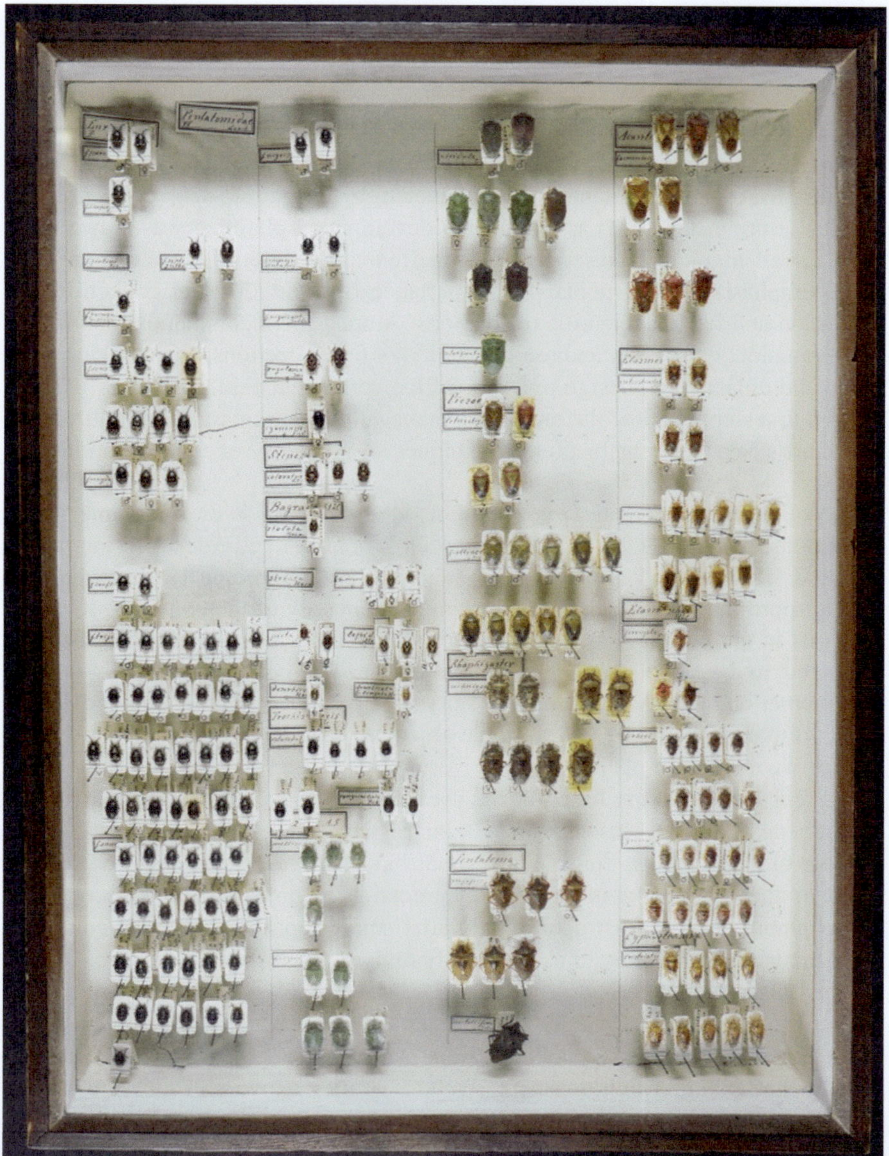

The entomological collection of Singer represents a centerpiece of the museum (Museen der Stadt Aschaffenburg, Ines Otschik)

7.4 High Potential of the Collections for a Prospective Natural History Museum

It has long been criticized, both by the public and among experts, that the obvious potential of the collections of the Museum of Natural Science is being greatly underused. An appraisal that was conducted by the zoological collections of Munich (Burmeister 2003) as well as the development plan that was drawn up in 2003 for all municipal museums of Aschaffenburg (Museum Development Plan 2003) emphasizes in particular the importance of the zoological collections for exhibition and further scientific evaluations. Among the recommendations were a reexamination, cataloging and reorganization of the collections, conservation measures, and didactic and design revisions of the permanent exhibition.

As a first step, in 2008, the museum's management commissioned employees of Frankfurt's Senckenberg Institute to conduct an inventory of the collections and exhibits.

In collaboration with museum staff, it was possible for the first time in the collections' history to identify all the museum's holdings and enter each into a new database. This resulted in nearly 3000 records for the zoological collection comprising more than 20,000 individual items. Due mainly to the convoluted species accounts in entomology, however, the actual number of preserved individuals is estimated to be about 120,000 (Malten 2009). The comprehensive systematic documentation and research of the history of the collection have been continued since, so that, in the future, every single object is readily available both for science and for the exhibition planning.

Current measures are aimed at the preservation of collections, improving the attractiveness of the Museum of Natural Science and the continued scientific analysis of the holdings. Considerable funds have therefore been invested in recent years in the renovation of buildings, protection against light, and the overall improvement of the infrastructure. Step after small step, the permanent collection now stands ready for being museologically revised.

The Museum of Natural Science has so far had to make do without its own staff, but has been cared for since 2008 by the technical, conservation, and curatorial staff of the city's museums. The collaborations with the Society of Natural Sciences, with Aschaffenburger Führungsnetz, Aschaffenburg's network of museum guides, with the children's group of the LBV (Landesbund für Vogelschutz, a Bird and Nature Conservation Society in Bavaria), and not least with the major museums and research institutions, are all indispensable for the operation and development of the museum and the collections.

Also single specimens like the extinct passenger pigeon illustrate the importance of the collection (Museen der Stadt Aschaffenburg, Ines Otschik)

With the development of Aschaffenburg's Museum Quarter (www.museen-aschaffenburg.de), great new opportunities have arisen for an entirely new concept of a regional natural history museum. Currently (2015), in the first phase of the Museum Quarter, next to the Art Gallery of the Jesuit Church, the new Christian Schad Museum is being constructed. The archaeological collections, the museum education, museum workshops, and storage facilities are also planned for this central museum campus in Aschaffenburg's upper end. With the construction of a new natural history museum at the site, the zoological collections will both provide a focus and teach vividly about local fauna as well as relate the history of collections.

References

Burmeister E-G (2003) Evaluation of the zoological collections of the Museum of Natural Science in Aschaffenburg, Munich 2003

Forestry communications, communications on the forestry and gamekeeping in Bavaria, published by the Royal Bavarian Ministerial Forestry Bureau, 2. Heft, Munich 1859

Malten A (2009) Inventory of collections and exhibits in the museum of natural science of Aschaffenburg (NMSA)
LORD Cultural Resources (2003) Museum development plan for the city museums of Aschaffenburg
Papius K (1820) The various modes of timber husbandry, a program for the opening of the lectures at the Royal Baier. Forestry Academy to Aschaffenburg for 1820/21, Aschaffenburg
Scherg T, SSAA, NL Scherg (No. 5), The natural sciences collection
Schmittner M (2007) Between birds of paradise and assassin bugs, Spessart, April 2007
von Herrlein A (1857) Aschaffenburg and its environs: a handbook for foreigners

Chapter 8
BAD DÜRKHEIM: The Zoological Collections at the Palatinate Museum of Natural History (Pfalzmuseum für Naturkunde – POLLICHIA-Museum), Bad Dürkheim (Germany)

Frank Wieland

Abstract The Palatinate Museum of Natural History—POLLICHIA-Museum in Bad Dürkheim (PMN)—houses the collections of the POLLICHIA, one of the oldest German societies of natural history and conservation. The zoological collections currently encompass approximately 210,000 specimens with a focus on Lepidoptera. The majority of the specimens was collected within the borders of today's federal state of Rhineland-Palatinate (Rheinland-Pfalz) throughout the past 200 years.

Keywords Palatinate Museum of Natural History • POLLICHIA • Rhineland-Palatinate • Insect collection • Mollusc collection • Vertebrate collection

8.1 POLLICHIA

The POLLICHIA is an active, registered society of natural history and conservation. Founded in 1840 on the initiative of the physician C. H. Schultz, it was named after and in honour of Johann Adam Pollich (1741–1780), physician and naturalist from Kaiserslautern, Germany (Müller 1990). Today, the society has approximately 2600 members. In the past 175 years, it has had a major impact on the study of the fauna and flora of Rhineland-Palatinate (Rheinland-Pfalz). Many of its members built collections of different organisms throughout their lifetime. The focus of the zoological collections was mostly on Lepidoptera and Coleoptera.

F. Wieland (✉)
Pfalzmuseum für Naturkunde – POLLICHIA-Museum, Hermann-Schäfer-Straße 17, 67098 Bad Dürkheim, Germany
e-mail: f.wieland@pfalzmuseum.bv-pfalz.de; http://www.pfalzmuseum.de

L.A. Beck (ed.), *Zoological Collections of Germany*, Natural History Collections,
DOI 10.1007/978-3-319-44321-8_8

8.2 History of the Collections

8.2.1 Origins

The natural history collections came to life in private hands and were later donated to the POLLICHIA. Some of them date back even before the foundation of the society and are more than 200 years old. In fact, the oldest specimens are lichen (*Xanthoria parietina*) collected in 1796 (Flösser et al. 2008) and 64 beetles without collection dates that survived as a remnant of a previously much larger collection dating back to ca. 1796 (Niehuis 2012). Some collectors only gathered a few drawers of specimens; others collected throughout their entire lives and accumulated hundreds of drawers of specimens.

8.2.2 The Early Years

Two years after the foundation of the society, the city council of Bad Dürkheim decided to provide space for the POLLICHIA to present their collections to the public. In 1842, two rooms in the north wing of the townhouse were used as display rooms, building the foundation for the Natural History Museum (Gettmann 1990). The POLLICHIA collections continued to grow and soon later they occupied further rooms until, in 1901, the entire first floor of the townhouse had turned into a natural history exhibition (Mehlis 1915). The need for additional space became more urgent. In 1872, a wealthy pharmacist from Bad Dürkheim had founded a trust to fund the construction of a natural history museum. In 1914, shortly before the cornerstone could be laid, World War I broke out and put an end to these plans. Most of the money was subsequently lost during the German hyperinflation between 1914 and 1923, thus ending the dream of a new museum building (Zumstein 1927; Spuhler 1965; Gettmann 1981).

A solution for the POLLICHIA collections had to be found as the conditions of the display rooms in the townhouse became worse. One of the options had been to transfer the collections to the nearby town of Speyer. However, it was soon recognized that this would mean a great cultural loss for Bad Dürkheim in the years to come. In 1920, the town council therefore decided to provide a building that had been acquired by the city during the hyperinflation (Spuhler 1965). However, it should take another 8 years to finish the preparation of the building for its new purpose (Gettmann 1990).

The building that became the POLLICHIA-Museum (then named "Naturwissenschaftliches Museum der Pfalz") for the next six decades was a former residential house, counting 34 rooms on three floors. The natural history collections were presented on the two upper floors, whereas the ground floor was occupied by the Museum of Local History (Walther 1953).

8.2.3 World War II

On March 18, 1945, only 48 h before the war ended in this part of the country, Bad Dürkheim suffered massive air raids. Nearly a third of the city was destroyed (Heiss 1995). At some point during the war, parts of the POLLICHIA collections had been transferred to safe locations or stored in secure basements where they survived the war more or less undamaged (Spuhler 1965; Kaiser 1967). The building itself was used to store the furniture of people who had lost their homes; the basement rooms were official air raid shelters for more than 200 people during the war (Kaiser 1967). The windows and many of the panes of showcases and dioramas were destroyed, but overall, the damage to the museum and its collections was minimal (Spuhler 1965). After the war, the collections were returned to the museum and the exhibitions were rebuilt.

The POLLICHIA-Museum occupied the building for more than half a century. By then, the building had become increasingly dilapidated. In the late 1970s, it was officially closed for the public, and the collections were in dire need of a new home.

8.2.4 The Palatinate Museum of Natural History - POLLICHIA-Museum

In 1981, the museum and its collections were transferred to a new locality. The building of the former "Herzogmühle" a historical mill first mentioned in a historical document in 1408, had been renovated and prepared for a new life as the Palatinate Museum of Natural History—POLLICHIA-Museum (PMN) (Fig. 8.1). An additional wing, added after World War II, today houses the collections and the offices as well as the library, while the main exhibition and usually one or two special exhibitions every year are presented to the public in the main building. The overall exhibition area encompasses about 2000 m^2.

Since 1989, the PMN has a branch on Lichtenberg Castle near the town of Kusel, Germany. At first, one of the buildings was opened for external PMN exhibitions. In 1998, the Primeval World Museum GEOSKOP (UGKU) opened its doors. While the zoological and botanical (and remnants of the Pleistocene fossil) collections and exhibitions are housed in the PMN main building in Bad Dürkheim, the main exhibition and special exhibitions at the GEOSKOP focus on the geological and palaeontological history of the Late Palaeozoic of Rhineland-Palatinate.

Since 2008 the PMN has successively been modernizing its exhibitions and added a new part to the building. The latter houses the new lobby, including an information desk and museum shop, a lecture room and, on the first floor, about 220 m^2 of floor space for special exhibitions (e.g. Flösser 2009).

There are currently 30 positions budgeted in the two branches of the museum.

Fig. 8.1 The Palatinate Museum of Natural History—POLLICHIA-Museum (PMN) housed in a historical mill in Bad Dürkheim

8.3 The Zoological Collections

The zoological collections comprise about 210,000 specimens. Although there are many specimens from all over the world, the majority was collected within the borders of today's Rhineland-Palatinate. The entire zoological collection is currently being evaluated; therefore, numbers may have to be adapted in the future. The inventory of non-lepidopteran insects was nearly completed in 2015. However, there are POLLICHIA collections that have not been transferred to the PMN yet (currently approximately 20,000 specimens, among them ca. 15,000 Hymenoptera). Old records in the annual reports list the specimens that were donated and added to the collections in the respective years, which today is a valuable basis for provenience research. The most recent overview of the invertebrate collections was published by Hallbach (1972).

Nonetheless, only small parts of the collections have so far been made available to the scientific community. Among them are parts of the Lepidoptera (see Sect. 3.1.1) and the Trichoptera collections (Neu 1999).

8.3.1 Entomology

8.3.1.1 Lepidoptera

The Lepidoptera constitute by far the largest part of the zoological collections housed in the PMN, consisting of many individual collections that were bought by or donated to the POLLICHIA and the museum by the respective collectors. Together, there are more than 130,000 specimens.

Parts of the Lepidoptera collection data are available in an online database (www.schmetterlinge-rlp.de). Others were published in faunistic papers and books (e.g. de Lattin et al. 1957, 1958, 1959, 1960, 1962, 1964, 1966, 1971). An overview of the Lepidoptera collections curated at the PMN at the time was published by van Gyseghem (1983) and Settele and van Gyseghem (1992).

8.3.1.2 Coleoptera

The beetles form the second largest taxon represented in the PMN collections, currently counting 44,000 specimens. There are many tropical specimens but the vast majority belongs to the local fauna. The largest family in this collection is the Carabidae, counting approximately 7250 specimens.

8.3.1.3 Other Insect Groups

The remaining insect orders are currently represented by little more than 14,000 specimens. Among them, Heteroptera is the largest group with about 4300 specimens, followed by Hymenoptera with about 4050 specimens (but see introduction to paragraph 3). Diptera and Orthoptera are represented by 1750 and 1700 exemplars, respectively. There are 800 Auchenorrhyncha and 670 Odonata, 350 Neuroptera, 320 Trichoptera and 140 Blattaria specimens. The collections of Mantodea, Phasmatodea, Ephemeroptera and Dermaptera each contain less than 100 specimens.

8.3.2 Mollusca

The mollusc collection encompasses 21,000 mostly Central European specimens, among them 18,100 Gastropoda and 2900 Bivalvia shells. Among the former is a rare abnormally sinistral (left-coiled, reverse-coiled) *Helix pomatia* specimen (Fig. 8.2; Röller 2008). All shells are stored in glass vials and plastic boxes in drawers. Their collection data have been completely digitized.

Fig. 8.2 An abnormally sinistral *Helix pomatia* specimen (a typical dextral specimen is shown on the *right-hand side*) in the mollusc collection of the PMN

8.3.3 Vertebrates

8.3.3.1 Aves

There are little more than 1000 dermoplastics in the exhibition and stored in the magazine. The greater part of the bird collection encompasses eggs (several thousand specimens), skulls (about 230 specimens), study skins and nests. Many of the bird specimens formed the basis for publications on the avifauna of Rhineland-Palatinate (e.g. Groh 1965).

The bird collection was critically reviewed by Groh (1966). There are many important records that are also historically interesting. For instance, the short-toed snake eagle (*Circaetus gallicus*) used to breed in Germany. The last pair that bred in Rhineland-Palatinate including its half-grown chick was shot for preparation in 1911. It was one of the last breeding pairs of this species in Germany (e.g. Groh 1965; Glunz von Blotzheim 1971). A wallcreeper (*Tichodroma muraria*), usually found in the high mountains of Eurasia and a rare vagrant in Germany, was collected in a quarry near today's museum in 1923 (Zumstein 1927; Groh 1966). A specimen of Manx shearwater (*Puffinus puffinus*) was found alive in 1922 in a town near Ludwigshafen and later donated to the museum. It has been assumed that this specimen might be the first physical record of this species in Germany (Zumstein 1927; Groh 1966). The dermoplastic of a rare subspecies of the hazel grouse (*Bonasa bonasia rhenana*; Fig. 8.3), shot in 1935, is possibly the only existing record from the Palatinate (the south-eastern region of Rhineland-Palatinate) (Schreiber et al. 2015).

Fig. 8.3 Dermoplastic of the hazel grouse subspecies *Bonasa bonasia rhenana*. This specimen, displayed in the main exhibition, is possibly the only physical record from the Palatinate (south-eastern Rhineland-Palatinate)

8.3.3.2 Mammalia

The mammal collection comprises about 270 dermoplastics in the exhibition and in the magazine. The greater part of the mammal species that used to occur in Rhineland-Palatinate is represented by dermoplastics, including the Eurasian beaver (*Castor fiber*), the wolf (*Canis lupus*, Fig. 8.4) and the Eurasian lynx (*Lynx lynx*, Fig. 8.5). The skull collection includes the first record of the raccoon dog (*Nyctereutes procyonoides*) for Rhineland-Palatinate. The specimen was shot and documented by a local hunter in 1983 and donated to the museum in 2003 (Röller and Müller 2004).

The skull collection comprises many local species including rodents, insectivores and carnivores. Among the exotic specimens are the skulls of dolphin, chimpanzee, lion, hippopotamus and giraffe.

8.3.3.3 Other Vertebrates

There are specimens of amphibians, non-avian sauropsids, lampreys, cartilaginous and bony fish (dermoplastics, parts of skeletons and ethanol-preserved specimens, replicas) in the collection and the exhibition, but their number is comparatively low.

8.3.4 Other Taxa

There are several hundreds of specimens of other taxa; among them are crustaceans, chelicerates and corals.

Fig. 8.4 Dermoplastic of the wolf (*Canis lupus*) presented in the main exhibition

Fig. 8.5 The Eurasian lynx (*Lynx lynx*) presented in the main exhibition

8.3.5 Wet Collections

There are more than 1000 specimens of many taxa stored in alcohol. This collection is currently being revised and successively transferred into new vials.

8.3.6 Type Material

The Palatinate Museum of Natural History currently houses 88 types of specimens. They encompass 75 Lepidoptera paratypes (18 for two new species and 57 for five new subspecies), ten Heteroptera paratypes (for one new species) and three Coleoptera paratypes (for two new species). A complete overview and evaluation of the type material has not been published so far but is currently in progress.

8.3.7 Current State and Future of the Collections

An accurate evaluation has so far only been done for parts of the zoological collections (e.g. Mollusca, Insecta except for Lepidoptera). A complete count is currently in progress in order to get an up-to-date number of all specimens. The digitization of specimen data is progressing constantly with the help of volunteers.

Many of the 1800 insect drawers are still those of the original collectors. Drawer formats vary considerably throughout the collection, and some do not close tightly enough to avoid damage by dermestid beetles. All specimens will subsequently be transferred into new drawers and cupboards to reduce potential damage. Furthermore, UV traps have been installed in all collection rooms for monitoring purposes.

The bird dermoplastics are being stored in a mobile aisle shelving system. As in many smaller museums, space is a crucial factor, and the PMN is no exception. Today's small collection storage rooms were not originally built for this purpose. Additionally, the climatic situation is problematic and cannot be properly controlled due to the architecture of the storage rooms.

8.4 Publications

The PMN publishes a house journal (*Perspektiven aus dem Pfalzmuseum*), covering annual reports and museum-related topics as well as special issues of general interest, including zoological and botanical topics and background information on special exhibitions. The POLLICHIA publishes two journals, the annual *Mitteilungen der POLLICHIA* and the quarterly *POLLICHIA-Kurier*, the latter of which is a popular scientific organ. Furthermore, the PMN scientists regularly publish their results in peer-reviewed national and international journals.

8.5 Research

The current focus of zoological research is on the evolution, morphology, phylogenetic systematics and biogeography of praying mantises (Mantodea). Studies are conducted in cooperation with several national and international museums and universities (e.g. Wieland and Schütte 2011; Wipfler et al. 2012; Wieland et al. 2014; Svenson et al. 2015). Smaller faunistic studies are focussing on the state and local distribution of neozoa in Rhineland-Palatinate (e.g. Wieland and Goldberg 2015) and the biology of rare and interesting species (e.g. Wieland 2014).

8.6 Education

The educational team of the PMN offers a varied and diverse spectrum of events for schools and visitors, ranging from a couple of hours up to several weeks (holiday programmes). There are also regular external programmes where the team visits schools for interactive events or organizes educational outdoor activities (e.g. Wolf 2013; Wolf and Bernd 2014).

Furthermore, all PMN departments supervise several student apprentices from schools and universities every year. The zoological department offers a course on scientific drawing in ink on a regular basis (Wieland 2015).

8.7 Social Media

The PMN maintains a website (www.pfalzmuseum.de) that informs on activities and events. A separate website provides information on the GEOSKOP (www.urweltmuseum-geoskop.de). The GEOSKOP is furthermore present on Facebook, a path that the PMN is currently planning to tread as well in the near future.

Acknowledgements For providing information, background data, literature and/or helpful comments on the manuscript, I would like to thank (in alphabetical order) Ernst Blum, Ronald Burger, Jan Fischer, Reinhard Flößer, Julia Goldberg, Roland van Gyseghem, Hans-Wolfgang Helb, Silke John, Volker John, Jürgen Ott, Maria Ritsch-Frenzel, Oliver Röller, Birte Schönborn, Sebastian Voigt and Ute Wolf.

References

de Lattin G, Jöst H, Heuser R, Roesler R (1957) Die Lepidopteren-Fauna der Pfalz. Mitt POLLICHIA III 4:51–167

de Lattin G, Jöst H, Heuser R, Roesler R (1958) Die Lepidopteren-Fauna der Pfalz. Mitt POLLICHIA III 5:82–90

de Lattin G, Jöst H, Heuser R, Roesler R (1959) Die Lepidopteren-Fauna der Pfalz. Mitt POLLICHIA III 6:169–244

de Lattin G, Jöst H, Heuser R, Roesler R (1960) Die Lepidopteren-Fauna der Pfalz. Mitt POLLICHIA III 7:245–321

de Lattin G, Jöst H, Heuser R, Roesler R (1962) Die Lepidopteren-Fauna der Pfalz. Mitt POLLICHIA III 9:322–390

de Lattin G, Jöst H, Heuser R, Roesler R (1964) Die Lepidopteren-Fauna der Pfalz. Mitt POLLICHIA III 11:391–526

de Lattin G, Jöst H, Heuser R, Roesler R (1966) Die Lepidopteren-Fauna der Pfalz. Mitt POLLICHIA III 13:527–538

de Lattin G, Jöst H, Heuser R, Roesler R (1971) Die Lepidopteren-Fauna der Pfalz. Mitt POLLICHIA III 18:11–85

Flösser R (2009) Bericht 2007/2008 über das Pfalzmuseum für Naturkunde – POLLICHIA-Museum, Bad Dürkheim und das GEOSKOP Urweltmuseum, Burg Lichtenberg bei Kusel. Mitt POLLICHIA 94:187–216

Flösser R, John V, Schönborn B, Schweiss D, van Gyseghem R, Wolf U (2008) Mit Tradition zur Innovation. Perspektiven aus dem Pfalzmuseum 1:1–64

Gettmann W (1981) Jahresbericht 1980/1981 des Pfalzmuseums für Naturkunde (POLLICHIA-Museum). Mitt POLLICHIA 69:276–289

Gettmann W (1990) Das Pfalzmuseum für Naturkunde (POLLICHIA-Museum) – Geschichte, Aufgaben, Ziele. In Gettmann W, John V (eds) Pfalzmuseum für Naturkunde Bad Dürkheim. Museen in Rheinland-Pfalz 4, pp 10–15, 136 pp

Glunz von Blotzheim UN (Hrsg) (1971) Handbuch der Vögel Mitteleuropas. Band 4: Falconiformes. Akademische Verlagsgesellschaft, Frankfurt a.M., 943 pp

Groh G (1965) Vogelfauna von Neustadt/Weinstraße und Umgebung. Mitt POLLICHIA III 12:69–129

Groh G (1966) Pfälzische Belegstücke in der Vogelsammlung der POLLICHIA. Mitt POLLICHIA 3(3):83–96

Hallbach L (1972) Die Wirbellosen Tiere im Museum Bad Dürkheim. Der Präparator 18:92–102

Heiss W (1995) Palmsonntag in Bad Dürkheim: Burgkirche steht in Flammen. In: Landkreis Bad Dürkheim (ed) Heimat-Jahrbuch 1995 des Landkreises Bad Dürkheim. Englram, Haßloch, pp 139–140, 285 pp

Kaiser K (1967) Zur Geschichte des naturwissenschaftlichen Museums der Pfalz und des Heimatmuseums Bad Dürkheim in der Kriegs- und ersten Nachkriegszeit. Pfälzer Heimat 18 (2):60–62

Mehlis C (1915) Unsere Sammlungen und ihre Bedeutung für die Heimatkunde. Mitt POLLICHIA 29:11–14

Müller GH (1990) Johann Adam Pollich (1741–1780), Medicinae Doctor und Pflanzengeograph der Pfalz. POLLICHIA-Buch Nr. 21. Bad Dürkheim, Selbstverlag der POLLICHIA, 177 pp

Neu PJ (1999) Revision der Köcherfliegen (Trichoptera) im Pfalzmuseum für Naturkunde, Bad Dürkheim. Mitt POLLICHIA 86:151–160

Niehuis M (2012) Pfarrer Philipp Wilbrand Jacob Müller (*4. Oktober 1771 - †31. März 1851) – ein bedeutender Insektenkundler aus Odenbach a. Glan. Mainzer Naturwissenschaftliches Archiv, Beiheft 33. 156 pp

Röller O (2008) Ein pfälzischer Schneckenkönig. POLLICHIA-Kurier 24(3):14–16

Röller O, Müller J (2004) Naturschätze aus der Pfalz. POLLICHIA-Sonderdruck 9. Meierdruck, Lingenfeld, 160 pp

Schreiber A, Wieland F, Weitz W (2015) Westliches Haselhuhn, *Bonasa bonasia rhenana* – eine dringliche Verantwortungsart für Rheinland-Pfalz. POLLICHIA-Kurier 31(2):37–43

Settele J, van Gyseghem R (1992) Die Schmetterlingssammlungen des Pfalzmuseums für Naturkunde (POLLICHIA-Museum) in Bad Dürkheim (Rheinland-Pfalz, Deutschland) (Lepidoptera). Nota lepidopterologica (Suppl 4):146–152

Spuhler L (1965) Zur Geschichte der POLLICHIA. Mitt POLLICHIA 12:5–21

Svenson GJ, Wieland F, Hardy NB, Cahill Wightman HM (2015) Of flowers and twigs, a total evidence phylogenetic revision of the plant-mimicking praying mantises (Mantodea: Empusidae and Hymenopodidae) with a new suprageneric classification. Syst Entomol. doi:10.1111/syen.12134

van Gyseghem R (1983) Schmetterlinge der Pfalz. Die Beleg- und Schausammlungen im Pfalzmuseum für Naturkunde (POLLICHIA-Museum) in Bad Dürkheim. Mitt POLLICHIA 71:123–130

Walther F (1953) Das naturwissenschaftliche Museum der Pfalz in Bad Dürkheim. Mitt POLLICHIA III 1:180–182

Wieland F (2014) Aktuelle Nachweise der Speispinne *Scytodes thoracica* in Bad Dürkheim. POLLICHIA-Kurier 30(3):19–20

Wieland F (2015) Auf den Punkt gebracht: Wissenschaftliches Zeichnen am Pfalzmuseum für Naturkunde. POLLICHIA-Kurier 31(3):49–50

Wieland F, Goldberg J (2015) Ein Überblick zur Biologie und Ausbreitung der Amerikanischen Kiefern- oder Zapfenwanze *Leptoglossus occidentalis* Heidemann, 1910 (Heteroptera: Coreidae) und erste Daten zur Verbreitung in Rheinland-Pfalz. POLLICHIA-Kurier 31 (1):17–20

Wieland F, Schütte K (2011) Unrecognized museum specimen expands distribution of *Mantoida* (Insecta: Mantodea) into the central Caribbean. Entomol Mitt Zool Museum Hamburg 15 (186):305–315

Wieland F, Schütte K, Goldberg J (2014) A review of the research on Canary Islands praying mantises (Insecta: Mantodea). Zootaxa 3797(1):78–102. doi:10.11646/zootaxa.3797.1.9

Wipfler B, Wieland F, DeCarlo F, Hörnschemeyer T (2012) Cephalic morphology of *Hymenopus coronatus* (Insecta: Mantodea). Arthropod Struc Dev 41:87–100. doi:10.1016/j.asd.2011.06.005

Wolf U (2013) Die Forschungswerkstatt auf Burg Lichtenberg – ein Rückblick. Nat Museum 3:18–20

Wolf U, Bernd C (2014) Pfalzmuseum unterwegs mit dem ArtenFinder. Nat Museum 4:42–46

Zumstein F (1927) Die Vogelsammlung der POLLICHIA zu Bad Dürkheim. Pfälzisches Museum/ Pfälzische Heimatkunde (1/2):6–8

Chapter 9
BAMBERG: The Zoological Collections at the Museum of Natural History in Bamberg

Matthias Mäuser

Abstract The Museum for Natural History in Bamberg dates back to 1791. It was established as a university "Naturalienkabinett." Since then, various collections were gathered, both for exhibitions and scientific purpose. Today the zoological collections comprise about 180,000 evidences. The focus is on entomology (140,000), especially on the special collections of the regional entomofauna. The collection of mounted birds (1550) holds numerous exotic specimens which are important with regard to historic respect. Due to the shortage of staff, only a small amount of the material has been recorded digitally up to now. Currently, the scientific focus of the museum lies on paleontology.

Keywords Natural History Museum Bamberg • Vogelsaal • Bird collection • Historic specimens • Quagga

The Museum of Natural History in Bamberg (Naturkunde-Museum Bamberg) and its collections are in the proprietary of the university-related Lyzeumstiftung Bamberg. In 1988, the foundation transferred the technical and scientific supervision to the Bavarian Natural History Collections, an administration of the Free State of Bavaria.

9.1 History

The Museum was founded in 1791 by prince-bishop Franz Ludwig of Erthal as a natural history collection, a so-called Naturalienkabinett (Mäuser 1995). The Naturalienkabinett, as well as a chair in natural history that was established at the same time, was supposed to be a teaching establishment for students of all faculties. For that purpose, the progressive regent had one wing of the former Jesuit monastery partially cored and had a two-story hall with gallery built in its inside. The

M. Mäuser (✉)
Museum of Natural History Bamberg, Fleischstr. 2, 96047 Bamberg, Germany
e-mail: info@naturkundemuseum-bamberg.de; http://www.naturkundemuseum-bamberg.de

L.A. Beck (ed.), *Zoological Collections of Germany*, Natural History Collections,
DOI 10.1007/978-3-319-44321-8_9

furnishing was done in the style of early classicism. Simultaneously, Erthal purchased several collections in order to create a basis for the exhibition and teaching materials—not only minerals and ore mounts but also stuffed animals.

During the following period of political unrest (The Napoleonic wars), the project stagnated, and most of the purchased collections got lost. The Naturalienkabinett finally reached completion and was opened as a museum between 1803 and 1810, when the Benedictine Father Dionysius Linder (1762–1838) brought his natural history collection of the secularized Banz Monastery to Bamberg, added it and acted as the museums first conservator until 1838 (Mäuser 2003). Linder added to the collection by purchasing from other museums and collections and by procuring gifts. Later conservators used the same methods, focusing on different groups of animals (Kolb 1983).

The main focus lies on the acquisition of showpieces for exhibitions as well as a broad presence of evidence from as many animal phyla as possible. When in mid-nineteenth century the collections had by far outgrown the holding capacities, new showrooms were added under the direction of conservator Andreas Haupt (Haupt 1893).

The initial main hall, nowadays referred to as the "Vogelsaal" (Hall of Birds), has to this day mostly been preserved in its original condition (Fig. 9.1). As a "museum inside a museum," it represents an original Naturalienkabinett. As a showroom it holds the greater part of the vertebrate collection with focus on ornithology. Furthermore, specimen of invertebrates as well as botanic, geological, and paleontological exhibits are shown as they fit the inventory of a historic Naturalienkabinett and were included in those in the nineteenth century as well.

9.2 The Zoological Collections

In total, about 180,000 evidences are stored in the Bamberg museum of natural history, most of them in the museums depot. 157,000 of those are zoological objects. Alongside a smaller collection of corals (200) collected in the nineteenth century, there are about 13,000 specimens of different invertebrates (without insects), most of which are mollusks (11,500). The latter comprise of a scientific collection of gastropods and bivalves from all over the world (4500), and a closed unit of East Asian mollusks (5500) that where collected between 1886 and 1921 by Joseph Schedel, a Bamberg pharmacist. Furthermore, the collection contains individual series of Upper Franconian freshwater bivalves that document their diversity of different stretches of water during the first half of the 20th century.

The main part of the invertebrate collection consists of insects (about 140,000). The collection of Theodor Schneid (80,000) is of particular scientific value (Dettner 2015). Schneid intensively collected in the greater area of Bamberg between 1930 and 1950 and had his catches properly dissected and documented and determinated by specialists. He included all orders of insects, though his main focus lies on Hymenoptera, *Coleoptera*, and Diptera. All in all, about two thirds of all insect

Fig. 9.1 The bird hall of the Natural History Museum of Bamberg is the prototype of a historic Naturalienkabinett. The showcases contain about 2200 exhibits, not only birds but also species from different phyla, botanical specimens, minerals, fossils and a rare historic collection of 200 fruit varieties made of wax

species known from Bavaria are documented in the Schneid collection, coming from the greater area of Bamberg. Precise data of the collection sites, the area-covering collection method and the huge numbers of individuals allow for conclusions on environmental changes in the regions analyzed at that time. The quality of the collection gave reason for intensive scientific efforts for the past 20 years and led to the data being included into the species protection file of the Bavarian state office for environmental protection.

Recently the museums entomological collection was complemented by Bernhard Stöckert's collection of regional Lepidoptera. 40,000 specimens document all Lepidoptera families that are found in the greater region of Bamberg.

Though the vertebrate collection (3000) appears weak in numbers, they naturally dominate the perception in the exhibition. The collections and exhibitions' main focus lies on ornithology (Fig. 9.2). First to deserve mention are the 1550 specimens of local and foreign birds (Steinheimer 2003). About 800 species represent almost 9 % of the worldwide avifauna. Most of the non-European species were acquired around the middle of the nineteenth century. Sources of supply were several in these days existing Naturalien stores and duplicate bought from museums and other collections. First to mention here is the State Collection in Munich, which provided numerous Brazilian species. Presumably those contain specimens that were collected during the famous Brazilian expedition by the German natural scientists Spix (1781–1826) and Martius (1794–1868). Other Brazilian species source from the collections of J. Natterer (1787–1843) and J.H.C. Sturm (1805–1862). The Godeffroy Museum in Hamburg provided several exemplars as well, including 80 species from Queensland, Australia, which were gathered by the professional

Fig. 9.2 Detail of a showcase with different species of Coraciiformes

collector Amalie Dietrich (1821–1891). The remaining specimens were given to the museum from private persons, most of them as gifts. The species-rich collection of middle European birds mostly is of younger origin. 552 eggs and 112 nests as well as some assembled skeletons complete the ornithological inventory.

The bird specimens of historic interest yet gain in value as they are excellently backed by archive records. To many specimens there is correspondence as well as invoices and receipts, which allows complete reconstruction of provenance.

Furthermore, the vertebrate collection contains about 1500 objects from all classes of vertebrates, mainly fish and mammals. Their visual condition matches their age, which in some cases goes back as far as the first half of the nineteenth century. Especially the fish specimens don't meet today's requirements for visually flawless exhibits. Nevertheless, or for this very reason, they fit the historic ambience rather well.

The specimen with certainly the highest value among the vertebrates is an excellently stuffed quagga (Fig. 9.3). Only 22 other specimens throughout the world exist to testify to the existence of this subspecies of a plains zebra eradicated by humans.

Fig. 9.3 Mounted specimen of *Equus quagga quagga*. Of this extinct subspecies, there are only 23 mounted skins housed in different museums worldwide

9.3 Data Situation

Georg Fischer, the museums conservator from 1885 to 1912, was given the task to inventory the significantly grown collections. Fischer diligently produced 55 inventory and cataloging tomes, covering all sectors of the collection, including the botanic inventory (the museum's herbarium was transferred to the Botanic State Collection in Munich in 1988). After Fischer's curatorship, the inventory sheets were continued, however not consequently. For example, there is no inventory sheet for the entomological collection of Schneid.

Due to the chronic understaffing of the museum, the much-needed digital registration of the collections could not happen until this day. Up to 2015, there has been no more than a single scientist responsible for the museum. Thankfully, since February of 2015, another part-time scientist post was established by the Free State of Bavaria, though that post is mostly devoted to educational and exhibition-related responsibilities. An appropriate inventory of these valuable collections as well as its inclusion into globally available digital networks is still due. A start on providing data for the scientific community is to be made by the bird collection, which is completely photographically captured and provided with all available data (thanks to voluntary help). The record sets of partitions of Schneid's entomological collection which were gathered in the year 1990 should be converted to compatible formats as soon as possible, in order to enable universal availability.

Unfortunately there is not enough manpower to allow a complete inventory of the museum's natural history collections.

9.4 Current Work Priorities

The "success" of museums nowadays primarily is measured in number of visitors. Hence, the main focus of work lies on the presentation of special exhibitions and the arranging of attractive events. Therefore, there is not enough time to fulfill the other major tasks of a museum—collecting and assessing—especially for small museums as the Naturkunde-Museum Bamberg. Thus, the gaining of zoological specimen primarily is limited to the receiving of gifts.

Furthermore, the current focus of collecting at the Naturkunde-Museum Bamberg lies in the paleontological sector. Since 2004, paleontological excavations take place in the quarry of Wattendorf near Bamberg where spectacular vertebrate fossils are found by a team of the Bamberg museum supported by volunteers.

References

Dettner K (2015) Dr. Theodor Schneid (1879–1958) – ein bayerischer Universalgelehrter. Ber Naturwiss Ges Bayreuth 27:321–350. Bayreuth

Haupt A (1893) Vermehrung des Kg. Naturalienkabinetts in Bamberg seit 50 Jahren. XVI. Ber Naturforsch Ges Bamberg, 135 S. Bamberg

Kolb A (1983) 180 Jahre Naturkunde-Museum Bamberg Linder'sche Stiftung. LVIII. Ber Naturforsch Ges Bamberg, pp 130–150. Bamberg

Mäuser M (1995) Zur Gründung des Bamberger Naturalienkabinetts durch Fürstbischof Franz Ludwig von Erthal. In: Baumgärtel-Fleischmann R (ed) Franz Ludwig von Erthal Fürstbischof von Bamberg und Würzburg 1779–1795. Bamberg

Mäuser M (2003) Der Benediktinerpater Dionysius Linder und seine zweite Karriere nach der Säkularisation. In: Baumgärtel-Fleischmann R (ed) Bamberg wird bayerisch. Bamberg

Steinheimer F (2003) Bamberg's natural history museum – the scientific significance of small collections. Bonner zoologische Beiträge 51(Heft2/3):141–146

Chapter 10
BERLIN: From Humboldt to HVac—The Zoological Collections of the Museum für Naturkunde Leibniz Institute for Evolution and Biodiversity Science in Berlin

Peter Giere, Peter Bartsch, and Christiane Quaisser

Abstract The zoological collections of the Museum für Naturkunde comprise historic specimens predating the founding of the institution, series of well-dated specimens collected over its 200-year history as well as modern additions such as tissue samples and sound recordings of animals. Overall, the zoological collections are estimated to hold more than 25 million specimens that are accessed by scientists from the Museum für Naturkunde and from around the world for research mainly in systematics and evolution. The zoological collections of the Museum für Naturkunde have their roots in the Berlin University, which—founded in 1810—had included a zoological museum from the start. After a period of growth as the principal zoological museum in Prussia, this and other museums from the Berlin University were united under the roof of a purpose-built building in 1889. This new "Museum für Naturkunde" underwent enormous growth in the following years, stemming both from expeditions and from acquisitions from the colonies. In World War II, the museum was affected by an air raid that left the eastern wing in ruins. This lasted until the bicentennial anniversary in 2010, when the eastern wing was reopened, now specially equipped for safely storing the vast wet collections in conjunction with a spectacular public insight into the collections. The reconstruction of other parts of the building will follow to provide up-to-date public galleries in conjunction with excellent storage for the invaluable collections. Being an institution that has combined scientific work with education and public outreach from the start, the zoological collections in the Museum für Naturkunde with its numerous international relations and projects will serve these purposes in the future as a backbone of an excellent research museum.

Collection Acronym: ZMB for most zoological collections

P. Giere (✉) • P. Bartsch • C. Quaisser
Museum für Naturkunde – Leibniz Institute for Evolution and Biodiversity Science, Invalidenstraße 43, 10115 Berlin, Germany
e-mail: Peter.Giere@mfn-berlin.de; http://www.naturkundemuseum-berlin.de/

L.A. Beck (ed.), *Zoological Collections of Germany*, Natural History Collections,
DOI 10.1007/978-3-319-44321-8_10

Keywords Museum für Naturkunde Berlin • MfN • Zoological collections • Research museum

The Collections of the Museum für Naturkunde at a Glance

30 million zoological, paleontological and geological specimens:
25,894,000 zoology:

more than *15,000,000 insects*, 10,000,000 other invertebrates
674,000 vertebrates, 80,000–100,000 embryological specimens
120,000 animal sound recordings

3,075,000 fossil specimens:

1,500,000 vertebrates
1,300,000 insects and invertebrates
268,000 plants
49,000 microfossils

331,000 minerals and rocks
6000 meteorites

10.1 The Museum für Naturkunde: An Overview

Inevitably, a visitor will look up when entering the Museum für Naturkunde. It is one of the historic buildings typical for the late nineteenth century: tall columns, huge windows, high ceilings and a large glass-roofed central hall producing a cathedral-like atmosphere (Fig. 10.1). "Ah!" is what follows when the visitor stands at the entrance to the central hall, having a first glance at the huge mounted *Brachiosaurus* and its companions presented there.

The Museum für Naturkunde is not the center of Berlin, but it is situated in the very heart of the city, just a 15-min walk from the main station and the Reichstag building. Following visitors' surveys, it is one of most popular museums and a must see in Berlin.

Not many visitors, however, are aware that behind the historic façade and behind the scenes of the public galleries, a modern research institution forms the backbone of the institution with expertise in fields like systematics, phylogeny, biogeography, evolution, impact of environmental changes and catastrophes, and citizen science. Its core research areas—biodiversity and evolutionary studies—are based on more than 30 million objects ranging from zoology, palaeontology and mineralogy. Comprising almost 26 million objects (including more than 15 million insects), the zoological collections are by far the largest part of it (see Box, http://www.naturkundemuseum-berlin.de/en/insights/collections/). However, no one will ever be able to count them, and an inventory of all specimens is in far reach. In total,

Fig. 10.1 Façade of the Museum für Naturkunde with main entrance (with permission from: Museum für Naturkunde—Leibniz Institute for Evolution and Biodiversity Science; photo: A. Dittmann)

280 staff are working at the museum (see Annual Report 2014, http://www.naturkundemuseum.berlin/sites/default/files/jahresbericht_mfn_2014.pdf). Many of these, ranging from technicians to scientists, take care of collections, curate them, digitize them, or provide service to the annually 700 external scientists and making sure that the collections remain accessible and are available for even more research.

Founded in 1810 as part of the university, research and academic education have been central tasks of the museum from the very beginning. With the first museum building and the quickly growing collections over the course of the nineteenth century, exhibitions became the third dimension, a process that can also be found in other natural history museums in Europe. In Berlin, however, history has it that a dispute between traditional and modern views left traces in the architecture of the building (see Sect. 10.3 for details).

Outstanding Specimens in the Zoological Collections of the Museum für Naturkunde

- *From notable collectors*: Collections of Marcus Elieser Bloch (1733–1799; fish collection) and Albertus Seba (1665–1736; herpetological collection), mineralogical specimens from Alexander von Humboldt (1769–1859)

(continued)

- *From extensive national expeditions*: German Deep Sea Expedition (1898–1899), Antarctic Expedition (1938–1939), Tendaguru Expedition (1909–1913)
- 177,000 *Type specimens*, including mountain gorilla and forest elephant

The most extensive growth of the collections was linked to the numerous expeditions starting in the early nineteenth century and the later colonial expansion in Imperial Germany (see Sect. 10.2). Consequently, focal areas of the collections were Africa and Southeast Asia. Until today, these collections build a solid ground for further research in these regions. Several of the museum's icons originate from this period (see box). The most famous is the mounted *Brachiosaurus* in the central dinosaur hall, dug out during the Tendaguru expedition to former German East Africa, today Tanzania (1909–1913, Fig. 10.2). Also, much of the scientifically important material dates back to this period. Others are even older than the founding date of the museum, e.g. objects from the collections of Albertus Seba or Marcus Elieser Bloch. Items collected by Alexander von Humboldt and Charles Darwin (1809–1882) show the close link to our cultural heritage (see box). Apart from outstanding individual objects found throughout the museum, some collections are exceptional, e.g. the embryological collection (see Sect. Vertebrate Collections in the appendix) and the animal sound archive which are unique for the zoological collections of the Museum für Naturkunde.

Key Networks and Consortia of Natural Science Collections

As an Integrated Research Museum of the Leibniz Association, the Museum für Naturkunde is both a research institution—as one of 89 within the *Leibniz Association* (http://www.leibniz-gemeinschaft.de/en/home/) that range from the natural, engineering and environmental sciences via economics, spatial and social sciences to the humanities—and a museum with public galleries and public outreach. As a museum, the Museum für Naturkunde is member of the *International Council of Museums (ICOM, http://icom.museum/*) and its German branch, the German Museums Association (*Deutscher Museumsbund, DMB, http://www.museumsbund.de/en/*). Whereas ICOM and DMB are associations concerned with different aspects of professional museum activities, the following are centered on the scientific collections housed in natural history museums.

The *Deutsche Naturwissenschaftliche Forschungssammlungen (DNFS*, http://www.dnfs.de/) is a German consortium of nine institutions holding the most comprehensive scientific research collections in Germany. It includes the working group on natural science collections in the DMB. Joint efforts and common approaches of the members of the DNFS help to

(continued)

answer the challenges and political and societal requests natural science collections are facing.

CETAF is the *Consortium of European Taxonomic Facilities*, http://cetaf.org/: a European network of 33 natural science museums, natural history museums, botanical gardens and biodiversity research centres with their associated biological collections and research expertise. CETAF aims to promote training, research and understanding in systematic biology and palaeobiology, and facilitate access to information (collections) and the expertise of its member institutions across Europe.

The mission of the *Scientific Collections International (SciColl, http://scicoll.org/)* is to increase the use and impact of scientific collections for interdisciplinary research and societal benefits to expand the access, awareness and appreciation of scientific collections. The aim of this international organization is to increase the return on investment that countries and institutions make in their scientific collections by catalyzing international and interdisciplinary collaboration, e.g. by research on major challenges and by new and more cost-effective management and use of collections in all disciplines.

For further information on networks, initiatives and projects see Quaisser and Woog (2011).

Although organized in separate institutional structures for most of its history, the zoological, palaeontological and mineralogical collections have always been hosted under one common roof. Only in 2009, almost 200 years after its foundation, the Museum für Naturkunde became an independent research institute and merged all management structures into one new organization led by a director general. In the same process, the museum became a member of the Leibniz Association, and as such it is financed jointly by federal and state funds. Today, the Museum für Naturkunde works towards an integrated research museum, emphasizing once again the close connection and finally the integration of research, collections and public engagement with science. It is a strong partner in many national and international consortia and initiatives, such as DNFS, CETAF and SciColl (see box). Though often closely cooperating in research initiatives, the Botanischer Garten und Botanisches Museum Berlin-Dahlem (BGBM) took quite a separate development in Berlin from the beginning. Only during the German separation, when the BGBM was in West Berlin (West Germany) and the Museum für Naturkunde in the East, Recent botany became part of the Museum für Naturkunde. After reunification, botanical specimens were transferred to the *Herbarium Berolinense* housed in the BGBM (Greuter et al. 1994; Köhler 2010).

Fig. 10.2 Skeleton of *Brachiosaurus brancai*, one of the icons of the museum (with permission from: Museum für Naturkunde—Leibniz Institute for Evolution and Biodiversity Science; photo: A. Dittmann)

10.2 Origins

The origins of the zoological collections reflecting the animal diversity known at the time is strongly linked to the opening of the *Alma Mater Berolinensis*, the Berlin University, in 1810. Suggested by Wilhelm von Humboldt (1767–1835) as part of his education reform following the new humanism ideal, it exemplified a new type of university with didactic *and* research components—as had been stipulated in von

Humboldt's text *Über die innere und äußere Organisation der höheren wissenschaftlichen Anstalten in Berlin* (http://edoc.hu-berlin.de/miscellanies/g-texte-30372/229/PDF/229.pdf). This unfinished and undated manuscript can likely be dated to 1810 (Paletschek 2007) and must be seen in a context of contemporary thinkers in education such as philosopher Johann Gottlieb Fichte and theologian Friedrich Schleiermacher. This university, endowed by King Frederick William III of Prussia, incorporated several previous institutions and organizations that were suitable to assist in the education of students. Collections were one essential part of this, as Article 1 of the preliminary regulations for the Berlin University indicates (http://edoc.hu-berlin.de/miscellanies/g-texte-30372/251/PDF/251.pdf). This article states, that "...the University is in connection with the two aforementioned Academies of Sciences and Arts as well as with institutions and collections with which it forms an organic whole" (translation by PG). The zoological collections now housed in the Museum für Naturkunde were no exemption and with reference to his much revered *Muséum National d'Histoire Naturelle* in Paris, Alexander von Humboldt supported the ideas to include a similar institution in the new university (Geus 1998). More explicitly, an outline of the museum was devised in a memorandum from 1810 by Johann Karl Wilhelm Illiger (1775–1813), which stated that a zoological museum was needed both for research and education (Jahn 1985). He was seconded by Johann Centurius Hoffmann Graf von Hoffmannsegg, an entomologist and botanist with a wide interest in zoology, who had visited the museum in Paris en route to his second collecting trip to Portugal (Eckert 2010). Like Alexander von Humboldt, he was impressed by the dual function of this museum (cf. Eckert 2010).

Von Hoffmannsegg was an avid collector of natural history specimens. After having combined his specimens with those from Braunschweig-based Professor Hellwig, they jointly owned the largest entomological collection of the time (Eckert 2010). Apart from this, Hoffmannsegg also owned many botanical specimens and a zoological collection including numerous animals collected by his servant and preparator Friedrich Wilhelm Sieber in Brazil (cf. Göllner-Scheiding 1972). He donated (according to Jahn 1985: sold) this collection of birds, mammals and amphibians to the Prussian State to form the base of the new museum, and, before moving on to Dresden in 1816, he also sold his entomological collection [Göllner-Scheiding (1972), but also see Jaeger and Uhlig (2010), who state an extended date for the transfer until 1820].

It was agreed to hire the entomologist Illiger as head of the new museum following a suggestion by von Hoffmannsegg, who had known Illiger from his time in Braunschweig. There, he had sorted and supervised the vast entomological collection of Hoffmannsegg and Hellwig (Eckert 2010). Other collections that formed the founding stock of the zoological museum were according to Jahn (1985) relocated from the "Academic Museum" of the Academy of Sciences with specimens that originally stemmed from the *Königlich Preußische Kunstkammer* (Royal Prussian Art Collection) and from several founding members of the *Gesellschaft Naturforschender Freunde zu Berlin* (Berlin Society of Friends of Natural

Science, founded in 1773). According to Geus (1998), these included the founder of the society, Friedrich Heinrich Wilhelm Martini (1729–1778, "conchyological" collection), Marcus Elieser Bloch (1723–1799, ichthyological collection), Friedrich Wilhelm Herbst (1743–1807, entomological collection) and Johann David Schoepf (1752–1800, herpetological collection). Until Bauer and Günther (2013) associated several reptilian specimens of the herpetological collection with Albertus Seba (1665–1736) and dated them to 1734 (possibly earlier, Fig. 10.3), the Bloch collection contained some of the oldest specimens of the zoological collections of the Museum für Naturkunde. This collection mainly is known for its fish specimens that soon were joined by Siberian fish specimens from Peter Simon Pallas (1741–1811), who had died shortly after the museum came into existence (Jahn 1985). Despite his support for inclusion of collections into the new university, Alexander von Humboldt did not contribute to the zoological collection initially. He remained mostly in Paris during the first decades of the nineteenth century to work on the report of his journey to South America and to get his biological specimens analysed by local experts. They then received most of his South American collection of botanical and zoological specimens. Nevertheless, apart from his pet parrot, which is now in the bird collection (Fig. 10.4), he gave his collection of minerals and the specimens of the later expedition to Russia to the Berlin museum.

The first head of the zoological collections (i.e. the "Zoological Museum Berlin", later and still using the acronym "ZMB" for most of its collections) in the new university, Karl Illiger, had devised a system for mammals and birds of his own (Illiger 1811), and the collections were ordered accordingly. After Illiger's early death in 1813 at age 37, Hinrich Martin Lichtenstein (1780–1857), the first

Fig. 10.3 *Python sebae* ZMB 1478, a specimen traced back to the collection of Albertus Seba. (with permission from: Museum für Naturkunde—Leibniz Institute for Evolution and Biodiversity Science; photo: F. Tillack)

Fig. 10.4 ZMB_Aves_ 14578, *Coracopsis vasa vasa*, the pet parrot of Alexander von Humboldt (with permission from: Museum für Naturkunde—Leibniz Institute for Evolution and Biodiversity Science; photo: C. Radke)

professor of zoology at the new university, succeeded him as head of the museum (cf. Lichtenstein 1816). To have a professor of zoology in its own right and not as part of medicine is another progressive feature of the *Alma Mater Berolinensis*. Lichtenstein had spent several years in South Africa as a private teacher (and physician) and explored the natural history of the area (Landsberg 2010). The position had been offered to von Hoffmannsegg, but he declined after his demands for more personnel, for an own library, and for funds to carry out collecting expeditions were turned down (Jahn 1985). If not realized for the hiring of von Hoffmansegg, these ideas were accomplished over time, and the Zoological Museum Berlin became the primary Prussian zoological collection; all others were entitled only to the acquisition of "duplicates" (Jahn 1985). However, the initial mistake not to include the Anatomical-Zootomical Collection, which held material for comparative vertebrate anatomy, was not corrected until 1889 (see Sect. 10.2.1).

Initially, the museum was expected to auction off "duplicates" from the collection and from collecting expeditions to finance both itself and further collecting trips. For this purpose, catalogues of *Doubletten* (duplicates) were printed and distributed, e.g. on mammals, birds, amphibians and fishes (Lichtenstein 1823). However, the surplus was soon used to buy missing species so that around 1850, the Zoological Museum Berlin could be compared to the largest European collections

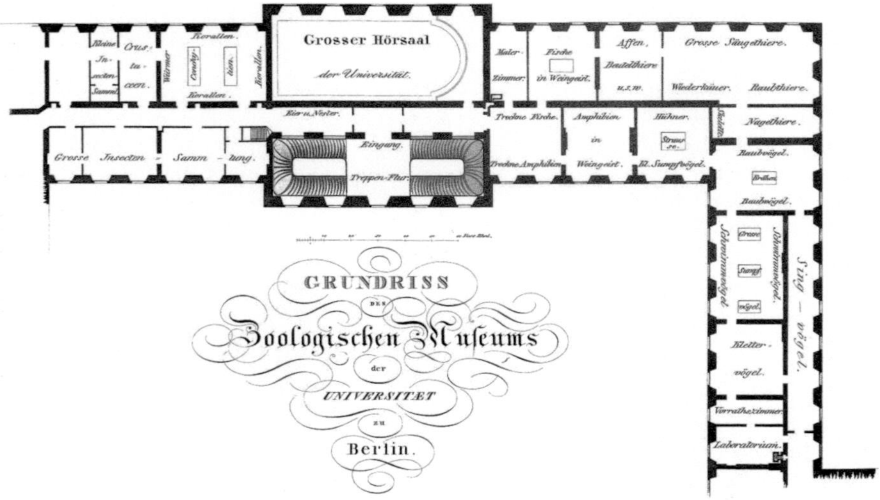

Fig. 10.5 Layout of the Zoological Museum Berlin in the building of the Berlin University around 1830 (with permission from Museum für Naturkunde Berlin) (© with Museum für Naturkunde—Leibniz Institute for Evolution and Biodiversity Science. Museum für Naturkunde Berlin, Historische Bildŭ. Schriftgutsammlungen (Sigel: MfN, HBSB), Bestand: Zool. Mus., Signatur: S I/Grundriss/1830)

of Paris, Wien, Leiden and London (Jahn 1985). This rapid growth can be seen in the number of rooms allocated to the museum in the university building (still the headquarters at *Unter den Linden 6*): it started out with nine rooms in the east wing, it occupied 22 rooms of the second floor after setting up the insect collection in 1825 (cf. Fig. 10.5), and in 1842 it extended over the entire third floor of the building (Jahn 1985).

The detrimental practice of selling off "duplicates" was continued until Wilhelm Carl Hartwig Peters (1815–1883) became director of the Zoological Museum Berlin in 1858, after Lichtenstein's death. He only allowed the sale of "duplicates" after his consent and generally reduced the number of sales considerably (Jahn 1985). Gradually, the storage facilities in the *Unter den Linden* location filled up, so that around 1870 most of the building was occupied with natural history collections, triggering complaints that teaching was impeded by collection objects obstructing access to lecture halls (Köstering 2010; Damaschun and Landsberg 2010). This led to a formal request for a different location for the vast collections and eventually, a decision was made to construct a new museum building at its present location, *Invalidenstraße 43*. This site, the former location of the Königliche Eisengießerei (Royal Prussian Foundry) had only been incorporated into the city boundaries of Berlin for a few decades, when the museum relocated there in 1889. The controversy about the new building with its dual arrangement in a traditional museum-type building is depicted in Sect. 10.3. Initially, this building with its total combined collection and exhibition space of 20,950 m^2 was large enough to house the collections of the *Unter den Linden* location. However, the collections grew steadily due to a worldwide network of collectors (see Angermann 1989) and due to an increase in the number of expeditions (e.g. Gazelle, Deutsche Tiefsee-Expedition, see Sect. 10.2.2) that in part were furbished by the museum. Another factor that considerably contributed to this growth around the time of the move to the new location was the shift in the policy towards the acquisition of colonies of the unified Germany in 1884. Germany then intensified government involvement in formerly purely trade activities of German companies in Africa and Southeast Asia and Oceania. Ensuing were colonies in what today is Namibia, Cameroon, Tanzania, New Guinea and several of the Oceanic Islands. These colonies provided numerous specimens, and the museum issued instructions for collectors (*Anleitung zum Sammeln, konservieren und Verpacken von Tieren für das Zoologische Museum Berlin*) for the zoological collections in several editions dated 1896, 1902 and 1907 (Jahn 1985). The numbers in the new building rose dramatically and within a few decades and despite the space-saving effect of the dual arrangement (specialized public exhibition and scientific collections locked away behind the scenes with study skins rather than mounted specimens, especially in birds and mammals), the need to build an extension for the collections became evident. The northern addition to the building was finished as one of only a few buildings in the war year 1917. At the time of completion of this wing, the colonies were no longer German, but the growth in the collections only slowed down, and up to now, many specimens are added to the zoological collections every year.

10.2.1 Zoology vs. Medicine

In its first decades, the Zoological Museum Berlin had difficulties to set itself apart from the medical faculty. It was intended by the university founders to include zoology in its own right in the faculty of philosophy—which succeeded by providing Hinrich Lichtenstein with the first chair of zoology in Berlin in the philosophical faculty (Damaschun and Landsberg 2010). However, despite Illiger's outline for the collections of 1810 (which explicitly included anatomical specimens) and the support of von Hoffmansegg in his concept of rules and regulations for the new museum, the collection profile was challenged for specimens by the collection of comparative anatomy ("Anatomisch Zootomische Sammlung") which demanded all skeletal material of vertebrates (Jahn 1985). Unfortunately, Illiger's and von Hoffmannegg's suggested outline of the collection was reviewed by Karl Asmund Rudolphi (1771–1832), first full professor of the medical faculty and head of the Anatomical-Zootomical Museum, who opposed these views (Jahn 1985) with regard to the Paris Muséum National d'Histoire Naturelle—where these objects are still in different locations today. In a decision by the Prussian administration dated 28 January 1811 (cited in Jahn 1985), a compromise was found in that both collections have separate administrations but remain together and could be used in the same manner. This separation resulted in the unfortunate situation that mammal specimens were divided between the two collections: the skeleton was given to the

Anatomical-Zootomical Museum and the skin to the Zoological Museum Berlin for mounting. The decision by the Prussian administration was kept up throughout the working lives of Rudolphi and his successors, Johannes Müller (1801–1858) and Karl Bogislaus Reichert (1811–1883) despite various attempts to incorporate the Anatomical-Zootomical Museum into the Zoological Museum. Only after a simultaneous vacancy of both the zoological and anatomical-zootomical chairs in 1883, a reorganization of the museums within the Berlin University was achieved, and the new director of the Zoological Museum Berlin was also appointed to supervise the Anatomical-Zootomical Museum in 1887–1888, and an integration of the latter into the Zoological Museum Berlin was decided in 1888 (Jahn 1985). With the move into the new building at the current location of the Museum für Naturkunde in 1889, the two collections were eventually combined (Jahn 1985), but the confusion stemming from the separation of the same specimen into two collections (along with two unrelated specimen numbers) remains until today, and the identification of matching skin and bones still can be an extremely time-consuming task for the collection personnel of the mammal collection (Angermann 1989).

10.2.2 Expeditions

The collections of the Museum für Naturkunde gained from several expeditions outfitted by government authorities and public funding. The first substantial expedition linked to the new museum took the young scientists Friedrich Wilhelm Hemprich (1796–1825) and Christian Gottfried Ehrenberg (1795–1876) to northern Africa and the Middle East. In 1820, the Prussian Academy of Sciences proposed to the Prussian Ministry that Hemprich and Ehrenberg accompanied an expedition of General Minutoli (1772–1846) to Egypt and Northern Africa in order to collect and record natural history objects. This expedition soon broke up and Hemprich and Ehrenberg continued to accomplish the tasks outlined in instructions by the Academy independently (cf. Stresemann 1954). Unfortunately, nine of the expedition members died for various reasons en route (see Ehrenberg 1828), including Hemprich, who died from malaria in June of 1825 (see Anonymus 1827). After the return of Ehrenberg to Europe late in the same year, Alexander von Humboldt gave a concise account of the voyage for the Academy in 1826 and recorded the scientific gains of the expedition beyond the 114 crates sent back to the museum with a total of 34,000 zoological and 46,000 botanical specimens (von Humboldt 1826).

Another Prussian expedition prior to German unification in 1871, the Royal Prussian Expedition to East Asia (1860–1862), was accompanied by Carl Eduard von Martens (1831–1904), curator of malacology and marine invertebrates at the museum, who besides his own interest in molluscs (von Martens 1867) collected animals in general, including mammals (von Martens 1876; Angermann 1989). Whereas the purpose of this expedition to Japan was mainly political, the focus shifted back to science in the following expedition of the "S.M.S. Gazelle"

(1874–1876), which was accompanied by zoologist Theophil Rudolf Studer (1845–1922) from Berne who included Berlin-based zoologists in the analysis of the material (Studer 1889). This frigate of the Prussian Navy was used for an expedition to the Kerguelen Islands in the southern Indian Ocean to contribute to the international effort in recording a transit of Venus at different remote stations. This was followed by two expeditions with a marine biological focus: the Plankton Expedition of 1889 (mainly funded by the Alexander von Humboldt Foundation for Nature Research and Travel) as a counterpart to the British Challenger Expedition (1872–1876) and the German Deep Sea Expedition (1898–1899) using the specially outfitted steamer "Valdivia" (Fig. 10.6). Whereas the Plankton Expedition was initiated by Kiel-based Victor Hensen (1835–1924), the head of the Prussian *Meereskommission* (Commission for the Seas), the German Deep Sea Expedition was led by Carl Friedrich Chun (1852–1914), whose scientific estate is kept at the Museum für Naturkunde in Berlin. Other noteworthy expeditions that contributed to the wealth of zoological specimens in the collections were the German Sunda Expedition (1929–1931), the German Expedition to Tibet (1938–1939) organized and led by Ernst Schäfer (1910–1992) and the Cuba Expedition (1967) led by Hans-Eckard Gruner (1926–2006), then in the middle of his 40 years as curator of the crustacean collection (Coleman 2007). The latter two exemplify expeditions that were constrained by ideological ties of their time which, in Schäfer's case, who was an SS officer, were prevalent in the participants themselves. The Cuba Expedition differs from other scientific expeditions carried out by staff members of the

Fig. 10.6 Steamer "Valdivia" on the German Deep Sea Expedition (1898–1899) (with permission from Museum für Naturkunde—Leibniz Institute for Evolution and Biodiversity Science. Museum für Naturkunde Berlin, Historische Bildŭ. Schriftgutsammlungen (Sigel: MfN, HBSB), Bestand: Zool. Mus., Signatur: B VI/3164)

museum in socialist German Democratic Republic as it was accompanied by media coverage that enabled participation of the general public in the exotic world which was out of reach for the vast majority of the population—not only for financial but also for political reasons. This is especially true for the coral reef that was disassembled and boxed up for transport to Berlin with the help of amateur divers who reported from the field (e.g. Wagner 1967). Using this material, a new diorama was created with the scientific expertise by the curator of marine invertebrates, Dietrich Hans Hermann Kühlmann (1927–2014), who continued to publicize these efforts (Kühlmann 1980). Parts of these corals are currently on display again. Today, expeditions are project based and, like the Cuba Expedition, in collaboration with host institutions in partner countries. In times of access and benefit sharing (see Chap. 5), this is a prerequisite for successful collecting of biological specimens (i.e. "genetic resources").

10.3 Spirit, Culture, Science and Technology

As outlined above, the Zoological Museum Berlin together with other natural history collections of anatomy, mineralogy, geology and palaeontology represented an integral part of the Berlin University in 1810. Two hundred years later, the Museum für Naturkunde has developed into one of the largest and most active integrated research museums of natural history in Europe and is an independent institution of the Leibniz Association. However, war action and inadequate building maintenance in the following 60 years have left their mark on the shell of the museum so that major reconstruction was needed. Following decades of limited repairs and provisional additions of infrastructure, the first phase of reconstructing the building was completed in 2010, still under the lead of the expert staff of the Humboldt University (Bartsch and Neuhaus 2011). After completion of the modernization of part of the galleries in 2007, which resulted in almost a doubling of the public visitor numbers to 500,000 per year, the first large scale reconstruction programme aimed particularly at rebuilding the eastern wing which had been destroyed by a bomb hit in February of 1945 (Fig. 10.7).

More space for the crowded collections was and still is needed urgently along with a higher security standard for the sensitive material part of zoological collections. From the very beginning in 2005, the project aimed at a safe storage of the 279,000 lots of the wet collection, the glass jars of alcohol-preserved zoological specimens. These amount to about 80 t of highly flammable fluid, and fire protection requirements demanded that they must be separated from the bulk of the dry collection. As such, it seemed to be shaped into a rather straightforward and clear-cut project with concordant aims of fire safety and an improvement of collection care. This included cooling and stable air-conditioning (HVac), an automatic fire-extinguishing system, laboratory space for handling, conservation and scientific use combined into a new replacement building of the eastern wing. The hope was that these measures would get rid of most of the immense and permanent task of refilling

Fig. 10.7 Ruined eastern wing in 2006 (with permission from: Museum für Naturkunde—Leibniz Institute for Evolution and Biodiversity Science; photo: C. Radke)

evaporated alcohol under the changing temperature conditions of the previous collection halls for wet specimens (Fig. 10.8), which ranged from 13 °C in winter to 30 °C in summer. At the same time, the internal, rather traditional procedures for conservation were reviewed, which was accomplished within a project for improving the curation of wet collections funded through the KUR—Programme for the Conservation of Moveable Cultural Assets by the German *Kulturstiftungen* (Neuhaus et al. 2012).

The logistics of disparate collections of hazardous materials originally forming historically grown entities and arranged according to biological systematics can be a problem for research availability and are a horror for the collection personnel and the fire marshal alike. To complicate matters, the remains of the eastern wing, the partly preserved façades and the head buildings of the building listed as a monument had to be preserved.

10.3.1 Building History

What about this monumental architectural heritage? It is simply marvellous. It is open, spacy, with large windows, inviting the public and with the stringency and Prussian austerity of a functional building at the same time. But, for a large museum of natural history with a huge influx of research collections during colonial times, it

Fig. 10.8 The fish collection in the old collection hall (with permission from Museum für Naturkunde—Leibniz Institute for Evolution and Biodiversity Science; photo: H.J. Götz)

had been an outdated near misconception already when it was opened in 1889. August Tiede (1834–1911), the architect of the building, a thoughtful specialist in museum construction (Tiede and Kleinwächter 1891; Tiede 1898), had originally envisioned a building with a separation of collection space and public access (dual arrangement), as it had been developed in Great Britain at that time. Thus, he seemed not to be wholly convinced of his construct that encompassed the demands of the director W.C.H. Peters and the authorities who had dismissed all his innovative plans from 1873 onward (Helbig 2010). They wanted to keep the entire collections open for public access as it had been in the original location in *Unter den Linden* and thus, the new building was designed along these lines. In a quirk of history, the successor of W.C.H. Peters, Karl August Möbius (1825–1908), who was the first director in the new building, supported the dual arrangement. Since the new building was devoid of capacious magazine areas for the ever growing zoological, palaeontological, geological and mineralogical collections, he directed that the grand staircases (Fig. 10.9) and the upper floors were closed to public access in order to gain collection space. Nevertheless, due to the growing collections, a northern wing soon had to be added—the construction lasted from 1913 to 1917. This was purposely built as collection and office space with a lower ceiling, thus accommodating two additional floors in the same building height. The dual arrangement also allowed Möbius to implement the new concept of public education with specially designed galleries incorporating didactic needs on the ground floor (Möbius 1884a, b; Jahn 1989).

Fig. 10.9 Internal stairs intended for visitors' access to the upper floors (with permission from Museum für Naturkunde—Leibniz Institute for Evolution and Biodiversity Science; photo: H.J. Götz)

What did this mean for the project of the reconstruction of the eastern wing and the ongoing restoration programme? The architect, Roger Diener, had immediately grasped the aesthetic potential of the thousands of glass jars of the zoological wet collections and the necessity of physically separating the walkway of visitors through the eastern wing on the ground floor from the collection space. After some disputes over the dominance of research and curation pragmatics over aesthetic demands, it was quickly understood that placement of an authentic research collection into the public gallery in a natural continuation of the old building concept is somewhat irritating yet highly attractive and appreciated by an educated citizen of today.

Accordingly, the high ceilings of the old building were continued into the new part, which essentially is a concrete box with wall-cooling set into the framework of the remnant façades of the original eastern wing (Fig. 10.10). The shelf system of the wet collection is 5–6 m high per floor, separated by integrated grid levels. At the intersection between original building and the later added northern wing, virtually no levels correspond. Even in the ground floor galleries, these parts are separated by several steps preventing wheelchair access. In the upper floors, internal ramps render internal transport of materials difficult; however, in 1913, with many cheap hands available, probably nobody thought of the internal logistics or barrier-free access. Now a large transport elevator with 13 stops provides access to all floors and is publicly accessible on the two ground floors. Since this transport facility had to be built in anyway, it was also decided to place the taxidermy, dissection and preparation laboratories on top of the new eastern wing. Instead of the usual basement situation for these facilities, this provides high ceilings with natural lighting.

Fig. 10.10 New eastern wing under construction in 2008 (with permission from Museum für Naturkunde—Leibniz Institute for Evolution and Biodiversity Science; photo C. Radke)

In the restored northern head building of the eastern wing, it was possible to construct offices and laboratories for most of the scientific and technical staff working with the larger wet collections, which at the same time are quite close to the corresponding dry collections. No work spaces are allowed in the wet collection itself for safety reasons. Other safety requirements of the new collection space include temperature control that cools it down to 15–18 °C (i.e. below the flashing point of 70 % ethanol), an explosion-protected electrical system, a high air exchange to prevent the accumulation of alcohol vapours and a gas fire-extinguishing system based on nitrogen. Access into the collection halls is only permitted for trained personnel. Work time is limited within the collections and examinations of the specimens, or conservational measures must be carried out in the laboratories of the northern head building.

Energy consumption of the new facility is moderate but much higher than the original building with its passive ventilation similar to that of a termite mound. Building large compartments reduced construction costs and optimized storage volume for the collections, thus allowing for 20 % growth. After overcoming initial technical problems, the museum staff are still quite content with the functionality of the building after five years—and this is perhaps the best compliment for everyone involved in the design and construction of this specialized collection space. With the fascinating insight for the public (Fig. 10.11) and the equally spectacular façade reconstruction (Fig. 10.12), the eastern wing represents a highly esteemed example of good architecture, almost—a piece of art.

Fig. 10.11 Eastern wing ground floor with visitor walkway (with permission from Museum für Naturkunde—Leibniz Institute for Evolution and Biodiversity Science; photo: C. Radke)

In essence, the close cooperation between architect, construction team and users, respecting the demands of the building monument as well as the functionality of the research facility, led to a synergetic effect rather than compromise (Diener 2010). For the remaining part of the highly needed reconstruction, it is planned to continue in this way by making use of the large room volumes for suitable collections,

Fig. 10.12 Partly destroyed façade of eastern wing supplemented by concrete casts (with permission from Museum für Naturkunde—Leibniz Institute for Evolution and Biodiversity Science; photo: H.J. Götz)

incorporating the high climate inertia of the thick walls of the building, applying loam rendering to the walls preventing drastic changes in relative humidity, tightening the old double box windows, diminishing energy consumption and carbon footprint by geothermic heating and cooling, and, at the same time, providing more room, more access and more insight for the interested public in the upper floors, rather than pressing all collections into the overloaded existing building (Bartsch et al. 2013). Taking into account the numerous collection storage areas outside the original building, it is not possible to accommodate all collections in the existing building effectively. This is especially true for the quickly growing and particularly fragile, yet frequently accessed entomological collections that according to current conservation standards (see standards) require strict pest management and climate control, and this is also true for the particularly heavy parts of mineralogy and palaeontology collections which currently amount to a load of 808 t. Thus, after 200 years of growth, a new on-site collection building is needed to reconcile the needs for increased space for public galleries in the original building and the need to accommodate the collections in an energy efficient building incorporating all conservation requirements.

10.4 Collections and Public Galleries

The collections were initially intended for student education and scientific research but soon were opened to the general public. As Lichtenstein (1816) notes, the ordering and preparatory work started by Illiger in 1810 had by 1814 advanced far enough that limited access to the halls of the museum could be granted. Since no differentiation between public displays and collection specimens was made at that time, the entire collection served as galleries. This only changed with the directorship of Möbius (see above), who separated collections from galleries with their specialized didactic displays, thus transforming the museum to one of the most advanced of its kind. This status slowly diminished over time, but, with the reconstruction of the public galleries that opened in 2007, international recognition was regained (Moldrzyk 2015). In his review, Moldrzyk (2015) explained that this was achieved by careful planning after an analysis of the shortcomings of the exhibition concepts in the previous decades. It was understood that the focus of the exhibitions should shift from education towards raising interest by focussing on the fascinating facts and remarkable stories rather than trying to explain biological processes or complex systems. This generates a comfortable situation for visitors and makes them more receptive for natural history issues. Therefore, object aesthetics, scenography and the display of the objects are in the focus of exhibition planning while content overload is avoided. Yet, in a second step, a wealth of information is offered to the interested public. Important conclusions drawn by Moldrzyk (2015) include the insight that original items from the collections should be preferred over casts, models or reconstructions and that content should focus on the research of the museum's scientists, especially on evolution and biodiversity. He states that more than 40 scientists were involved in creating the new permanent exhibitions that display thousands of collection specimens. A special case in this context is the fish wet collection in the east wing which serves as active collection space but nevertheless can be viewed by the public (see above). Following the insight that original specimens from the collections are the key to success, future concepts must allow visitors to have a closer look at the collections. In a way, this resembles the initial concept of the Zoological Museum Berlin as implemented 200 years ago by Lichtenstein (1816).

10.5 Collections and Research

Scientific curiosity has always been the driving force behind the development of natural history collections, especially zoological collections, and the collections of the Museum für Naturkunde reflect this. In the same way as research moved from descriptive biology to morphological studies, molecular genetics and ecological research over the course of time, collecting and consequently the character of collections changed. Starting with "cabinets of curiosities", the focus moved to

Fig. 10.13 Series of study skins in the bird collection (with permission from Museum für Naturkunde—Leibniz Institute for Evolution and Biodiversity Science; photo: H.J. Götz)

"complete" collections at the beginning of the nineteenth century and further to whole series of specimens of specific taxa in order to understand their morphological variability (see Sect. 10.2 and Fig. 10.13).

Today, the Museum für Naturkunde takes an interdisciplinary approach to study biological and geological/palaeontological issues. With partners around the world, zoologists, palaeontologists and geoscientists work closely together on the discovery of microevolutionary mechanisms of population differentiation, on speciation, evolutionary genetics and biodiversity. The work covers the entire time frame from the birth of the solar system via the present to the modelling of future scenarios. Processes of diversity dynamics are also investigated based on changes of ecosystems in time and space.

10.5.1 New Collections

The collections support this research and, in turn, are complemented by newly acquired materials. Along with the arrival of new technologies like DNA sequencing, new collection types such as DNA and tissue collections arose. Especially in vertebrate zoology, scientists now take blood and tissue samples instead of whole organisms, or, if so, they take special care that the preservation of voucher specimens will not destroy the genetic resources. The DNA and tissue collection of the Museum für Naturkunde is one of the youngest collections of the institution with a history of only 30 years. In 2015, about 30,000 DNA samples and another 3000 tissue samples were stored in freezers at the Museum für Naturkunde, mainly from mammals, amphibians, molluscs, insects and crustaceans. Each new research question is translated into the growth of this collection.

Another relatively new facet in natural history collections are the fast-growing digital collections. Based on both digital access to specimen information and to digital representations of it, digitization and new scanning facilities facilitate access to the specimens, speed up processes and enable new ways of research. On the other hand, this leads to an impressive amount of digital data that need their own management strategies, access and curation. In 2015, about 4 % of all collections objects were accessible through central databases and management systems.

Outstanding in this context is the animal sound archive. It is one of the largest collections of animal sounds, consisting of about 120,000 analogue and digital bioacoustical recordings comprising almost all groups of animals. The animal sound archive is basis for a whole range of research projects and educational programmes, from behavioural studies to bioacoustic monitoring of birds and the detection of bioacoustical patterns. Almost all recordings are digitized and available online, see www.tierstimmenarchiv.de.

10.5.2 Collection-Based Research

Collections are especially relevant for taxonomic research and the entire field of biodiversity discovery, and they play an important role in many areas of evolutionary studies. There is an impressive record of good examples of collection-based research at the Museum für Naturkunde. Snapshots from recent years will give an impression of the variety of this research conducted by scientists of the museum.

Between 2009 and 2011, scientists of the Museum für Naturkunde discovered and described almost 500 new taxa which is about 1 % of all taxa described in this period. One of the most impressive findings is probably the giant wasp, *Megalara garuda*, a new genus and species of larrine wasps from Indonesia (Larrinae, Crabronidae, Hymenoptera). Collected on the Indonesian island of Sulawesi in 1930, it remained hidden in the collection for about 80 years. Only then the curator Michael Ohl stumbled over two specimens in his collection that caught his eye by their sheer size. Very quickly it became clear that this extraordinary sphecid species had never been described. The description of the Giant Wasp (Kimsey and Ohl 2012) caught the attention of hundreds of newspapers and journals all around the world (Fig. 10.14).

How to Name a Wasp

In a unique process of describing a new species, public visitors to the museum had the chance to vote and consequently decide on the scientific name of a

(continued)

new digger wasp (Ampulicidae, Hymenoptera). As guidance, they were provided with a description of biology and behavior of the new species and a selection of four names based on peculiarities of the species. As the result most visitors selected the name "dementor" (full name *Ampulex dementor* Ohl, 2014). This name, derived from the "soul-sucking" dementors from the popular Harry Potter books is an allusion to the wasps' behavior to selectively paralyze its cockroach prey. In this example, public voting on a scientific name has been shown to be an appropriate way to link museum visitors emotionally to biodiversity and its discovery (Ohl et al. 2014).

In another example, molecular as well as anatomical analyses revealed new evolutionary relationships between very similar-looking African rainforest frogs specialized on waterfalls and rapids. It turned out, that Eastern and Central African frog species were closely related. However, the West African frogs represented an evolutionary lineage of their own, branched off from all other extant frogs as early as the Cretaceous, a period when dinosaurs still roamed the earth. Furthermore, the new discovery has anatomic features not found in other frogs (long, pointed and backward-bending teeth in the upper and massive fangs in the lower jaw, Fig. 10.15), underpinning the evidence for the discovery of a new frog family. This finding has not only academic and scientific value but is also relevant for conservation and underlines the necessity to protect the unique and species rich forests of West Africa (Barej et al. 2014).

Fig. 10.14 Pinned specimens of the giant wasp *Megalara garuda* and other species of Hymenoptera (with permission from Museum für Naturkunde—Leibniz Institute for Evolution and Biodiversity Science; photo: C. Radke)

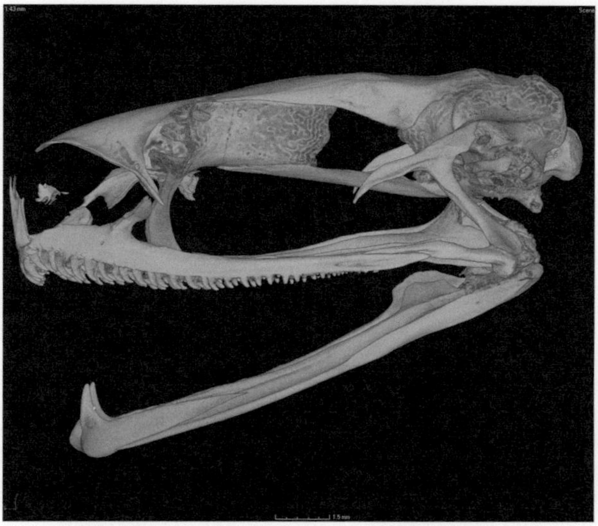

Fig. 10.15 Visualization of the skull and teeth of *Odontobatrachus natator* (ZMB 78203) from μCT data, lateral aspect (with permission from Museum für Naturkunde Berlin—Leibniz Institute for Evolution and Biodiversity Science, visualization by M. Barej, μCT scanning by K. Mahlow)

The combination of taxonomic expertise, comprehensive collections and complementary fieldwork builds a solid ground for the documentation and description of biodiversity and biogeography. Not surprisingly, several scientists from the Museum für Naturkunde are authors of comprehensive monographs on specific taxonomic groups, including recent works on bugs (Wachmann et al. 2004–2008; Deckert and Göllner-Scheiding 2006), poisonous snakes (Sharma et al. 2013) and sawflies (Koch 2005; Koch et al. 2015a, b). Others are involved in national programmes for marine exploration on research vessels such as the "Sonne" which results in the description of new species from the deep sea (e.g. Logan et al. submitted). Some of this work is linked to long-term national research projects such as BIOTA and maritime research programmes, e.g. with the research vessel "Sonne" in the Indian and Pacific Oceans.

In 2014, the German National Academy of Sciences Leopoldina published a review of taxonomic research in the era of OMICS technologies. This milestone document on the future of taxonomic research in Germany was coauthored by colleagues of the Museum für Naturkunde (C. Häuser, C. Lüter in Nationale Akademie der Wissenschaften Leopoldina 2014, http://www.leopoldina.org/uploads/tx_leopublication/2014_Stellungnahme_Taxonomie_LANG_final.pdf). As part of the recommendations, three areas for future collaborative projects were proposed: the description of all species in Central Europe supported by new high-throughput methods and technologies, an internal revision of the collections and the future development of taxonomy especially with regard to new standards and rules.

Recommendations and results have a direct influence on strategies and priorities of taxonomic research at the Museum für Naturkunde. The development of efficient methodology for recording and analysing biodiversity changes (taxa, ecosystems and timescales) is one of the strategic aims of the museum (https://www.naturkundemuseum.berlin/en/insights/about-us) and is realized, e.g. in the joint project Indonesian biodiversity discovery and information system (INDOBIOSYS, http://www.indobiosys.org/). This project develops and provides core components for a knowledge-based functional screening approach employed to the discovery of new anti-infective compounds from Indonesian organisms. This comprises (1) a novel integrated high-throughput biodiversity discovery pipeline for sampling, identification and provision of target groups from areas with a high level of biodiversity and (2) setting up a digital Indonesian Biodiversity Information System. The combination of primary biodiversity data and relevant metadata supporting an innovative approach towards the discovery of active compounds creates a novel platform that allows a targeted, efficient and sustainable exploitation of biological resources in Indonesia. The approach is accompanied by an internal revision of existing material in the collections, e.g. of the Museum für Naturkunde.

The research on microevolution focuses on gradual evolutionary changes within organisms that can lead to the emergence of new forms over extended periods of time. The research on evolutionary morphology is concerned with the evolution of genetic blueprints and characteristic complexities and builds a bridge between deep time and Recent biodiversity. Various methods are used, including modern imaging procedures and different types of digitization, as well as molecular biological approaches. Collections are playing a key role in many of the research projects in these fields. Outstanding in this context is the Embryological Collection with approximately 600 vertebrate species in more than 3000 alcohol jars and around 80,000 histological preparations of developmental stages of vertebrates and their reproductive organs (cf. Richardson and Narraway 1999). Originating mainly from the historical collections of Ambrosius Arnold Willem Hubrecht (1853–1915) and James Peter Hill (1873–1954), it is the largest and most significant collection of its kind. Recent publications on the basis of this specific collection include Ashwell et al. (2012) and Koyabu et al. (2014).

One of the focal research areas at the Museum für Naturkunde is the evolution of tetrapods and more specific the evolution and development of the vertebrate body plan and the relationship between ontogeny and phylogeny throughout the evolutionary history of tetrapods and patterns of phylogenetic and morphological diversification. In an integrative approach data from both fossil records and extant taxa are combined including fresh material as well as collections objects (Müller et al. 2010; Fröbisch et al. 2015).

In many other cases, collection specimens support the work on specific research questions by providing essential details or missing links. At the same time, collections benefit from the research input and the acquisition of new collection material through expeditions and research work. This is not only true for research fields mainly based on collections like taxonomy, systematics and biogeography but also for evolutionary studies and research on diversity dynamics, covering the entire time frame from deep time to the present situation. In 2014, research at the Museum für Naturkunde resulted in 208 scientific publications, 115 of them in ISI-listed journals (see Annual Report 2014, http://www.naturkundemuseum.berlin/sites/default/files/jahresbericht_mfn_2014.pdf).

10.5.3 Training

The collections also build the ground and are the resource for manifold educational activities, ranging from student courses to training workshops for collection management staff.

Well-trained collection management staff is one of the key elements for the management and development of natural history collections. The Museum für Naturkunde is aware of this and has put a special emphasis on staff training. On the way to a career development plan and life-long learning for collection management staff, a lot has already been realized with the support of national and EU projects as well as institutional funding. With focus on the staff of the Museum für Naturkunde, projects and initiatives have always taken a broader approach, including national and international partners and networks. Major achievements are:

- Regular training courses on relevant issues, e.g. Integrated Pest Management, disaster preparedness, collection techniques, legal aspects (partly funded by EU, Synthesys project, http://www.synthesys.info/)
- Staff exchange with EU partner institutions (funded by Leonardo da Vinci, Daubenton project)
- Development of a competency framework for collection management (funded by Leonardo da Vinci, EUColComp project, http://eucolcomp.myspecies.info/)
- Development of a wiki platform for collection management as tool for collaborative work and knowledge pool (funded by BMBF)

10.5.4 Beyond the Traditional Use

Besides the core research areas such as systematics, phylogeny and evolutionary studies, scientists are also facing new research questions. Many of them are linked to the increasingly demanding societal and environmental challenges, e.g. questions regarding food security, natural resources, emerging diseases, biodiversity loss, etc. With more than 200-year history the zoological collections of the Museum für Naturkunde have already proven in many cases that they can assist in answering these questions, and they have the capability to demonstrate their usefulness in these and many other in the future. However, to do so and to remain relevant for science and society, it is essential that institutions holding collections address these challenges in their research strategies and build new and fruitful partnerships not only within the scientific community but also beyond, with the industry and the general public. This is even more important since the valorization of collections and public engagement with science are becoming more and more driving factors in the world of zoological collections.

Acknowledgements We would like to thank Lothar Beck, the editor of this volume, for enabling our inclusion despite several delays in the preparation of the manuscript. We would like to thank everyone from the Museum für Naturkunde who contributed to the manuscript in one way or another: Uwe Moldrzyk provided his new paper on the exhibitions highlighting the close links between collections and public galleries in Berlin, Sabine Hackethal double-checked the historical facts, Manuela Bauche contributed articles on the Cuba Expedition, Frank Tillack, Michael Barej, Kristin Mahlow, Antje Dittmann, Carola Radke, and Hwaja Götz each provided images or improved their appearance, Nadia Fröbisch and Johannes Müller provided examples for research related to collections, Petra Ebber and Gesine Steiner assisted in the paperwork, Edda Aßel collated the collection statistics, and both Ronja Fröhlich and Selma Lindert assisted in proofing.

Appendix

1. National and international networks
2. Collection use
3. Collection statistics
4. Standards

1. National and International Networks

The collections of the Museum für Naturkunde are part of a dispersed international scientific infrastructure that is linked by joint initiatives, research projects, exhibitions and of course the shared desire among the institutions' directors and scientists that these collections be maintained, utilized, displayed and augmented. These efforts include national and international projects such as the Berlin-Brandenburg Institute of Advanced Biodiversity Research (BBIB, http://www.bbib.org/), Synthesys (an integrated European infrastructure for natural history collections, http://www.synthesys.info/), Biota (Biodiversity Monitoring Transect Analysis in Africa, http://www.biota-africa.org/) and EDIT (European distributed Institute of Taxonomy, http://www.e-taxonomy.eu/).

2. Collections Use

The zoological collections of the Museum für Naturkunde are not only used by internal scientists but they also provide a huge international research infrastructure for scientists from all over the world. During the 6-year period from 2009 to 2014, an average of 650 external scientists visited the Museum für Naturkunde annually. In 2014, they spent a total of 2408 days working in the collections. Overall, 1764 collection-related enquiries were recorded in 2014, resulting in 47,921 objects on loan. On average, about 33,000 objects were sent on loan to partner institutions and scientists abroad in each of year between 2009 and 2014, and about 1600 scientific inquiries were answered by the collection staff of the museum.

3. *Zoological Collections Statistics*

a. Vertebrate Collections

The oldest stock of the Vertebrate collection are several newly identified reptilian specimens bought from Albertus Seba and donated to the museum in 1817 by Graf Friedrich Heinrich von Borcke (1776–1825, Bauer and Günther 2013) followed by more than 800 fish specimens from M. E. Bloch and P.S. Pallas (late eighteenth century). The vertebrate collections have a global geographic scope with some emphasis on Africa (e.g. Peters, Pascha, Hemprich and Ehrenberg, Stuhlmann), Japan (Döderlein, Hilgendorf) and Southeast Asia (Day, v. Bork and v. Martens). Considerable contributions derive from the early expeditions of exploration vessels like the "S.M.S. Gazelle" (1874–1876) or "Valdivia" (Deutsche Tiefsee-Expedition 1898–1899), from the former colonies at the end of the nineteenth and beginning of the twentieth century and from land expeditions (e.g. Hemprich and Ehrenberg, v. Humboldt, Temminck, and Mayr). Recent activities include additions to the herpetological collection by R. Günther (New Guinea) and M. O. Rödel (West and Central Africa). As depicted above, the Embryological Collection is the largest of its kind and is specialized in vertebrate developmental stages and reproductive organs (histology and wet specimens). Primary collections are not only from A.A.W. Hubrecht and J.P. Hill but also from A. Dohrn, L. Bolk and experimental work by Mangold, Spemann, Grüneberg and others.

Collection	Skeletal preparation	Wet specimens	Skins and mounted specimens	Other collection material	Types	Estimated number of individuals
Ichthyological collection	1100	130,640	1750		>1,700	134,000
Herpetological collection	500	118,000	1500		2600	155,000
Bird collection	7000	5000	155,000	40,000 eggs 1500 nests	6000	207,000
Mammal collection		35,000 lots			2210	150,000–180,000
Embryological Collection	0	3000 lots	0	80,000–100,000 histological slides	0	ca. 30,000 databank entries

b. Invertebrate Collections (Without Entomology)

The collections comprise all recent invertebrate groups with the exception of insects. Some 6.2 million specimens are systematically arranged and allow for efficient retrieval and curation. The majority of the specimens are preserved in ethanol complemented by dry and microscope slide collections. Most taxa are covered worldwide with rich material from the expeditions of the nineteenth and early twentieth century and from colonial origin. Zoologists such as Kükenthal, von Martens, Döderlein, Arndt, Ehrenberg, Haeckel, Philippi, Dunker, Rudolphi, Blumenbach, Esper, Plate and even Darwin deposited at least part of their material in the Berlin collection. Many of the current marine expeditions by research vessels such as the "Polarstern", the "Meteor" or the "Sonne" are accompanied by the curators of these collections who thus or in other field work contribute to the growth of the collections.

Collection	Taxa	Curatorial units	Types	Estimated number of individuals
Vermes	Platyhelminthes, Nemathelminthes, Nemertini, Sipuncula, Echiuroidea, Chaetognatha, Pogonophora, Annelida	41,159	2150	100,000
Crustaceans	Crustacea	36,500	800	500,000
Chelicerates/ Myriapods/ Onychophorans	Chelicerata, Myriapoda, Onychophora	52,300	5098	250,000
Molluscs	Mollusca	110,000	9000	7,000,000
Marine Invertebrates	Porifera, Cnidaria, Ctenophora, Phoronida, Brachiopoda, Bryozoa, Echinodermata, Pterobranchia, Enteropneusta, Tunicata, Acrania	130,000	3000	350,000

c. Entomology

The insect collections are estimated to count around 15 million predominantly pinned specimens. They have a global geographic coverage with an emphasis in the western Palearctic, Central Asia and the former German colonies and cover approximately 10–30 % of the known insect species. Apart from vast collections of pinned material and a smaller portion of wet specimens, there are several tens of thousands of microscope slides. The oldest specimens date back from at least 1775 and have been part of collections by A. von Humboldt, Pallas, Herbst, Illiger, Gravenhorst, Hellwig, v. Hoffmansegg, Fabricius, Panzer and other eminent entomologists of that time. However, the major part of the collections originates from the nineteenth and early twentieth century and due to collection activities by the curators, they are growing. The insect collections hold an enormous number of type specimens, particularly from the nineteenth century.

Collection	Prepared specimens	Types
Lepidoptera	4,000,000	10,700
Coleoptera	6,000,000	100,000
Hemimetabola	680,000	8800
Hymenoptera	2,227,880	11,700
Diptera and Siphonaptera	1,300,000	21,000
Neuropterida, Orthopteroidea, Sphecidae	350,000	9200
Unprepared material	1,000,000	
Entomology total	15,500,000	161,400

4. Standards

MGC Museum and Galleries Commission Standards 2. in the Museum Care of Biological Collections. 1992. pp. 1–55.

NHM Life and Earth Sciences and Library Collections. Curatorial Policies and Collections Management Procedures 2003—Revised edition. The Natural History Museum, London 2003. pp. 1–47 www.nhm.ac.uk

National Museum of Natural History Department of Entomology Collections Management Policy. Smithsonian, Washington, May 1998. pp. 1–82.

Sammlungsrichtlinien des Museums für Naturkunde Berlin MfN. – Aberhan, M., Leinfelder, R. & Bartsch, P. (eds.) Museum für Naturkunde, Berlin, April 2012. pp. 1–62, 15 ff. Appendices.

References

Angermann R (1989) Die Säugetierkollektion des Museum für Naturkunde der Humboldt Universität zu Berlin. Säugetierkd Inf 3(13):47–68

Annual Report (2014) http://www.naturkundemuseum-berlin.de/fileadmin/startseite/institution/publikationen/jahresberichte/Jahresbericht_2014_a.pdf

Anonymus (1827) Neuer Nekrolog der Deutschen, Dritter Jahrgang 1825. Erstes Heft. Druck und Verlag von Bernhard Friedrich Voigt, Ilmenau

Ashwell KWS, Hardman CD, Giere P (2012) Distinct development of peripheral trigeminal pathways in the platypus (*Ornithorhynchus anatinus*) and short-beaked echidna (*Tachyglossus aculeatus*). Brain Behav Evol 79:113–127. doi:10.1159/000334469

Barej MF, Schmitz A, Güther R, Loader SP, Mahlow K, Rödel M-O (2014) The first endemic West African vertebrate family – a new anuran family highlighting the uniqueness of the upper Guinean biodiversity hotspot. Front Zool 11:8. doi:10.1186/1742-9994-11-8

Bartsch P, Neuhaus B (2011) Der Wiederaufbau des Berliner Museums für Naturkunde. Museum Aktuell 179:13–16

Bartsch P, Quaisser C, Giere P, Theuer-Kock A, Feck N (2013) New meets old – the requirements and limits of new collection facilities at the Museum für Naturkunde Berlin. In: Ashley-Smith J, Burmester A, Eibl M (eds) Climate for collections. Standards and uncertainties. Postprints of the Munich climate conference 7th–9th Nov 2012. Archetype Publications, London, pp 163–177. http://www.doernerinstitut.de/downloads/Climate_for_Collections.pdf

Bauer A, Günther R (2013) Origin and identity of the von Borcke collection of amphibians and reptiles in the Museum für Naturkunde in Berlin: a cache of Seba specimens? Zoosyst Evol 89 (1):167–185

Coleman CO (2007) Nachruf auf Herrn Professor Dr. Hans-Eckhard Gruner. Mitt Mus Nat Berl Zool Reihe 83(2):99–104

Damaschun F, Landsberg H (2010) "...so bleiben dem materiell Gesammelten und geographisch Geordneten fast allein ein langdauernder Wert" – 200 Jahre Museum für Naturkunde Berlin. In: Damaschun F, Hackethal S, Landsberg H, Leinfelder R (eds) Klasse Ordnung Art – 200 Jahre Museum für Naturkunde Berlin. Basilisken Presse, Rangsdorf, pp 11–22

Deckert J, Göllner-Scheiding U (2006) Lace bugs of Namibia (Heteroptera, Tingoidea, Tingidae). Denisia 19, neue Serie 50:823–856

Diener R (2010) Kontinuität und Neufassung – Die Renovierung des Ostflügels. In: Damaschun F, Hackethal S, Landsberg H, Leinfelder R (eds) Klasse Ordnung Art – 200 Jahre Museum für Naturkunde Berlin. Basilisken Presse, Rangsdorf, pp 60–61

Eckert C (2010) Blüten, Kerfe und Brasiliens Schätze – Graf von Hoffmannsegg als Museumsgründer. In: Damaschun F, Hackethal S, Landsberg H, Leinfelder R (eds) Klasse Ordnung Art – 200 Jahre Museum für Naturkunde Berlin. Basilisken Presse, Rangsdorf, pp 112–115

Ehrenberg CG (1828) Naturgeschichtliche Reisen durch Nord-Afrika und West-Asien in den Jahren 1820–1825 von Dr. W.F. Hemprich und Dr. C.G.: Ehrenberg. Ernst Siegfried Mittler, Berlin

Fröbisch NB, Bickelmann C, Olori JC, Witzmann F (2015) Deep-time evolution of regeneration and preaxial polarity in tetrapod limb development. Nature. doi:10.1038/nature15397

German National Academy of Sciences Leopoldina (2014) Challenges and opportunities of integrative taxonomy for research and society – taxonomic research in the era of OMICS technologies. Deutsche Akademie der Naturforscher Leopoldina e.V. Nationale Akademie der Wissenschaften (German National Academy of Sciences). Halle/Saale, 54 p. http://www.leopoldina.org/uploads/tx_leopublication/2014_Stellungnahme_Taxonomie_EN_final_01.pdf

Geus A (1998) Zoologische Disziplinen. In: Jahn I (ed) Geschichte der Biologie – Theorien, Methoden, Institutionen, Kurzbiographien, 3rd edn. Nikol Verlag, Hamburg, pp 324–355

Göllner-Scheiding U (1972) Hoffmannsegg, Johann Centurius Graf von. p. 440 in Neue Deutsche Biographie 9. http://www.deutsche-biographie.de/ppn116942630.html

Greuter W, Breitwieser I, Vogt R (1994) Bericht über den Botanischen Garten und das Botanische Museum Berlin-Dahlem (BGBM) für das Jahr 1993. Wildenowia 24:5–31. http://www.bgbm.org/sites/default/files/documents/3996675.pdf

Helbig J (2010) Entwurfs- und Baugeschichte des Museums für Naturkunde. In: Damaschun F, Hackethal S, Landsberg H, Leinfelder R (eds) Klasse Ordnung Art – 200 Jahre Museum für Naturkunde Berlin. Basilisken Presse, Rangsdorf, pp 28–32

Humboldt Av (1826) Bericht über die Naturhistorischen Reisen der Herren Ehrenberg und Hemprich durch Ägypten, Dongola, Syrien, Arabien, und den östlichen Abfall des Habessinischen Hochlandes in den Jahren 1820–1825. Druckerei der Königlichen Akademie der Wissenschaften, Berlin

Illiger K (1811) Prodromus Systematis Mammalium et Avium. Sumptibus C. Salfeld, Berlin

Jaeger B, Uhlig M (2010) Zur frühen Geschichte der Käfersammlung des Zoologischen Museums von 1810–1880. In: Damaschun F, Hackethal S, Landsberg H, Leinfelder R (eds) Klasse Ordnung Art – 200 Jahre Museum für Naturkunde Berlin. Basilisken Presse, Rangsdorf, pp 132–135

Jahn I (1985) Zur Vertretung der Zoologie und zur Entwicklung ihrer institutionellen Grundlagen an der Berliner Universität von Ihrer Gründung bis 1920. Wiss Z Humboldt-Univ Math-Nat Reihe 34(3/4):260–280

Jahn I (1989) Der neue Museumsbau und die Entwicklung neuer museologischer Konzeptionen und Aktivitäten seit 1890. Wiss Z Humboldt-Univ Berl Math-Nat Reihe 38:287–307

Kimsey LS, Ohl M (2012) *Megalara garuda*, a new genus and species of larrine wasps from Indonesia (Larrinae, Crabronidae, Hymenoptera). Zookeys 177:49–57. doi:10.3897/zookeys.177.2475

Koch F (2005) Biogeography, diversity and ecology of sawflies in the Afromontane region (Insecta: Symphyta). In: Huber B, Sinclair J, Lampe K-H (eds) African biodiversity: Molecules, organisms, ecosystems. Springer, Berlin, pp 159–166

Koch F, Goergen G, van Noort S (2015a) The sawflies of Namibia and western South Africa (Symphyta, Hymenoptera). ABC Taxa, Bruxelles 15:8+1–262

Koch F, Pauly A, Hora ZA, Boevé J-L (2015b) Sawflies of Ethiopia (Hymenoptera: Argidae, Tenthredinidae). Zootaxa 3998

Köhler E (2010) Kuba, Expeditionen und weitere Zusammenarbeit. In: Damaschun F, Hackethal S, Landsberg H, Leinfelder R (eds) Klasse Ordnung Art – 200 Jahre Museum für Naturkunde Berlin. Basilisken Presse, Rangsdorf, pp 244–247

Köstering S (2010) Eine "Musteranstalt" "naturkundlicher Belehrung" – Museumsreform im Berliner Naturkundemuseum 1810–1910. In: Damaschun F, Hackethal S, Landsberg H, Leinfelder R (eds) Klasse Ordnung Art – 200 Jahre Museum für Naturkunde Berlin. Basilisken Presse, Rangsdorf, pp 37–45

Koyabu D, Werneburg I, Morimoto N, Zollikofer CPE, Forasiepi AM, Endo H, Kimura J, Ohdachi SD, Truong Son N, Sánchez-Villagra MR (2014) Mammalian skull heterochrony reveals modular evolution and a link between cranial development and brain size. Nat Commun 5 (3625):1–9. doi:10.1038/ncomms4625

Kühlmann DHH (1980) Das Korallenriff-Diorama. Neue Museumskunde 3:209–215

Landsberg H (2010) Vom "Hottentottenarzt" zum Museums- und Zoodirektor – Hinrich Lichtenstein. In: Damaschun F, Hackethal S, Landsberg H, Leinfelder R (eds) Klasse Ordnung Art – 200 Jahre Museum für Naturkunde Berlin. Basilisken Presse, Rangsdorf, pp 118–121

Lichtenstein H (1816) Das zoologische Museum an der Universität zu Berlin. Ferdinand Dümmler, Berlin, http://reader.digitalesammlungen.de/de/fs1/object/display/bsb10308061_00005.html

Lichtenstein H (1823) Verzeichnis der Dubletten des Zoologischen Museums der königlichen Universität zu Berlin nebst Beschreibung vieler bisher unbekannter Arten von Säugetieren, Vögeln, Amphibien und Fischen. Trautwein, Berlin, http://www.archive.org/download/verzeichnissderd00lich/verzeichnissderd00lich.pdf

Logan A, Hoffmann J, Lüter C (submitted) Checklist of Recent thecideoid brachiopods from the Indian Ocean and Red Sea, with a description of a new species of Thecidellina from Europa Island and a re-description of *T. blochmanni* Dall from Christmas Island. Zootaxa

Möbius KA (1884a) Rathschläge für den Bau und die innere Einrichtung zoologischer Museen. Zool Anz 7:378–383

Möbius KA (1884b) Rathschläge für den Bau und die innere Einrichtung zoologischer Museen. Zentralbl Bauverwaltung 4:420–421

Moldrzyk U (2015) Renovació de grans museus clàssics. Nous punts de vista sobre naturalesa? Renovació dels museus d'història natural segons l'exemple de Berlín. Mnemòsine. Revista Catalana de Museologia 8:29–37. (English translation: New views on Nature? Renewal of Natural History Museums by the Berlin example, pp 71–75. http://www.museologia.cat/wp-content/uploads/2014/03/Traduccions-dossier-Mnemòsine-8.pdf)

Müller J, Scheyer TM, Head JJ, Barrett PM, Werneburg I, Ericson PGP, Pol D, Sánchez-Villagra MR (2010) Homeotic effects, somitogenesis and the evolution of vertebral numbers in recent and fossil amniotes. Proc Natl Acad Sci USA 107(5):2118–2123. doi:10.1073/pnas.0912622107

Nationale Akademie der Wissenschaften Leopoldina (2014) Herausforderungen und Chancen der integrativen Taxonomie für Forschung und Gesellschaft—Taxonomische Forschung im Zeitalter der OMICS-Technologien. Deutsche Akademie der Naturforscher Leopoldina e.V. Nationale Akademie der Wissenschaften, Halle/Saale. http://www.leopoldina.org/uploads/tx_leopublication/2014_Stellungnahme_Taxonomie_LANG_final.pdf

Neuhaus B, Allspach A, Bartsch P, Burckhardt D, Coleman CO, Fries I, Fuchs R, Gudo M, Kotrba M, Mentjes M, Moore S, Neumann D, Obererl C, Potthast A, Riedel J, Rudolf R, Schnalke T, Schönbohm D, Schuda M, van Dam A, Widulin N (2012) KUR-Projekt: Aufbau und öffentliche Kommunikation eines wissenschaftsbasierten Sammlungsmanagements für naturkundliche Nasssammlungen. – Abschlussbericht und Projektdokumentation Kulturstiftungen des Bundes und der Länder. http://www.hornemann-institut.de/german/epubl_projekte116.php

Ohl M, Lohrmann V, Breitkreuz L, Kirschey L, Krause S (2014) The Soul-Sucking Wasp by popular acclaim – museum visitor participation in biodiversity discovery and taxonomy. PLoS One 9(4):e95068. doi:10.1371/journal.pone.0095068

Paletschek S (2007) Zurück in die Zukunft? Universitätsreformen im 19. Jahrhundert. In: Jäger W (ed) Das Humboldt-Labor: Experimentieren mit den Grenzen der klassischen Universität. Albert-Ludwigs-Universität, Freiburg, pp 11–15

Quaisser C, Woog F (2011) Sammlungsrichtlinien am Beispiel naturkundlicher Sammlungen. In: Weber C, Mauersberger K (eds) Universitätsmuseen und -sammlungen im Hochschulalltag. Aufgaben. Konzepte. Perspektiven. Beiträge zum Symposium vom 18.–20. Februar 2010 an der Humboldt-Universität zu Berlin, pp 33–42. http://edoc.hu-berlin.de/conferences/ums2010

Richardson MK, Narraway J (1999) A treasure house of comparative embryology. Int J Dev Biol 43:591–602

Sharma SK, Pandey DP, Shah KB, Tillack F, Chappuis F, Thapa CL, Alirol E, Kuch U (2013) Venomous snakes of Nepal – a photographic guide. Schweizerische Eidgenossenschaft, Embassy of Switzerland in Nepal, 77 pp, http://www.bik-f.de/files/publications/kuch_venomous_snakes_of_nepal_-_english_edition.pdf; http://www.bik-f.de/files/publications/kuch_venomous_snakes_of_nepal_-_nepali_edition.pdf

Stresemann E (1954) Hemprich und Ehrenberg, Reisen zweier naturforschender Freunde im Orient geschildert in ihren Briefen aus den Jahren 1819–1826. Abhandlungen der deutschen Akademie der Wissenschaften zu Berlin, Klasse für Mathematik und allgemeine Naturwissenschaften (1):1–177

Studer T (1889) Vorwort. In: Hydrographisches Amt des Reichs-Marine-Amts, (ed) Die Forschungsreise der S.M.S. “Gazelle” in den Jahren 1874–1876 unter Kommando des Kapitän zur See Freiherr von Schleinitz. Ernst Siegfried Mittler und Sohn, Berlin, pp 5–6. http://edoc.hu-berlin.de/ebind/mfn/3_ZoologieundGeologie/XML/

Tiede A (1898) Museumsbaukunde. Wilhelm Greve, Berlin

Tiede A, Kleinwächter F (1891) Das Museum für Naturkunde der Universität Berlin. Sonderdruck aus der Zeitschrift für Bauwesen, pp 1–7, 6 Pl

von Martens E (1867) Die Preussische Expedition nach Ost-Asien : nach amtlichen Quellen. Zoologischer Theil / bearbeitet von Eduard v. Martens. Verlag der Königlichen geheimen Ober-Hofbuchdruckerei, Berlin. (R. v. Decker). doi:10.5962/bhl.title.14188

von Martens E (1876) Die preussische Expedition nach Ost-Asien. Nach amtlichen Quellen. Zoologischer Teil, Erster Band. Verlag der Königlichen geheimen Ober-Hofbuchdruckerei, Berlin. (R. v. Decker). http://www.deutschestextarchiv.de/book/images/berg_ostasienzoologie01_1876

Wachmann E, Melber A, Deckert J (2004–2008) Wanzen, vols 1–4. Tierwelt Deutschlands, vols 75, 77, 78, 81

Wagner J (1967) In den Riffen von Arroyo Bermejo. Poseidon 68(8):346–350

Chapter 11
BERLIN: Zoologische Lehrsammlung (Zoological Teaching Collection)

Gerhard Scholtz

Abstract The zoological teaching collection of the Berlin University (now Humboldt-Universität zu Berlin) was founded in 1884 by the zoologist Franz Eilhard Schulze. From the onset, the concept for the collection was the creation of a didactic tool for teaching zoology. After a continuous growth for several decades, dramatic losses had to be bemoaned during World War II and in the 1970s due to the new orientation of biology in the GDR. After German reunification most of the remaining collection was transferred to its current placement on the science campus Philippstraße. Since 1995 the collection has a new director. The last 20 years were characterized by consolidation, expansion, repair, and cataloging. Currently, the teaching collection encompasses about 30,000 objects, among them are wax, glass, and papier-mâché models from the nineteenth century, wet and dry anatomical specimens, insect boxes, microscope slides, and educational wall charts from Leuckart, Pfurtscheller, and the Zoological Institute's own production. Many of the objects are still heavily used for teaching purposes. In addition, parts of the collection have been presented to the public during various exhibitions. Furthermore, the Internet, several publications, and public talks led to an increasing public awareness of the zoological teaching collection. Nevertheless, concerning infrastructure such as room conditions, finances, and personnel, the collection is in a suboptimal shape.

Keywords Didactics • Anatomical models • Wall charts • Microscopic slides • Anatomical preparations • Public awarness

11.1 A Short History

11.1.1 The Beginning

The zoological teaching collection was founded by the sponge specialist Franz Eilhard Schulze (1840–1921) (Richter 2000). With his appointment at the

G. Scholtz (✉)
Institut für Biologie/Vergleichende Zoologie, Humboldt-Universität zu Berlin, Philippstr. 13, 10115 Berlin, Germany
e-mail: gerhard.scholtz@rz.hu-berlin.de

L.A. Beck (ed.), *Zoological Collections of Germany*, Natural History Collections,
DOI 10.1007/978-3-319-44321-8_11

Friedrich-Wilhelm-Universität in the year 1884, Schulze established the Zoological Institute (Zoologisches Institut) and the teaching collection in addition to the already existing Zoological Museum and the Zootomical-Anatomical Museum (Jahn 1985; Tembrock 1985; Richter 1999, 2000). All these collections were housed in the university building on the street Unter den Linden in close proximity to each other and to the lecture halls and laboratories. It has been documented that material was exchanged between these three institutions (Richter 1999). Due to space restrictions, in 1888 the zoological teaching collection moved to the west wing of the new natural history museum building (Museum für Naturkunde) on the Invalidenstraße 43. This part of the building combined the department of zoology, the collection, and the private house of the director Schulze.

From the onset Schulze's concept of the collection was devoted to creating a powerful tool for teaching zoology. Hence, the focus was laid on didactic material such as different kinds of anatomical and developmental models, wall charts, and anatomical and microscopic preparations. During the phase of his directorship, the collection grew very quickly and constantly. This is documented in the still existing inventories of contents. The objects were either bought from commercial suppliers or made by technicians, craftsmen, and illustrators in the department.

11.1.2 Cuts, Changes, and Losses

During World War II, large parts of the collection were destroyed. In 1943 a bomb that hit the west wing of the museum led to the loss of 100 vertebrate skeletons, 500 mounted animals, 200 invertebrate specimens, 5 wax model series, and 150 wall charts. Further losses were caused by fights during the final stage of the war in Berlin in May 1945 (Tembrock 1985).

The third reform of the higher education system in the GDR in 1968 led to a fusion of the Zoological Institute and the Institute of Zoology of the agricultural faculty (Herbst 1985; Tembrock 1985; Richter 1999). This also led to an integration of parts of the zoological collection of the Institute of Zoology into the zoological teaching collection. The zoologist Hans Georg Herbst (1920–1991) was in charge of this combined collection. He introduced the complex numbering system which is still in use today. In the end the department of zoology was closed and transformed into Behavioral Studies in the Section of Biology (Tembrock 1985). In 1970 the zoological teaching collection became dramatically reduced. Many objects were given or thrown away, buried in the yards of the former department (Wolfgang Dreier, pers. comm.), or transferred to the collections of the natural history museum (Hackethal and Hackethal 1999). In particular, numerous historical wax, glass, and papier-mâché models and anatomical preparations are now integrated into the Museum für Naturkunde. Nevertheless, a large part was retained and continuously used in teaching by dedicated zoologists.

11.1.3 New Developments

During the early 1990s, most of the remaining objects were transferred from the Invalidenstraße to the old veterinary anatomy building on the campus Philippstraße (Campus Nord). This building originally housed the zoological collection of the veterinarians (Grünbein 1997). Hence, a number of showcases and cabinets, and enough space, provided a reasonable infrastructure for the zoological teaching collection of the new department of biology (Fig. 11.1). However, no one was officially in charge of the collection.

Fig. 11.1 A characteristic view into one of the collection showcases. The mixture of models and wet and dry anatomical preparations is clearly visible

This changed with the appointment of Gerhard Scholtz (1954 -) to the professorship of Comparative Zoology in 1995. In addition to the director, there is a half position for a technician and a half position for a scientist. The first actions were to unify all the bits and pieces of the collection from the various places and to combine them. The precious Blaschka glass models were repaired, cleaned, and displayed. Parts of a forgotten collection of pests were included. Later objects such as skulls, skeletons, and eggs from the Tierpark Berlin were added. In turn some insect collections of scientific, as opposed to teaching, value were transferred to the German Entomological Institute (Ziegler and Menzel 2000), the Julius Kühn Institute, and the Museum für Naturkunde. Some parts are on semipublic display in the corridors of the zoology building. For several years now, the collection has been financially supported by the biology department.

11.1.4 Drawbacks

The scientific position has been vacant for several years, and there is currently no money to fill it. The rooms have not been renovated since the 1960s, and they do not correspond even to the most basic safety regulations (e.g., fire) for this kind of collections. Furthermore, the climatic and light conditions are inappropriate for many objects. More manpower and a large amount of money are needed to restore all the objects, to digitize them professionally, and to renovate the rooms in an appropriate way.

11.2 Current Status

Currently the zoological teaching collection harbors about 30,000 objects. Most of them date back to the end of the nineteenth century. The collection comprises more than 27,000 microscopic slides (Fig. 11.2); about 2500 anatomical preparations, skeletons, and insect boxes; 600 wall charts; and more than 100 models. Among the anatomical models, there are glass models of Leopold (1822–1895) and Rudolf (1857–1939) Blaschka (Hackethal 2008) mainly of radiolarians, cnidarians, and annelids (Fig. 11.3), Adolf Ziegler (1820–1898) wax models (Hopwood 2002) such as the development of the lancelet, and some worm larvae, and several wax models of Rudolf Weisker (1845–1887)/Paul Loth which represent mostly parasitic worms such as tape worm larvae, heads and proglottids, trichinae and other nematodes, and acanthocephalans. In addition, parasitic mites like *Demodex* and the itch mite scabies and a series of developmental stages of the noble crayfish (Scholtz 2014). Papier-mâché models of a silk worm caterpillar, human female reproductive organs including embryonic stages, and some bees from the Paris manufacturer Louis Auzoux (1797–1880) (Grob 2000). More recent models include large insect wax models from Alfred Keller (1902–1955) (Hackethal 2008) such as moths and

Fig. 11.2 Historical microscopic preparation of a thin section through a bivalve shell showing its various layers

Colorado beetle life stages, and SOMSO® plastic models of a crayfish, a mussel, a sea star, and an earthworm, among others. The wall charts include about 30 of the very detailed Rudolf Leuckart (1822–1898) charts of invertebrates (Redi et al. 2000) and about 15 Paul Pfurtscheller (1855–1927) charts as well as numerous original charts produced in the Zoological Institute from the late nineteenth century to the 1960s (Fig. 11.4). Vertebrate skeletons, mounted animals, and anatomical dry and wet specimens include rarities and interesting preparations of internal structures which could not be acquired again, if broken (Fig. 11.5), for instance, large tape worms, the eye of a blue whale, sea lilies from the deep sea, a collection of leaf mimicry animals, and series of crayfish wet specimens showing separately all their internal organs, and the development and in situ latex casts of the blood vascular system of vertebrates, to name but a few.

Fig. 11.3 Glass model of the radiolarian *Thalassicolla nucleata* from Leopold and Rudolf Blaschka (1885)

11.3 Teaching

The collection is still widely used for teaching at all levels. Wall charts, models, and anatomical preparations accompany lectures and practical lab courses. In some of the latter, the comprehensive collection of microscope slides is central. Likewise, dried arthropod specimens, mollusk shells, skulls, and mounted vertebrates are used for courses on species identification. These more traditional teaching devices are combined with modern tools such as digital presentations and videos. This mix seems attractive for many students, and in particular, wet preparations showing exotic specimens or the internal anatomy and skeletons always catch people's attention and lead to further questions.

In advanced level bachelor and master courses, objects from the collection are integrated in project-oriented studies. In these, the students use the zoological objects autonomously for the reconstruction of evolutionary and phylogenetic processes. For instance, series of preparations such as the bone patterns of mammalian forelimbs or the branching patterns of blood vessels in vertebrates in combination with plastic models of vertebrate hearts serve as basis for projects on comparative morphology and phylogenetics.

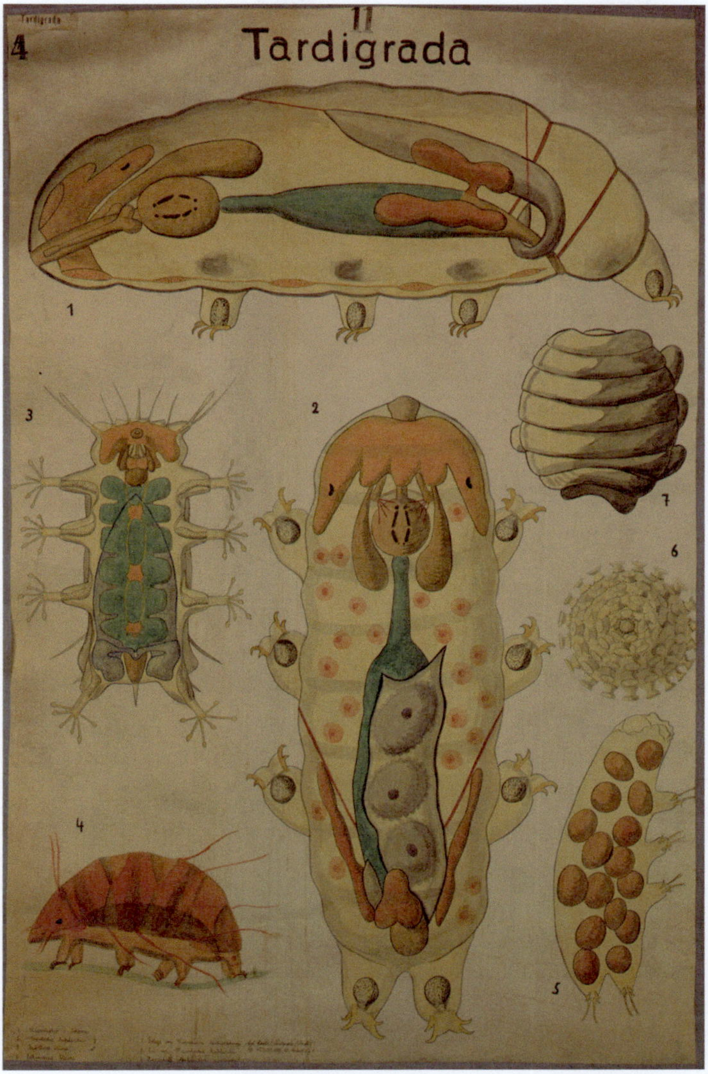

Fig. 11.4 Wall chart depicting the anatomy and ontogeny of water bears (Tardigrada) after Ernst Marcus. Drawn at the Zoological Institute (1934) probably by the illustrator Erika von Bruchhausen

11.4 Science

Due to the main focus of the collection for teaching purposes, the actual role of the collection for scientific questions is only marginal. Some objects have been used as valuable material for comparative studies, but in most cases, images of specimens,

Fig. 11.5 Wet specimen of the beef tapeworm *Taenia saginata*. This specimen is about 6 m long and comprises 600 proglottids coiled around a black glass pane (late nineteenth century). Anatomical preparation of a domestic cat. The animal's right side shows the natural habit with the original fur and a natural posture; the animal's left side shows the internal organization with the original skeleton and plaster models of the inner organs (1960)

models, wall charts, and microscopical and anatomical preparations served as illustration in scientific articles. Yet, for this purpose the collection is well suited as has been exemplified in several publications (e.g., Scholtz 2003; Reindl et al. 2015).

11.5 Public Awareness

For a number of reasons, it is crucial for university collections to be presented to the public: (1) the public has the right to know about them because they pay for them, (2) the aesthetic and scientific values of the objects are interesting even to a wider audience, (3) public awareness protects collections from being neglected or destroyed, and (4) the public may be a source of funding and donations.

Visibility and awareness always create further visibility and publicity. For instance, the Internet visibility of the zoological teaching collection leads to many requests for loan of material for exhibitions in Germany and other countries

Fig. 11.6 The large installation by Mark Dion at the exhibition "WeltWissen" (World Knowledge) at the Martin Gropius Bau, Berlin (2010/2011). Many objects from the zoological teaching collection were used

and for the right to publish images of wall charts, other objects in school books, etc. This often implies donations. Furthermore, a visible collection attracts artists, who in turn enlarge the degree of visibility of the collection through their work. Photographers Heidi and Hans-Jürgen Koch (Koch and Koch 2007) and Mark Kessell produced picture series of Blaschka glass models and anatomical preparations (Kessel 2013). Mark Dion used skeletons, models, wall charts, cabinets, and preparations for his installation in the "WeltWissen" exhibition (Fig. 11.6). Wolf von Kries, Andrea Roe, and Lucy Powell choose the zoological teaching collection for their contributions to the exhibition "On the Edge. Artists in Dialogue with Humboldt University Collections."

11.5.1 Publications

There are some publications on the history of the collection (e.g., Herbst 1985; Tembrock 1959, 1961, 1985; Richter 1999, 2000) and about individual objects (Scholtz 1997, 2010, 2014, 2015; Scholtz and Müller 2000; Hackethal 2008).

11.5.2 Exhibitions

The zoological teaching collection has been presented to the public through a variety of means such as the Internet, printed articles, exhibitions, and public talks.

To mention a few examples: In 2000/2001 the collection was shown to the public for the first time within the framework of the great exhibition of the collections of the Humboldt-Universität "Theatrum naturae et artis." There was a whole room dedicated to the zoological teaching collection. Wall charts, glass and wax models, anatomical preparations, etc., were not just presented, but arranged under various headings such as "The aesthetics of learning," "Collection and science," "Zoology and mythology," and so on (Scholtz and Müller 2000).

Models and anatomical preparations are relatively easy to present. Their sometimes exotic charm even leads to reactions from laypeople who do not know much about the scientific value of these objects. In contrast to this, the sometimes very beautiful and appealing structures of microscopic slides are difficult to show. The exhibition "Die Ästhetik des Kleinen" (The aesthetics of the small) shown in Berlin (2011) and Jena (2011/2012) addressed this problem by an enlarged and transparent reproduction of scans of microscopic preparations that were arranged in an aluminum scaffold (Scholtz 2011) (Fig. 11.7). Other exhibitions displaying numerous objects and with conceptual contribution of the collection were "Tiermodelle-Modelltiere—Die Zoologische Lehrsammlung" (2008), "WeltWissen. 300 Jahre

Fig. 11.7 View of the display of pictures of historical microscopic slides from the exhibition "Die Ästhetik des Kleinen" (The Aesthetics of the Small) at the Humboldt-Universität zu Berlin (2012)

Wissenschaften in Berlin" (2010/2011), "ModellSchau" (2015), and "On the Edge. Artists in Dialogue with Humboldt University Collections" (2015).

11.5.3 Public Lectures

Public lectures about the collection were given by the author on several occasions such as exhibitions, seminar series, meetings, and workshops. These lectures were entitled "Not just a tasty snack: The zoological teaching collection and the evolution of freshwater crayfish" (2001), "How to create public awareness for a university teaching collection" (2012), "Worm stories—From the zoological teaching collection of the Humboldt-Universität" (2012), and "Why are old zoological collections so attractive?" (2015).

11.5.4 Internet

The zoological teaching collection is present in the Internet via the homepage of the Vergleichende Zoologie https://www2.hu-berlin.de/biologie/zoologie/sammlung.html and also through the database Kabinette des Wissens http://www.sammlungen.hu-berlin.de/kdw/ and other databases.

References

Grob B (2000) The world of Auzoux: models of man and beast in Papier-Mâché. Museum Boerhaave, Leiden

Grünbein D (1997) Im Museum der Mißbildungen. In: Trautner T, Sabinski J (eds) Sammeln! Sammeln! Sammeln! Geschichten von einer besitzergreifenden Leidenschaft. Aufbau Taschenbuch Verlag, Berlin, pp 125–132

Hackethal S (2008) The Blaschka models of the Humboldt University and their historical context. Hist Biol 20:19–28

Hackethal S, Hackethal H (1999) Modelle als Zeugnisse biologischer Forschung und Lehre um 1900 – Neuzugänge in der Historischen Arbeitsstelle des Museums für Naturkunde Berlin. In: Geus A, Junker T, Rheinberger HJ, Riedl-Dorn C, Weingarten M (eds) Repräsentationsformen in den biologischen Wissenschaften. Beiträge zur 5. Jahrestagung der DGGTB in Wien 1996 und zur 7. Jahrestagung in Neuburg a.d. Donau 1998. Verhandlungen zur Geschichte und Theorie der Biologie, vol 3, pp 177–190

Herbst HG (1985) Zur Geschichte der Zoologie an den landwirtschaftlichen Hochschuleinrichtungen in Berlin. Das Institut für Zoologie von der Gründung der Landwirtschaftlichen Hochschule bis zur Auflösung der Landwirtschaftlich-Gärtnerischen Fakultät in der Hochschulreform von 1968. Wiss Z Humboldt-Univ Berlin Math Nat Reihe 34:330–341

Hopwood N (2002) Embryos in wax. Models from the Ziegler Studio, Cambridge

Jahn I (1985) Zur Vertretung der Zoologie und zur Entwicklung ihrer institutionellen Grundlagen an der Berliner Universität von ihrer Gründung bis 1920. Wiss Z Humboldt-Univ Berlin Math Nat Reihe 34:260–280

Kessel M (2013) Sparrowness – or inspiration from nature. In: Bruhn M, Scholtz G (eds) Morphologien. Bildwelten des Wissens, Kunsthistorisches Jahrbuch für Bildkritik (eds Bredekamp H, Bruhn M, Werner G), 9.2, pp 97–105. Akademie Verlag, Berlin

Koch H, Koch H-J (2007) Blachka. Gläserne Geschöpfe des Meeres. Dölling und Galitz Verlag, München

Redi CA, Garagna S, Zucotti M, Capanna E, Zacharias H (2000) Visual zoology. The Pavia collection of Leuckart's zoological wall charts (1877). Ibis, Como

Reindl A, Strobach T, Becker C, Scholtz G, Schubert T (2015) Crab or lobster? Mental principles underlying the categorization of crustaceans by biology experts and non-experts. Zool Anz 256:28–35

Richter S (1999) Die Lehrsammlung des Zoologischen Instituts der Berliner Universität – ihre Geschichte und ihre Bedeutung. Sitzber Ges Naturf Freunde Berlin (N.F.) 37:59–76

Richter S (2000) Franz Eilhard Schulze und die Zoologische Lehrsammlung der Berliner Universität. In: Bredekamp H, Brüning J, Weber C (eds) Theater der Natur und Kunst/Essays. Henschel, Berlin, pp 119–127

Scholtz G (1997) Glasmodell eines marinen Polychaeten aus der Werkstatt von Leopold und Rudolf Blaschka. In: von Campenhausen C (ed) Andenken Zoologischer Vergangenheit, pp 78–79, Mainz

Scholtz G (ed) (2003) Evolutionary developmental biology of Crustacea. Lisse

Scholtz G (2010) Die Lehrsammlung der Zoologie. In: Hennig J, Andraschke U (eds) Weltwissen 300 Jahre Wissenschaften in Berlin. Hirmer Verlag, München, pp 286–290

Scholtz G (2011) Mikroästhetik – Bilder von historischen histologischen Präparaten der Zoologischen Lehrsammlung der Humboldt-Universität zu Berlin. Mikrokosmos 100:280–285

Scholtz G (2014) *Astacus fluviatilis*. Wachsmodellserie zur Entwicklung des Flusskrebses. In: Ludwig D, Weber C, Zauzig O (eds) Das materielle Modell – Objektgeschichten aus der wissenschaftlichen Praxis. Wilhelm Fink, Paderborn, pp 41–51

Scholtz G (2015) Morphologische Modelle von Tieren und Pflanzen – veraltet oder auf dem Weg zu neuer Blüte? In: Grotz K (ed) Modellschau – Perspektiven auf Botanische Modelle (Ausstellungskatalog). Botanischer Garten und Botanisches Museum, Berlin, pp 60–65

Scholtz G, Müller H (2000) Traditionsreich und höchst aktuell – Die Zoologische Lehrsammlung des Instituts für Biologie. In: Bredekamp H, Brüning J, Weber C (eds) Theater der Natur und Kunst/Katalog, Henschel, Berlin, pp 182–196

Tembrock G (1959) Zur Geschichte der Zoologie in Berlin. Wiss Z Humboldt-Univ Berlin Math Nat Reihe 8:185–196

Tembrock G (1961) Die Geschichte des Zoologischen Instituts. Wiss Z Humboldt-Univ Berlin Beiheft zum Jubiläumsjahr 9:107–125

Tembrock G (1985) Das Zoologische Institut der Humboldt-Universität von 1945 bis zur Gründung der Sektion Biologie (1968). Wiss Z Humboldt-Univ Berlin Math Nat Reihe 34:38–47

Ziegler J, Menzel F (eds) (2000) Die historische Dipteren-Sammlung Carl Friedrich Ketel. Nov Suppl Entomol 14:1–268

Chapter 12
BERLIN: The Zoological Collection of the Stadtmuseum Berlin

Manfred Graefe

Abstract The zoological collection of the Stadtmuseum Berlin was founded in 1978 as part of the Naturwissenschaftliche Sammlung Berlin by a sponsorship association, the "Fördererkreis der naturwissenschaftlichen Museen Berlins e.V.". In 1995 the collection was integrated into the newly founded Stadtmuseum Berlin, Landesmuseum für Kultur und Geschichte Berlins.

The small collection mainly contains lifelike mounted specimens of birds and mammals as well as casts of fishes, reptiles and amphibians created for exhibition purposes. There are only few scientific collections, especially of insects.

According to the focus of the Stadtmuseum, the collection is concentrated on the regional reference. Besides specimens living in Berlin in our days, there are also some examples of formerly native species of the region, like the wolf and the brown bear.

Some formerly prominent inhabitants of the Zoological Garden Berlin also belong to the collection, for example, the giant panda Tien Tien, the male hippopotamus Knautschke and the American alligator Swampy.

Keywords Naturwissenschaftliche Sammlung • Stadtmuseum Berlin • Berlin • Fauna of Berlin • Giant panda • Hippopotamus • American alligator

The zoological collection of the Stadtmuseum Berlin concentrates upon regional animals from small insects up to large mammals. Lifelike mounted animals, eggs, nests and feeding traces illustrate the plenty of fauna in Berlin and Brandenburg.

M. Graefe (✉)
Stiftung Stadtmuseum Berlin, Landesmuseum für Kultur und Geschichte Berlins, Berlin, Germany

Fachbereich Alltagskultur | Naturwissenschaftliche Sammlung, Hans-Poelzig-Str. 20
13587 Berlin, Germany
e-mail: graefe@stadtmuseum.de; http://www.stadtmuseum.de

L.A. Beck (ed.), *Zoological Collections of Germany*, Natural History Collections,
DOI 10.1007/978-3-319-44321-8_12

12.1 History

After the establishment of the Berlin Wall in 1961, the access to the Naturkundemuseum in East Berlin with its comprehensive collections and large exhibition spaces was restricted for the residents of West Berlin. In the west part of the town merely remained the Botanisches Museum and the Botanical as well as the Zoological Garden as publicly accessible natural history institutions. Besides, university collections were available for research and teaching.

To preserve and expand the collections of natural history items existing in West Berlin and in perspective to establish a new Museum of Natural History in the west part of the city, a sponsorship association, the "Fördererkreis der naturwissenschaftlichen Museen Berlins e.V." was founded in July 1978.

12.2 The Naturwissenschaftliche Sammlung

Financial contributions of the Berlin senate permitted the Fördererkreis to hire six employees: A Biologist, earth scientist, zoological taxidermist, librarian and secretary as well as a depot manager.

In 1989 several rooms in a building provided by the senate were opened to the public near Schloss Charlottenburg. In the so-called Naturwissenschaftliche Sammlung, offices and well-equipped preparation studios were available for zoology and geological purposes also like storage areas and exhibition spaces. The restricted place permitted no attractive permanent exhibition, but special exhibitions on various zoological and geoscientific subjects could be shown. Among others exhibitions were presented on the subject bear (Berliner Wappentier—Bären weltweit), Antarctic (Lebensraum Antarktis), oviparous animals (Rund um das Ei), bees (Bienen—Blüten—Bienenzucht) and ice age (Die Eiszeit in Berlin). Some of these exhibitions were loaned to other museums all over Germany.

Training events for teachers, students and children rounded off the didactic offer of the Naturwissenschaftliche Sammlung.

After the fall of the Berlin Wall, there was no more need to build another natural history museum in the city. Nevertheless, the high attendance of the special exhibitions and the educational offering at that time convinced the Berlin senate of the importance of the small Naturwissenschaftliche Sammlung as a natural history education centre.

Therefore it was not closed down yet, but was integrated into the newly founded Stadtmuseum Berlin, Landesmuseum für Kultur und Geschichte Berlins in 1995.

Work continued and several well-frequented special exhibitions were developed, but the additional educational offering for groups at times had to be reduced due to staff shortage.

In 2012 the Naturwissenschaftliche Sammlung was closed down as a publicly accessible museum. The collections were moved and now are stored in the central depot building of the Stadtmuseum.

12.3 The Zoological Collection

From the beginning the zoological collection mainly contained lifelike mounted specimens. Vertebrates were exclusively created for exhibition purposes; scientific preparations did not occur.

However, the Fördererkreis took over several extensive insect collections, among others rose chafers and butterflies. They partly came from private hand, partly from public facilities, in particular the universities who had to separate for spatial reasons from parts of their collections

According to the focus of the Stadtmuseum, a concentration of the collections on the regional reference became necessary. Therefore since 1995 most of the non-European preserved specimens had to be passed to other museums as well as most of the scientific collections.

Today the zoological collection encloses around:

450 mounted vertebrates, models and casts
70 specimens in alcohol
600 bird eggs and nests
20,000 insects
100 scull preparations

The collection of mounted birds, mammals and casts of fishes, reptiles and amphibians contains beside specimens living in Berlin in our days also some examples of formerly native species of the region, like the wolf and the brown bear.

The Berlin heraldic animal, the bear, is represented in several subspecies. The most precious specimen is the female giant panda Tien Tien (Figs. 12.1 and 12.2). This animal reached Berlin Zoo besides its male companion Bao Bao in 1980 as a

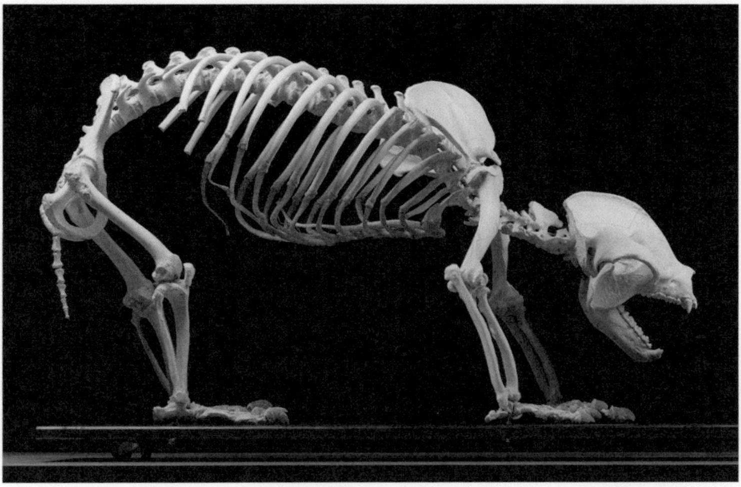

Fig. 12.1 Giant panda Tien Tien, skeleton (with permission from: Stiftung Stadtmuseum Berlin)

Fig. 12.2 Giant panda Tien Tien, dermoplastic (with permission from: Stiftung Stadtmuseum Berlin)

gift of the People's Republic of China to German chancellor Helmut Schmidt. Unfortunately, in spite of intensive veterinarian medical efforts, Tien Tien passed away of viral infection in 1984.

A formerly prominent inhabitant of the Zoological Garden Berlin besides Tien Tien also was the male hippopotamus Knautschke (Fig. 12.4), one of only 91 animals which survived the Second World War in Berlin Zoo. Its life-size gypsum sculpture belongs to the zoological collection of the Stadtmuseum Berlin as well as the American alligator Swampy. This reptile, in its younger years, reached post-war Germany as a mascot of the sixth US infantry regiment and was a gift to the Zoo Aquarium in 1952 (Fig. 12.3).

The preserved amphibians, reptiles and snails not only remind of historical museum presentations but are also a document for the occurrence of these species in Berlin 100 years ago. Today most of them are rare or even have become extinct. The majority of the specimens were collected in Berlin at the beginning of the twentieth century.

A permanent loan of the Zoologischer Garten Berlin is the bird's egg collection of Dr Oskar Heinroth (1871–1945). The ornithologist was the founder of ethology and long-standing manager of the Aquarium of the Berlin Zoo. After hand rearing the offspring of most bird species of Germany in cooperation with his first wife Magdalena, he published the outstanding volumes Die Vögel Mitteleuropas (Birds of Central Europe) from 1924 to 1934.

Several ancient insect boxes from the first half of the twentieth century show various local types of insects with their stages of development in a reproduced favourite habitat.

Fig. 12.3 American alligator Swampy, dermoplastic (with permission from: Stiftung Stadtmuseum Berlin)

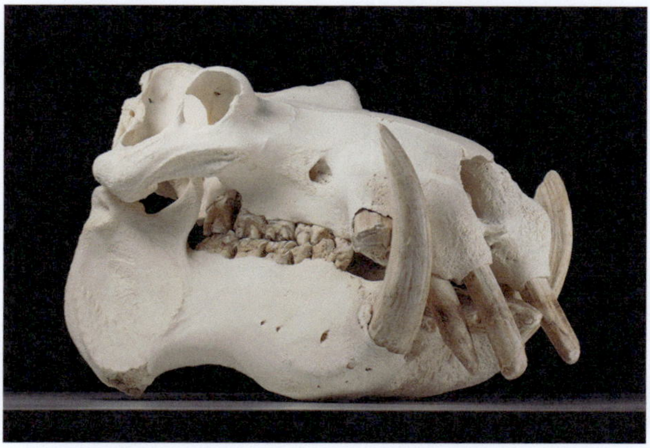

Fig. 12.4 Male hippopotamus Knautschke, skull (with permission from: Stiftung Stadtmuseum Berlin)

12.4 The Actual Situation

Since the Naturwissenschaftliche Sammlung was closed down in 2012, special exhibitions on zoological subjects have not taken place any more till today.

In the exhibition area "Frag Deine Stadt!" (Ask your town) of the Märkisches Museum, some species which were found in the region when people first settled at the place of Berlin were shown in a setting. Moreover, the exhibition picked out as a central theme the actual animal life in the river Spree in connection with environmental pollution. This small exhibition area was closed down in 2016 and for the next times, there probably will be no exhibition on zoological subjects in the Stadtmuseum Berlin.

The Märkisches Museum is the permanent exhibition building of the Stadtmuseum Berlin.

Presently the collection is supervised by a zoological taxidermist. A small workroom is available for conservation and restoration purposes.

Chapter 13
BERLIN: The Veterinary Collection of the Institute of Veterinary Anatomy, Department of Veterinary Medicine, Freie Universität Berlin

Johanna Plendl, Janet Weigner, Juliane Rieger, and Klaus-Dieter Budras

Abstract The Institute of Veterinary Anatomy at the Department of Veterinary Medicine, Freie Universität Berlin, houses several veterinary collections, ranging from various sections with a historical background to sections showcasing veterinary plastinates, which have only been produced in recent years. For example, the Gurlt collection of the institute, which was founded by the veterinary anatomist Ernst Friedrich Gurlt (1794–1882), showcases unique preserved skeletons as well as wet specimens of malformations of domestic animals. The Ziegler collection comprises more than 160 wax models, which were used for teaching during the nineteenth and twentieth centuries and are mainly based on drawings of the anatomist Wilhelm His (1831–1904). The collection of wax models consists of 48 models, mainly heads of dogs, horses, and cattle, which were produced between 1975 and 1995. Moreover, the institute houses 42 corrosion cast specimens, which for a large part have been produced from the mid-1970s until today. Also, approximately 40 anatomical models that can be taken apart and are made of gypsum and wood, originating from various epochs of twentieth-century Germany, are on display. The institute also houses a comprehensive collection of large-format wall charts, originating from the years before 1945 and from the years between 1949 and 1985, which were almost exclusively custom-made.

Keywords Veterinary medicine • Gurlt • Ziegler • Wax models • Anatomical models • Corrosion casting • Comparative anatomy • Plastination • Wall charts

---

J. Plendl (✉) • J. Weigner • J. Rieger • K.-D. Budras
Department of Veterinary Medicine, Institute of Veterinary Anatomy, Freie Universität Berlin, Koserstraße 20, 14195 Berlin, Germany
e-mail: johanna.plendl@fu-berlin.de

L.A. Beck (ed.), *Zoological Collections of Germany*, Natural History Collections,
DOI 10.1007/978-3-319-44321-8_13

13.1 Introduction

The collections of the Institute of Veterinary Anatomy at the Department of Veterinary Medicine, Freie Universität Berlin, reflect a history that is steeped in tradition, shaped by losses during the Second World War, by relocations due to Germany's separation, and finally by merging the respective departments of the Humboldt and the Freie Universität.

The different sections are largely located in the anatomical museum of the institute, as well as in glass cabinets in the dissecting room and other rooms of the institute. Most of the exhibits are used as visual aids for teaching the degree courses of veterinary medicine, agricultural sciences, and equine sciences. The methods of preparation used range from the fixation of organs and body parts in various liquids, such as formalin, to the impregnation with paraffin, polyethylene glycol, and other types of wax and to the method of plastination according to Dr. Gunther von Hagens (1979).

In recent years, there has been a reduction in practical, anatomical dissection courses in favor of virtual simulations and teaching concepts based on e-learning. Because of this development, anatomical models have become increasingly important due to their durable and reusable nature. The use of realistic, three-dimensional models allows for a tactile experience and underlines their significance in the history of science as well as the history of developing anatomical specimens.

13.2 The Gurlt Collection

The Gurlt collection of the institute showcases unique preserved skeletons as well as wet specimens of malformations of domestic animals. It has its roots in the collection of preserved specimens of the "Berliner Tierarzneischule", which was founded by the veterinary anatomist Ernst Friedrich Gurlt (1794–1882) and was focused on malformations. In 1841, the collection comprised 3358 specimens (Smollich 1988), most of which Gurlt described in his textbook "Lehrbuch der pathologischen Anatomie der Haus-Säugethiere" (Fig. 13.1) (Gurlt 1832). The part that survived World War II consists of 143 skeletons and skulls as well as 105 wet specimens displaying a range of malformations of the head (Fig. 13.2), torso, entire body (Fig. 13.3), limbs, and various organs. To this day, the collection is an important part of the curriculum of veterinary embryology and teratology, which are both taught at the institute. The specimens are stored in historical cabinets in the anatomical museum of the institute. The website of the Institute of Veterinary Anatomy displays most of them in close-up photographs with captions giving further information on the individual exhibit (http://www.vetmed.fu-berlin.de/einrichtungen/institute/we01/gurltsche_sammlung_startseite/index.html).

Part of this historical collection is "Conde". According to the enamel board that belongs to the exhibit, it is the skeleton of Prussian king Frederik the Great's

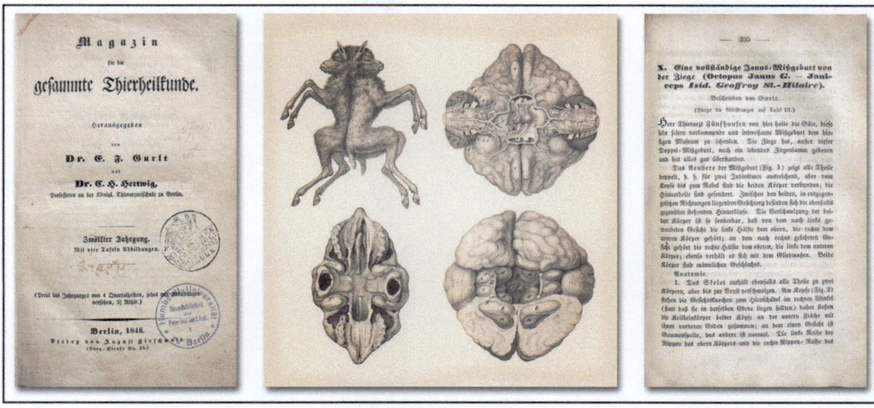

Fig. 13.1 This picture has been taken from the "Magazin für die gesamte Thierheilkunde," 1846, which Gurlt published along with Prof. Carl Heinrich Hertwig. Pictured is the entire janus-shaped, deformed body of a goat (Octopus janus G.)

Fig. 13.2 Dry specimens of the Gurlt collection: Head of a calf with separated double face (Diprosopus sejunctus). Head of a foal, cyclopia, brachygnathia superior. Distorted spine of a foal (Campygnathia, Kyphosis und Skoliosis cervicalis). Bovine hydrocephalus

(1712–1786) personal riding horse, which died in 1804 at the age of 38. After its death, the skeleton of the horse was exhibited at the Anatomical Museum of the Langhansscher Kuppelbau (Königliche Thierarzneischule). In the wake of the relocation of the Anatomical Theater in 1902, "Conde's" skeleton was moved to the newly built Veterinary Anatomy, only a few meters away from the Langhansscher Kuppelbau. After the merge of the two educational institutions, the skeleton was moved to the Veterinary Anatomy of the Freie Universität Berlin in Dahlem, where it is exhibited in the entrance area (Budras and Berg 1998).

13.3 The Ziegler Collection

The Ziegler collection of the institute comprises more than 160 wax models (Fig. 13.4), which were used for teaching during the nineteenth and twentieth centuries and are mainly based on drawings of anatomist Wilhelm His (1831–1904), who set up the collection to be able to compare the embryology of

Fig. 13.3 Wet specimen from the Gurlt collection: Immature fetus of the horse without front legs (Peromelus achirus), ca. 1846

the skull as well as facial and brain development of vertebrates. For the most part, these wax models were built in A. Ziegler's (1820–1889) Freiburger Atelier für wissenschaftliche Praxis and in a few other workshops. The highlights are two series of models, one demonstrating the development of the lancelet (*Amphioxus lanceolatus Y.*, *Branchiostoma lanceolatum*) in 25 parts (Fig. 13.5) and the other one showing "the development of the chicken in the egg," as well as models on the "anatomy of human embryos." The models are often presented in embryology courses as historic visual aids and are also useful to investigate preparation techniques of wax models.

13.4 Collection of Wax Models Focusing on Comparative Anatomy

This collection consists of 48 models, which were produced between 1975 and 1995 by Berlin preparator Dietrich Seifert (Witte 2010), who was working at the Institute of Veterinary Anatomy at the time. The models are mainly heads of dogs, horses, and cattle cut in the median into halves to present various levels of preparation with

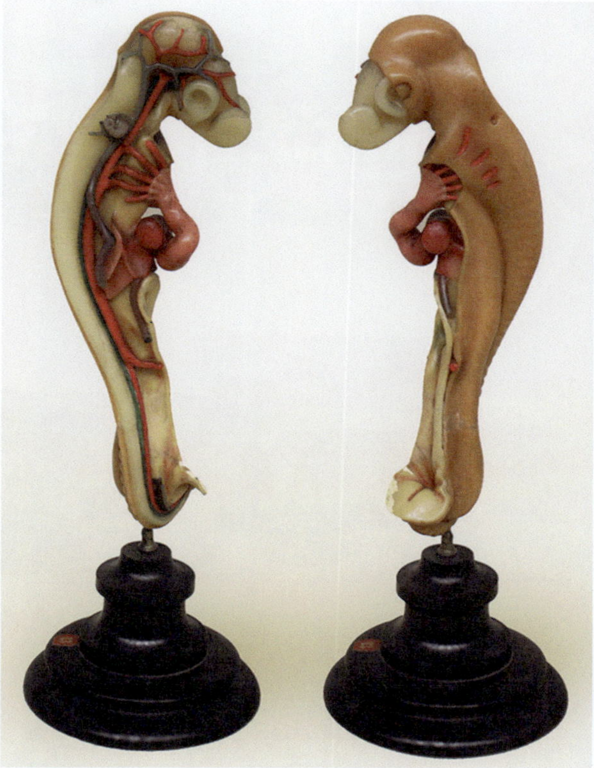

Fig. 13.4 Wax model from the Ziegler collection showing the embryo of a pig

a focus on vascular structures such as veins and arteries as well as nerves against the backdrop of bone structure and muscular system. The original material was fixated and prepared heads and their anatomical structures, which were dehydrated and impregnated in a vacuum using a mixture of paraffin and wax. In order to present these models in an elaborate and artistic but also realistic way, each anatomical structure of the respective specimen was remodeled using colorful wax mixtures (Fig. 13.6). These exhibits are displayed in the cabinets of the anatomical museum and the dissecting room of the institute and are still used for teaching as visual aids to demonstrate comparative anatomical structures.

13.5 Anatomical Models That Can Be Taken Apart

A part of the collection consists of approximately 40 anatomical models made of gypsum and wood, originating from various epochs of twentieth-century Germany. These anatomical large models of the cow, horse, sheep, and chicken, as well as anatomical models of body parts and organs, can be taken apart. For example, one specimen from 1943 that is approximately 50 cm in height shows the midsagittal

Fig. 13.5 Wax models from the Ziegler collection showing the development of the lancelet (*Branchiostoma lanceolatum*) in 25 parts

section of the pelvis of the cow without an embryo, with a removable uterus. These models are still very popular among students, as their tactile nature makes them ideally suited for comprehending anatomical structures (Fig. 13.7).

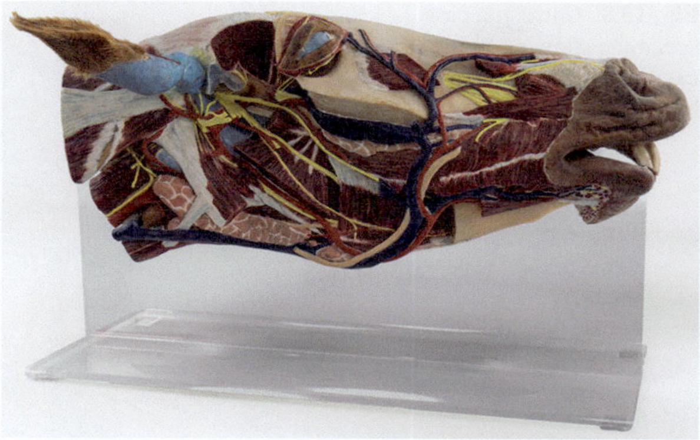

Fig. 13.6 Specimen from the collection of wax models focusing on comparative anatomy: This is the combination of an original specimen and the wax modeling technique focussing on the vessels and nerves of the head of a horse

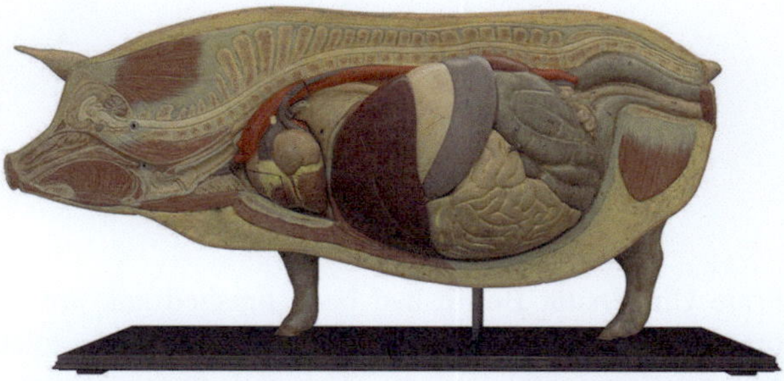

Fig. 13.7 Anatomical model of a pig that can be taken apart

13.6 Cast Specimens

The anatomical collection of the institute houses approximately 42 corrosion cast specimens, which for a large part have been produced from the mid 1970s until today. These display anatomical cavity systems which were filled with a corrosion-resistant mixture (e.g., synthetics) and subsequently cleaned from remaining tissue (maceration). The specimens are mainly three-dimensional, bronchial casts of the lung as well as vascular casts of the kidneys, cephalic arteries, and heart cavities (Fig. 13.8) as well as casts of the cerebral ventricles and the paranasal sinuses.

13.7 Comparative Collection Used for Teaching

The institute also has a comprehensive collection on the comparative anatomy of domestic and farm animals, in particular dog, cat, pig, small and large ruminants, and horse, which is continuously expanded and updated. Among these exhibits are a number of specimens of domestic and exotic wild animals, including birds and reptiles, at various ages. Most skeletons, dry and wet specimens have been produced from 1960 after the construction of the Institute of Veterinary Anatomy at the

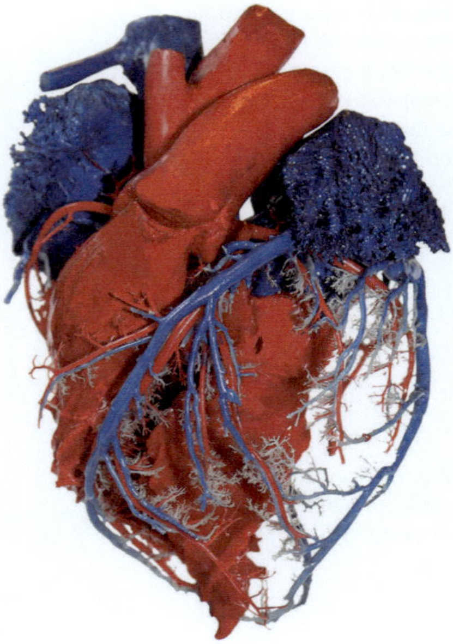

Fig. 13.8 Corrosion cast specimen showing the cavities and vessels of the heart of a sheep

Freie Universität Berlin had been finished and are still extensively used during practical anatomical courses to demonstrate issues related to comparative anatomy and developmental biology. The collection consists of 6500 individual bones; 500 mounted parts of skeletons; 115 whole, mounted skeletons as well as 700 skulls; and 100 dry specimens. A part of the collection is exhibited in glass cabinets in the anatomical museum and constantly updated. The specimens are arranged according to which organ system they belong to. Last but not least, 650 wet specimens, which are partly exhibited in specimen jars (ca. 250), also belong to the collection although the biggest part of them is stored in closed specimen tubs.

13.8 Plastinates

Plastination, the preservation method invented by Dr. Gunther von Hagens (1979), was introduced at the institute in order to give students the possibility of studying multidimensional specimens outside the dissection courses and be flexible with regard to time and space. The method allows for creating specimens that are very similar to their natural models, durable in light and at room temperature, harmless in contact with the skin, odor-free, and non-hazardous, for example, for pregnant students. The collection, which is continuously expanded, currently consists of 360 whole-body plastinates (Fig. 13.9) as well as plastinates of body parts and organs of all domestic animals. Apart from those, the collection comprises 25 series of plastinates (e.g., the head of a horse sliced into 25 disks), which are produced using various methods based on silicon or epoxy resin. Plastinates of wild animals

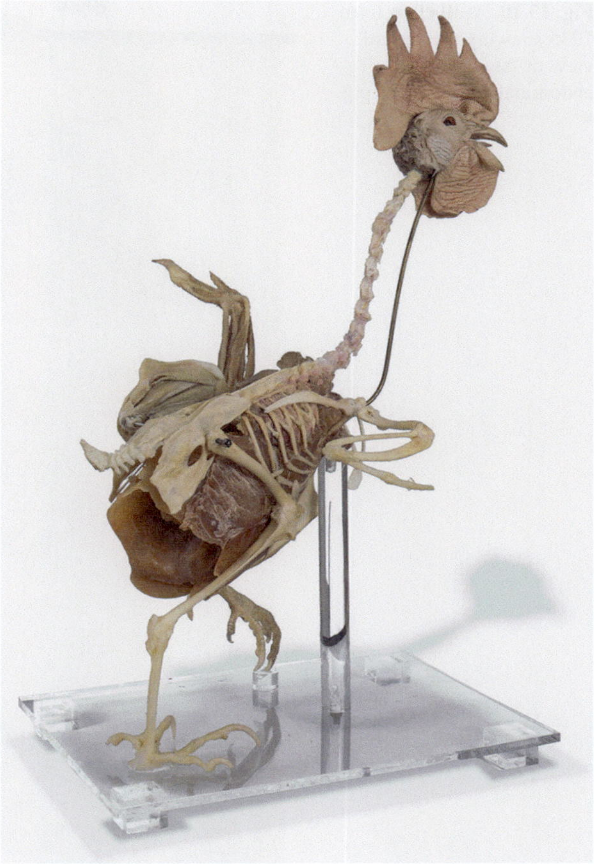

Fig. 13.9 Whole-body plastinate of a rooster

are also among the collection, such as disks of an elephant's trunk or reptiles and fish. Most plastinates are stored in glass cabinets in the dissection hall and are accessible to students at all times.

13.9 Collection of Wall Charts

The institute also houses a comprehensive collection of large-format wall charts in different sizes (most of them are around 2 m wide and between 2 and 3 m long). According to the three subjects taught at the institute, the collection comprises approximately 500 registered wall charts made specifically to be used in the lecture hall, among those around 370 charts on veterinary anatomy (Fig. 13.10), around 110 charts on veterinary histology, and around 20 charts on comparative embryology. These wall charts originate from the years before 1945 and from the years between 1949 and 1985 and are almost exclusively custom-made. Most charts are

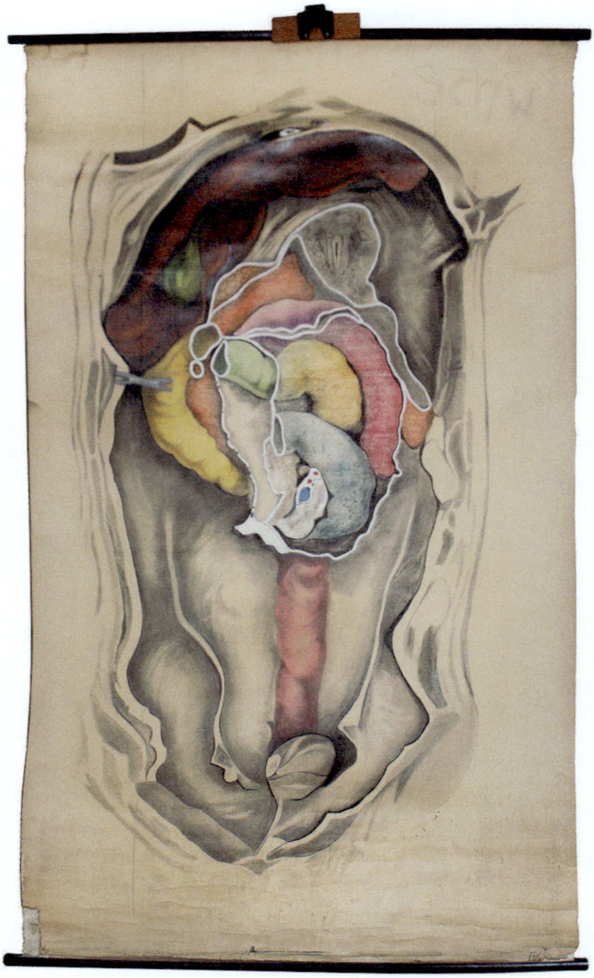

Fig. 13.10 Wall chart from 1935 showing the ventral view of the dorsal abdominal cavity of the pig

made of primed linen that was hand painted in different colors (tempera paint, watercolor, the later ones also with airbrush painting) and annotated. The charts are rolled up, stored in a safe place, and only rarely used for teaching these days.

Acknowledgments This book chapter is the result of a collaboration of members of staff of the Institute of Veterinary Anatomy at the Freie Universität Berlin. We would like to give special thanks to Ms. Wiebke Gentner for translating the original German text into English.

We would also like to thank our preparators Ms. Harriet Wendel and Mr. Florian Grabitzky for producing a great many specimens including plastinates and bone specimens and also restoring a considerable number of historical specimens to their former glory.

Furthermore, special thanks go to our graphic designers Ms. Diemut Starke and Mr. Martin Werner for producing the images accompanying the various text passages.

References

Budras K-D, Berg R (1998) Einmal quer durch den runden Salon von Sanssouci. Condé – das letzte Leibreitpferd Friedrich II von Preußen als Zeitzeuge der Geschichte Preußens und der Veterinärmedizin in Berlin. Reiten Zucht Brandenburg 1:20–22

Gurlt EF (1832) Lehrbuch der pathologischen Anatomie der Haus-Säugethiere, vol 2. Reimer, Berlin

Smollich A (1988) Ernst Friedrich Gurlt. Gegenbaurs morphol Jahrb 134(4):575–583

von Hagens G (1979) Impregnation of soft biological specimens with thermosetting resins and elastomers. Anat Rec 194(2):247–255

Witte W (2010) Vom Diener zum Meister. Zur Geschichte des Anatomischen Präparators in Berlin 1852–1959. In: Kunst B, Schnalke T, Bogusch G (Hrsg) Der zweite Blick: Besondere Objekte aus den historischen Sammlungen der Charité. De Gruyter Berlin, New York, pp 185–217

Chapter 14
BONN: Zoologisches Forschungsmuseum Alexander Koenig in Bonn: Transformation of a Classical Natural History Museum of the Nineteenth Century into a Biodiversity Research Institution

Jonas J. Astrin, Sabine Heine, Claudia Koch, Dennis Rödder, Till Töpfer, and J. Wolfgang Wägele

Abstract Natural history museums have their origin in times when scientists started to explore unknown landscapes where they collected specimens for scientific study and to demonstrat the general public the wonders of our world. Researchers usually were also hunters and gatherers. The "Zoologisches Forschungsmuseum Alexander Koenig" in Bonn was built on this tradition by a wealthy naturalist at the end of the nineteenth century. Today, this institute has a very different profile. It is a research institute with a focus on biodiversity studies at species level, with a strong taxonomic tradition and a special expertise in molecular studies that are relevant for phylogenetic inference, population genetics, speciation research, and molecular taxonomy. A major challenge for the future is to bridge the gap between knowledge about species taxonomy and biology and the application of this knowledge in ecological research and in applied conservation biology.

In this contribution, we explain the historical background and the use of collections for the case of the bird and reptile departments, the challenges for molecular research, the necessity to build new types of collections (biobanks), the use of collection data for macroecological research, and some examples for knowledge transfer to the general public.

Keywords Voucher specimens • Dry collections • Wet collections • Biobanking • Molecular samples • Biodiversity research • Taxonomy

---

J.J. Astrin • S. Heine • C. Koch • D. Rödder • T. Töpfer • J.W. Wägele (✉)
Zoologisches Forschungsmuseum Alexander Koenig—Leibniz-Institut für Biodiversität der Tiere, Adenauerallee 160, 53113 Bonn, Germany
e-mail: W.Waegele@leibniz-zfmk.de; http://www.zfmk.de

L.A. Beck (ed.), *Zoological Collections of Germany*, Natural History Collections,
DOI 10.1007/978-3-319-44321-8_14

14.1 Research Built on Sugar: Short History of the "Museum Koenig"

In the Rhineland around Cologne, Bonn, and Koblenz, our institute is known as "Museum Koenig." The founder, Alexander Koenig (1858–1940), was a passionate hunter, whose inherited fortune allowed him to dedicate his life to the study of animals, especially birds. The family's fortune had been amassed by his father Leopold Koenig, who owned large estates in Eastern Europe (mainly in Ukraine) to grow sugar beets and several sugar factories located in St. Petersburg and in Ukraine (Hutterer 2008). His enterprise produced about 10 % of the sugar consumed in Eastern Europe. The family possessed several large villas in St. Petersburg, estates in Germany, and the villa later known as "Villa Hammerschmidt" in Bonn, which after the Second World War became Germany's White House. Alexander Koenig went to school in Bonn and as a boy already started hunting and collecting birds. He studied zoology in several German universities and inaugurated in 1900 a small museum close to his parents's villa in Bonn. He never worked to earn money. After his father's death, Alexander Koenig started to construct in 1912 the large museum we know today (Fig. 14.1), enlarging the first building. However, during the Russian Revolution (1917), all possessions in Eastern Europe and the family's main income got lost, forcing A. Koenig to negotiate with the Prussian Government a takeover of the museum, which since then was a governmental institution until it became in 2013 a public law foundation. The Zoological Research Museum Alexander Koenig (ZFMK) is today an institute of the Leibniz Association, jointly funded by federal states and the federal government (via the Federal Ministry for Education and Research). Among German natural history museums, the ZFMK started very early with the implementation of molecular biodiversity research and the informatics needed to analyze sequence data.

Fig. 14.1 The main building of the Zoological Research Museum Alexander Koenig in Bonn harbors all exhibitions and part of the collections of the Vertebrate Department. The architecture was designed in the first decade of the twentieth century. The institute consists of several buildings, e.g. for the arthropod collection or for the Center for Molecular Biodiversity Research (Photo: © with C. Koch)

Today, the Center for Molecular Biodiversity Research of ZFMK is unique in Germany.

14.2 Preparing for the Future: A Transformation Process

Classical natural history museums show the "wonders of the world," classified into separate collections for animals, plants, fossils, and minerals. The scientific collections are the result of expeditions and of project-dependent research (like a study on the phylogeny of birds in Papua New Guinea). In the future, these museums will be the main providers for information on species, the living components of ecological systems. To understand the consequences of land use, desertification, drainage, climate change, etc., data at species level are needed to model scenarios and predict ecological consequences and the probability that species will survive. This is an essential basis for well-informed responses to the questions policy makers need to answer. To be able to provide this information, the Museum Koenig developed a strategic process that will change the institute's profile. New tasks are:

- To speed up inventorying in species-rich habitats, new tools for automated sampling and species identification have to be developed. An important aspect is the development of workflows for rapid and cost-efficient barcoding of mass samples.
- To speed up the access to species-specific information, we have:
  - to digitize collection-based data. ZFMK is using the Diversity Workbench (diversityworkbench.net) as repository and to provide data for international databases.
  - to link various databases and to build user-friendly portals (such as BiNHum: wiki.binhum.net) for rapid access to information from different sources.
- To make use of species-specific information for ecological studies. Modeling tools used to predict geographical distribution and dispersal abilities have to be completed with more niche parameters that are relevant for the survival of a species (e.g. biotic interactions and conditions for breeding success).
- Taxonomists have to enlarge their field of activity to document a species' biology.
- Taxonomists have to be trained to enable them to accelerate species discovery and description.
- To make use of the full potential of available knowledge, it is necessary to bridge the gap between (1) citizen scientists, who possess important data on the occurrence of species, (2) well-funded academic research, and (3) conservation biologists and governmental agencies that have to report on the state of nature.

These are new tasks that should not replace the classical museum. Instead, they require additional resources. *It is a grave mistake to replace taxonomic collections* and research departments with more "modern" working groups. Taxonomy will always be the basis for biodiversity research and for conservation biology. Without the ability to discern species and to protect them, we will transform our planet into a poor and desolate place in the universe.

14.3 Linking the Past and the Future: The Ornithological Collection of the ZFMK

A major part of the ZFMK's vertebrate collections, including the ornithological collection, consists of dry-preserved specimens. These specimens comprise study skins (Fig. 14.2), mounted specimens, and skeletons as well as bird eggs and nests. Separately stored and cross-referenced subsamples of organ or tissue material link the dry collections with the museum's alcohol and tissue collections.

Scientific bird collections are available for a wide range of research approaches and methods (faunistics, ecology, molecular systematics, conservation, etc.) of which many were unknown when the collections were established (Joseph 2011; Frahnert et al. 2013). The increasing availability of databased information on collection holdings and associated meta-data facilitates both access to and use of collections for a broad scientific audience. Therefore, the scientific potential even of historic collections is greatly enhanced, particularly with a combination of a continued contemporary extension and scientific advertisement (Viscardi 2013). Using the example of the ZFMK's ornithological collection, it shall be demonstrated how modern research approaches can be implemented in sustainably curated zoological collections.

14.3.1 The Bird Collection of the ZFMK

Dating back to the very beginning of the museum's founder and director Alexander Koenig's private activities in the 1870's, the bird collection, and notably the bird egg collection, became the cradle of the recent Museum Koenig (Rheinwald 1984). Alexander Koenig, an acknowledged ornithologist of his times, invested both his wealth and his personal dedication to the extension of the bird collection and the growth of the museum as a research institution. Hiring able assistants who performed the necessary custodial work as well as assigning major scientific tasks to them, Koenig himself already defined the use of the bird collection primarily for scientific purposes right from the start. Once the museum's public displays and iconic dioramas were finished in 1934 (Hutterer 2015), the focus of the bird collection swung even more into a research direction. The acquisition of the collection of Otto Kleinschmidt in 1935, a huge private bird collection with more than 10,000 study skins, boosted the size and the scientific importance of the bird collection. Subsequently, the overall appearance of the ornithological collection changed from Koenig's aesthetically motivated array of mounted specimens

Fig. 14.2 The two lectotype specimens of the Canary Island Kestrel (*Falco tinnunculus canariensis*), a taxon described by Alexander Koenig in 1889. They represent an example for the numerous type specimens kept in the ZFMK's ornithological collection, making them a valuable resource for taxonomic research (Photo: © with J.F. Struwe)

(Fig. 14.3) to a more prosaic, yet practical arrangement of a contemporary study skin collection.

The curators succeeding Koenig continuously added scientifically meaningful material to the ornithological collection. There were stages of extensive growth resulting from acquisitions of large private collections, exchange of material, and own collecting activities. Prominent examples are the purchase of the collection of Karl Dernedde in 1944 that contained about 3500 hummingbird skins and the acquisition of a part of the collection of Christian Ludwig Brehm in 1964 by exchanging 2800 European bird skins with the American Museum of Natural History in New York. Currently the ornithological collection holds about 76,000 study skins, 9000 mounted specimens, 3000 skeleton and alcohol specimens, some thousand feather sheets plus Germany's largest egg collection with about 60,000 eggs and clutches, and additional 1700 bird nests. Since a few years, cross-referenced tissue samples, stored in the museum's blood and tissue bank, are taken routinely with every incoming bird specimen. Last but not least, there are numerous type specimens, securing the physical basis of 322 scientific bird names

Fig. 14.3 Mounted specimen of a Sacred Ibis (*Threskiornis aethiopicus*), collected in 1910. This bird not only exemplifies the fine taxidermic quality of many early bird specimens of the ZFMK but also demonstrates their enduring scientific value because of the meticulous labeling (Photo: © with J.F. Struwe)

(Rheinwald and van den Elzen 1984; van den Elzen 2010). Considering the countless names given by C.L. Brehm, whose collection is held in part at the ZFMK, the total number of types would increase substantially, even though many of these names were never in use to a greater extent.

Although important parts of the present collection are considerably older than the foundation of the Museum Koenig, a striking feature of the ZFMK's bird collection is the high amount of data-rich specimens. This refers not only to skins but, in contrast to many other museums, also to mounted specimens both in the collection and in the exhibition. This holds true also for specimens shown in the dioramas, making them not only historically interesting but also scientifically valuable. Such a high degree of meticulous documentation still allows for outstanding research approaches.

Of particular scientific importance are local specimen series (Fig. 14.4) that represent insightful cross sections of the variation of populations at a certain place and time. For example, in the ZFMK's ornithological collection, there are remarkable local series of North Atlantic seabirds from the early 1900s (coll. A. Koenig) and songbirds from Thuringia collected during the first half of the nineteenth century (coll. C.L. Brehm). Series like these allow comparisons of genetic diversity with extant populations from the very same localities using modern analytical methods.

Another notable part of the museum's bird collection is its large number of eggs and clutches (Fig. 14.5). Compiled and labeled with great care, Alexander Koenig himself made the biggest contribution to this outstanding part of the collection. Because egg collecting has meanwhile ceased due to serious legal restrictions, the egg collection archives unique species records and sometimes even local series of populations that are now becoming of great interest again. Regarding the most recent development in scientific collecting, the ZFMK currently extends its feather collection, mainly through incorporation of bequeathed private material. Feather

Fig. 14.4 A major strength of the ZFMK's ornithological collection lies in the existence of local series of study skins of many bird species. Being the most requested preparation type in bird collections, study skins are not only useful for a wealth of morphological approaches but have also gained tremendous interest as resources for historical DNA during the last two decades. Combined methods thus form the basis for the current research foci in the Section Ornithology of the ZFMK (Photo: © with J.F. Struwe)

Fig. 14.5 Clutches of Golden Eagles (*Aquila chrysaetos*). The ZFMK's egg collection is the largest of its kind in German museums. It offers a wide range of study opportunities, including ongoing three-dimensional assessments of intrapopulational size and shape variability in avian eggs (Photo: © with J.F. Struwe)

collections have proved to be valuable for an array of studies, ranging from identification of bird strike remains to stable isotope analyses (Töpfer 2010).

14.3.2 Current Research Approaches in Museum Ornithology

The advent of affordable broad-scale applications of molecular genetics, the profound technical advances in three-dimensional visualizations (e.g. computed tomography), and the accompanied bioinformatics and statistical tools have substantially altered the appreciation of zoological collections. The new techniques have created new demands on the use and treatment of specimens (e.g. destructive sampling). Simultaneously, the self-conception of scientific museums has changed in many places toward becoming multi-methodological research institutions. This trend is in accordance with the current scientific agenda of the ZFMK that defines the museum's main duties in the fields of original zoological research, archiving and collecting physical vouchers of biodiversity and academic teaching.

The section's predominant research strategy is defined as integrative eco-phylogeny, i.e., the study of trait evolution in space and time. Focusing on the population level, integrative methodological approaches allow an up-to-date exploitation of the scientific potential of the collection, and the research outline clearly revolves around the individual bird specimens. By doing so, valuable data can be generated that complement contemporary macroecological approaches. This is particularly interesting because community patterns are influenced by the evolutionary history and the respective adaptations of the taxa involved which can only be disentangled on a fine scale using populational approaches. Taken together, such data sets become highly informative on very different biological levels without losing their intrinsic coherence.

Fig. 14.6 A glimpse into the collection of Christian Ludwig Brehm held at the ZFMK. These Bluethroats (*Luscinia svecica*) illustrate the enormous scientific value of this collection since most of their specimens were collected close to just a single locality in Thuringia. Compiled in the first half of the nineteenth century, such rare local series allow for unique comparisons of population structures across times scales that usually cannot be covered due to the poor preservation of specimens from these days (Photo: © with J.F. Struwe)

One of the best examples for the multifarious usefulness of bird collections for a wide range of scientific questions and methods can be seen in the bird collection of C.L. Brehm (Fig. 14.6). Compiled between 1808 and 1858, the Brehm collection is an icon of ornithology, representing one of the largest and oldest local collections of birds from those days. Because of their excellent preservation and data richness, the specimens are of paramount relevance for population studies in space and time since they allow a huge variety of modern study opportunities ranging from reflectance spectrophotometry over computed tomography to genomics. In an ongoing project, we make use of the manifold technical possibilities in order to comprehensively compare the structure of selected bird populations over a time span of 200 years.

Moreover, new technologies open up opportunities for the reexamination of natural history objects that were previously considered fully explored. For example, bird egg collections were rarely studied systematically during the last five decades but experienced a revival of interest with the application of reflectance spectrophotometry and molecular methods. One of such new methods employed at the ZFMK is the three-dimensional assessment of egg shape and coloration using a white light scanner. With the aid of highly resolved three-dimensional data, not only size and proportions can be assessed on a very fine scale, but also the absolute and relative distribution of egg shell patterns can be analyzed very well.

However, contemporary museum research is not limited to the museum itself but also includes fieldwork. Regarding the substantial research and collection tradition in Afrotropical ornithology, we currently focus on research in the Horn of Africa. The main interest lies on endemic and often only small-scaled distributed birds of the region. These taxa are studied in the field and in the museum, including molecular population studies. Of course, such research can only be realized in close collaboration with regional partners, ranging from universities to nongovernmental organizations, and it is hoped to establish joint long-term activities.

14.3.3 Scientific Perspective and Future Growth

Regarding the increasing awareness of the importance of voucher specimens for population studies, biochemical or molecular analyses (Clemann et al. 2014), museums might become unable to provide adequate contemporary material in the very near future. In their own interest, museums need to prevent being reduced to merely repositories of past biodiversity. Even if historic collection holdings currently experience a scientific renaissance, the widening gaps in contemporary material will unavoidably impede similar future studies if the current temporal and geographic void cannot be filled immediately. Therefore, coordinated collecting strategies (Viscardi 2013) are needed to strengthen museums as actively working scientific institutions with inevitable research resources of data-rich material.

Research with Dry Collections

- Documentation for taxonomy, species descriptions based on morphology and DNA data.
- Voucher specimens for DNA samples.
- Analyses of variations within species (biogeography, evolution of subspecies, sexual variation etc.).
- Comparative morphology of whole organisms, skeletons, structure of feathers and skins, bird egg shapes.

14.4 A Taxonomic Collection in Alcohol: Herpetology

It is a tradition to store specimens of fragile arthropods, fish, amphibians, and reptiles in alcohol (Fig. 14.7). It is almost impossible to elucidate a species on the whole only by studying live animals. The herpetological collections of the Museum Koenig in Bonn are a typical example, which is the result of and the basis for intense research. Specimens preserved in alcohol have the advantage that many details can be studied much later after preservation. Besides only documenting the presence of species, collections can be used to describe diversity at many different levels. The examination of voucher specimens stored in alcohol allows researchers to compare the morphology, stomach contents, parasitic infestation and genetic differences, among many other disciplines of science, which would often not be possible with live or stuffed animals.

One of the primary tasks of a scientific collection is a professional and safe storage of primary types. These so called name-bearing specimens are irreplaceable

Fig. 14.7 A small part of the very large collection of Lacertidae of the Museum Koenig conserved in alcohol (Photo: © with C. Koch)

documents of the registered biodiversity. They are indispensable for the identification of newly discovered species and allow a recheck if results from a study lead to questions about the identification of a specimen/species. Taking recourse of type specimens for comparative studies must be warranted in the long term and at any time (Böhme 2003).

Type specimens together with non-name-bearing voucher specimens document the presence of an animal at a specific place and time and document intraspecific and interspecific diversity.

In the past, diversity has often been overlooked as for identification researchers primarily studied the outer morphology of the animals. In the last years, the use of newer methods such as DNA sequencing or micro-CT scanning resulted in the discovery of higher species diversity in some taxa than previously expected. The application of these methods does not require fresh material and is also possible with voucher specimens stored in a collection for many years, allowing the reassessment of specific diversity.

Collections further allow researchers to assess how populations, species, or communities of animals have changed over time. Some collections have excellent historical series from certain areas and thus are very good representatives of past communities. Those areas can be revisited to see if the same species still exist in the same places, and if so, historical and current information on habitat, morphology, genetics, diet and other traits of the species can be compared.

Research with Wet Collections

- Species descriptions based on morphology and DNA data.
- Studies of population genetics with DNA extracted from preserved specimens.
- Comparative anatomy based on X-ray images and 3D-reconstructions of skeletons and soft tissues.
- Analysis of geographical variation (evolution of races, range shifts).

14.5 ZFMK's Herpetological Collection

Thanks to the noteworthy collectors such as Herrmann Grün (1892–1963) and Jost H. Jokisch (1881–?) and his first herpetological curator Karl F. Buchholz (1911–1964), ZFMK holds remarkable sample sizes of some Mediterranean squamate species with a particular focus on Portugal, Spain including Balearic Islands, and Greece and most of its islands and islets. This material led to numerous taxonomic revisions and new descriptions (e.g., Buchholz 1954a, b, 1955, 1960, 1961, 1962a, b, 1963, 1964; Gruber 1971; Gruber and Schultze-Westrum 1971; Beutler and Gruber 1977) and was to a large part the basis of studies on the diet (Maragou et al. 1996; Adamopoulou et al. 1999; Perez-Mellado et al. 2011), caudal autotomy (Pafilis and Valakos 2008), reproductive biology (Pafilis et al. 2011), and gastrointestinal parasites of lizards (Pafilis et al 2009; Jorge et al. 2011).

After Buchholz' sudden death in 1967, Ulrich Gruber became the second curator of herpetology at the ZFMK but returned to Munich already in 1971 to become the herpetological curator of the Bavarian State Collection of Zoology (ZSM). Subsequently, only 6 weeks after his oral doctoral examination, Wolfgang Böhme (born 1944) became the third curator of the collection which at that time consisted of only about 9500 specimens. He restructured and reorganized the collection and published the first type catalogue of ZFMK's herpetological collection, representing vouchers for 34 taxon names, including 22 primary types (Böhme 1974). The integration of private collections, together with Böhme's own collecting activities and those of his many diploma and doctoral students led to a massive increase in ZFMK's specimen numbers over the last decades with representatives from every continent. It is especially strong in its holdings from South America (Venezuela, Bolivia, Peru), Africa (Ethiopia, Cameroon, Senegal, Morocco, Mauretania, Guinea, Benin, Kenya, Zambia, Madagascar), and Asia (Caucasus, Cambodia, Iran, Vietnam, Indonesia). Many newly discovered taxa were among the collected material, leading to a considerable increase in type material (Fig. 14.8). The monotypic New Caledonian Skink (*Geoscincus haraldmeieri*), for example, was described by Böhme in 1976 (originally placed in the genus *Eugongylus*), but despite intensive searches at the type locality, it has never been found again and is meanwhile probably already extinct. The two type specimens of the ZFMK seem to be the only existing voucher specimens (Böhme 2014).

The transfer of university-based collections from Kiel, Göttingen, Heidelberg, and Saarbrücken/Trier to the ZFMK further augmented the herpetological material and comprised collections of several historically important zoologists such as Wilhelm F.G. Behn (1808–1878), Friedrich Boie (1789–1870), Heinrich Kuhl (1797–1821), Karl Möbius (1825–1908), Johann Friedrich Blumenbach (1752–1849), Adolph Arnold Berthold (1803–1861), Wilhelm Moritz Keferstein (1833–1870), Moritz Wagner (1813–1887), and Jacques von Bedriaga (1854–1906). The material collected by Paul Müller (1940–2010), formerly stored in the collections of the universities of Saarbrücken and Trier, added about 7.000 Brazilian and 3.000 Tyrrhenian amphibians and reptiles (Böhme 2014).

With these transfers, also some important type specimens (e.g., *Bothriechis schlegelii, Montivipera wagneri*) and other very interesting materials were passed into ZFMK's collection, such as an individual of the draconine agamid *Pseudocophotis sumatranus* which is extremely rare in museum collections. Or a series of the probably already extinct Chile Darwin's Frog (*Rhinoderma rufum*) that could, e.g., be examined in order to determine if an infection with the Chytrid Fungus (*Batrachochytrium dendrobatidis*) is responsible for the decline of the species. The likewise extinct Yunnan Lake Newt (*Cynops wolterstorffi*) was described by the famous European herpetologist George A. Boulenger in 1905, and investigations have shown that the specimen, which ZFMK received through the integration of the Göttingen collection, most probably stems from the same

Fig. 14.8 Holotypes of some Varanidae described by scientists of the ZFMK. *Varanus yemenensis* (*middle*) was discovered during a television documentary in 1985 (Photo: © with C. Koch)

imported series as the six individuals which were used in Boulenger's original description (Böhme 2014).

Recently Böhme (2014) published a detailed overview of the herpetological activities in Bonn including information on the composition of the herpetological collection of the ZFMK and a critically commented type catalogue. By that time about 97.000 specimens had been catalogued at the ZFMK, and a further about 8.000 specimens are still waiting to be integrated and will bring the entire collection to considerably more than 105.000 specimens.

Turtles and crocodiles play a minor role in the herpetological collection of the ZFMK, whereas amphibians and squamates are well represented, although, as mentioned before, they are unevenly distributed based on the geographical foci. With respect to taxonomic groups, the Lacertidae are by far the best represented group with more than 25.000 individuals, including 41 primary types representing 9 species and 32 subspecies. Among this material are also some large series of certain species representing more than 20 generations and thus being of particular interest for genetic population studies. Altogether, ZFMK's herpetological collection currently houses type material of nearly 670 taxa from more than 70 countries, including 340 primary types representing 244 species and 95 subspecies. One fifth of the primary types (representing 48 species and 16 subspecies) originated from Madagascar, emphasizing the great effort of ZFMK's scientists regarding biodiversity research in this gravely threatened island. With respect to the number of described species within those groups, the Varanidae are especially well represented with 13 primary types (representing ten species and three subspecies). Among these is, e.g., the holotype of the Yemen Monitor (*Varanus yemenensis*, Fig. 14.8) which Wolfgang Böhme accidentally discovered in 1985 while watching a television documentary about the Yemen. Not to be sneezed at is also the number of primary types in the families Chamaeleonidae (16: representing 11 species and 5 subspecies), Agamidae (29: representing 23 species and 6 subspecies), and Gekkonidae (45: representing 27 species and 18 subspecies). Some of the most famous amphibian and reptile species in the world are also represented among the type material, such as two paratypes of the Golden Poison Frog (*Phyllobates terribilis*), the probably most poisonous amphibian and one of the most poisonous animals worldwide, whose poison is used in the hunting darts of indigenous people in the Colombian rainforest. Other examples of spectacular specimens housed in ZFMK's collection are the neotype of the famous Asiatic Reticulated Python (*Malayopython reticulatus*), one of the largest snake species in the world, and the precious holotype of the gecko *Sphaerodactylus dommeli* from the Dominican Republic, which is encased in amber and probably derived from the Middle Oligocene about 25 million years ago (Fig. 14.9).

Most specimens are alcohol-preserved, but the collection also contains about 1500 dry skeletal preparations (Fig. 14.10) which are of value not only for comparative anatomical studies on extant species (e.g., de Buffrénil et al. 2005, 2008) but also in paleontological respect as shown, e.g., in the description of a new snake genus from the upper Oligocene (Szyndlar and Böhme 1996) or in an investigation of evolution of snakes based on the presence or absence of certain skull bones, as supposed homologies among snakes (Palci and Caldwell 2013, 2014). Particularly noteworthy are the about 150 skulls and skeletons of the Galapagos Marine Iguana (*Amblyrhynchus cristatus*) collected and donated by Fritz Trillmich of the

Fig. 14.9 25 million-year-old holotype of *Sphaerodactylus dommeli* encased in amber (Photo: © with C. Koch)

Fig. 14.10 Part of the collection of herpetological dry skeletal preparations (Photo: © with C. Koch)

University of Bielefeld. Some of these preparations were used in a comparative study on differences in the microstructure and the bone compactness values compared to its terrestrial iguanid relatives (Hugi and Sánchez-Villagra 2012). Additionally there are several hundreds of glycerine-stored cleared and stained preparations, which served in numerous studies on comparative osteology of certain reptile and amphibian taxa (e.g., Lang 1991; Glaw et al. 1998; Vences et al. 1998, 1999).

Several dozens of lung preparations in the collection demonstrated species-specific characters and were used to hypothesize the phylogenetic relationship of varanid lizards (Becker et al. 1989).

Furthermore, the collection houses nearly hundred everted and prepared hemipenes of different snake and lizard species which have not only been used as characters for descriptive morphology but also for studies on phylogenetics, sexual selection, and copulatory behavior (e.g., Ziegler and Böhme 1997; Böhme and Ziegler 2009).

Fig. 14.11 The two-story gallery hall with cabinets for snake specimens plays an important role in guided visitor tours behind the scenes (Photo: © with C. Koch)

Exceptional—compared to other collections and thus worth mentioning—is also the arrangement of ZFMK's herpetological collection, since it is (1) very close to the offices of the scientists who work with the collection and (2) the cabinets with the snake specimens are arranged in an attractive gallery hall on two levels, which is big enough to provide space for visitor groups of 10–15 people for guided tours behind the scenes where visitors can learn much about snakes and herpetological research (Fig. 14.11).

After Böhme's retirement in 2010, Dennis Rödder (born 1979) received the position of the curator of herpetology at the ZFMK. Fortunately and contrary to a recent trend to ax taxonomic positions due to budget cutbacks, a second curatorial position was established in 2014 as a consequence of two main facts: (1) the considerable size of the herpetological collection and the respective scientific output and (2) the fact that amphibians and reptiles belong to different animal classes and were just traditionally grouped together, rooting back to Linnaeus' classification of animals. As Rödder's focus is mainly on the evolution and macroecology of amphibians and reptiles, the second position was filled with Claudia Koch (born 1980), who has a clear focus on taxonomy of amphibians and reptiles.

14.6 Molecular Biodiversity Research: Preparing for the Future

Research based on DNA, RNA or proteins is directly linked to biological specimens. Today, transparency, reproducibility, and credibility require the safe deposition of vouchers, samples, primary data, and open access to corresponding metadata. Therefore, natural history museums have to enlarge their scope to include new types of collections and data.

The Center for Molecular Biodiversity Research of the ZFMK was implemented in 2004 and is today one of the leading European centers for DNA barcoding, molecular phylogenetics, genomics, and biobanking.

Molecular Biodiversity Research at the Museum Koenig

- Coordination of a genetic inventory of the German fauna and flora based on DNA barcodes (GBOL project).
- Development of new algorithms and software for phylogenetic analysis of sequence data.
- Generation of large transcriptome data sets to study animal phylogeny.
- Genomic research to better understand the evolution of genomes and their relation to phenotypic evolution.
- Molecular phylogenies as a basis for studies on biogeography, speciation, classification.

14.7 DNA Barcoding: Securing Taxonomic Knowledge for Large-Scale Ecological Studies

The engagement of the Museum Koenig for large-scale DNA barcoding (Hebert et al. 2003) is motivated by the fact that taxonomy has been losing its anchorage in universities over the last decades, and today most experts able to discern and discover species are surviving among citizen scientists and in natural history museums. In view of both the rapid loss of biodiversity and decreasing number of specialists, taxonomic inventories and monitoring at species level are urgently needed. DNA barcoding is therefore a tool that can improve the availability of data drastically:

- DNA barcode libraries of correctly identified species preserve the taxonomic knowledge of experts and make this knowledge available for everybody and everywhere.
- Linking barcode libraries with databases that offer species-specific information (geographical data, images, biological data) enables rapid identification and access to information.

- Metabarcoding (mass-identification of mixed samples) using automated workflows is currently the only option for efficient and sufficiently precise analysis of environmental samples.
- Barcodes allow the precise characterization of unknown species, even when no taxonomic description is available.

A prerequisite for linking barcodes with already available knowledge is the construction of validated barcode libraries. For biodiversity in Germany, this is currently being pursued in a ZFMK-led consortium of several institutes and many citizen scientists: the German Barcode of Life (GBOL) project. The following table contains data on results of the first 3 years of this project:

Taxon (examples)	Barcoded specimens (2012–2014)	Barcoded species (2012–2014)
Vertebrata	1287	173
Coleoptera	7546	~3031
Diptera	1181	577
Lepidoptera	4758	2141
Hymenoptera	5671	2137
Arachnida	3420	670

At finalization of the first 3 years, about one third of the German fauna has been barcoded. The completion of the inventory depends mainly on the availability of specimens, which are mostly *provided by qualified citizen scientists.*

14.8 A Frozen Bank Within the Museum: The ZFMK Biobank

Within their genomes, organisms contain an immense wealth of information. These genomes have been shaped by billions of years of evolution in changing habitats. They conserve the code to understand, with high granularity and in a unified "language", adaptational processes in a multitude of different life forms. They also lend themselves ideally to identify organisms, comprehend population structures, etc. Today, progress in molecular technologies dramatically speeds up research on genetic biodiversity and finally makes it possible to tap into this rich resource. Consequently, the demand is rapidly increasing for professionally preserved and managed genome-quality tissue, DNA, RNA, and compound samples from throughout the whole tree of life. Evolution, phylogenetics, taxonomy, ecology, and conservation biology are all in need of ample sources of molecular data, as are many other disciplines.

Biodiversity biobanks ('biodiversity' as opposed to human medical biobanks) form an emerging collection type that caters to these needs specifically and complements classical natural history collections. A biobank is "a curated collection/ repository of biological materials that warrants long-term integrity at molecular level, authenticity, availability, and rights management of its samples by adhering to standard operating procedures (SOPs)" (Astrin et al. 2013). While often conveniently associated with "traditional" collections, biobanks are much better suited for modern downstream molecular analysis, as their focus lies explicitly on conserving the molecular structure of their samples. They blend in perfectly into an integrated natural history collection environment, but they contribute a new specialization to the scope of curatorial roles. Biobank curators usually operate with *sub*samples and are very aware of the distinction between specimen vouchers (e.g., morphological vouchers) and molecular vouchers, as of the general importance to deposit tandem molecular *plus* morphological vouchers. They operate in a network of interconnected samples and interconnected data, where the data is expanded by many additional (molecular) data categories and relations, including several public databases. Standardization is of special importance in biobanks due to (1) the fragility of the samples, (2) the influence of preanalytical variables on sensitive downstream applications, (3) the physically challenging environment of ultracold storage, and (4) the increased storage cost. Sample types and formats, storage conditions, labeling, operating procedures, monitoring, security, metadata, and many other aspects are important elements to standardize.

Biobanking is a rapidly growing field, and an increasing number of natural history collections are installing dedicated facilities to serve as repositories for molecular-grade samples. This is an indicator for the globally increased awareness of how relevant molecular data has become in biodiversity studies. From the pragmatic perspective, it is an indicator also for the progressive realization that holding molecular samples in individual laboratories or freezers is a storage form sustainable only for a very short term and that delegating those samples to a dedicated central facility offers a standardized, documented, secure, and sustainable form of storage.

For more than 1.5 decades, ZFMK has placed a strong focus on DNA-based applications. A central biobank for the museum, run by dedicated zmb staff, was established in 2010, building on some of the structures that had been developed since 2007 within a third-party funded project (Gemeinholzer et al. 2011): the "DNA Bank Network." While initially focused on Germany only, it later gathered international partners as well. Its data architecture and philosophy eventually (2011/2012) led into the Global Genome Biodiversity Network, GGBN (Droege et al. 2014), whose secretariat is based at the Smithsonian Institution in Washington DC. GGBN, of which ZFMK is an active partner, aims at virtually unifying the scattered collections of biodiversity biobanks worldwide in order to facilitate centralized sample access and to identify taxonomic or geographic gaps among collections, thus building a global molecular biodiversity resource. GGBN further strives to harmonize biodiversity biobank activities (through various working groups), including the development of standards and best practices.

The ZFMK Biobank archives animal samples, in accordance with the institute's profile. The biobank is structured into a tissue bank (institute and collection code: ZFMK-TIS), a DNA bank (ZFMK-DNA), and soon also an RNA bank (ZFMK-RNA) as well as a molecular compound library (ZFMK-MOL). A biobank curator

Fig. 14.12 Ultra-frozen DNA samples in the ZFMK Biobank. For each sample, a voucher specimen is kept in the taxonomic collections (Photo: © with J.J. Astrin)

and a curatorial assistant on salaried positions plan and develop the biobank, carry out daily duties, perform experiments to optimize standards, and interact intensively with various biobanking networks and associations.

Sample storage at the time of writing is mostly at −80 °C for buffered DNA and for ethanol-fixed or snap-frozen tissues in 2 ml vials at −140 °C and −180 °C, i.e., below glass transition temperature of water (Fig. 14.12). Temperatures are being monitored externally. While samples are scattered in freezers over five different rooms at the time of writing, a new ZFMK building for molecular biodiversity research is planned and gives hope to unify the freezers and to implement a new infrastructural vision for the biobank.

External users are welcome to donate samples for long-term storage that have been or can be used for molecular analysis. To date, however, most sample depositions result from ZFMK projects. The most copious source of samples so far is the German Barcode of Life (GBOL) project (see above). Accordingly, through this project, a very strong biobank focus lies on species occurring in Central Europe.

The ZFMK Biobank grows quickly, currently containing more than 100,000 samples. Almost 10,000 species are represented.

At the ZFMK, data is kept in the Diversity Workbench database. From there, specimen and molecular data are transferred by wrappers to the GGBN web portal and to the Global Biodiversity Information Facility (GBIF), partly also to the Barcode of Life Data Systems (BOLD) and to EMBL/GenBank/DDBJ (the databases of the International Nucleotide Sequence Database Collaboration).

The website of the ZFMK Biobank can be found at https://www.zfmk.de/en/biobank.

Research Based on Biobank Material

- Phylogenetic or evolutionary analyses based on molecular data including complete genomes.
- Population genetics, ecology, and analyses for conservation biology.
- Addition of sequence information to complement earlier studies (e.g. extended 'DNA barcodes').
- Validation of previously published data based on the same material, especially in the context of integrative taxonomy.
- Discovery of parasites and symbionts.

14.9 Collection Data in Macroecology, Conservation Biology, and Evolutionary Biology

Increasing availability of computation power and of large-scale data sets including both gridded environmental information (e.g., temperature, precipitation, land cover, remote sensing data) and specimen-based data offer nowadays the unique opportunity to integrate very different sets of information to answer exciting novel research questions (Graham et al. 2004; Kozak et al. 2008). Digitalization efforts of natural history collections provide now not only information on the availability of specific taxa in research collections but also snapshots of the spatiotemporal distribution of species, including their biological attributes (morphology, genetics, ecology) (e.g., Graham et al. 2004; Lavoie 2013; Violle et al. 2014, Powney and Isaac 2015). For example, the Global Biodiversity Information Facility (www.gbif.org), a network collecting and providing information from numerous natural history institutions, harbors more than 500 million species records today, >470 million including exact GPS coordinates (as of July 6th 2015). The data availability is steadily increasing. However, digitalization of museum collections is a very time- and cost-consuming process which is still ongoing. Therefore, the utilization of their full value may require allocation of more resources (Johnson et al. 2011). The potential of these data sets is further increased by adding tissue and/or DNA samples and/or morphological data in specialized repositories. This opens novel fields of research and promotes the development of novel methods such as species distribution modelling (SDM) or environmental niche modelling (ENM), which have emerged only during the last decades (e.g., Guisan and Zimmermann 2000; Zimmermann et al. 2010). The popularity of these methods is steadily increasing, and today they represent an integral part of numerous further applications in biological sciences (Ahmed et al. 2015).

By combining georeferenced species records with gridded environmental data, a number of SDM algorithms have been developed to characterize a species' realized niche in environmental space (e.g., Guisan and Zimmermann 2000; Peterson et al. 2011). Once those environmental combinations, which are likely to provide suitable

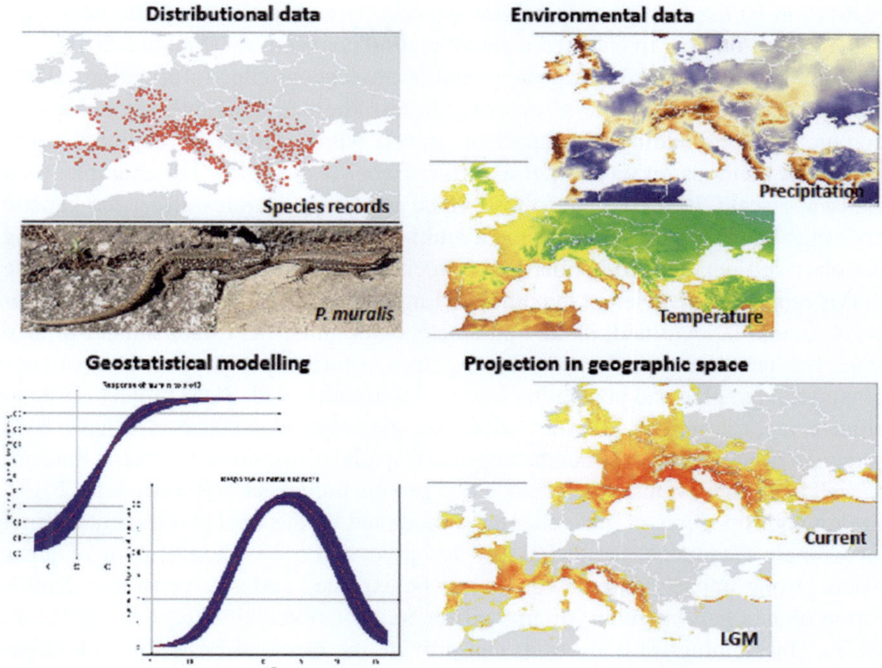

Fig. 14.13 Generalized workflow of how a species' potential distribution can be estimated based on georeferenced museum specimens. As first step, a comprehensive set of species records is compiled (*upper left*), as well as environmental predictors, which are biologically important for the target taxa (e.g., temperature, precipitation, solar radiation, etc.) (*upper right*). Extracting the environmental conditions at the species records, it is possible to derive statistical relationships between the occurrence probability of the species given the environmental conditions (or combinations thereof) in the general area of occurrence (*lower left*). This model can now be projected from environmental space back into geographic space. Using alternative environmental conditions for model projection, it is also possible to estimate the species' potential distribution for, e.g., the last glacial maximum (LGM, *lower right*) or future climate change scenarios as proposed by the Intergovernmental Panel on Climate Change

conditions for infinite population persistence, are identified, it is possible to project this realized niche space back into geographic space as well as across time in order to assess a species' potential distribution under current, past, and future conditions (for a recent review, see Peterson et al. 2011) (Fig. 14.13). Although the general concepts of correlative SDMs is well settled in the meanwhile, this relatively new field of research is rapidly developing (Ahmed et al. 2015). There are numerous direct applications of SDMs (Powney and Isaac 2015): these comprise the identification of previously unknown populations guiding further field surveys or the identification of potential areas of risk of biological invasion (Peterson 2003). Impacts of past or future environmental changes on biodiversity can be estimated (e.g. Ihlow et al. 2012; Rödder et al. 2013), which in turn can be used for spatial explicit conservation planning and risk management. Furthermore, fine-scale

SDMs can be used to quantify habitat availability and quality to guide landscape management on how to enhance metapopulation connectivity (McShea 2014).

Combined with population genetic and/or phylogeographic information, SDMs provide an independent line of evidence to test alternative hypotheses in biogeographic analyses adding an important aspect, which cannot be captured using molecular methods only (Chan et al. 2011; Habel et al. 2015). For example, on a landscape scale, it is possible to test which landscape features affect the genetic exchange between populations using a multi-model inference approach comparing the observed genetic differentiation between populations and a set of alternative SDMs representing different hypotheses (Engler et al. 2014). Once the most appropriate model is identified, its key parameters can guide the establishment of step stone habitats during landscape planning. In a historical perspective, comparisons of phylogeographic and population genetic information with SDM projections onto climate reconstructions as expected during glacial-interglacial cycles have been shown to be able to reconstruct the origin and spatial distribution of genetic lineages as well to predict events which have led to genetic bottlenecks (Rödder et al. 2013). Facing anthropogenic climate change as predicted by the Intergovernmental Panel on Climate Change, these techniques are pivotal for the identification of areas where strong impacts on biodiversity can be expected and vice versa the identification of safe harbors requiring immediate conservation action (e.g., Keppel et al. 2012). Only biological collections can provide the essential basis for such large-scale analyses.

Information on the morphology of specimens as, e.g., obtained via geometric morphometrics can be combined with georeferenced records of the respective specimens and analyzed in a geographic information system (GIS) framework (e.g., Tomovic et al. 2010). The main advantage using this approach is that spatial interpolation techniques such as kriging allow the identification of areas with significant break points in morphological characters. This information may help to delimit taxa and allows to test for ecological and/or geographic factors leading to diversification, especially when combining them with phylogeographic/population genetic analyses and SDM-based reconstructions of historic changes in potential distributions (e.g., Habel et al. 2015).

In times of ubiquitous data availability and steadily increasing computation power, biological collections including the associated metadata are very valuable. Accumulated over centuries, only data provided by natural history collections allow detailed assessments of temporal changes in global biodiversity. Next to its immense value for basic science, the information stored in natural history collections is crucial to develop effective strategies to mitigate the most pressing effects of habitat loss/modification and climate change to safeguard global biodiversity.

The New Use of Biodiversity Collections

- Collections document the presence of species in space and time and are the basis for biological global change research.
- The knowledge of species experts is the basis for biodiversity monitoring and for the sustainable use of biodiversity.
- Collection and monitoring data are the basis for the development of scenarios for the future of biodiversity in a changing world.
- Collections provide a wealth of genomic information for basic and applied research.

14.10 The ZFMK as a Partner of the "Bonner Museums curriculum"

The "Bonner Museums curriculum" is an outstanding project that shows the role of collections for teaching and education. In this project, the primary schools of the city of Bonn cooperate with seven museums by developing a program with a variety of modules. These modules are coordinated and built upon each other, combining their contents. An increasing number of schools signed an agreement which ensures that every single class visits one museum in Bonn at least once a year. All partners benefit from their contributions to school education. Depending on how many programs the students have already gone through, at their next visit of a museum, they will carry on with the appropriate lesson.

The Museumscurriculum increases the competence of school staff for teaching and learning incorporating museums. Students and teachers can use museums independently, learning outside the classroom. Furthermore, museums and primary schools improve their image during their cooperation. A continuous progress results from these learning loops, and this process links scholar curricula contents with museum- and collection-based education.

At ZFMK, zoological specimens are used by school teachers as a basis for subjects like evolution, biogeography, species diversity, and conservation biology. However, in the Museumscurriculum, the collections per se are in focus. Young students can learn why and how specimens are preserved. They are able to show initiative and to actively participate in the coursework of collecting and cataloguing. The principles ensuring a consistent acquisition policy in a natural history museum are also mentioned.

14.10.1 School Programs at ZFMK

First grade classes: get to know the museum, its exhibitions and collections. Explore and discover the objects. Learn about the tasks of the museum, such as collecting, preservation, research, and the museum as information provider. The museum is a "castle" for animals. Some parts of the exhibition are older than the founder Alexander Koenig himself. Each century has its own requirements for the documentation of the collected objects. During a guided tour through the museum, students discover different specimens (e.g. vertebrates, insects, or tissues of animals). Also, they get to know all about preparation techniques and learn about the meaning of the objects for the museum.

Second grade classes: experiencing original objects, approach, and methods (Part I). Children will perceive original objects with all senses. As a student, Alexander Koenig was an enthusiastic collector. By exploring a photo album about his life, students are going to find out why the egg collection was on focus for his research and a passion. Furthermore, children discover the egg collection and have the opportunity to experiment with chicken eggs to create their own egg collection, which then opens their view to modern collection methods.

Third grade classes: experiencing original objects, approach, and methods (Part II). The children will describe and compare and discover historical time-varying and content-based relationships. What are the changing tasks of a museum with a growing collection? By playing games, drawing, painting, sculpturing, as well as discovering and comparing original specimens, students learn the methods of identifying organisms to some taxonomic level and determining how they might be related to each other. The pupils will understand common methodologies and guidelines of collection development through educational games and experiments.

Fourth grade classes: work with the exhibits, understand their contents, and present them. Students work with objects and exhibits on their own. Ask employees of the museum: What do visitors expect from an exhibition? What kinds of information about the exhibits are necessary? Students create teams, choose specific objects for shoebox dioramas, and create mediating and easily comprehensible texts with their earned knowledge.

The "Bonner Museumscurriculum" is a trendsetting project for cultural education of children and was awarded with the first prize by "Medienberatung NRW." Classes take around 90 minutes.

References

Adamopoulou C, Pafilis P, Valakos ED (1999) Diet composition of *Podarcis milensis, Podarcis gaigeae* and *Podarcis erhardii* (Sauria: Lacertidae) during summer. Bonn Zool Beitr 48(3/4): 275–282

Ahmed SE, McInerny G, O'Hara K, Harper R, Salido L, Emmott S, Joppa LN (2015) Scientists and software—surveying the species distribution modelling community. Divers Distrib 21: 258–267

Astrin JJ, Zhou X, Misof B (2013) The importance of biobanking in molecular taxonomy, with proposed definitions for vouchers in a molecular context. ZooKeys 365:67–70

Becker H-O, Böhme W, Perry SF (1989) Die Lungenmorphologie der Warane (Reptilia: Varanidae) und ihre systematisch-stammesgeschichtliche Bedeutung. Bonn Zool Beitr 40(1): 27–56

Beutler A, Gruber U (1977) Intraspezifische Untersuchungen an *Cyrtodactylus kotschyi* (Steindachner, 1870); Reptilia, Gekkonidae. Beitrag zu einer mathematischen Definition des Begriffs Unterart. Spixiana 1(2):165–202

Böhme W (1974) Die Typusexemplare der Herpetologischen Sammlung des Museums Alexander Koenig, Bonn. Bonn Zool Beitr 25(1/3):165–176

Böhme W (1976) Über die Gattung *Eugongylus* Fitzinger, mit Beschreibung einer neuen Art. Bonn Zool Beitr 27:245–251

Böhme W (2003) Biodiversität: Zoologische Sammlungen und die globale Erfassung der Vielfalt des Lebens. Rundgespr Komm Ökol 26 ("Biologische Vielfalt: Sammeln, Sammlungen und Systematik"), München (Bayer Akad Wiss):43–54

Böhme W (2014) Herpetology in Bonn. Mertensiella 21:I–VI + 1–256

Böhme W, Ziegler T (2009) A review of iguanian and anguimorph lizard genitalia (Squamata: Chamaeleonidae; Varanoidea, Shinisauridae, Xenosauridae, Anguidae) and their phylogenetic significance: comparisons with molecular data sets. J Zool Syst Evol Res 47:189–202

Boulenger GA (1905) Description of a new newt from Yunnan. Proc Zool Soc Lond 1905:277–278

Buchholz KF (1954a) Ein neuer *Hemidactylus* von den Balearen (Rept. Geckonidae (sic)). Bonn Zool Beitr 5(1/2):68

Buchholz KF (1954b) Zur Kenntnis der Rassen von *Lacerta pityusensis* Bosca (Reptilia, Lacertidae). Bonn Zool Beitr 5(1/2):69–88

Buchholz KF (1955) Herpetologische Ergebnisse zweier Cycladenreisen. Bonn Zool Beitr 6(1/2): 95–110

Buchholz KF (1960) Zur Kenntnis von *Lacerta peloponnesiaca* (Rept.: Lacertidae). Bonn Zool Beitr 11(1):87–107

Buchholz KF (1961) Über *Elaphe quatuorlineata* von den Zykladen und das *Elaphe*-Taxon von der Insel Amorgos (Rept.: Colubridae). Bonn Zool Beitr 12(1/2):68–86

Buchholz KF (1962a) Die Mauereidechse von Ananes, Milos-Archipel. Bonn Zool Beitr 13(1/3): 216–218

Buchholz KF (1962b) Die Smaragdeidechse der Kykladeninsel Tinos. Bonn Zool Beitr 13(4): 341–352

Buchholz KF (1963) Die Perleidechse der Sierra Nevada (Reptilia: Lacertidae). Bonn Zool Beitr 14(1/2):151–156

Buchholz KF (1964) Zur Kenntnis des Genus *Algyroides* (Reptilia: Lacertidae) in Spanien. Bonn Zool Beitr 15(3/4):239–246

Chan LM, Brown JL, Yoder AD (2011) Integrating statistical genetic and geospatial methods brings new power to phylogeography. Mol Phylogenet Evol 59:523–537

Clemann N, Rowe KM, Rowe KC, Raadik T, Gomon M, Menkhorst P, Sumner J, Bray D, Norman N, Melville J (2014) Value and impacts of collecting vertebrate voucher specimens, with guidelines for ethical collection. Mem Mus Victoria 72:141–151

de Buffrénil V, Ineich I, Böhme W (2005) Comparative data on epiphyseal development in the family Varanidae. J Herpetol 37(3):328–335

de Buffrénil V, Houssaye A, Böhme W (2008) Bone vascular supply in *Monitor lizards* (Squamata: Varanidae): influence of size, growth, and phylogeny. J Morphol 269:533–543

Droege G, Barker K, Astrin JJ, Bartels P, Butler C, Cantrill D, Coddington J, Forest F, Gemeinholzer B, Hobern D, Mackenzie-Dodds J, O Tuama E, Petersen G, Sanjur O,

Schindel D, Seberg O (2014) The global genome biodiversity network (GGBN) data portal. Nucleic Acids Res 42:D607–D612
Engler JO, Balkenhol N, Filz K, Habel JC, Rödder D (2014) Comparative landscape genetics of three closely related sympatric Hesperid butterflies with diverging ecological traits. PLoS One 9:e106526
Frahnert S, Päckert M, Tietze DT, Töpfer T (2013) Aktuelle Schwerpunkte sammlungsbezogener Forschung in der Ornithologie. Vogelwarte 51:185–191
Gemeinholzer B, Droege G, Zetzsche H, Haszprunar G, Klenk H-P, Guentsch A, Berendsohn WG, Waegele J-W (2011) The DNA Bank network: the start from a German initiative. Biopreserv Biobank 9:51–55
Glaw F, Vences M, Böhme W (1998) Systematic revision of the genus *Aglyptodactylus* Boulenger, 1919 (Amphibia: Ranidae), and analysis of its phylogenetic relationships to other Madagascan ranid genera (Tomopterna, Boophis, *Mantidactylus*, and *Mantella*). J Zool Syst Evol Res 36:17–37
Graham CH, Ferrier S, Huettman F, Moritz C, Peterson AT (2004) New developments in museum-based informatics and applications in biodiversity analysis. Trends Ecol Evol 19:497–503
Gruber U (1971) Dier Inselpopulationen der Cycladen-Eidechse (*Lacerta erhardii*) in der Ägäis. In: Strid A (ed) Evolution in the Aegean. Opera Bot 340:71–79
Gruber U, Schultze-Westrum T (1971) Zur Taxonomie und Ökologie der Cycladen-Eidechse (*Lacerta erhardii*) von den Nördlichen Sporaden. Bonn Zool Beitr 22:101–130
Guisan A, Zimmermann NE (2000) Predictive habitat distribution models in ecology. Ecol Model 135:147186
Habel JC, Zachos FE, Dapporto L, Rödder D, Radespiel U, Tellier A, Schmitt T (2015) Population genetics revisited—towards a multidisciplinary research field. Biol J Linn Soc 115:1–12
Hebert PD, Cywinska A, Ball SL, deWaard JR (2003) Biological identifications through DNA barcodes. Proc R Soc Lond Ser B Biol Sci 270:313–321
Hugi J, Sánchez-Villagra MR (2012) Life history and skeletal adaptations in the Galapagos Marine Iguana (*Amblyrhynchus cristatus*) as reconstructed with bone histological data—a comparative study of Iguanines. J Herpetol 46(3):312–324
Hutterer R (2008) Auf Zucker gebaut—Alexander Koenig und sein Museum in Bonn. Koenigiana 2:3–8
Hutterer R (2015) Habitat dioramas as historical documents: a case study. In: Tunnicliffe SD, Scheersoi A (eds) Natural history dioramas. Springer, Dordrecht, pp 23–32
Ihlow F, Dambach J, Engler JO, Flecks M, Hartmann T, Nekum S, Rajaei H, Rödder D (2012) On the brink of extinction? How climate change may affect global Chelonian species richness. Glob Chang Biol 18:1520–1530
Johnson KG, Brooks SJ, Fenberg PB, Glover AG, James KE, Lister AM, Michel E, Spencer M, Todd JA, Valsami-Jones E, Young JR, Stewart JR (2011) Climate change and biosphere response: unlocking the collections vault. BioScience 61:147–153
Jorge F, Roca V, Perera A, Harris DJ, Carretero MA (2011) A phylogenetic assessment of the colonisation patterns in *Spauligodon atlanticus* Astasio-Arbiza et al., 1987 (Nematoda: Oxyurida: Pharynodonidae (sic)), a parasite of lizards of the genus *Gallotia* Boulenger: no simple answers. Syst Parasitol 80:53–66
Joseph L (2011) Museum collections in ornithology: today's record of avian biodiversity for tomorrow's world. Emu 111:i–xii
Keppel G, Van Niel KP, Wardell-Johnson GW, Yates CJ, Byrne M, Mucina L, Schut AGT, Hopper SD, Franklin SE (2012) Refugia: identifying and understanding safe havens for biodiversity under climate change. Glob Ecol Biogeogr 21:393–404
Kozak KH, Graham CH, Wiens JJ (2008) Integrating GIS-based environmental data into evolutionary biology. Trends Ecol Evol 23:141–148
Lang M (1991) Generic relationships within Cordyliformes (Reptilia: Squamata). Bull Inst R Sci Nat Belg (Biol) 61:121–188

Lavoie C (2013) Biological collections in an ever changing world: herbaria as tools for biogeographical and environmental studies. Perspect Plant Ecol Evol Syst 15:68–76

Maragou P, Valakos ED, Giannopoulos Z, Stavropoulou A, Chondropoulos B (1996) Spring aspect of feeding ecology in *Podarcis peloponnesiaca* (Bibron & Bory, 1833) (Squamata: Sauria: Lacertidae). Herpetozoa 9(3/4):105–110

McShea WJ (2014) What are the roles of species distribution models in conservation planning? Environ Conserv 41:93–96

Pafilis P, Valakos ED (2008) Restriction of caudal autotomy during life in Balkan green lizard (*Lacerta trilineata*). J Nat Hist 42(5/8):409–419

Pafilis P, Foufopoulos J, Valakos ED, Pafilis P (2009) Gastrointestinal helminth communities of the Aegean wall lizard (*Podarcis erhardii ruthveni*, Lacertidae) from the Northern Sporades Islands (Northwestern Aegean Sea, Greece). Amphibia-Reptilia 30(4):493–503

Pafilis P, Foufopoulos J, Sagonas K, Runemark K, Svensson E, Valakos ED (2011) Reproductive biology of insular reptiles: marine subsidies modulate expression of the "island syndrome". Copeia 2011(4):545–552

Palci A, Caldwell MW (2013) Primary homologies of the circumorbital bones of snakes. J Morphol 274:973–986

Palci A, Caldwell MW (2014) The upper cretaceous snake *Dinilysia patagonica* Smith-Woodward, 1901, and the crista circumfenestralis of snakes. J Morphol 275:1187–1200

Perez-Mellado V, Pérez-Cembranos A, Garrido M, Luiselli L, Corti C (2011) Using faecal samples in lizard dietary studies. Amphibia-Reptilia 32:1–7

Peterson AT (2003) Predicting the geography of species' invasions via ecological niche modeling. Q Rev Biol 78:419433

Peterson AT, Soberón J, Pearson RG, Anderson RP, Martínez-Meyer E, Nakamura M, Araújo MB (2011) Ecological niches and geographic distributions. Princeton University Press, Princeton, NJ, p 328

Powney GD, Isaac NJB (2015) Beyond maps: a review of the applications of biological records. Biol J Linn Soc 115:532–542

Rheinwald G (ed) (1984) Die Wirbeltiersammlungen des Museums Alexander Koenig. Bonn Zool Monogr 19:1–239

Rheinwald G, van den Elzen R (1984) Die Wirbeltiersammlungen des Museums Alexander Koenig und ihre Typusexemplare. II. Vögel. In: Rheinwald G (ed.) Die Wirbeltiersammlungen des Museums Alexander Koenig. Bonn Zool Monogr 19:49–150

Rödder D, Lawing AM, Flecks M, Ahmadzadeh F, Dambach J, Engler JO, Habel JC, Hartmann T, Hörnes D, Ihlow F, Schidelko K, Stiels D, Polly PD (2013) Evaluating the significance of paleophylogeographic species distribution models in reconstructing quaternary range-shifts of Nearctic chelonians. PLoS One 8:e72855

Szyndlar Z, Böhme W (1996) Redescription of *Tropidonotus atavus* von Meyer, 1855 from the Upper Oligocene of Rott (Germany) and its allocation to *Rottophis* gen. nov. (Serpentes, Boidae). Palaeontogr Stuttg (A) 240(4/6):145–161

Tomovic L, Crnobrnja-Isailovic J, Brito JC (2010) The use of geostatistics and GIS for evolutionary history studies: the case of the nose-horned viper (*Vipera ammodytes*) in the Balkan Peninsula. Biol J Linn Soc 101:651–666

Töpfer T (2010) Scientific importance and management of feather collections in natural history museums. J Afrotrop Zool Special Issue:93–99

van den Elzen R (2010) Type specimens in the bird collections of the Zoologisches Forschungsmuseum Alexander Koenig, Bonn. Bonn Zool Bull 59:29–77

Vences M, Glaw F, Mausfeld P, Böhme W (1998) Comparative osteology of Malagasy poison frogs of the genus *Mantella* (Amphibia: Ranidae: Mantellinae). Bonn Zool Beitr 48:205–215

Vences M, Glaw F, Böhme W (1999) A review of the genus *Mantella* (Anura, Ranidae, Mantellinae): taxonomy, distribution and conservation if Malagasy poison frogs. Alytes Paris 17(1/2):3–72

Violle C, Reich PB, Pacala SW, Enquist BJ, Kattge J (2014) The emergence and promise of functional biogeography. Proc Natl Acad Sci 111:13690–13696

Viscardi P (2013) A survival strategy for natural science collections: the role of advocacy. J Nat Sci Collect 1:4–7

Ziegler T, Böhme W (1997) Genitalstrukturen und Paarungsbiologie bei squamaten Reptilien, speziell den Platynota, mit Bemerkungen zur Systematik. Mertensiella 8:1–210

Zimmermann NE, Edwards TC, Graham CG, Pearman PB, Svenning J-C (2010) New trends in species distribution modelling. Ecography 33:1–5

Chapter 15
BRAUNSCHWEIG: Staatliches Naturhistorisches Museum Braunschweig

Ulrich Joger

Abstract The Braunschweig State Natural History Museum, founded in 1753/1754, in the period of enlightenment, is one of the oldest museums worldwide. Its collections are of medium size, but they contain a considerable number of eighteenth- or nineteenth-century specimens of rare and even extinct species. The bird collection is especially remarkable. It contains about 60,000 specimens, representing about half of the world's bird species. The museum has been renovated recently. Current research activities center around vertebrates, focusing on reptiles, both recent and fossil.

Keywords Braunschweig • Bird collection • Reptile collection • Extinct species • Blasius • Enlightenment

15.1 History of the Museum and Its Collections

The Staatliches Naturhistorisches Museum in Braunschweig is the second oldest natural history museum on the European continent that was accessible for the public. It was founded in 1754 by Duke Carl I. of Braunschweig and Lüneburg (1713–1780) in the historical city center of Braunschweig, which Carl had recently predestined as the capital of his duchy (Ahrens 2004). It was a universal museum combining art and natural history collections. The museum is predated only by the foundation of the "Kunstkammer" in St. Petersburg, Russia, in 1724, which also holds natural history objects. The British Museum in London was founded in 1753 but opened in 1759. As in most natural history museums of the eighteenth century that are based on the cabinets of natural curiosities and fine arts of noblemen and aristocrats, also parts of the collections in Braunschweig are considerably older. They date back to the private cabinets of Duke Ferdinand Albrecht I. (1636–1687) at Bevern and Duke Anton Ulrich of Braunschweig and Lüneburg (1633–1714) at Salzdahlum, both located in the surrounding hinterland of Braunschweig. From this

U. Joger (✉)
Staatliches Naturhistorisches Museum, Pockelsstraße 10, 38106 Braunschweig, Germany
e-mail: u.joger@3landesmuseen.de; info.snhm@3landesmuseen.de; http://www.naturhistorisches-museum.de/

L.A. Beck (ed.), *Zoological Collections of Germany*, Natural History Collections,
DOI 10.1007/978-3-319-44321-8_15

Fig. 15.1 Turtle carapace decorated with the dukes' of Braunschweig and Luneburg coat of arms (seventeenth or eighteenth century) (reproduced with permission from Staatliches Naturhistorisches Museum Braunschweig)

period of the mid-seventeenth century, single specimens are still present today. Among the oldest surviving specimens are, for instance, the skeleton and the shell of a European pond turtle (*Emys orbicularis*) that was kept by Ferdinand Albrecht I. as a pet. The carapace is decorated with the ducal coat of arms (Fig. 15.1). Since the amount of natural history and art items in the collection increased considerably during the first years of its existence, the museum was moved to new locations several times.

In 1857, however, the universal arts and natural history collections were separated following the general progress of the natural sciences and the specialization of institutions during the nineteenth century. In that year, the famous vertebrate taxonomist Johann Heinrich Blasius (1809–1870, Fig. 15.2) became director of the independent natural history museum. Since 1836, he was already appointed professor of natural sciences at the Collegium Carolinum, the predecessor institution of the current Technical University at Braunschweig, the Carolo-Wilhelmina. Under his directorship, the collections grew substantially due to his connections and exchange with scholars and collectors from around the world. When H. H. Blasius died, his younger son Wilhelm Blasius (1845–1912, Fig. 15.3) became professor for zoology and botany at the university and, consequently, also director of the natural history museum. Wilhelm Blasius continued the work of his father and further acquired many collections for the museum, mainly birds and their eggs. The most important ornithological collections that were gathered during his directorship were the voucher specimens from Carl Constantin Platen (1843–1899), who collected on various islands of the Indo-Australian Archipelago between 1878 and 1894.

After World War I, the future of the natural history museum was publicly discussed, and due to the political upheaval in Germany, space was now available in Braunschweig palace, the residence of Ernst August (1887–1953), the last reigning Duke of Braunschweig.

Fig. 15.2 Johann Heinrich Blasius (1809–1870) (reproduced with permission from Staatliches Naturhistorisches Museum Braunschweig)

Fig. 15.3 Wilhelm Blasius (1845–1912) (reproduced with permission from Staatliches Naturhistorisches Museum Braunschweig)

15.2 The Current Building

In 1933, director Gerhard von Frankenberg (1892–1969), a zoologist and politician, was dismissed. The museum was united with the institute for biology, and for both institutions, a new complex of buildings was erected in the Pockelsstraße, which until today houses the museum with its various collections and exhibitions (Fig. 15.4). During the years of World War II, the museum was spared from the bombings of Braunschweig, but parts of the collections and of the archives got lost. The plan of transforming it into a center for genetics and racial education was—luckily—never realized. In the early 1950s, the natural history museum became an independent institution again and gained more space at the building since the zoological institute moved out. Together with the general economic boom in Germany, also the natural history museum flourished. New positions were opened, and the equipment of offices, carpentry, and the taxidermic and paleontological

Fig. 15.4 The actual museum building, completed 1937 (reproduced with permission from Staatliches Naturhistorisches Museum Braunschweig)

ateliers improved considerably. Also the store rooms under the roof of the museum were reorganized and offered more space and better conditions for the collections.

The modern concept of the museum is inspired by its traditional role as a "museum of enlightenment." As in the period of enlightenment in the eighteenth century, its main goal is to teach the general public about nature and natural history, with emphasis on evolution and biodiversity, thereby using mainly original specimens for display. Another principle is that all services without public participation have to leave the museum building, in order to gain more space for public exhibition. Recently, this space has been significantly enlarged by moving magazines and offices to other buildings.

15.3 The Exhibitions of the Museum

15.3.1 The Dioramas

On the occasion of the 200th anniversary of the museum, the construction of a new series of *dioramas* was commenced in 1954, while the first of such showcases had already been finished in the late 1930s. Today, 25 of the historical dioramas are still extant. Four were only recently dismantled in the course of the rearrangement and enlargement of the entrance hall of the museum. All dioramas show exclusively native mammals and birds in their natural or human-influenced habitats. Often they have a close relation to Braunschweig and its hinterlands, such as the once last lynx of the Harz Mountains that was shot in 1818 (Fig. 15.5). Since the year 2000, lynxes are native to the Harz again due to a successful resettlement project.

15.3.2 A Time Journey Through the Collections

Since 2015, the new exhibitions at the ground floor offering 240 m^2 more space are accessible to the visitors, while a first room was already opened in December 2013. The new concept realizes a time journey through the collections and the history of the museum from its very beginnings to the twenty-first century. The first room, the *treasure hall* ("Schatzkammer"), houses some of the oldest and most valuable objects from the cabinets of curiosities of the Braunschweig dukes. Among these valuables are, for instance, richly ornamented shells of mussels, nautilus, and turtles and the world's only pair of stockings made of golden mussel silk from the noble pen shell (*Pinna nobilis*). A fluid-preserved embryo of a Ceylonese elephant (*Elephas maximus*) from the eighteenth century arouse such a broad attention at that time that even Johann Wolfgang von Goethe (1749–1832), the polymath, wanted to examine the specimen for his comparative anatomical studies (Fig. 15.6). The planned dissection, however, never took place.

The next room represents—in a modern transcription—a *traditional storeroom* of the nineteenth century displaying more than 500 specimens, representing the diversity of the museum's natural history collections of animals from colorful beetles to a taxidermy replica of the famous Brunswick Lion, a Romanesque statue of a lion which was built in 1166 during the reign of Henry the Lion (1129/1130–1195), Duke of Bavaria and Saxony, as a symbol of his ducal authority and jurisdiction. The huge showcases also demonstrate the different kinds of preservation from dried specimens of insects, corals, and shells to the alcohol-based wet collections mainly for other invertebrates, fish, amphibians, and reptiles, to stuffed specimens of mammals and birds (Fig. 15.7). The most prominent specimen is a skeleton of the extinct aurochs (*Bos primigenius*) found in 1870. Another remarkable specimen on display is the historical name-bearing holotype of the bird *Mergus anatarius*, which was scientifically described by the former museum inspector

Fig. 15.5 Last lynx of the Harz mts. (shot in 1818) (reproduced with permission from Staatliches Naturhistorisches Museum Braunschweig)

Fig. 15.6 "Goethe's elephant embryo" (first mentioned 1783) (reproduced with permission from Staatliches Naturhistorisches Museum Braunschweig)

Friedrich Eimbeck in 1829 (Fig. 15.8). Later, it turned out that the specimen is a rare hybrid between the goldeneye duck (*Bucephala clangula*) and the smew (*Mergus albellus*) (Blasius 1887). Two multimedia screens in the middle of the room provide detailed information about each individual on display.

The third room at the ground floor is a *discovery hall* with many interactive museum educational elements which are a pleasure not only for children. One issue treated is animals' teeth with their enormous variation in size, shape, and function. Furthermore, mimicry and camouflage are easily explained with the aid of several taxidermic specimens. The discovery hall is dominated by a 13-m-long showcase which demonstrates life above and below the ground (Fig. 15.9). It is corresponding with the historical dioramas which border the showroom.

15.3.3 Paleontology and Extinct Animals

Another highlight of the exhibitions is the *dinosaur hall*. Here, a 13-m-long skeleton and a life-sized reconstruction of the African sauropod *Spinophorosaurus nigerensis* are shown, a new genus discovered in 2006 during an expedition of

Fig. 15.7 Modern version of a traditional show and store room (reproduced with permission from Staatliches Naturhistorisches Museum Braunschweig)

Fig. 15.8 "*Mergus anatarius*", a hybrid bird (shot in 1829) (reproduced with permission from Staatliches Naturhistorisches Museum Braunschweig)

museum members and later scientifically described (Remes et al. 2009; Fig. 15.10). In modern computer-based animations, the extinct giants are revived, and aspects of their extraordinary biology are visualized for the visitors. The successful paleontological project of the natural history museum in Braunschweig was the first German dinosaur expedition to Africa since nearly a hundred years, and further

Fig. 15.9 Partial view of the large showcase in the discovery hall (reproduced with permission from Staatliches Naturhistorisches Museum Braunschweig)

expeditions to the Republic of Niger were conducted the following years. In a *second paleontological hall*, the geological histories of the Braunschweig area and the Harz Mountains are illustrated and explained. Among the many fossils exhibited are also some excellently preserved ichthyosaurs (note that 175 million years ago northern Germany was covered by a Jurassic sea) that were found and excavated near Braunschweig by scientists of the natural history museum.

Special treasures of the scientific collections are on display in the *great light hall* on the second floor of the museum's building. There, specimens of iconic and now extinct species, such as the great auk (*Alca impennis*) and Steller's sea cow (*Hydrodamalis gigas*), are shown (Fig. 15.11). Of the latter species, which was already extirpated by seafarers in the eighteenth century, only few complete skeletons still exist today.

Further key issues of the great exhibition hall are the Pleistocene fauna of the Harz Mountains, as well as the evolution of our own species, *Homo sapiens*. Visitors are especially fascinated by the lifelike replica of a Neanderthal man.

15.3.4 The Aquarium

Next to taxidermically preserved specimens, the natural history museum also has living animals on display since more than 50 years. Hence, in 1968 an aquarium was opened in the basement of the museum's building. Also several terraria for amphibians and reptiles were constructed, which, until today, show a variety of tropical aquatic and terrestrial habitats, such as coral reefs and the Amazonian

Fig. 15.10 Reconstructed skeleton of the sauropod *Spinophorosaurus nigerensis*, discovered in 2006 (reproduced with permission from Staatliches Naturhistorisches Museum Braunschweig)

Fig. 15.11 Front view of the skeleton of Steller's Sea cow (*Hydrodamalis gigas*) (reproduced with permission from Staatliches Naturhistorisches Museum Braunschweig)

rainforest with their characteristic fauna and flora. Due to its naturalistic and attractive enclosures, the aquarium, which was modernized in 1991, is one of the highlights of the exhibitions. For many years, several species, such as the white-blotched river stingray (*Potamotrygon leopoldi*) from South America or the green tree python (*Morelia viridis*) from New Guinea, have regularly been bred at the museum. In addition, a colony of honeybees can be observed by the visitors in the insect room on the first floor.

15.4 The Scientific Collections

15.4.1 Vertebrates

The collections of the natural history museum in Braunschweig amount to nearly 500,000 specimens in total. Those collections of international importance are the *ornithological* and the paleontological section.

The former is also the largest vertebrate collection of the museum with more than 62,000 specimens, 10,300 of which are eggs. This is the largest bird collection in Lower Saxony and one of the most comprehensive collections in Germany. About half of the world's known bird species are represented in the Braunschweig collection. The oldest specimens date back to the foundation of the museum in the mid-eighteenth century. Noteworthy is the high number of specimens of extinct bird species, such as great auk (*Alca impennis*), passenger pigeon (*Ectopistes migratorius*), laughing owl (*Sceloglaux albifacies*), and Carolina parakeet (*Conuropsis carolinensis*). Another rare specimen is the Faroese white-speckled raven, a color aberration of the extinct Faroese raven (*Corvus corax varius*), of which merely 26 specimens still persist in natural history collections worldwide (van Grouw and Bloch 2015). The Braunschweig specimen is probably the oldest of them and was collected before 1755.

The *mammal collection* contains more than 8000 specimens, about half of which are skulls and skeletons (not counting the bone collection from Pleistocene caves, also amounting to the thousands). The most valuable specimen is the skeleton of Steller's sea cow (*Hydrodamalis gigas*), described by famous Braunschweig scientist F. A. W. Zimmermann (1783). It is the most complete skeleton of this iconic extinct species outside of Russia.

The natural history museum keeps the only noteworthy *herpetological* collection in Lower Saxony. It comprises reptiles from all around the world. As in the other sections of the museum, the oldest specimens are from the seventeenth and eighteenth century. Among these historical specimens are, for instance, a water monitor lizard (*Varanus salvator*) from Southeast Asia and some turtles (see above). In the nineteenth century, the collection was increased by snakes gathered by J. H. Blasius in southern Europe. Around the turn to the twentieth century, further specimens from the German colony of Cameroon and from Madagascar came to Braunschweig. In addition, Fritz Grabowsky (1857–1929) collected amphibians and reptiles on Borneo between 1880 and 1884. Among the specimens was also a new snake species, *Calamaria grabowskyi* (Fischer 1885). In recent years the collection has been enlarged with specimens from Northern Africa and the Middle East by the museum's director Ulrich Joger.

Although the smallest vertebrate collection, the *ichthyology section* has a worldwide scope. It comprises about 1700 specimens including 12 historical type specimens of five fish species collected during the German Deep Sea Expedition in 1898–1899 (Fricke 1991). The ichthyological collection originated in the eighteenth century as is evident from the first catalogue of the cabinet that was written in 1754, which mentions several unlabeled alcohol-preserved and a few dried specimens like a diodontid porcupinefish.

15.4.2 Invertebrates

The invertebrate collections are the most numerous of the museum. In total, they comprise about 200,000 insects and 100,000 mollusks and other invertebrates such as spiders.

The insect collections have a regional emphasis and were mainly acquired during the twentieth century (Hevers 2006). The oldest entomological collections in town, however, namely, those by Johann Hellwig (1743–1831) and Hermann von Heinemann (1812–1871), did not reach the natural history museum in Braunschweig but were sold to the respective institutions in Berlin and Hannover. Especially the latter collection of Microlepidoptera contained many type specimens.

The oldest parts of the malacological collection date back to the cabinets of natural curiosities of the dukes of Brunswick due to the fascinating variety in form and colors of their shells and their good suitability for long-term preservation. Hence even today, painted shells of mussels and nautilus from the seventeenth century still persist in the museum. Indigenous mollusks were mainly collected by Victor von Koch (1840–1915). His voucher specimens are important historical documents since they reflect the ecological situation 100 years ago.

15.4.3 Paleontology

Next to the bird collection, the paleontological section houses the most important collection of the museum with about 45,000 specimens. They are mainly fossil reptiles of the Mesozoic era. Another focus is Pleistocene mammal remains from the Rübeland caves in the Harz Mountains and from (Salzgitter-)Thiede only about 12 km from Braunschweig. These are the oldest paleontological specimens in the collection; some were already described in the early nineteenth century and belong to the earliest paleontological collections worldwide. In addition, one of the most species-rich collections of fossil sea urchins in Germany belongs to the museum. About 3000 specimens were only recently donated in 2003.

Notably, the museum in Braunschweig is the first German natural history museum to conduct paleontological excavations for dinosaur fossils in Africa after World War I. The huge dinosaurs found are on display in the exhibition (see above). There are, however, also regional fossils of remarkable scientific notice in the collection. Since a few years, excavations are conducted in Cremlingen, Hondelange, and Schandelah, where well-preserved ichthyosaur remains were found.

15.4.4 Type Specimens

According to Hevers (2005), the museum houses 184 type specimens of 76 taxa. They are mostly ornithological specimens described by J. H. Blasius (Europe), W. Blasius (South East Asia), and Eugen Ferdinand von Homeyer (1809–1889; Palaearctic region), whose collection was acquired for the museum after his death. This number of original type specimens, however, represents an underestimation of the museum's holdings since more and more historical name-bearing type specimens have been discovered in recent years.

15.4.5 Infrastructure

The museum has a permanent staff of four scientists, two educators, and nine technicians. Nonpermanent positions amount to about equal numbers. Volunteers are mainly active in the paleontology department. Administration and public relations are shared with two other museums, which together amount to a staff of about 120 persons in the administrative unit "Niedersächsische Landesmuseen Braunschweig" (www.3landesmuseen.de). This is the largest museum unit in Lower Saxony, totaling more than 120 employees, an annual budget of about 8 million euros and using more than 10 buildings. The State Natural History Museum alone spreads over five buildings in town. Three of them are collection magazines. There are workshops for taxidermy and for paleontological preparation, a graphics atelier, and a 3-D scanning unit. Molecular genetic work is not done in own laboratories, but at Braunschweig Technical University. The museum has its own library and produces a scientific periodical, "Braunschweiger Naturkundliche Schriften", enabling a regular exchange with about 300 other institutions.

15.5 Scientific Research

Research is done at the State Natural History Museum in three different fields:

(A) Vertebrate paleontology. The dinosaur bones collected in Niger (see above) are still being examined, actually for pathological effects such as deformations. This is done in collaboration with the Berlin Museum für Naturkunde (Witzmann et al. 2016).
Newly discovered Jurassic fossils from the Braunschweig region such as dinosaurs, ichthyosaurs, and pterosaurs are also being studied.
The enormous collection of Pleistocene mammal bones is used for paleogenetic and paleoecological studies, in cooperation with universities and other museums.

(B) Vertebrate zoology. The current focus of research is on reptiles and amphibians from North Africa, the Middle East, and Europe. There is a collection of frozen samples which is continuously enlarged during expeditions and which is the basis for molecular genetic studies in order to reconstruct phylogenetic trees, phylogeographic patterns, and speciation processes. This is done in collaboration with the Institute of Zoology at Braunschweig Technical University and with foreign partners (e.g., Stümpel and Joger 2009; Vences et al. 2014). The large bird collection is also often in the focus of research activities, nowadays with molecular genetic methodology, too.

(C) Entomology and research on collection data base management. There is an ongoing research project led by the Braunschweig Natural History Museum but with four other nature museums as partners, all of which pursue the goal of linking their entomological collection data into a single network of metadata. This is executed with the collections of Coleoptera first. The partners are fully aware that this is the most numerous group—both in species and in individuals—the magnitude of data is the special challenge of this pioneering project. The metadata will also be accessible via GBIF.

It is a principle of the museum that scientific research and public exhibitions do not simply coexist, but one is the basis of the other. Therefore, many temporary exhibitions (on mammoths, dinosaurs, snakes, and other subjects) have been designed by incorporating own research. Visitors are thus enabled to follow up and understand the progress in scientific work.

15.6 Educational Work

Cooperation with schools is based on the pedagogic program of the museum offering diverse courses which have been designed to meet the biology curricula of different grades of schools. Materials needed for those courses (which the schools do not have) are being produced in the workshops of the museum and are designed to fit the specific requirements, respectively. This includes modern IT-based approaches.

For many schools in the region, a regular visit to the Natural History Museum is obligatory. Pupils are also accepted in the museum for executing project work.

Special teachers focus on the integration of disabled pupils and immigrants. Other programs are designed for elderly people. The programs developed for these groups are welcomed very well.

There is also a program of seminars for teachers.

The *didactic concept of the museum* is primarily based on enthusiasm which should be developed in all visitors and participants of didactic programs. This is achieved by working with original objects of nature. Esthetics play a dominant role in awaking sympathy for nature. The traditional dioramas, the aquarium, and the giant showcase in the new "discovery hall" (Fig. 15.7) mediate a positive emotional approach to biology and encourage to spend time to discover nature's secrets in its amazing details. In addition to the pedagogic team, the scientists also offer insights into their work.

As part of the "3landesmuseen", the State Natural History Museum is present on Facebook, Twitter, and Wikipedia.

References

Ahrens S (2004) 250 Jahre Naturhistorisches Museum in Braunschweig. Braunschweig: Staatliches Naturhistorisches Museum. 407 S

Blasius R (1887) *Mergus anatarius*, Eimbeck, ein Bastard zwischen *Mergus albellus*, Linné und *Glaucion clangula*, Linné. Monatsschr Deutsch Ver zum Schutze Vogelwelt 12(14):1–30

Fischer JG (1885) Ueber eine Kollektion von Amphibien und Reptilien aus Südost-Borneo. Arch Nat 51(1):41–72

Fricke (1991) Catalogue of the Fish Collection of the Staatliches Naturhistorisches Museum in Braunschweig. Braunschweiger Nat Schriften 3(4):1021–1049

Hevers J (2005) Die Typusexemplare des Staatlichen Naturhistorischen Museums in Braunschweig. Braunschweiger Nat Schriften 7(2):443–490

Hevers J (2006) Die entomologischen Sammlungen des Staatlichen Naturhistorischen Museums in Braunschweig. Braunschweiger Nat Schriften 7(3):697–757

Remes K, Ortega F, Fierro I, Joger U, Kosma R, Ferrer JMM (2009) A new basal Sauropod Dinosaur from the middle Jurassic of Niger and the early evolution of Sauropoda. PLoS One 4 (9):e6924

Stümpel N, Joger U (2009) Recent advances in phylogeny and taxonomy of Near and Middle Eastern Vipers—an update. ZooKeys 31:179–191. doi:10.3897/zookeys.31-138

van Grouw H, Bloch D (2015) History of the extant museum specimens of the Faroese white-speckled raven. Arch Nat Hist 42(1):23–38

Vences M, De Pous P, Nicolas V, Diaz-Rodriguez J, Donaire D, Hugemann K, Hauswaldt JS, Amat F, Barnestein JAM, Bogaerts S, Bouazza A, Carranza S, Galan P, de la Vega JP G, Joger U, Lansari A, El Mouden H, Ohler A, Sanuy D, Slimani T, Tejedo M (2014) New insights on phylogeography and distribution of painted frogs (Discoglossus) in northern Africa and the Iberian Peninsula. Amphib Reptil 35:305–320

Witzmann F, Hampe O, Rothschild B, Joger U, Kosma R, Schwarz D, Asbach P (2016) Subchondral cysts at synovia vertebral joints as analogies of Schmorl's nodes in a sauropod from Niger. J Vert Paleontol. doi:10.1080/02724634.2016.1080719

Chapter 16
BREMEN: Geosciences Collection of the University of Bremen: The Recent Mollusk Compilation

Jens Lehmann

Abstract The Geosciences Collection of the University of Bremen is a midsize collection of estimated about 800,000 specimens that are housed at the Faculty of Geosciences of Bremen University, focused on the field of paleontology. Nevertheless, the collections include more than 300,000 individual neontological specimens; most of these are gastropods, a focal point that fits well into the mollusk focus of the paleontology collection. Among the recent gastropods, the most important part are species of the unranked clade Stylommatophora, terrestrial gastropods comprising about one third of the total number of specimens in the collection. The gastropod collection contains originals to the nineteenth-century collections of Borcherding, Gerhard von dem Busch, Hartmann, Jantzen, and Schmacker among others, including important historical collections of achatinellids from Hawaii and clausilids from China. A few hundred types or potential types are registered in the database; to illustrate the research potencies, some type material is presented exemplarily. Technically, the material and the collection are in an excellent condition, and thus preconditions for scientific revisions are excellent. A permanent preservation and curation are ensured by a contract between the Übersee-Museum, the institution formerly holding these city-owned collections, and the University of Bremen. As a consequence the collection has a good staffing today, including one technical assistant and one scientist handling collection affairs. The lively exhibition projects include major exceptional exhibits on inland water vessel and shopping centers, with the conceptional focus to present zoological specimens in comparison with fossil representatives with respect to constructional morphology, evolution, and diversity.

Keywords Mollusks • Marine • Terrestrial • Historical collection • Exceptional exhibits

J. Lehmann (✉)
Faculty of Geosciences, University of Bremen, Klagenfurter Strasse, 28357 Bremen, Germany
e-mail: jens.lehmann@uni-bremen.de

L.A. Beck (ed.), *Zoological Collections of Germany*, Natural History Collections,
DOI 10.1007/978-3-319-44321-8_16

16.1 Introduction

"One of Germany's most comprising mollusk collections is that [...] in Bremen. This was almost unknown to the professional world" (translated from Knipper 1954). This quotation dates back more than 60 years, and since then the collection owned by Germany's smallest federal state underwent a couple of complete relocations and one full rearrangement and a refurbishment of data. Originally it was placed in the Übersee-Museum, an internationally highly recognized Natural History and Ethnographic museum in Bremen, Germany, that is integrating topics of nature, culture, and distribution across the world in its exhibitions. In the 1990s, the opinions differ significantly across the country when a relocation of this compilation altogether with geoscientific specimens of the same museum to the University of Bremen as one uniform, the "Geosciences Collection of the University of Bremen," took place. It became part of the university note before 1994, when its former curator finally moved her place of work from the Übersee-Museum to the University of Bremen. This major and out of the ordinary move from a public museum to an university was suspiciously watched for several years; museum's unions and others, before the conditions, facilities, and staffing convinced critics that this change was all around positive for the development of the collection, its accessibility and visibility on all fields.

The Geosciences Collection of the University of Bremen became a lively place of research and international academic exchange during the last decade on the field of history of life. It is belonging to the Faculty of Geosciences, one of Germany's newly founded universities, established in 1971. Nevertheless, the collection has its pre-Übersee-Museum roots in the seventeenth century. The collection is mainly housing paleontological objects, and according to an overview on all paleontology collections in Germany, it belongs to Germany's midsize academic collections (Jansen and Steininger 2002). However, the more than 300,000 individual zoological object, mainly gastropod shells, are at least as important as the paleontological specimens. Knipper's (1954) statement is still right that the main cadre of the modern mollusk collection is based on the historical private collection of Schmacker, a tradesman and private scholar devoted to researching the terrestrial gastropods of Asia. Schmacker's collections comprise a significant amount of type material. However, several other important collections need to be mentioned as well, since quite many types came from these compilations, among these some nineteenth-century collections (von dem Busch, Borcherding, Hartmann, Jantzen, Cox, Eylmann, Hasshagen, Isenberg, Meyer, Philippi, Schmacker, Suter as well as expedition results from Cohn, Krause (1881–1882) and Schauinsland. Some of these are presented in more detail in Sect. 16.4.1. General accounts on the mollusk collectors and the zoological specimens are given by Kuster-Wendenburg (1999, 2002, 2003, 2008).

This chapter aims to give an overview about the zoological content to an international community, since the collection comprises scientifically significant gastropod shells including a few hundred types and potential types. The most valuable scientific assemblages are listed, and the key papers are cited. Not all of the type material is identified with certainty yet, and thus this attempts to encourage specialists to work with the collection. After giving an historical overview, this contribution is presenting the essential information, including specific aspects of collection management and conditions. After a presentation of focal points, the state of the art of research is given, including its potencies. The Geosciences Collection presents some specimens to the public, among others by touring exhibitions. In the final paragraphs of this chapter, the didactic conceptions are briefly discussed, that is dominated by the concept of integrating zoological with paleontological topics.

16.2 Historical Background

Systematic collection of natural history specimens started in the fifteenth to seventeenth century, with the founding of rarity and curiosities collections across Europe by peers or social societies. In the nineteenth century, they vanished again; many of these form the base for newly founded natural history museums. In the Hanseatic City of Bremen, particularly its long tradition of trading triggered to collect natural history specimens from all over the world. Here systematic collection started around the seventeenth century, and this is also the beginning of the collection known as Geosciences Collection today. The historical development of the geoscientific collection was described by Kuster-Wendenburg (1999), including information on the collections of recent mollusks. This contribution is written in German language as are other contributions on specific neontological aspects of the collection (Kuster-Wendenburg 2002, 2003, 2008). An overview of the historical milestones is given by Lehmann (2003a). These sources are the base for the brief historical development presented as follows; particularly Kuster-Wendenburg (2008) is considered. The Bremen collection did not undergo much damage directly by destruction in World War II, but a significant loss of information needs to be constituted for parts of the material by various moving during times of war but also before and after, a feature typical also for other important German mollusk collections (e.g., Zilch 1971).

16.2.1 Origin and Early Development of the Collection

The "Gesellschaft Museum" (Museum Society) of Bremen (until 1783 known as "Physikalische Gesellschaft," = Physical Society) decided to build a "Naturalien Kabinett" (nature cabinet) for their members in 1784. Recent mollusks, originating from all over the world, have been housed since then in the nature cabinet in the Hanseatic City of Bremen—respectively in the succeeding public collections. The administration of the natural history collections has been done by the fluctuating heads of the society and later by various teachers. The first fully professional conservator Otto Finsch has been employed in 1864, who was later the first director of the collections (until 1878) and who was mainly working on the field of ornithology (Abel 1970; Kuster-Wendenburg 1999, 2008). Since the founding of the city collections for Natural History and Ethnography ("Städtische Sammlungen für Naturgeschichte und Ethnographie") in 1876–1890, the specimens were owned by the city of Bremen and are still today. In the framework of the Hanseatic history, the Bremen collection was much benefitting by sailors, tradesmen, and scientists in the first half of the nineteenth century who were in close contact to the Gesellschaft Museum. Main additions came, for example, from captain Johann Willem Wendt who sailed the Earth three times and donated his collection to the collection in 1836 (Kuster-Wendenburg 2002). Gerhard von dem Busch, a physician who collected mollusks on a scientific base and who kept contact with many scientists on a supraregional base, transferred his material at around the middle of the nineteenth century (Kuster-Wendenburg 1999, 2008). The tradesman and consul Johannes A. Jantzen donated his huge collection in 1888 (Kuster-Wendenburg 2008). In 1887 Dr. Hugo Schauinsland became director of the city collections and achieved to build a new building for display purposes in 1895/1896 (Abel 1970).

16.2.2 History During World Wars I and II

In the year 1911, an extension of the museum's building made more space available for the zoological collection (Abel 1970). Particularly the important Schmacker collection, acquired in 1896, has been still wrapped into boxes until these years. It has been gradually integrated into the systematic collection during World War I, since the curator for mollusks at that time (F. Borchering; see Lehmann 2003b) has been too old to be forced to join the army (Abel 1970). However, the build-up of zoological exhibits and that of further cabinets was stopped in 1914 by recruiting the preparators to the military service.

After the World War I, the name of the museum changed twice during the 1930s. The name "Deutsches Kolonial- und Übersee-Museum," introduced in 1935, is indicating the intention to establish Bremen as "Stadt der Kolonien" (city of colonies; Abel 1970). This was embedded in a new political framework, since the demands to pass back the German protectorates were increasing shortly after the National Socialists took over the force. Thus, the reinforcement of the collections from the former colonies was in the focus during the period around World War II; material has been purchased mainly focusing on Africa (Abel 1970). With respect to the input into the neontological mollusks collection during this time, Franz Stapelfeldt (born 1877, deceased 1954) needs to be mentioned. During the 1930s until the end of World War II, he was a prominent person involved in the special circumstances of the 3rd Reich history of modern Geosciences Collection of the Bremen University. He has been head of the "Deutsche Schiff- und Maschinenbau Aktiengesellschaft Bremen" (German ship- and machine construction incorporated company Bremen), and he has been "Wehrwirtschaftsführer" (Chairman of the Reich Association of Industry/Industrialists). His role was double edged, with connections to the resistance, supported haunted financially but also cultivated relationships to major people of the National Socialism. Specimens to more than 70 catalogue entries in the database of the Geosciences Collection were acquired into the collection in 1940, associated with a number of ethnological specimens of African origin. According to a letter by Stapelfeldt, these specimens were purchased from Mrs. von Müldner in Berlin. Stapelfeldt states that he supported Mrs. von Müldner financially by purchasing the material. These information are the result of provenance research by Bettina von Briskorn (Übersee-Museum Bremen) that is unpublished yet. She is currently leading a research project to clarify the origin of the incomings initiated by Stapelfeldt and if the salesperson has been hunted by the National Socialism or if the sale was differently motivated.

With respect to the destruction during World War II, 11 events of destruction of the museum building took place between 1940 and 1945 (Abel 1970). Parts of the collection were removed from the museum for safety reasons. Fortunately, the mollusk collection was not heavily damaged, if damaged at all. Although there is no detailed knowledge about what happened exactly with this part of the collection, some unique zoological specimens were sheltered in the cellar during bombings (Abel 1970), and possibly this was true for the mollusks as well. In fact almost no material did get lost during the periods of war according to Kuster-Wendenburg (2008).

16.3 General Information

16.3.1 Official Label, Conditions, and Contact

The Geosciences Collection of the University of Bremen, in German "Geowissenschaftliche Sammlung der Universität Bremen," is an administrative part of the "Fachbereich Geowissenschaften" (Faculty of Geosciences) with its own annual budget available for collection affairs. The acronym is GSUB. All biological specimens are given a smaller letter before the running number; fossils bear a large letter (e.g., G = fossil gastropods and g = recent gastropods). The collection is housed on the ground floor of the Marum building on the campus of the University of Bremen, Leobener Strasse, in 28357 Bremen, Germany. Specimens are housed in a total number of 144 individual units; these are either individual rows of drawers within a coherent cupboard system made of metal or rows in a manual mobile shelving system.

One chemical laboratory and one workshop for technical preparation and cleaning that are separated from the collection room are exclusively associated to the collection. Additionally, there is an outpost in the "Verfahrenstechnik" building in the central area of the campus. The later is currently housing mainly the recent bivalves among the zoological objects. It should be noted that in some cases, specimens including both shell and soft body exist; however, the latter are not housed at the university. All of the collection of alcohol preparations is still housed at the Übersee-Museum in the city center of Bremen.

All enquiries to the collection should be made to PD Dr. Jens Lehmann, Faculty of Geosciences, University of Bremen, Klagenfurter Strasse, 28357 Bremen (e-mail: jens.lehmann@uni-bremen.de).

16.3.2 Management Techniques Used

16.3.2.1 Registration

The holdings until 2005 were fully registered in a relational database based on Microsoft Access that was adapted to the specific needs (Kuster-Wendenburg 2008). Paper labels were printed out directly from this computer program, limited by the fact that not all fields potentially included in the relational database fitted on the printout, and consequently a selection of database fields has been printed for space reasons. Avoiding these limits, the registration of new entries to the collection today is realized in a Microsoft word file for the biology object, and the full set of information is printed on one or, in case of too many information, on subsequently numbered labels. All of these entries are continuously amended in Microsoft Access also. The language of registration changed from German to English at the beginning of the year 2005 for paleontological and neontological specimens of the collection.

16.3.2.2 Techniques of Labeling

Between the years 1995 and 2005, the main part of the zoological section of the collection has been inventoried. During this time, the acronym-number combination has been directly written onto the specimens using customary waterproofed ink, in order to achieve a long-lasting labeling. The contemporary paper labels associated with the specimens during this time are laser printer hardcopies. Since 2010 on-specimen labeling is improved to enhance permanence by first applying a basic coat, and secondly, after drying of the first, an India ink is used for the acronym-number labeling, and at last a finish is coated, principally following the method described by Davidson et al. (2006). A couple of years before 2010, paper labeling was improved by laminating the hardcopies produced by a laser printer. After a short experimental stage with glossy laminating foils that are disadvantageous with respect to photography works, matt surface laminating foil is used today.

Table 16.1 Modern gastropods of the geosciences collection partitioned by main habitats

Gastropod habitat	Database entries	Specimens	Types
Terrestrial	13,487	113,832	413
Limnic	3989	84,256	113
Marine	8528	89,145	29
Brackish	539	13,558	28
Total	26,543	300,791	583

Note that row "types" include types as well as potential type material which status is not clarified yet

16.3.3 Focal Points

The most important main group of the zoological compilation are mollusks; however, the zoological part of the collection comprises less than a hundred specimens of scattered groups also, mainly echinoderms and crustaceans. All of these are unregistered. Additionally, there are 136 database entries (176 specimens) of scleractinian corals, mostly collected for research purposes by the scientists Exoperanzio and Nemenzo (ded. Gerold Wefer in the 1990s). A systematic collection is that of recent fish otoliths; the 262 database entries represent 509 specimens, assembled by Menzel (Lehmann 2010) collected for a comparison with fossil statoliths assembled by the same collector (Lehmann and Menzel 2005).

Among the main group of the zoological compilation, the mollusks, only the bivalves are not fully registered (currently 33 database entries and 51 specimens only of an estimated number of at least 5000 specimens). The gastropods are fielded as follows (Table 16.1). Following the taxonomic concept used in the data base, currently a total number of 163 families and 1071 genera are represented (Kuster-Wendenburg 2008).

The main fields of the individual gastropod compilations and its association with people and corresponding publications are listed in the next chapter.

16.4 Research

Material that is unified in the Geosciences Collection today has been considered in the scientific literature since the nineteenth century (e.g., Philippi 1842–1845, 1845–1847; von dem Busch 1858; Boettger 1879a, b, c, 1880, 1881a, b; Moellendorff 1884, 1887; Boettger and Schmacker 1894). Pulmonate gastropod shells were in the focus of research interest (Fig. 16.1). In the twentieth century a few scientific publications have been added, including a substantial-type catalogue for a part of the pulmonate collection (Knipper 1954, 1956, 1958; Boettger 1956; Schilder 1958). These research activities are almost exclusively focused on gastropods; only a couple of papers are dealing with the bivalves of the collection (Meyer 1958, 1970). In the current century, the revision of parts of the pulmonate

Fig. 16.1 Selection of type Gastropoda in the Geosciences Collection of the University of Bremen. (**a–c**) *Acrotoma* (*Acrotomina*) *semicincta*, paralectotype of *Clausilia* (*Acrotoma*) *semicincta* (Boettger 1881b); GSUB g31801. Northwest Caucasus, Georgia. (**d–f**) *Phaedusa* (*Phaedusa*) *hainanensis*, paralectotype of *Clausilia* (*Pseudonenia*) *hainanensis* (Moellendorff 1884); GSUB g31798. Isle of Hainan, southeast China. (**g**) dito, original illustration of the lectotype by Yen (1939). (**h–j**) *Likharevia gustavi*, paralectotype of *Clausilia* (*Oligoptychia*) *gustavi* (Boettger 1880); GSUB g31803. Astara spring, northwestern Iran. (**k–l**) *Georissa regularis*, SEM image of paralectotype; GSUB g31814 and g31815. Isle of Busuanga Philippines. (**m–o**) *Macrophaedusa frankei*, holotype of *Clausilia* (*Hemiphaedusa*) *frankei* (Boettger and Schmacker 1894; GSUB g10295. Kiangsi province, China. (**p–r**) dito, original illustration refigured. (**s–u**) *Quadriplicata lederi gradata*, syntype of *Clausilia* (*Euxina*) *gradata*

gastropods was the most important contributions, due to its outstanding importance among the recent mollusks of the collection (e.g., Köhler and Glaubrecht 2002, 2003; Nordsieck 2001, 2003). Due to the full electronic accessibility of the collection (see Sect. 16.3.2) and permanent staff, all technical requirements for further research are given.

16.4.1 Research Collections

Here the main data of the most important research compilations of modern mollusks in the Bremen collection is given. As far as known, the year of birth and death of the collectors are given in brackets:

1. Gerhard von dem Busch (1791–1868): Stylommatophora, Germany
2. Friedrich Borcherding (nineteenth century): Stylommatophora, Germany
3. Erhard Eylmann (nineteenth/early twentieth century): Various, nonsystematic approach, Australia
4. William Del Hartman (nineteenth century): Stylommatophora from Hawaii, USA
5. Johannes A. Jantzen (1810–1897): Nonsystematic approach, worldwide collecting
6. Helmut Knipper (1914–1974): Stylommatophora from Germany and particularly a large number of achatinellids from Tanzania
7. Philipp Bernhard Schmacker (1852–1896): Stylommatophora from China
8. Frank Schröder (twentieth century): Stylommatophora (Cochlicopidae, Orculidae, Clausiliidae), the vast majority from northwest Germany.
9. Adolf Wagner (twentieth century): Cypraeidae worldwide
10. Johann Wilhelm Wendt (1802–1847): Nonsystematic approach, worldwide collecting

Fig. 16.1 (continued) (Boettger 1879a); GSUB g9989. upper River Kura area, southern Russia. (**v–x**) original illustration refigured. (**y–aa**) *Leptachatina* (*Leptachatina*) *brevicula*, paratype of *Heliceras* (*Leptachatina*) *brevicula* (Pease 1869); GSUB g31813. Kauai County, Hawaii, USA. (**ab–ad**) *Palaina* (*Palaina*) *quadrasi*, SEM image of paralectotype of *Diplommatina* (*Palaina*) *quadrasi* (Moellendorff 1887); GSUB g31817, g31818, g31819. Rizal province NE of Manila, Isle of Luzon, Philippines. (**ae–af**) dito, original illustration refigured. (**ag–ai**) *Paludina pyramidata* von dem Busch in Philippi, 1844, holotype; GSUB g13012. Probably Isle of Palawan, Sri Lanka. (**aj**) dito, original illustration refigured. If not stated otherwise all photographic images are compiled by focus stacking technique (Johnson 2008); GSUB is the acronym of Geosciences Collection of the University of Bremen

16.4.2 Research Potencies and Publications

The list of collectors and individual compilations given above already suggests that the research focus of the collection is and always has been the pulmonate gastropods. This includes a total number of 583 types or potential types registered in the current database. All of these are gastropods. The regeneration of the type material, with refiguring type material, has been started by Knipper (1954, 1958), including the clausilids, neritids, and thiarids only. However, particularly in the compilations by von dem Busch and Schmacker, the labeling suggests that there is a much larger number of types or potential types (Kuster-Wendenburg 2008). During the years of the modern inventory from 1994 to 2005, by Elisabeth Kuster-Wendenburg and many helpers, limited time has been spent on checking background and validity of gastropods indicated as type material according to Kuster-Wendenburg (2008). The publicity of the collection, its international visibility, among others, and its presence in the Internet before the year 2000 have been in need of improvement, and thus a couple of revising scientists were not aware of the material. This is true with respect to the revision of achatinellids by Johnson (1996); compare Kuster-Wendenburg (1999, 2008) (here erroneously 1994 has been given as the date of publication). In a few cases material has been considered during the past two decades since 1994 when the collection moved to the university (e.g., Nordsieck 2001, 2003; Köhler and Glaubrecht 2001, 2002, 2003; Kuster-Wendenburg 2008). Thus, today the research potencies of the material are high and the largest part of the historical collection in need for revision. In sharp contrast to the gastropod collections, the recent bivalve compilation does not contain any types, and there are a couple of scattered publications about this group only (Meyer 1958, 1970).

Exemplarily a selection of type material is presented in Fig. 16.1 to illustrate the research foundations, potencies, and difficulties. Many of the species among the material housed in the Geosciences Collection are in need for revision. Furthermore, it illustrates that many nineteenth-century collectors and scientists intensively exchanged material. In some cases this included specimens of type series, and this is how some specimens of the Bremen collection are types due to exchange activities as exemplified below. In some cases labels in the collection help to locate type localities from a modern point of view, for example, the type material of *Palaina quadrasi* is given as "Montalban" in the original publication (Moellendorff 1887), and the additional information on the label "near Manila" (translated from German) was able to confirm that this is most probably referring to Rodriguez in the Rizal province northeast of Manila in the Philippines.

16.5 Educational Work

The Geosciences Collection of the University of Bremen is designed as a research collection. There is neither a permanent public exhibition associated with the storage rooms nor planned. However, a limited exhibit is presented for guided tours presenting the cast of a dinosaur skeleton and that of a subfossil Irish elk among the main attractions but also including some recent gastropods, like the giant gastropod *Syrinx aruanus*, the Australia trumpet or false trumpet, from the Indian Ocean (Lehmann 2014). A recently invented booster club plans to expand offers to the public (Lehmann and Liebenberg 2015), and a part of these activities is to obtain new recent mollusk material for the collection and to present this in the Internet in a fossil-recent context (Lehmann 2015). Outside the campus some material is permanently on display in the Übersee-Museum; specimens are not concentrated but embedded in the individual topics presented by the museum (e.g., parts of the world like Oceania, Africa). Furthermore the recent mollusk compilation of the collection was also published as a whole or in parts on the field of public relation, mostly for documentation of their importance (Kuster-Wendenburg 1999, 2002, 2003) but also with respect to fossil-modern comparisons (e.g., Lehmann and Friedrich 2007).

16.5.1 Public Presentation

In the field of public understanding of science, neontological material of the Bremen collection is on display in special exhibitions (e.g., Schmieder and Lehmann 2011). Occasionally, there have been special presentations of the modern gastropod collection during scientific meetings as well (e.g., Lehmann 2001).

The Geosciences Collection is often cooperation with the Center of Marine Environmental Sciences at the University of Bremen (Marum), and this incorporates exhibitions at exceptional places. An example is the "Geoschiff" (geoship) project realized in the "Year of Geosciences" in Germany in 2002. This was a 105 m long inland water vessel altered to contain a 600 m^2 special exhibition "Abenteuer Meeresforschung" (adventure marine science). During almost six months of exhibition, 117,000 visitors have been visiting the exhibition in 62 cities across Germany (Bundesministerium für Bildung und Forschung 2006). Due to the success, the federal government financed exhibitions in the succeeding years until 2009, displaying further topics off geobiological sciences (Münder 2009). In 2002 the Geosciences Collection issued one sixth of the total exhibition area with combined displays of fossil and recent mollusks (Krogmann and Lehmann 2002).

Another distinguished cooperation is "MeerErleben"("Experience Ocean")—a traveling exhibition initialized by the Marum at the University of Bremen and financed and organized by the ECE (=Einkaufs-Center-Entwicklung) shopping center project management company currently managing 183 centers across Europe and the Near East (Fig. 16.2; Gerdes and Pätzold 2012; Schmieder 2014; Anonymous 2015).

Fig. 16.2 A display board of the diversity and evolution module featured in the current public exhibit "Experience the sea" (German title "MeerErleben") on display in a shopping mall in Hamburg, Germany, in 2012

16.5.2 Didactic Conceptions of Exhibitions and Public Talks

The striking feature of exhibits drafted by the Geosciences Collection is the combination of paleontological and neontological specimens, covering the key topics evolution, biodiversity, and constructional morphology. Examples are the

Fig. 16.3 Gastropod shell association testifying the major tsunami in Asia in December 2004. Terrestrial, limnic, and marine species are originating from different habitats, having been transported into the sea by the tsunami wave first, and subsequently its empty shells were washed upon the beach. An example of a current topic that was touching people's life has been involved in the public understanding of science activities of the collection. All specimens belong to a series from Polhena Beach near Matara, Sri Lanka. Registered as (*upper row*, from *left* to *right*): GSUB g31824 (*Achatina* cf. *marginata*), g31828 and g31832 (both *Acavus haemastoma*); and (*lower row*, from *left* to *right*): g31821 (*Pomacea* cf. *paludosa*: *L*), g31826 (*Cyclophorus involutus*), g31834 (*Bullinidae ampulla*: *M*), g31794 (*A.* cf. *marginata*), g31820 (*Pythia reeveana*: *M*), g31793 (*A.* cf. *marginata*). Geosciences Collection of the University of Bremen. Leg. Martha Meyer. Environmental assignment: *M* marine, *L* limnic, all others are terrestrial taxa

display "Evolution" during the geoship tour (Krogmann and Lehmann 2002) and that on diversity and evolution in the currently running presentation "Experience Ocean" (Schmieder and Lehmann 2011). In this general conception, specimens are embedded in "display landscapes," e.g., by arranging an evolutionary sequence on the background of a quarry wall with layered sediment or surrounding small display cases with bivalves into a wall mural of their habitat (Fig. 16.3). Captions to individual specimens are preferably very short, inconspicuously embedded and refer to daily subjects everybody is familiar with, a concept that is owed by the fact that some exhibits attract visitors at places untypical for displays. In other words people are getting involved that are usually not systematically visiting museums or public displays. These ideas are strongly influenced by the Tübingen paleontology school of the twentieth century (e.g., Seilacher and Gishlick 2014). Particularly aspects of constructional morphology of recent and fossil specimens, including its analogies with daily life, are often picked up in the didactical conceptions (Fig. 16.2).

Regularly public talks are offered for a broad audience, including a series of lectures organized ten times a year by the Geosciences Collection and that is announced in the local press and on various websites. These talks also refer to analogies with daily life or up-to-date topics, like the disastrous tsunami in Asia in 2004 (Fig. 16.3).

Acknowledgments I am thankful to Martin Krogmann for the photographic and Petra Witte for the SEM work (both Bremen). Bettina von Briskorn (Bremen) gave valuable information about the history of collection material during World War II and Hartmut Nordsieck (Frankfurt) as well as Frank Köhler (Sydney) provided important information on the material they worked on. Hans Konrad Nettmann (Bremen) is thanked for discussion about taxonomic difficulties.

References

Abel H (1970) Vom Raritätenkabinett zum Bremer Überseemuseum. Die Geschichte einer hanseatischen Sammlung aus Übersee anläßlich ihres 75jährigen Bestehens. In: Röver, Bremen, p 352

Anonymous (2015) ECE Projektmanagement. http://de.wikipedia.org/wiki/ECE_Projektmanagement. Accessed 31 May 2015

Boettger O (1879a) Kaukasische Mollusken. Gesammelt von Herrn Dr. G. Sievers in Tiflis. Jahrb deutsch malakozoologischen Gesellschaft 6:388–412

Boettger O (1879b) Monographie der Clausiliensection Albinaria v. Vest. Novit Conchol 5:39–173

Boettger O (1879c) Neue recente Clausilien IV. Jahrb deutsch malakozool Gesellschaft 6:385–387

Boettger O (1880) Diagnoses molluscorum novarum ab ill. Hans Leder in regione caspia Talysch dicta lectorum. Jahrb deutsch malakozool Gesellschaft 7:379–383

Boettger O (1881a) Sechstes Verzeichnis transkaukasischer, armenischer und nordpersischer Mollusken aus Sendungen der Herren Hans Leder, z. Z. in Kutais, und Dr. G. Sievers in St. Petersburg. Jahrb deutsch malakozool Gesellschaft 8:167–261

Boettger O (1881b) Sectiones speciesque novae Clausiliarum Caucasiarum. Jahrb deutsch malakozool Gesellschaft 8:341–346

Boettger CR (1956) Im tropischen Ostafrika eingeschleppte mediterrane Landschnecken. Veröff Übersee-Museum Bremen 2:331–340

Boettger O, Schmacker PB (1894) Descriptions of some Chinese land-shells. Proc Malacol Soc London 1:169–174

Bundesministerium Für Bildung Und Forschung (2006) Jahr der Geowissenschaften 2002. http://planeterde.de/Members/redaktion/Fotos/02_2006/JDG_in_Zahlen.pdf. Accessed 31 May 2015

Davidson A, Alderson S, Fox M (2006) Assembling an archival marking kit for paleontological specimens. In: Society of Vertebrate Paleontology, Abstracts Poster Presentations 66th Annual Society of Vertebrate Paleontology Meeting, Oct 2006, Ottawa. Society of Vertebrate Paleontology, Ottawa, pp 1–6

Gerdes A, Pätzold M (2012) Wie das Meer an Land kommuniziert wird. Zielgruppenorientierte Öffentlichkeitsarbeit am MARUM. In: Dernbach B, Kleinert C, Münder H (eds) Handbuch Wissenschaftskommunikation. VS Verlag für Sozialwissenschaften, Berlin, pp 127–132

Jansen U, Steininger FF (2002) Die paläontologischen Sammlungen in Deutschland. Inhalte, Erfassung und Gefährdung. E Schweizerbart'sche Verlagsbuchhandlung, Stuttgart, p 101

Johnson RI (1996) Types of land and freshwater mollusks from the Hawaiian islands in the Museum of Comparative Zoology. Bull Museum Comp Zool 155:159–214

Johnson D (2008) How to do everything: digital camera. McGraw-Hill Osborne Media, New York City, p 428

Knipper H (1954) Die Typen und Typoide des Überseemuseums Bremen, 1: Mollusca: Clausiliidae. Veröff Übersee-Museum Bremen 2:251–274

Knipper H (1956) Neue Achatiniden aus Ostafrika (Moll. Gastrop. Stylomm.). Veröff Übersee-Museum Bremen 2:341–356

Knipper H (1958) Die Typen und Typoide des Überseemuseums Bremen, 5: Mollusca (Gastrop. Prosobranch.): Neritidae und Thiaridae. Veröff Übersee-Museum Bremen 3:39–74

Köhler F, Glaubrecht M (2001) Toward a systematic revision of the Southeast Asian freshwater gastropod *Brotia* H. Adams, 1866 (Cerithioidea: Pachychilidae): an account of species from around the South China Sea. J Molluscan Stud 67:281–318

Köhler F, Glaubrecht M (2002) Annotated catalogue of the nominal taxa of Southeast Asian freshwater gastropods, family Pachychilidae Troschel, 1857 (Mollusca, Caenogastropoda, Cerithioidea), with an evaluation of the types. Mitt Nat Berl Zool Reihe 78:121–156

Köhler F, Glaubrecht M (2003) Morphology, reproductive biology and molecular genetics of ovoviviparous freshwater gastropods (Cerithioidea, Pachychilidae) from the Philippines, with description of a new genus Jagora. Zool Scripta 32:35–59

Krogmann M, Lehmann J (2002) Evolution. In: Bremen U (ed) Das Geoschiff: Abenteuer Meeresforschung, 11. April (Potsdam) bis 6. Oktober (Würzburg), Bremen, p 2–7

Kuster-Wendenburg E (1999) Der Bremer Stein und die Dinosaurier. Die Geschichte einer Sammlung. Aschenbeck & Holstein, Bremen, p 95

Kuster-Wendenburg E (2002) Entdeckungsfahrten im Auftrag Preussens. Der Bremer Kapitän Wendt 1802 bis 1847. Aschenbeck & Holstein, Delmenhorst, p 110

Kuster-Wendenburg E (2003) Marine Gastropoda und Lamellibranchiata der Sammlung von Johann Wilhelm Wendt (1802–1847). Aschenbeck & Isensee, Oldenburg, p 47

Kuster-Wendenburg E (2008) Die rezenten Gastropoden in der Geowissenschaftlichen Sammlung der Universität Bremen. Abh herausgegeben Nat Verein Bremen 46:143–272

Lehmann J (2001) Biodiversität – historisches Material der Geowissenschaftlichen Sammlung der Universität Bremen. In: Wehrmann A (ed) 3,5 Milliarden Jahre Biodiversität, Gemeinsame Jahrestagung der Paläontologische Gesellschaft und der Gesellschaft für Biologische Systematik, 17.-21. September 2001, Tagungsführer. Terra Nostra, Schriften der Alfred-Wegener-Stiftung. Alfred-Wegener-Stiftung, Berlin, Oldenburg, p 184

Lehmann J (2003a) Chronology of the collection history. http://www.geosammlung.uni-bremen.de/sammlung/geschich/chrono_en.php3. Accessed 22 May 2015

Lehmann J (2003b) People of the collection. http://www.geosammlung.uni-bremen.de/sammlung/geschich/personen_en.php3. Accessed 22 May 2015

Lehmann J (2010) Geologie und Paläontologie des nordwestdeutschen Raumes – die Beiträge von Herbert Menzel. Abh herausgegeben Nat Verein Bremen 46:393–396

Lehmann J (2014) Wikipedia-*Iguanodon* und -Riesenhirsch nun in der Geowissenschaftlichen Sammlung der Universität Bremen. GMit. Geowiss Mitt 58:84–85

Lehmann J (2015) Neues Objekt durch die Freunde der Geowissenschaftlichen Sammlung finanziert. http://www.fgsub.de/neues-objekt-von-den-freunden-der-geowissenschaftlichen-sammlung-finanziert/. Accessed 31 May 2015

Lehmann J, Friedrich O (2007) *Dentalium decussata* – Ein Scaphopode aus der norddeutschen Kreide. Fossilien 24:93–96

Lehmann J, Liebenberg W (2015) Neuer Förderverein der Geowissenschaftlichen Sammlung der Universität Bremen. GMit. Geowiss Mitt 59:75–76

Lehmann J, Menzel H (2005) Die Otolithen der Geowissenschaftlichen Sammlung der Universität Bremen – Provenienzen und Katalog der Originale. Abh herausgegeben Nat Verein Bremen 45:657–678

Meyer KO (1958) Notizen zur Anatomie von *Iridina* (*Cameronia*) *spekii*. Ein Beitrag zur Kenntnis der Muteliden (Moll. Bivalvia). Veröff Übersee-Museum Bremen 3:25–31

Meyer KO (1970) Unioniden und Muteliden aus Tansania (Moll. Bivalvia). Veröff Übersee-Museum Bremen 4:45–49

Moellendorff OF (1884) Diagnosen neuer chinesischer Arten. Nachr deutsch malakozool Gesellschaft 16:169–174

Moellendorff OF (1887) Von den Philippinen. Jah deutsch Malakozool Gesellschaft 14:85–97

Münder H (2009) Das schwimmende Science Center. In: (BMBF) BfBuF (ed) Wissenschaftsjahre 2000 bis 2009. Erfahrungen und Perspektiven der Wissenschaftskommunikation. Kessler Druck + Medien, Bonn, p 25

Nordsieck H (2001) Revision of the system of the Phaedusinae from mainland China with the description of new taxa. Arch Molluskenkunde 129:25–63

Nordsieck H (2003) New zaptychoid Phaedusinae from Taiwan (Gastropoda: Stylommatophora: Clausiliidae). Arch Molluskenkunde 132:105–119

Pease WH (1869) Description d'especes nouvelles du genre Helicter, habitant les îles Hawaii. J Conchyliologie 17:167–176

Philippi RA (ed) (1842–1845) Abbildungen und Beschreibungen neuer oder wenig gekannter Conchylien. Erster Band. Theodor Fischer, Kassel

Philippi RA (ed) (1845–1847) Abbildungen und Beschreibungen neuer oder wenig gekannter Conchylien. Zweiter Band. Theodor Fischer, Kassel

Schilder FA (1958) Eine fast unbekannte Porzellanschnecke der Hawaii-Inseln. Veröff Übersee-Museum Bremen 3:32–38

Schmieder F (2014) MARUM-Wanderausstellung MeerErleben. https://www.marum.de/meererleben.html. Accessed 31 May 2015

Schmieder F, Lehmann J (2011) Modul 2. Vielfalt und Evolution. In: Wefer G (ed) Meer Erleben. Die Erlebnisausstellung von ECE und MARUM. Begleitheft zur Ausstellung. Marum, Bremen, pp 6–9

Seilacher A, Gishlick AD (2014) Morphodynamics. CRC Press, Boca Raton, p 551

Von Dem Busch G (1858) Neue Melanien. Malakozool Blätter 5:33–37

Zilch A (1971) Die wissensschaftliche Bedeutung alter Molluskensammlungen. Nat Museum 101:437–439

Chapter 17
BREMEN: Zoological Collection of the University of Bremen

Hans Konrad Nettmann

Abstract The Zoological Collection of the University of Bremen is a very young collection of arthropod and vertebrate material collected in different ecological research projects between 1960 and 2010. Carabid and staphylinid beetle and spider communities of bogs and salt marshes are the greatest part of the collections which includes time series within different habitats and documented communities of formerly undisturbed bogs. Other arthropod groups are included as far as pitfall traps will get them. As climate change was one aspect of most salt marsh projects, the material was collected as baseline documents for further studies.

The collection is housed in rooms of the university but has only a retired scientist as staff.

Keywords Insects • Spiders • Bogs • Salt marshes

17.1 Introduction

Within the institute of ecology and evolutionary biology at Bremen University, several projects with intensive field research were performed. Zoological material was collected mainly in the ecosystems group (Prof. Dr. Gerd Weidemann) (focus on soil biology) and the evolutionary biology group (Prof. Dr. Dietrich Mossakowski) (focus on arthropod communities in bogs and coastal marshes and on phylogeography of Carabid beetles). These groups housed their material within their rooms, but when the leading persons retired and new groups with new themes were founded, the task to save the old research material has to be solved. Thus based on an agreement between Prof. Dr. Mossakowski and the University administration, the zoological collection of the university was founded and started to work in 2012. Perhaps it is the youngest university collection in Germany, and it is still in a rather primordial state because it has only some room but no real staff except the

H.K. Nettmann (✉)
Faculty 2, Biology and Chemistry, University of Bremen, Leobener Strasse, 28359 Bremen, Germany
e-mail: nettmann@uni-bremen.de

L.A. Beck (ed.), *Zoological Collections of Germany*, Natural History Collections,
DOI 10.1007/978-3-319-44321-8_17

author in a post retirement job. Cooperation with the Übersee-Museum in Bremen is good, but the capacities there are limited.

17.2 Status, Label, etc.

The Zoological Collection of the University of Bremen, in German "Zoologische Sammlung der Universität Bremen," is an independent unit, attached but not part of the Faculty of Biology and Chemistry" with no annual budget. The catalogue is in preparation, and up to now no acronym is defined because taxonomically important specimens will be transferred to the Übersee-Museum in Bremen or to the ZFMK Bonn.

The collection is housed in the central area of the campus.

All enquiries to the collection should be made to Dr. Hans Konrad Nettmann [Fachbereich 2 (Biologie/Chemie) University of Bremen, Leobener Strasse, 28359 Bremen. e-mail: nettmann@uni-bremen.de].

17.3 Collections

The most important part is the material collected by Mossakowski during ecological research in raised bogs all over Germany in the late 1960s (Mossakowski 1970a, b, 1971, 1973, 1977). As in the other projects, it is a material from year cycles of pitfall traps, placed in different habitat types or along ecological gradients and changed monthly. Carabids and staphylinids are completely analyzed and mostly published; spiders were given to other specialists in some cases, while other groups are still unsorted but available. Most of the bogs may have changed dramatically since that time; thus comparisons over longer time scales can be made. Other bog projects were made later in Iceland and northern Finland (Hoffmann 2002), in Scandinavia and the Baltic States (Främbs 1994, 1988; Främbs et al. 2002; Schikora 2002; Mossakowski et al. 2003), and in Northwest Germany (Mossakowski and Främbs 1993, 2003; Främbs and Mossakowski 2001). The second and largest part of the collection (some thousand probes, partly sorted) is the material from several large projects about climate change effects in coastal salt marshes (1996–2002). This is pitfall trap material from several stations along the German coast of North and Baltic sea (Vagts et al. 2000; Kinder et al. 2003; Dormann et al. 2002, 2008, 2010). Also pitfall material from smaller projects in forest (Dülge 1992) and grassland habitats (Andretzke 2002) is present as well as material from military heathland projects (Mossakowski et al. 1990) and other studies.

Some material from the former Solling project in Göttingen, collected by G. Weidemann (Weidemann and Schauermann 1986), is also present, while most soil arthropod material of that group is still housed in the rooms of the still existing group of theoretical ecology (Prof. Dr. Juliane Filser).

In 1980–1983, a mapping project on fauna and flora of Bremen took place, and the carabid material as well as some small mammal material is present in the collection (Mossakowski 1991; Nettmann et al. 1991). Also small mammal bones from different owl pellet probes from Bremen are stored. A large lizard collection as result of several long-lasting hybridization and breeding experiments (several thousand individuals) (Rykena 1991, 1996, 2001) has been transferred to the ZFMK Collection in Bonn, due to the very good conditions there.

Last but not least there is a collection of frozen tissue probes, mainly of Carabus species from the western Palaearctic, used in phylogeographic studies (publications not listed here).

It is expected that more material from projects of other groups, which is distributed in the faculty, will be integrated in the collection.

17.4 Research Potencies and Publications

The main value of all the preserved material is that aspects of real arthropod communities in defined places at ecological conditions of a certain time are documented. This can be used to check changes in later studies as well as to study former distribution pattern following new taxonomic findings. But it is necessary to register all this in a new database, together with the original field protocols which exist for many places. As shown above several publications are based on the existing material (not all are listed), but there is also much unpublished and even not analyzed material from defined places.

17.5 Educational Work

There is also a collection of mammalian skulls formerly used in determination courses. It is planned to organize courses in vertebrate determination as cooperation between University Collection, Übersee-Museum, and Hochschule Bremen. Also cooperation with the specialist groups of the "Naturwissenschaftlicher Verein Bremen" will be intensified.

References

Andretzke H (2002) Erfolgskontrolle von Naturschutzmaßnahmen in der Wümmeniederung bei Bremen anhand der Carabidenfauna. Angew Carabidol 4(5):3–17

Dormann W, Kamer N, Hochkirch A, Mossakowski D (2002) Innerartliche und zönotische Diversität der Carabiden von Salzgrasländern der westlichen Vorpommerschen Boddenküste. Verh Ges Ökol 32:351

Dormann W, Metzing D, Mossakowski D (2008) Klimafolgen und Küsten: Auswirkungen auf die Verteilung und Verbreitung von Laufkäfern der Salzwiesen (Carabidae). Insecta 11:51–62

Dormann W, Mißkampf R, Kamer N, Mossakowski D (2010) Die Dynamik der Diversität von Tierarten und Zönosen in einer sich wandelnden Umwelt unter besonderer Berücksichtigung von Dispersionsprozessen. B. Westliche Bodenlandschaft. In: Schickhoff U, Seiberling S (Hrgs) Entwicklung der Biodiversität in Salzgrasländern der Vorpommerschen Boddenlandschaft. Naturschutz und Biologische Vielfalt vol 102, pp 161–226

Dülge R (1992) Die Carabidenfauna (Coleoptera: Carabidae) ausgewählter Geestwälder nördlich von Bremen. Abh Naturwiss Ver Bremen 42(1):95

Främbs H (1988) Habitatpräferenz von Carabiden (Coleoptera) auf Palsmooren in Torne-Lappmark. Schweden – Mitt dtsch Ges allg angew Ent 6:379–390

Främbs H (1994) The importance of habitat structure and food supply for carabid beetles (Coleoptera, Carabidae) in peat bogs. Mem Entomol Soc Can 126 (Suppl S169):145–159

Främbs H, Mossakowski D (2001) Dynamik von Laufkäfer-Gemeinschaften (Coleoptera, Carabidae) nach Wiedervernässungsmaßnahmen im Leegmoor. Angew Landschaftsökol 38:107–162

Främbs H, Dormann W, Mossakowski D (2002) Spatial distribution of Carabid beetles on Zehlau Bog. Baltic J Coleopterol 2(1):7–15

Hoffmann J (2002) Habitat preferences of spiders (Araneae) in palsa mires in central Iceland and northwestern Finland, 247 S

Kinder M, Främbs H, Hielen B, Mossakowski D (2003) Regeneration von Salzwiesen in einem Sommergroden an der Nordseeküste: E + E Vorhaben "Salzwiesenprojekt Wurster Küste". Nat Landsch 78(8):343–353

Mossakowski D (1970a) Ökologische Untersuchungen an epigäischen Coleopteren atlantischer Moor- und Heidestandorte. Z wiss Zool 181:233–316

Mossakowski D (1970b) Das Hochmor-Ökoareal von *Agonum ericeti* (Panz.) und die Frage der Hochmoorbindung. Faunistisch-ökol Mitt 3:378–392

Mossakowski D (1971) Ökologische Untersuchungen der Coleopterenfauna salzbeeinflußter Torfe. Verh Ver naturwiss Heimatforsch Hamburg 38:117–131

Mossakowski D (1973) Programmierte Auswertung faunistisch-ökologischer Daten. Faunistisch-ö kol Mitt 4:255–272 (Daten: Hochmoor Wurzacher Ried)

Mossakowski D (1977) Die Käferfauna wachsender Hochmoorflächen in der Esterweger Dose. Drosera 77:63–72

Mossakowski D (1991) Zur Verbreitung der Laufkäfer (Carabidae) im Lande Bremen. Abh Naturwiss Ver Bremen 41(3):543–639

Mossakowski D, Främbs H (1993) Carabiden als Indikatoren der Auswirkungen von Wiedervernässungsmaßnahmen auf die Fauna im Leegmoor. Natursch Landsch Niedersach 29:79–114

Mossakowski D, Främbs H (2003) Arthropoden intakter Hochmoore und regenerierender Abtorfungsflächen. In: Aschemeier C, Rückriem C, Ikemeyer D (Hrsg) Naturschutz in Moor und Heide. Ergebnisse der Tagung in der Biologischen Station Zwillbrock vom 16–17. 4. 2002, pp 35–41. Biologische Station Zwillbrock e. V. Vreden

Mossakowski D, Främbs H, Baro A (1990) Carabid beetles as indicators of habitat destructions caused by tank movements. In: Stork NE (ed) The role of ground beetles in ecology and environmental studies. Intercept Ltd, Andover, pp 237–243, 424 pp

Mossakowski D, Främbs H, Lakomy W (2003) The Carabid and Staphylinid fauna of raised bogs. A comparison of Northwest Germany and the Baltic region. Baltic J Coleopterol 3(2):137–144

Nettmann HK, Hellbernd-Tiemann L, Roschen A (1991) Zur Verbreitung der Säugetiere (mit Ausnahme der Fledermäuse) im Land Bremen. Abh Naturwiss Ver Bremen 41(3):641–660

Rykena S (1991) Kreuzungsexperimente zur Prüfung der Artgrenzen im Genus Lacerta sensu stricto. Mitt Zool Mus Berlin 67:55–68

Rykena S (1996) Experimental interspecific hybridization in the genus Lacerta. Isr J Zool 42 (2):171–184

Rykena S (2001) Experimental hybridization in green lizards (*Lacerta s. str.*), a tool to study species boundaries. Mertensiella Rheinbach 13:78–88

Schikora H-B (2002) Spinnen (Arachnida, Araneae) nord- und mitteleuropäischer Regenwassermoore entlang ökologischer und geographischer Gradienten. 567 S. Mainz-Verlag, Aachen

Vagts I, Cordes H, Weidemann G, Mossakowski D (2000) Auswirkungen von Klimaänderungen auf die biologischen Systeme der Küsten (Salzwiesen und Dünen). Abschlußbericht eines Verbundprojektes, gefördert durch BMBF, Land Mecklenburg-Vorpommern und DLR. Teil A. Synthese. 199 pp. TIB Hannover

Weidemann G, Schauermann J (1986) Die Tierwelt, ihre Nahrungsbeziehungen und ihre Rolle. In: Ellenberg H, Mayer R, Schauermann J (eds) Ökosystemforschung - Ergebnisse des Solling-Projektes. Ulmer, Stuttgart, pp 179–266

Chapter 18
BREMEN: The Zoological Collections of the Übersee-Museum Bremen (UMB)

Michael Stiller, Fernanda M. Rodrigues, and Volker Lohrmann

Abstract The Übersee-Museum Bremen, which was founded in 1896 as the "Städtisches Museum für Natur-, Völker- und Handelskunde", embraces three departments: "natural history", "ethnology", and "trade history". With this combination the Übersee-Museum is unique not only in Germany but also in Europe. It houses a collection of about 1.2 million plants, fungi, and animals as well as ethnographic and trade objects collected over the past 350 years. The origins of the zoological collections date way back into the seventeenth century. Today, the zoological collection of the Übersee-Museum comprises about 700,000 specimens including more than 1500 types and rare samples of extinct animals like Steller's sea cow. The collection is particularly strong with respect to the birds and insects. Current research at the Übersee-Museum includes the study of selected groups of aculeate Hymenoptera, but most research on the collections is carried out by external scientists. The official acronym for the zoological collection of the Übersee-Museum Bremen is UMB.

Keywords Hugo Schauinsland • Entomology • Ornithology • Africa • Asia • America • Europe • Oceania • Types • Extinct

18.1 Then and Now

Today, the Übersee-Museum Bremen embraces three departments: "natural history", "ethnology", and "trade history". With this combination the Übersee-Museum is unique not only in Germany but also in Europe. It houses a collection of about 1.2 million plants, fungi, and animals as well as ethnographic and trade objects collected over the past 350 years.

The origins of the zoological collections date way back into the seventeenth century and have been part of the collection of the "Gymnasium Illustre"

M. Stiller (✉) • F.M. Rodrigues • V. Lohrmann
Übersee-Museum Bremen, Stiftung öffentlichen Rechts, Bahnhofsplatz 13, 28195 Bremen, Germany
e-mail: m.stiller@uebersee-museum.de; http://www.uebersee-museum.de; f.rodrigues@uebersee-museum.de; v.lohrmann@uebersee-museum.de

L.A. Beck (ed.), *Zoological Collections of Germany*, Natural History Collections,
DOI 10.1007/978-3-319-44321-8_18

Fig. 18.1 Hugo Schauinsland in Qufu (China), 1913 (with permission from: Übersee-Museum Bremen)

(Abel 1970; this publication provides valuable information on the history of the Übersee-Museum Bremen). The first society in Bremen with a particular focus on natural history was the "Physikalisch ökonomische Gesellschaft" established in 1776 which was later renamed the "Gesellschaft Museum". Here the fundamentals of the collections were established, e.g., the entomological collection. Due to the changing objectives of the "Gesellschaft Museum", a new society was founded, named the "Naturwissenschaftlicher Verein zu Bremen". One of the aims of the new society was to preserve, curate, and enlarge the natural history collections and to concentrate its research on the fauna and flora from the northwestern region of Germany.

The museum was founded in 1896 as "Städtisches Museum für Natur-, Völker- und Handelskunde" with the aim to provide public access to the valuable regional collections and those brought back from overseas. These collections, now housed in the new museum, were accumulated by merchants, missionaries, scientists, and people interested in science, as well as the Norddeutscher Lloyd (a shipping company from Bremen). The first director Hugo H. Schauinsland (Fig. 18.1) focused on the exploration of Southeast Asia, Oceania, and Hawaii with a particular personal interest on isolated islands, e.g., the Chatham Islands or Laysan.

18.2 "Deutsches Kolonial und Übersee-Museum" and World War II

In 1933 Schauinsland was forced by the national socialists to resign, followed by a reorganization of the focal points of the museum in accordance with the new ideology. Thus the former colonies became more and more important, and the museum was subsequently renamed the "Deutsches Kolonial und Übersee-Museum" in 1935.

From 1939 the museum was temporarily closed during World War II. For safety reasons most of the collections were evacuated and stored in the countryside nearby or even more distantly in eastern Germany. Because of its location near the main train station, the Übersee-Museum was bombed by allied airplanes in 1943. The first glazed atrium of the museum was completely destroyed—including the exhibition and collections still stored in this part of the museum. Most affected by the war was the collection of the department of trade history, which was totally destroyed, and large parts of the documentation. Whereas the ethnographic collection survived the war without major harm, the damage to parts of the natural history collection was devastating. Some went up in flames while others never returned to the museum after the war.

18.3 Reorganization After 1945

The end of the war resulted in a complete exchange of the scientific staff of the department of natural history. Scientists were suspended by the American military administration and the museum was again renamed the "Museum für Natur-, Völker- und Handelskunde". The first glazed atrium and the main entrance which were destroyed during the war were rebuilt and reopened in 1951. Helmuth O. Wagner (1897–1977), an ornithologist, became the new director of the museum before the museum was finally renamed the "Übersee-Museum" in 1952. The museum was again closed for renovations from 1976 to 1978 and reopened in 1979. Finally, in 1999 a new air-conditioned building for large parts of the collection, the so-called Übermaxx, was completed giving the museum the appearance that it has today.

18.4 The Collections

A skeleton of a minke whale formerly belonging to the collection of the "Gymnasium Illustre" is the oldest known part of the natural history collection. The whale was hunted in the Lesum near Bremen in 1669. The museum is one of only two collections worldwide that holds skin fragments of the extinct Steller's sea cow (*Hydrodamalis gigas*). Besides the collections accumulated by the aforementioned "Gesellschaft Museum", numerous expeditions have helped to increase the collection. Some of them are listed here to provide a feel for the international scope of the collection: Hugo Schauinsland (Australia, Bismarck Archipelago, China, Egypt, Hawaii Archipelago, Hong Kong, India, Indonesia, Japan, Korea, New Guinea, New Zealand, Samoa, Sri Lanka, the USA, 1896/1897, 1905/1906, 1907/1908, 1913/1914, and 1926), Ludwig Cohn (New Guinea, 1908, 1912), Herbert Abel (Namibia, 1952), Helmut Knipper (East Africa, 1952), Herbert Hohmann

Table 18.1 Overview of the zoological collection of the Übersee-Museum Bremen (UMB)

Taxon	Min. # of objects	Min. # of types	Geographical focus	Main collecting period
Mammalia	6000	3	E Africa, Argentina, Asia, Oceania	~1800–1960
Aves	30,000	264	E Africa, Mexico, Paraguay, Asia, Oceania, Hawaii	~1850–1950
"Reptilia"	3300	4	Asia, Oceania incl. New Zealand	~1890–1935
Amphibia	700	2	Madagascar, Asia, Oceania	~1850–1935
"Pisces"	10,000	9	E Asia, Oceania	~1890–1935
Tunicata/Agnatha/ Hemichordata	200	–	E Asia, Oceania	~1890–1915
Echinodermata	1000	–	–	~1890–1915
Mollusca	2500	~200	E Africa, N America, E Asia, Indo-Pacific, Oceania	~1870–1955
Arthropoda				
Insecta				
Lepidoptera	75,000	~760	Europe, SE Asia, Oceania, S America	~1840–present
Coleoptera	300,000	15	Europe, SE Asia, Oceania, S America	~1840–present
Diptera	35,000	2	Europe, SE Asia, Oceania	~1840–present
Hymenoptera	100,000	~200	Europe, SE Asia, Oceania, S America	~1840–present
Neuroptera	3500	–	Europe, SE Asia, Oceania	~1840–present
Hemiptera	26,000	2	Europe, SE Asia, Oceania	~1840–present
Orthoptera	33,000	14	Europe, SE Asia, Oceania	~1840–present
Odonata	3000	10	Europe, W Africa, SE Asia, Oceania	~1840–present
Other Insecta	7000	–	–	~1840–present
Other Arthropoda				
Myriapoda	300	13	Europe, E Africa	1890–1950
Chelicerata	1000	7	Africa, Asia, Oceania	1890–2005
Crustacea	5000	21	–	1890–1915
Other invertebrates	1200	13	–	1890–1915

(New Guinea, 1972/1973), and Thorwald Kruckow, Herbert Hohmann, Heinrich Kuhbier, Horst Gnettner, and Eberhard Focke (Costa Rica, 1971). Information about some of these expeditions can be found in Schauinsland (1999) and Kruckow (1974). Additional important material was obtained from the first German Deep Sea

Expedition in 1898/1899 or purchased from natural history collectors, e.g., Alan Owston (1853–1915, Japan), Hans Sauter (1871–1943, Taiwan), Fritz Plaumann (1902–1994, Brazil), and Th. Andersen. Since the 1970s most additions to the collection were obtained through donations, purchases, and exchanges with other museums.

Today the zoological collection of the Übersee-Museum comprises about 700,000 specimens (Table 18.1) and is particularly strong with respect to the birds and insects. These two most important collections are discussed below in more detail.

18.4.1 Ornithological Collection

The most significant parts of the ornithological collection were compiled by the Ornithologists Gustav Hartlaub (1814–1900) and Otto Finsch (1839–1917). In addition to material collected during Finsch's expeditions, e.g., into the west Siberian region along with Alfred Brehm and Karl Graf Wartburg-Zeil in 1876, both were involved in intensive exchanges of zoological material with the international scientific community. Because of their excellent relationships to, e.g., the so-called Emin Pascha (Sudan, 1840–1892), Philip Lutley Sclater (UK, 1829–1913), and the collectors of the Museum Godeffroy (Oceania), the bird collection became an important part of the zoological collection. The material donated to the museum by international institutions or scientists demonstrated the outstanding reputation of the collection and its curators during that period. More recent material was added by Alberto Schulze and Helmuth O. Wagner. The latter added his collection of Mexican and North American birds, and his excellent contacts to South American colleagues helped to complement the collection.

Today, the bird collection of the Übersee-Museum contains roughly 30,000 objects; about 10,000 of them are mounted specimens, while 16,000 are dried skins only. The collection is complemented by an extensive collection of skulls and skeletons and a large collection of egg shells and nests as well as about 100 specimens stored in alcohol (Becker 2008). The real gems of the collection are the 280 types (Fig. 18.2) and 100 individuals of several now extinct bird species—among them a series of specimens sampled by Schauinsland on Laysan and one of the last sampled specimens of the great auk.

18.4.2 Entomological Collection

The most continuity in growth and research was within the entomological collection which has a tradition going back about 200 years. The oldest parts date back in the early half of the nineteenth century. The most significant collection from these early days still kept at the Übersee-Museum is the one from the merchant Adam

Fig. 18.2 Parts of the ornithological type collection (with permission from: Übersee-Museum Bremen; photo: M. Haase)

Heinrich Norwich (1771–1851) which comprises over 4000 species of insects from around the world (Hohmann 1980; this publication provides valuable information on the history of entomology in Bremen). Johann Dietrich Alfken (1862–1945) was the first entomologist employed by the museum, followed by Hans Georg Amsel (1905–1999), Eberhard Jaeckh (1902–1993), Herbert Hohmann, and Helmut Riemann. In addition to the material accumulated by those individuals, the collection holds important material from, e.g., Karl Pfankuch (1871–1924), Edward Meyrick (1854–1938), Adalbert Brinkmann (1860–1943), Karl Schawerda (1869–1945), and Gustav Harttig (1871–1965) (Hohmann 1980). More recent additions to the collection include material from, e.g., Adolf Wagner (1915–2000), Hubert Pschorn-Walcher (1926–2006), Fritz Kelschinske (1939–2006), Hans-Joachim Van Loh (1948–2007), Gerd Weidemann (1934–2013), Jens-Hermann Stuke, Rolf Niedringhaus, and Werner Bode.

Today the collection holds about 600,000 specimens, including nearly 1000 types (Fig. 18.3), and is housed in "Cornell drawers" in a new air-conditioned building in a modern compactor system. Most specimens are mounted and more or less catalogued in a preliminary database. Nevertheless, there are ten thousands of insect specimens preserved in alcohol. The collection is particularly strong in Hymenoptera, Lepidoptera, and Coleoptera. Whereas the first two groups received continuous attention over the years, the beetle collection is in urgent need of a re-curation.

The entomological collection is still growing, e.g., in 2000 alone two collections were handed over to the Übersee-Museum comprising about 30,000 insect specimens (Riemann 2001). Furthermore, it was agreed that several collections are to be

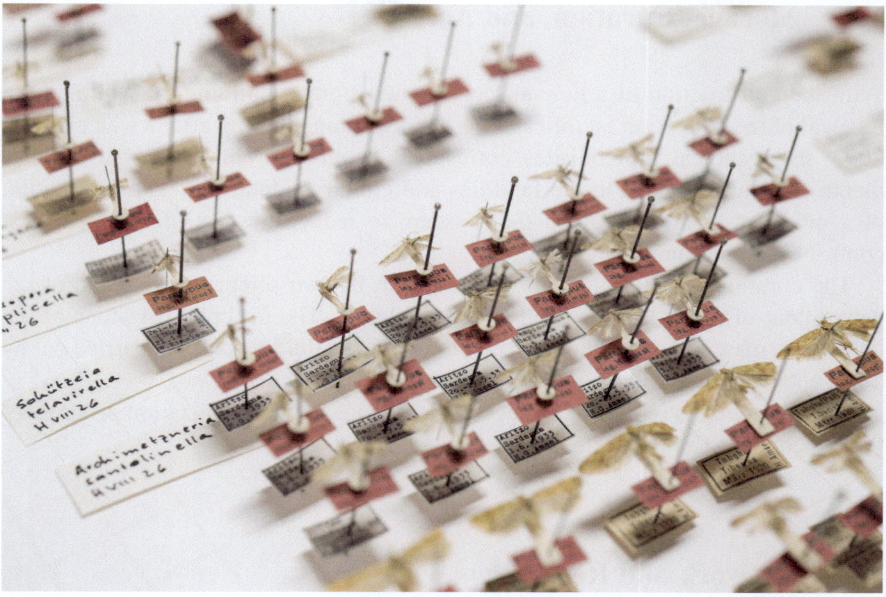

Fig. 18.3 Parts of the entomological type collection (with permission from: Übersee-Museum Bremen; photo: M. Haase)

transferred to the Übersee-Museum in the near future, e.g., the ant collection of Ursula Winter and the Lepidoptera collection of Georg Lakmann.

18.4.3 Lost and Transferred Collections

Apart from damage caused by World War II, the museum lost parts of its collection under different circumstances, such as the majority of the birds received through exchange with Emin Pascha which were destroyed even before the war due to lacking curatorial care.

Furthermore, the Alfken collection did not stay at the Übersee-Museum but was transferred to the "Museum für Naturkunde" in Berlin in 1945 although Alfken served the entomological collection for more than 40 years. Therefore, the Übersee-Museum holds only a comparable small series of duplicates from this well-known bee collection.

In 1994 the geological collection, including large parts of the mollusk collection, was relocated to the University of Bremen. During the same process, the position of curator of the collection was transferred to the university as well.

18.5 Storage, Curation, and Access

With the construction of a new air-conditioned storage building in 1999 and the renewal of the collection facilities in the historic main building after a water leak in 2007, the technical prerequisites correspond to modern conservational requirements. This also applies to the facilities and climatic conditions. The installation of a nitrogen chamber enabled the museum to reduce the use of chemical pest control to a minimum.

The curation of the collection is carried out by two scientists, the head of the department of natural history and an entomologist. The different divisions of the collection, mammalia, ornithology, entomology, and fluid storage are carried out by a respective technical staff member.

The collection is registered in several catalogues but the digitization progress of the collection differs considerably between the divisions.

18.6 Network and Research

The Übersee-Museum Bremen is well connected in the German and European museum landscape. It is a member of the "Deutsche Museumsbund" (DMB), the "Museumsverbund der Nord- und Ostsee Region" (NORe e.V.), and the "Consortium of European Taxonomic Facilities" (CETAF).

Current research at the Übersee-Museum includes the study of selected groups of aculeate Hymenoptera (e.g., Lohrmann et al. 2014; Ohl et al. 2014; Perrichot et al. 2014). Most research on the collections of the Übersee-Museum, however, is carried out by external scientists. The museum supports these efforts by welcoming visiting researchers from around the world in our collection or by providing information, photos of specimens, and loans. Even DNA samples are provided when the preservation of the specimen allows, as was recently done for the specimen of the extinct great auk (*Pinguinus impennis*) and Steller's sea cow (*Hydrodamalis gigas*).

One of the last projects included investigations of dried skins belonging to several historic bird specimens in the Übersee-Museum and the discovery of no less than 14 new species of parasitic feather mites (Acari) (e.g., Mironov et al. 2014 and therein cited references), emphasizing the importance of "old and dusty" museum collections. But new species have also been discovered in other parts of the collection, e.g., a new species of Diptera (Stuke and Clement 2008). In 2014 a visiting scientist rediscovered a forgotten series of odonates which has been donated to the Übersee-Museum in 1875 by Baron Michel Edmond de Sélys Longchamps (1813–1900). The series includes several types and a few specimens collected by Alfred Russel Wallace (1823–1913) and Henry Walter Bates (1825–1892) (Seehausen 2014).

18.7 Publications

Well before the museum published its own periodical, many publications dealing parts of the Übersee-Museum collection have been published in the journal "Abhandlungen des Naturwissenschaftlichen Vereins zu Bremen". Beginning in 1935 the museum published its own periodical named "Veröffentlichungen aus dem Deutschen Kolonial und Übersee-Museum in Bremen" until 1942. The journal was relaunched as the "Veröffentlichungen aus dem Übersee-Museum/Naturwissenschaften" in 1949. These volumes include many publications on the zoological collection of the Übersee-Museum, e.g., a series of publications dealing the type specimens of the Übersee-Museum (e.g., Knipper 1956; von Wahlert 1955). This periodical appeared until 1990 and was replaced in 1992 by the "Tendenzen" that still appears once a year. Until 2000 this newest title functioned as a yearbook of the activities of the museum. Since this time, in terms of its focus, this publication annually rotates between the three departments of the Übersee-Museum. With respect to natural history, the most important contributions are a supplement published in 1999 and volumes XIII and XVI with catalogues of, e.g., the Mantidae, Phasmidae, Decapoda, Odonata, Trichoptera, and the bird type specimens of the Übersee-Museum (Übersee-Museum Bremen 1999; Becker 2006, 2010).

18.8 Exhibitions

On 10,000 m^2 the Übersee-Museum presents in its main exhibitions topics related to natural history, ethnology, and trade history using an interdisciplinary approach (Fig. 18.4). Focal points are humans in the twenty-first century, their culture, their interactions with nature, and their economical dependencies in non-European regions. In a second building, the museum presents about 30,000 objects from the three departments on 4800 m^2—the so-called Übermaxx (Fig. 18.5) was the first storage accessible for museum visitors in Europe. In terms of zoological objects, it functions as an open reference collection as part of the permanent exhibition and, among other things, is used for training biology students from the University of Bremen. The educational service of the museum offers a wide range of activities for a broad audience across all age groups/cohorts. In terms of visitors, the museum ranges among the top 5 % in Germany.

Fig. 18.4 In the exhibition of the Übersee-Museum: Rice cultivation in Southeast Asia (with permission from: Übersee-Museum Bremen; photo: M. Haase)

Fig. 18.5 In the Übermaxx: A reference collection as part of the permanent exhibition (with permission from: Übersee-Museum Bremen; photo: M. Haase)

Acknowledgment We sincerely thank Helmut Riemann (Bremen) for providing information on the entomological collection, Matthias Haase (Bremen) for providing the photographs, and Jason Dunlop (Berlin) for improving the English.

References

Abel H (1970) Vom Raritätenkabinett zum Bremer Überseemuseum: Die Geschichte einer hanseatischen Sammlung aus Übersee anlässlich ihres 75 jährigen Bestehens. Verlag Friedrich Röver, Bremen

Becker P-R (ed) (2006) Tendenzen 2005 Bestimmtes Sammeln, Jahrbuch XIII. Übersee-Museum Bremen, Bremen

Becker P-R (2008) Bedeutende Sammlung von Exoten Überseemuseum Bremen. Der Falke 55: 28–32

Becker P-R (ed) (2010) Tendenzen 2008/1 Unbestimmtes Erforschen, Jahrbuch XVI. Übersee-Museum Bremen, Bremen

Hohmann H (1980) Zur Geschichte der Entomologie in Bremen. Jb Wittheit Bremen 24:121–152

Knipper H (1956) Die Typen und Typoide des Übersee-Museums Bremen, 3: Spongiaria. Veröff Übersee-Museum Bremen Reihe A 2(6):357–361

Kruckow T (1974) Ergebnisse der Costa-Rica-Expedition des Übersee-Museums, Nr. 1. Veröff Überseemuseum Bremen Reihe A 4(15):131–140

Lohrmann V, Falin ZH, Bennett DJ, Engel MS (2014) Recent findings of *Olixon banksii* in the Nearctic with notes on its biology (Hymenoptera: Rhopalosomatidae). J Kansas Entomol Soc 87(2):258–260. doi:10.2317/jkes130820.1

Mironov SV, Ehrnsberger R, Dabert J (2014) New species of the feather mite genus *Protolichus* Trouessart, 1884 (Astigmata, Pterolichidae) from lories and lorikeets (Aves: Psittaciformes). Zootaxa 3774(2):131–151

Ohl M, Lohrmann V, Breitkreuz L et al (2014) The soul-sucking wasp by popular acclaim – museum visitor participation in biodiversity discovery and taxonomy. PLoS One 9(4):e95068. doi:10.1371/journal.pone.0095068

Perrichot V, Engel M, Nel A et al (2014) "False Ants" from the Cretaceous (Hymenoptera: Chrysidoidea: Falsiformicidae). Mitt Entomol Ver Stuttgart 49(1):26

Riemann H (2001) Neue Insektensammlungen im Übersee-Museum. Abh Naturwiss Ver Bremen 45(1):187–188

Schauinsland HH (1999) Unterwegs in Übersee: aus Reisetagebüchern und Dokumenten des früheren Direktors des Bremer Übersee-Museums. H.M. Hauschild, Bremen

Seehausen M (2014) Forgotten duplicates from the Odonata collection of Edmond de Selys Longchamps rediscovered at the Übersee-Museum Bremen (Germany). Int Dragonfly Fund 70:1–15

Stuke J-H, Clements DK (2008) Revision of the *Myopa testacea*-group in the Palaearctic region (Diptera: Conopidae). Zootaxa 1713:1–26

Übersee-Museum Bremen (1999) Tendenzen Supplement 1999. Übersee-Museum Bremen, Bremen

von Wahlert G (1955) Die Typen und Typoide des Übersee-Museums Bremen, 2: Pisces. Veröff Übersee-Museum Bremen Reihe A 2(5):323–326

Chapter 19
CHEMNITZ: Museum of Natural History Chemnitz: Identity Through Tradition

Thorid Zierold, Sven Erlacher, and Ronny Rößler

Abstract Museums are more than a warehouse or storage place; museums are places of communication and education. The knowledge evolves from comparative studies of historical and current exhibits. Although numbers of exhibits are relevant, it is the potential of each single object which helps understanding natural and cultural history. Highlights of our zoological collection are regional collections of Geometridae, Mollusca, and Avifauna as well as juvenile vertebrates. In the context of new technologies and citizen science, it is now time for transparent magazines and international cooperation to prepare the path for tomorrow's museums.

Keywords Citizen science • Ecological shift • Geometridae • Juvenile • Vertebrate

19.1 Introduction

In the second half of the nineteenth century, Chemnitz was an upcoming and growing industrial city. However, contradictory to its commercial relevance, science and culture were underrepresented in public life. Chemnitz had neither held a patriarchal university nor baronial collections or scientific libraries as intellectual centers. Under those circumstances, 16 engaged citizens founded the "Chemnitz Society of Natural Science" on the 17th October 1859. The association evolved from the initial internal reader circle of scientific papers to a society being more open to the public by inviting to talks, conferences, and excursions. They even published their own scientific journal. The register of members contained teachers, medical doctors, scientists, merchants, and businessmen (Barthel 2001). Soon the society thought about developing and preserving the growing natural history collection and bibliographical inventory. In 1868 they donated the collections to the city of Chemnitz with the obligation to make them available to the public. That was the beginning of the very first civic museum in Chemnitz (Rössler and Zierold 2014). Due to the work of volunteering curators and immense public engagement,

T. Zierold (✉) • S. Erlacher • R. Rößler
Kustodin Naturwissenschaften, Museum für Naturkunde Chemnitz, Das TIETZ, Moritzstraße 20, 09111 Chemnitz, Germany
e-mail: zierold@naturkunde-chemnitz.de; http://www.naturkunde-chemnitz.de

L.A. Beck (ed.), *Zoological Collections of Germany*, Natural History Collections,
DOI 10.1007/978-3-319-44321-8_19

Fig. 19.1 Interior of the zoological collection as part of the King Albert Museum, Chemnitz, about 1909 (with permission from: Museum für Naturkunde Chemnitz; photo: MfNC)

biological and geological collections were developed and completed (Zierold 2014). They illustrated both the local specific features and the exotic nature worldwide, such as kiwi, duck-billed platypus, giant turtle, whale vertebra, and also an Egyptian mummy (Fig. 19.1).

Today the zoological collection comprises about 250,000 single objects subdivided into the areas of Entomology, Malacology, other Invertebrates, and Vertebrates (Mieth et al. 2006).

19.2 Small but Powerful: Insights into the Specific of the Zoological Collection of the Natural History Museum Chemnitz

The *entomological collection* distinguishes itself through comprehensive objects of Western Saxony and Eastern Thuringia butterflies and local beetles (Fig. 19.2). The collections highlight the ecological shift of the entomofauna due to growing urbanization in the area of Western Saxony. This allows scientific questions in the field of taxonomy and evolutionary ecology. Providing digital data of the entomological collection in combination with recent distribution patterns is a huge challenge and requires both personnel engagement and technical know-how. Our start-up project "Online database of Geometridae of Saxony, Thuringia and Saxony-Anhalt" is not only an efficient research tool but also an application to engage the public.

The collection of the two lipped door snail (*Balea biplicata*) of West Saxony is particular for the *Malacological section*. The research of corresponding historic manuscripts and field data and the determination and digitalization are challenging future tasks (Clemens Kleindienst 1945/1946). In cooperation with amateur researchers and volunteers, the regional collection of other terrestrial and freshwater Mollusca was documented and determined (Enzenross and Enzenross 1996). Their studies provide new insights into the zoogeography of this group. Furthermore Trübsbach published results on the distribution of Gastropods along the

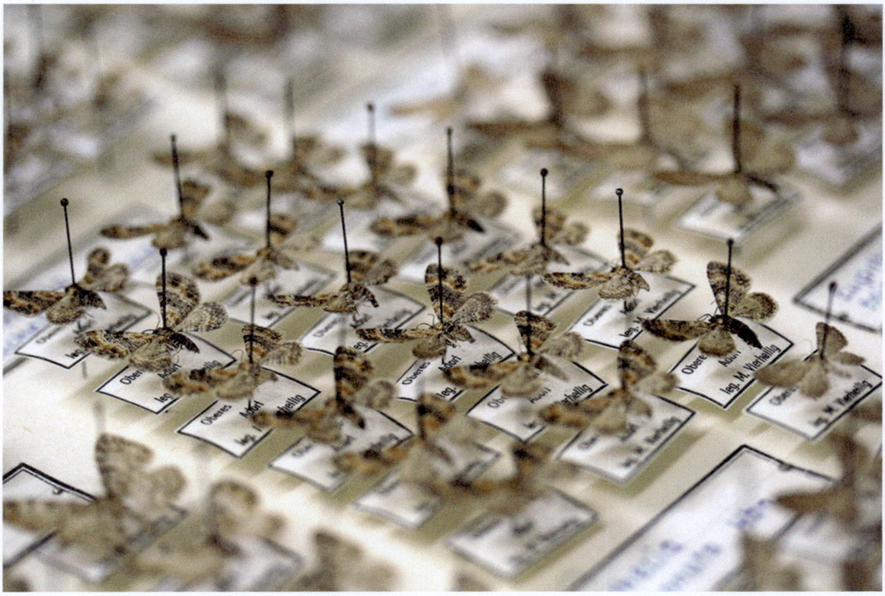

Fig. 19.2 Historical collections highlighting the ecological shift of the entomofauna due to growing urbanization in the area of Western Saxony (with permission from: Museum für Naturkunde Chemnitz; photo: MfNC)

Fig. 19.3 Great argus in courtship with hen. The exhibits prepared for the temporary exhibition "1 + 1 = SEX – The love life of animals" by taxidermist Holger Rathaj (with permission from: Museum für Naturkunde Chemnitz; photo: MfNC)

Zschopau valley and physiological experiments of the carbonate metabolism in freshwater Mollusca (Trübsbach 1934, 1947).

Historic *vertebrate* object are often called stuffed exhibits. Prepared with seaweed or wooden wool, they hardly represent their natural anatomy and often prepared in excessively aggressive position. However those objects are relevant for the history of science in general and of taxidermy in specific, they act as cultural accelerator as mentioned by the communication theoretician Derek de Kherkove (1995). Investigating the material and displays, stuffed exhibits can tell a story of its designer—the taxidermist. Trends in preparation techniques and materials can also be deduced. Thus contemporary taxidermy including dermoplastic, freeze-dried, and plastination established to present an authentic object to visitor (Fig. 19.3). The vertebrate collection of the museum provides an insight into the passion of collectors in the early nineteenth century and allows the investigation of several preparation techniques with a special focus on local and regional avifauna. Traditionally juvenile vertebrates, local to regional avifauna including 1,350 eggs, are in the strategic focus of our collection (Fig. 19.4). They are fundamental for educational purposes and for taxonomic determination. Insurance documents from 1899 onward report about the growing vertebrate collection. However, the inventory number system in the vertebrate section did not start before 1963. Thus, digitalization and provenance research are challenging tasks in the vertebrate collection.

19.3 Digital Registration

The digital documentation is supported by a web-based data management system. It has been implemented in cooperation with CD-LAB Bonn and Saxony Regional Office of Museums. It is our short-term aim to increase transparency and to engage

Fig. 19.4 Bird egg collection with about 1350 exhibits goes back to 1920 (with permission from: Museum für Naturkunde Chemnitz)

citizens. Therefore we are working on the presence of our collection on the Web portal "museum digital."

Literature

Barthel M (2001) Vom alten Schulschrank über die Kunsthütte ins König-Albert-Museum. In: Rößler R (ed) Der Versteinerte Wald von Chemnitz. Chemnitz, Museum für Naturkunde Chemnitz, vol pp 52–65

Derek DK (1995) The skin of culture: investigating the new electronic reality. Kogan Page, Toronto

Enzenross R, Enzenross L (1996) Historische Kollektionen des Museums für Naturkunde Chemnitz – Sammlung Trübsbach (Land- und Süßwassermollusken). Veröff Museums Naturkunde Chem 19:23–26

Mieth KM, Sächs, Landesstelle für Museumswesen (2006) Museum für Naturkunde Chemnitz. Dößel, Sächsische Landesstelle für Museumswesen, Freistaat Sachsen

Rössler R, Zierold T (2014) Der Weg zum Erfolg – Das Museum für Naturkunde Chemnitz im Eigenbetrieb "Das Tietz". Jahrestagung des Sächsischen Museumsbundes e.V., 22.03.14. Chemnitz, Sächsische Landesstelle für Museumswesen & Sächsischer Museumsbund e.V. Informationen des Sächsischen Museumsbundes e.V. Heft 47/2014

Trübsbach P (1934) Geographische Verbreitung der Gastropoden im Gebiete der Zschopau nebst biologischen Untersuchungen. Ber naturwiss Gesellsch Chem 24:15–98

Trübsbach P (1947) Der Kalk im Haushalte der Mollusken 2, mit besonderer Berücksichtigung des physiologischen Vorganges der Schalenbildung. Arch Molluskenkunde 76:145–162

Zierold T (2014) 1+1= SEX – Das Liebesleben der Tiere – Eine Ausstellung des Museums für Naturkunde Chemnitz. Veröff Museums Naturkunde Chem 37:137–147

Chapter 20
DARMSTADT: The Zoological Collections of the Hessische Landesmuseum Darmstadt

Jörn Köhler and Francisco Hita Garcia

Abstract The origins of the zoological collections of the Hessische Landesmuseum Darmstadt (HLMD) are rather old and can be traced back to the early eighteenth century. In past times, it was mainly two curators responsible for the growth of the collection, namely, Johann Jakob Kaup (1803–1873) and Gottlieb von Koch (1849–1914). The latter was involved in the construction of the museum building opened in 1906 and was the scientific designer of Darmstadt's unique and internationally important zoological dioramas. Both curators acquired several rare specimens of recently extinct species for the museum. After World War II, it was Georg Scheer who significantly contributed with collections originating from the so-called Xarifa expeditions led by diving pioneer Hans Hass, among them an internationally very important collection of stony corals. Today, the HLMD holds a representative collection about 15,000 vertebrate and 380,000 invertebrate specimens from all continents, among them several historical and recent type specimens.

Keywords Johann Jakob Kaup • Gottlieb von Koch • Georg Scheer • Historical collection • Natural history diorama • Extinct species

20.1 General Information

The Hessisches Landesmuseum Darmstadt (Fig. 20.1) is one of the very few museums conserving and displaying an unusually wide array of objects of arts, ethology, paleontology, geology, and zoology. Visitors will face historical paintings, sculptures, and Eocene fossils but also the taxidermy of a polar bear. The collections in the zoology section of the Hessisches Landesmuseum Darmstadt (HLMD) have their origins in the "Natural History Cabinet" of Duke Ludwig X (1753–1830), who later became Grand Duke Ludwig I. This "Cabinet," formerly part of a kind of "Wunderkammer," continued growing and was later transferred to a more technically managed collection (Krause 1972).

J. Köhler (✉) • F. Hita Garcia
Hessisches Landesmuseum Darmstadt, Friedensplatz 1, 64283 Darmstadt, Germany
e-mail: joern.koehler@hlmd.de; http://www.hlmd.de; fhitagarcia@gmail.com

L.A. Beck (ed.), *Zoological Collections of Germany*, Natural History Collections,
DOI 10.1007/978-3-319-44321-8_20

Fig. 20.1 Southern view of the museum building of the Hessische Landesmuseums Darmstadt in 2014 (with permission from HLMD; photo: Steffen Harms)

Johann Jakob Kaup (1803–1873), who was natural history curator in the nineteenth century, was keen to extend the collection and regularly acquired new objects for research and exhibition. These objects were either bought from professional collectors or exchanged with other museums. This way, the HLMD acquired a considerable collection of remarkable specimens, among them numerous Australian marsupials and rare birds, as well as several specimens of recently extirpated species. These, for example, include a quagga (*Equus quagga quagga*), a couple of huia (*Heteralocha acutirostris*), skull of Steller's sea cow (*Hydrodamalis gigas*), and the very rare bone remains of the Rodrigues solitaire (*Pezophaps solitaria*).

The growing collection required a new museum building which was officially opened in 1906. This period was mainly characterized by activities of Kaup's successor, Gottlieb von Koch (1849–1914). As zoological curator, von Koch was deeply involved in the planning and construction of the new museum building. His focus was on education and didactics. He placed major emphasis on an architectural layout that would create a shell around the displays and as a result maximize the effect of the planned dioramas. Comprising nearly 1000 animal specimens, the ten dioramas of the Hessisches Landesmuseum Darmstadt ranked among the first displays that ever focused on zoogeography and systematic zoology (Fig. 20.2). Displaying animal specimens in an abstracted environment that merely provided an idea of the natural habitat and never aimed at reproducing it true to every detail was very modern indeed then and could rightfully be termed revolutionary from a historical perspective. In fact, there is no other museum in the world where a comparative concept has been realized, rendering in particular the dioramas "Africa," "Asia," and "Australia" exceptionally important historical documents

Fig. 20.2 View of the right part of the historical Asia diorama in 2014, briefly after its restoration (with permission from HLMD; photo: Wolfgang Fuhrmannek)

on an international scale as they still exist in their original conditions (Schulze et al. 2014; Munsch et al. 2015).

World War II caused considerable damage to parts of the museum building, but fortunately only small portions of the collections were affected.

In the second half of the twentieth century, some far-reaching research expeditions led to the acquisition of further specimens. Most notable are the two so-called Xarifa expeditions in the 1950s directed by famous diving pioneer Hans Hass. These expeditions reached the Caribbean and Galapagos and later several localities in the Indian Ocean, respectively. The curator Georg Scheer (1910–2004) was one of the crew members with his research focusing on stony corals. Consequently, apart from birds, squamates, and mollusks, these expeditions brought an internationally important collection of corals to the HLMD.

Later, expeditions by the curators Wolfgang Schneider and Ulrich Joger to the island of Socotra and northern Africa contributed to HLMD zoological collections. These collections mainly consist of different groups of arthropods and squamates. Since recently, the HLMD also holds an important collection of ants with a geographical focus on Africa, Madagascar, and Southeast Asia.

The zoological collections today cover a representative range of mammal, bird, mollusk, arthropod, and coral species. Together, the HLMD today holds about 15,000 vertebrate and approximately 380,000 invertebrate specimens from all continents. Among them are several historical and recent type specimens. Staff of the zoology department includes two research curators, two scientific volunteers, and two taxidermists. Apart from administrative staff, there are third-party funded postdoc researchers.

20.2 Research

Historical research by Johann Jakob Kaup focused on systematics and taxonomy of phasmids (Fig. 20.3) and beetles, whereas Gottlieb von Koch, a former student of Ernst Haeckel in Jena, mainly worked on the biology of marine Cnidaria and their alteration of generations. Georg Scheer was an expert on stony corals and published many important articles on their systematics and taxonomy, but also he did some research on birds. Ulrich Joger did most of his research on systematics of lizards and snakes, whereas Wolfgang Schneider mainly worked on dragonflies and damselflies in the Near East.

Today's research is characterized by broad international collaborations and joint projects with other research institutions. Its main competencies are classical museum research on systematics, taxonomy, phylogeny, and biogeography by application and integration of modern methodology like computer-based bioacoustic analyses, molecular genetics, and microCT scanning. The senior author mainly works on phylogeny, systematics, natural history, and biogeography of African and Neotropical anurans and squamates (e.g., Köhler et al. 2015). In recent years his studies had a geographical focus on Amazonia and Madagascar. The junior author is specialized in ants. His work consists in the taxonomy, biogeography, evolution, and ecology of ants from Africa, Madagascar, and, to a lesser extent, China and the Western Pacific (e.g., Hita Garcia et al. 2014).

The HLMD natural history department edits an own publication named *Kaupia*. *Kaupia* publishes international research on geology, paleontology, and zoology and usually appears on an annual basis. Whereas some issues are to be treated monographs, others constitute proceedings volumes of scientific meetings held at the museum.

Fig. 20.3 Historical holotype specimen of *Eurycantha rosenbergii* (Phasmatodea) described by Johann Jacob Kaup in 1871 (with permission from HLMD; photo: Jürgen Krebs)

20.3 Educational Work

Historically, the HLMD zoological collections were used as a basis for several important book publications. The most striking example is probably *Teutsche Ornithologie oder Naturgeschichte aller Vögel Teutschlands in naturgetreuen Abbildungen und Beschreibungen*. Published between 1800 and 1817 in 22 parts, it gives descriptions and realistic drawings of the birds of Germany and includes lifelike illustrations by Johann Conrad Susemihl (1767–1847). These drawings were based on specimens from the Darmstadt collection, such as the northern gannet (*Sula bassana*) collected in the Odenwald, Hesse, in 1796 (Fig. 20.4).

Being a curator and at the same time professor at the University of Darmstadt, Gottlieb von Koch was involved in teaching and preparation of didactic materials. He is one of the authors of the so-called *Jung-Koch-Quentell'schen Lehrtafeln* which displayed anatomical details of different organisms and were used in lectures at university and schools until the late twentieth century (Fig. 20.5). Furthermore, von Koch together with his taxidermists prepared some of the first "plastinated" specimens showing inner organs for teaching medical students.

Whereas the former permanent zoological museum exhibition was generally organized in a strictly systematic approach, showing protozoans first and then guiding through different groups of organisms reaching the primates, the entirely new exhibition, opened to the public in September 2014, by contrast, is organized into topic groups with particular emphasis on evolutionary connections and mechanisms. There are exhibition areas addressing species diversity, the structure of organisms, research, and evolution itself (Fig. 20.6). Showcases are partly accompanied by interactive media and a QR code-based level of information accessible through handheld devices. All exhibitions are frequently visited by classes and guided through certain relevant topics by a specialized pedagogic team.

Fig. 20.4 Painting of the northern gannet (*Sula bassana*) by Johann Conrad Susemihl for the publication *Teutsche Ornithologie* (1800–1817) at the left and historical original specimen used as model for the respective painting at the right (HLMD-A-528) (with permission from HLMD; photos: Wolfgang Fuhrmannek)

Fig. 20.5 Original painting of the "Jung-Koch-Quentell'sche Lehrtafel" on the anatomy of the honey bee (with permission from HLMD; photo: Wolfgang Fuhrmannek)

Fig. 20.6 View of the huge new showcase (16 m × 4.6 m) on species diversity on display in the permanent zoological exhibition since September 2014 and containing about 800 specimens (with permission from HLMD; photo: Wolfgang Fuhrmannek)

References

Hita Garcia F, Hawkes PG, Alpert GD (2014) Taxonomy of the ant genus *Proceratium* Roger (Hymenoptera, Formicidae) in the Afrotropical region with a revision of the *P. arnoldi* clade and description of four new species. Zookeys 447:47–86

Köhler J, Glaw F, Pabijan M, Vences M (2015) Integrative taxonomic revision of mantellid frogs of the genus *Aglyptodactylus* (Anura: Mantellidae). Zootaxa 4006(3):401–438

Krause R (1972) Zur Geschichte der Zoologischen Abteilung des Hessisches Landesmuseum Darmstadt 1795-1914. Institut für Naturschutz, Darmstadt, 64 pp

Munsch M, Schmiese H, Angelov A, Riedel G, Köhler J (2015) Conservative restoration and reconstruction of historical natural history dioramas. In: Tunnicliffe SD, Scheersoi A (eds) Natural history dioramas—history, construction and educational role. Springer, Dordrecht, pp 115–130

Schulze A, Kurbel M, Köhler J (2014) Zur Historie der zoologischen Dioramen im Hessischen Landesmuseum Darmstadt. In: Schulze A, Köhler J, Gruber G (eds) Naturkundliche Dioramen. Darmstädter Beiträge zur Naturgeschichte. Kaupia 19, pp 7–16

Chapter 21
DARMSTADT: The Zoological Collection at the Department of Biology/TU Darmstadt History and Present State of the Zoological Collection

Berend Koch

Abstract As the main part of the zoological collection at the Department of Biology/TU Darmstadt had been destroyed during World War II, a rebeginning started in the 1950s and 1960s. After a slow expansion during the following 25 years, the collection has experienced a permanent and substantial increase since 1988. Nowadays, it consists of more than 1500 vertebrate preparations, mainly European birds and mammals, completed by more than 20,000 invertebrate preparations, mainly arthropods and molluscs. Scientific models and a large bird egg collection are further parts of it. Several vertebrate exhibits are shown in display cases, but a large part is stored in a shelf system and in steel lockers. The collection plays an important role for the education of bachelor students, but it also offers traineeships for apprentices in taxidermy profession and for interested pupils.

Keywords TU Darmstadt • Zoological collection • Vertebrate and invertebrate preparations • Scientific models • Egg collection • Exhibition

Today, the zoological collection with exhibition and taxidermy is a central institution at the Department of Biology of the Technische Universität Darmstadt. The old Institute for Zoology with the collection was located in the loft of the university's main building before it was moved to a casern building, also under the roof, in 1940. In World War II, the main air raid on Darmstadt was in the night of the 11/12 September 1944, when over 70 % of the buildings were totally destroyed, including the Institute for Zoology (Information from: W. E. Ankel, 1945: Das Zoologische Institut Darmstadt. Nekrolog, der Erinnerung an Margit Gaydoul, unseren Lehrling, gewidmet). After the war, most of the Darmstadt Institutes were rebuilt. So, although the collection includes some older exhibits (nineteenth

B. Koch (✉)
Fachbereich Biologie—Zoologische Sammlung, Technische Universität Darmstadt, Schnittspahnstrasse 10, 64287 Darmstadt, Germany
e-mail: koch-b@bio.tu-darmstadt.de; http://www.bio.tu-darmstadt.de/zoologischesammlung

L.A. Beck (ed.), *Zoological Collections of Germany*, Natural History Collections,
DOI 10.1007/978-3-319-44321-8_21

century), it is relatively young, especially in comparison with other zoological collections. Huge parts of the collections are dated in the 1950s and 1960s of the twentieth century when the Institute for Zoology of the Technische Hochschule Darmstadt (the renaming in Technische Universität Darmstadt took place in 1997) was located in the Roßdörfer Straße 140.

In 1976, the Institute for Zoology with the collection moved into the current location, a new building at the campus "Botanical Garden", Schnittspahnstraße 10, 64287 Darmstadt, Building B2/03, first floor. After the move, the collection expanded only partially, especially in the part of invertebrates and models. But also skeletons, skulls and mounted birds and mammals were purchased or made by Bernd Amelunxen, the employed taxidermist from 1974 to 1987. In the middle of the 1980s, a small exhibition for supporting the zoology studies was established.

Since 1988, with the employment of the current taxidermist and collection manager, Berend Koch, the collection expanded permanently, especially the bird and mammal collections. In addition, the workshop of the Department of Biology designed and produced seven modern large display cases between 1998 and 2014 (Figs. 21.1 and 21.2). Together with eight commercially produced display cases (Figs. 21.3 and 21.4), it is a worth seeing exhibition. With the substantial renovation 2010/2011, the attractiveness of the exhibition was obviously heightened. With the displayed exhibits of the zoological collection, our students and the interested public as well have the possibility for a fascinating view into biodiversity. The exhibition is also a very useful support for the biology studies' module "Biodiversity and Phylogeny", showing not only animals but giving also information about the natural range of the displayed species in Europe and worldwide.

The collection room (42 m^2) and the taxidermy laboratory (29 m^2) with an included office cabin were also substantially renovated in 2010. After installation of a mobile shelving system, the collection can be used more effectively and the service is much easier. Materials for the practical courses are stored in steel lockers on the ground floor.

Today, the collection consists of about 25,000 exhibits, including ca. 1500 vertebrate preparations and a bird egg collection with ca. 7000 items. The main focus is on European birds (full mounts, study skins, skulls/skeletons) and mammals (full mounts, study skins, skulls/skeletons), but also wet preparations of reptiles, amphibians and fishes. The invertebrates are mainly dry or wet preserved arthropods and molluscs. Apart from the local insect fauna, we have small collections from alpine and Mediterranean regions. A collection of scientific models completes the educational aspect.

The main part of the collections (vertebrates, wet preparations, microscopic preparations) has been digitally recorded and has been transferred to a database.

Because of the departments' main focus "Synthetic Biology", internally the collection is mainly used for education in the bachelor studies but we promote some extern research projects on taxonomy and biodiversity from time to time. It is also possible for museums or institutions to get items out of the collection for research, educational purposes or special exhibitions temporary as a loan.

The zoological collection and especially the taxidermy play an important role in the public relation of the Department of Biology.

Fig. 21.1 Shorebird Mounts

Independent on the lack of infrastructural circumstances like constant temperature and humidity, the quality of conservation of the exhibits is the most important component of a collection. To be up to date with conservation techniques or to develop new ones, e.g. new methods for micro mammal taxidermy, is essential. It is always the goal to improve the quality of the mounts (gold medals in World and European Taxidermy Championships). To share this knowledge and to keep the future of the collection quality in mind, traineeships are offered for apprentices in taxidermy profession and also for interested pupils. Presentations about taxidermy are offered for whole school classes.

Fig. 21.2 Owl Mounts

Fig. 21.3 Finch Mounts

Fig. 21.4 Canid Mounts

References

Publications

Koch B (1991) Eine Methode zur Umarbeitung alter Vogelpräparate. Der Präparator 37 (3):116–121

Koch B (1993) Die Präparation von Vogelfüßen—ein oft vernachlässigter Arbeitsschritt. Der Präparator 39(3):101–104

Koch B (1998) Die Herstellung der Replik des Mecklenburger Pfeilstorches. Der Präparator 44 (3/4):99–104

Koch B (2001) Dermoplastik—auch eine Alternative in der Vogelpräparation. Der Präparator 47 (3):97–109

Koch B (2002) Mounting a European white-tailed eagle. Breakthrough Magazine Issue 67 (Spring):60–71

Koch B (2005) Keine Angst vor kleinen (Säuge)Tieren—Kleinsäugerpräparation am Beispiel von Fledermäusen (Chiroptera). Der Präparator 51:54–61

Koch B (2006) No fear of mounting small or micro-mammals. Breakthrough Magazine Issue 85 (Winter):61–69

Oral Presentations

Koch B (1991) Eine Methode zur Umarbeitung alter Vogelpräparate. 30. Internationale Arbeitstagung des Verbandes Deutscher Präparatoren e. V., Marburg

Koch B (1998) Die Herstellung der Replik des Mecklenburger Pfeilstorches. 37. Internationale Arbeitstagung des Verbandes Deutscher Präparatoren e. V., München

Koch B (2001) Dermoplastik—auch eine Alternative in der Vogelpräparation. 40. Internationale Arbeitstagung des Verbandes Deutscher Präparatoren e. V., Manderscheid

Koch B (2003) Seminario su calco e realizzazione di Aquila di mare. 4° Concorso Nazionale di Tassidermia, Longarone/Italy

Koch B (2004) Conservazione delle pelli degli uccelli, tramite concia (Seminar 1), Creazione di una replica della cicogna con freccia di Mecklenburg (Seminar 2). 5° Concorso Nazionale di Tassidermia, Longarone/Italy

Koch B (2005) Keine Angst vor kleinen (Säuge)Tieren—Kleinsäugerpräparation am Beispiel von Fledermäusen (Chiroptera). 44. Internationale Arbeitstagung des Verbandes Deutscher Präparatoren e. V., Augsburg

Koch B (2007) Tecnica di preparazione di un pipistrello con metodo PEG. 7° Concorso Nazionale di Tassidermia, Longarone/Italy

Koch B (2008a) Don't be afraid mounting small or micro-mammals. World Taxidermy Championships WTC®, Salzburg, Austria

Koch B (2008b) Fondamenti di anatomia nella preparazione degli uccelli. 8° Concorso Nazionale di Tassiderma, Longarone/Italy

Koch B (2009) Consigli per la scelta e l'applicazione di occhi artificiali nella preparazione degli uccelli. 9° Concorso di Tassidermia, Longarone/Italy

Koch B (2011) (New) techniques to solve common problems in the field of taxidermy. Convention "The Art of Taxidermy in Natural History Museums", Museo Civico di Zoologia di Roma

Koch B (2014a) Grundlagen der Vogelanatomie in der zoologischen Präparation. Arbeitstreffen Landesgruppe Bayern im Verband Deutscher Präparatoren e. V., München

Koch B (2014b) Die Kunst der zoologischen Präparation. Eröffnungsvortrag beim Kulturhistorischen Verein Roßdorf im Südhessischen Handwerksmuseum

Chapter 22
DETMOLD: Lippisches Landesmuseum Detmold

Dorothee Suray

Abstract The zoological collection at the Lippisches Landesmuseum Detmold is the only collection in Eastern Westphalia (Ostwestfalen-Lippe) with an emphasis on non-European animal species. Many native species were only incorporated into the collection in the second half of the twentieth century. The museum was founded in 1835, and within a few decades, an extensive collection comprising zoological and botanical preparations, fossils and minerals from Lippe and overseas was created. Today's zoological collection still contains extremely valuable and irretrievable old mounted animals, altogether over 4000 species. Among them are species which are already or will be extinct in the foreseeable future. The natural history permanent exhibition was reopened in 2009 and comprises five segments: collection's history, non-European animal world, central European animal world, palaeontology and mineralogy containing also the collection of meteorites.

Keywords Non-European animals • Central European animals • Carl Weerth • Philipp Leopold Martin • Eastern Westphalia • Detmold

22.1 General Information

The museum was brought to life in 1835 as a foundation of the scientific association (Naturwissenschaftlicher Verein für das Fürstenthum Lippe) founded in the same year by 40 citizens of Detmold. Through private bequests the "Naturaliensammlung für das Fürstenthum Lippe" grew especially over the course of the nineteenth century and the first two decades of the twentieth century. Owing to the excellent contacts of the first museum leader Dr. Carl Weerth (1812–1889), an extensive collection comprising zoological and botanical preparations, fossils and minerals from Lippe and overseas was created within a few decades. From the early 1880s onwards, particularly diplomats, administrative officials and officers serving in the colonies donated single objects or whole collections to the museum in Detmold. Additionally, some pieces were purchased such as some Indonesian birds from the

D. Suray (✉)
Ameide 4, 32756 Detmold, Germany
e-mail: mail@lippisches-landesmuseum.de; http://www.lippisches-landesmuseum.de

L.A. Beck (ed.), *Zoological Collections of Germany*, Natural History Collections,
DOI 10.1007/978-3-319-44321-8_22

expedition yield by Dr. Platen (1843–1899). Due to the collection's history, the zoological collection at the Lippisches Landesmuseum is the only collection in Eastern Westphalia (Ostwestfalen-Lippe) with an emphasis on non-European animal species. Many native species were only incorporated into the collection in the second half of the twentieth century.

The museum sustained no damages throughout World War I. Nevertheless, after the war the scientific association applied for state aid to maintain the collections. Since the 1880s, these had grown considerably through art-historical and cultural-historical "antiquities". The negotiations concluded with the transfer of all collections to the Free State of Lippe which was established after World War I in 1919. Thus the association's museum became a state museum in 1919 and a district museum in 1949 when the Free State of Lippe became a part of the newly founded federal state North Rhine-Westphalia. In comparison, World War II had a more substantial influence on the museum's routine. Between 1935 and 1945, the museum had to cede showrooms to national-socialist institutions (Lehr- und Forschungsanstalt für Germanenkunde), and in 1940 the house was closed to the public. The director at the time, Oskar Suffert, had been obliged to join the military service. When the air raids in the area increased (e.g. on nearby Bielefeld and Lage), the zoological collection was stored in the museum's cellar and in the air-raid shelter of the grammar school. Fortunately, Detmold and the museum were mostly spared from bombs. However, the collections suffered some damages from lootings and, above all, pests. It was impossible to keep those voracious insects in check due to the tightly packed and improper storage. As a result, the 1960s saw a huge number of mounted animals destroyed due to severe damages caused by insects, including specimens mounted by the well-known taxidermist Philipp Leopold Martin (1815–1885). But nevertheless today's collection still contains extremely valuable and irretrievable old mounted animals. Among them are numerous species which are already or will be extinct in the foreseeable future. Examples include the passenger pigeon (*Ectopistes migratorius*), the Pitt shag (*Phalacrocorax featherstonii*), the kakapo (*Strigops habroptilus*), the Juan Fernandez firecrown (*Sephanoides fernandensis*), the aye-aye (*Daubentonia madagascariensis*), the southern pudú (*Pudu puda*) and the platypus (*Ornithorhynchus anatinus*).

As an up-to-date compilation of the collection's inventory is absent, the following table (Table 22.1 and 22.2) contains predominantly estimated numbers:

Together with the botanical, palaeontological and mineralogical collections, the zoological collection is housed in its own building within the building complex of the Lippisches Landesmuseum which is sustained by the regional association of Lippe (Landesverband Lippe). The natural history permanent exhibition was reopened in 2009 after 2 years of extensive refurbishments. Apart from installing air-conditioning in the exhibition and magazine rooms, interactive elements and a lecture theatre were added. The new exhibition comprises five segments: collection's history, non-European animal world, central European animal world, palaeontology and mineralogy containing also the collection of meteorites.

Table 22.1 Extent of the zoological collection of the Lippisches Landesmuseum

	Number
Insecta	6000 specimens
Conchifera	1000 species (from which ~ 120 European Gastropoda)
Echinodermata	100 species

Table 22.2 Extent of the zoological collection of the Lippisches Landesmuseum

	Species number of mounted specimen	Species number of skulls/skeletons
"Pisces"	15	6
Lissamphibia	8	3
"Reptilia"	50	6
Aves	900	50
Mammalia	200	125

22.2 Research

The zoologist Dr. Friedrich Goethe was the first scientist employed exclusively for the zoological collection. He served as a scientific curator and nature conservation representative for the former administrative district of Detmold from 1946 to 1951. Subsequently he took up the position as the scientific director of the Institute of Avian Research in Wilhelmshaven (Institut für Vogelforschung "Vogelwarte Helgoland"). The palaeontologist Prof. Dr. Rainer Springhorn followed Goethe as a curator of the natural history collection in 1977 and served as the museum's director from 1986 to 2011. From 2009 to 2011, there was a biologist as a trainee at the natural history department. Since 2011 the Lippisches Landesmuseum does not have a resident scientist curating the natural history department. Hence, the latest research and publications on the zoological collection date back some years:

- Springhorn R (2012) Die Naturkundesammlung des Lippischen Landesmuseums Detmold blickt auf eine 175-jährige Geschichte zurück. Natur im Museum 2:38–39
- Springhorn R, Suray D (2009) Älteste Naturkundesammlung Nordrhein-Westfalens im neuen Glanz: Das Lippische Landesmuseum Detmold vor seinem 175. Geburtstag. Museum Aktuell 155:105–106
- Springhorn R, Suray D (2008) Eine Arche Noah im Teutoburger Wald: Das Lippische Landesmuseum Detmold bereitet sich auf seinen 175. Geburtstag vor. Lippische Mitteilungen aus Geschichte und Landeskunde 77:262–287
- Springhorn R (ed) (2007) Lippisches Landesmuseum Detmold: Die Schausammlungen. Deutscher Kunstverlag, München, Berlin
- Springhorn R (2004) A skull of the Caspian Tiger from the estate of the Russian Scholar Constantin von Scharnhorst (1846–1908) in the collection of the Lippisches Landesmuseum Detmold: Research history and a review of the

subspecie's extinction history. Lippische Mitteilungen aus Geschichte und Landeskunde 73:323–350

- Springhorn R (2003) A wild horse (Equus przewalskii Poliakov 1881) of Mesolithic Age from Kempen (Germany, Northrhine-Westfalia, Lippe-County). Eiszeitalter und Gegenwart 52:40–46, Stuttgart
- Springhorn R (1985) 150 Jahre Naturwissenschaftliche Sammlung und Naturhistorisches Museum in Detmold. Heimatland Lippe 78(8/9):287–314
- Springhorn R (1985) Die frühe Phase der Naturhistorischen Sammlung des Lippischen Landesmuseums: Stiftungen aus der Zeit von 1835–1891. Lippische Mitteilungen aus Geschichte und Landeskunde 54:85–107
- Springhorn R (1980) Führer durch die Naturhistorische Abteilung des Lippischen Landesmuseums Detmold. Topp + Möller, Detmold
- Springhorn R (1980) Pferdereste eines wikingerzeitlichen Siedlungsplatzes bei Müssen (Lippe). Zeitschrift für Tierzüchtung und Züchtungsbiologie 97 (2):116–126
- Fleischhack E (1974) Aus der Frühgeschichte des Lippischen Landesmuseums: Ferdinand Freiligrath und Georg Weerth als Aufbauhelfer. Lippische Mitteilungen aus Geschichte und Landeskunde 43:145–166
- Trumler E (1958) Das Burchell-Zebra des Lippischen Landesmuseums in Detmold. Lippische Mitteilungen aus Geschichte und Landeskunde 27:239–256
- Goethe F, Zippelius HM (1951) Ein dreibeiniger Fischreiher (Ardaea c. cinera L.). Columba 3:29–30
- Goethe F (1950) Die Vogeleiersammlung des Lippischen Landesmuseums. Mitteilungsblatt des Verbandes Lippische Heimat 1:105–106
- Weerth O (1922) Führer durch die zoologische Sammlung des Lippischen Landesmuseums. Meyer, Detmold
- Thorbecke H (1875) Detmold und der Teutoburger Wald mit dem Hermannsdenkmale. Meyer, Detmold, pp 7–9
- Martin PL (1870) Die Praxis der Naturgeschichte: Ein vollständiges Lehrbuch über das Sammeln lebender und toter Naturkörper; deren Beobachtung, Erhaltung und Pflege im freien und gefangenen Zustande; Konservation, Präparation und Aufstellung in Sammlungen, 2. Teil: Dermoplastik und Museologie oder das Modelliren der Thiere und das Aufstellen und Erhalten von Naturaliensammlungen. Voigt, Weimar, p 6
- Weerth C (1862) Das naturwissenschaftliche Museum in Detmold. Programm des Gymnasium Leopoldinum
- Weerth C (1840) Bemerkungen über den gegenwärtigen Zustand der Sammlungen des naturwissenschaftlichen Vereins im Fürstenthum Lippe. Lippisches Magazin 5(18):314–318

22.3 Educational Work

Both the non-European animal world and those of central Europe are displayed within spacious close to nature scenarios. The non-European wildlife is introduced under the motto "Exotic and Dramatic". The animals are displayed within naturally designed landscapes of the polar regions, the tropical rain forests and the grass ecosystems. A physical map of the world forms the floor of the showroom. On this map pictures assign some of the displayed animals to their respective biogeographic region. A hidden chainsaw between a jaguar and tapir and pictures of a melting icicle shown on a screen in the polar region serve as potential triggers for considerations of possible effects of environmental destruction and climate change on natural habitats. The fauna of central Europe is presented under the motto "Wild and Familiar" and includes native species still—or again—living in our woods and on our lakes as well as species deliberately or unintentionally naturalised in our region. The lifelike presentation of the native and exotic flora and fauna is supported by interactive media applications. These enable the visitor to further explore the displayed animal species through more detailed information about their biology and their habitats such as the Senne in Lippe or tropical rainforests. Once the user chooses an animal on the touchscreen she/he hears the animal's voice simultaneously to lighting effects emphasising the chosen animal in the glass cabinet. Short films about special topics of interest are offered on big screens. Smaller screens within the glass cabinets provide further information through slide shows. Besides the mounted animals in the glass cabinets, there is a large aquarium with living cichlids in the entrance area. The aquarium is maintained by the German cichlid association (Deutsche Cichliden-Gesellschaft e.V.). The collection's history is presented in a room fitted out like the showroom described in a tourist's guide of Detmold from 1875 (Thorbecke 1875). This enables the visitor to imagine the beginnings of the museum at a time when the world was full of strange animals and exotic landscapes to be explored. With the help of a globe and a screen (controlled via touchscreen) the visitor may discover the remote areas all over the world from which the specimens originated and the people who have so effectively supported the museum at home.

The Lippische Landesmuseum co-operates with two elementary schools in Detmold. Within the scope of the all-day school (Offene Ganztagsschule, OGS), the pupils are able to partake in museum-educational offers on different subject areas. In addition, the museum offers several worksheets and guided tours for children to explore special topics of the exhibition.

Chapter 23
DRESDEN / THARANDT: Zoological Collections Housed at the Institute of Forest Botany and Forest Zoology of Technische Universität Dresden

I. Brunk, C. Scheibner, J. Zschille, and M. Roth

Abstract The objectives and principles of zoological collecting at the Institute of Forest Botany and Forest Zoology of Technische Universität Dresden are presented. Since nearly 200 years, zoological objects have been systematically collected in Tharandt. The history of the foundation, as well as information about acquisitions, important sources and donations, and the scientific and educational significance of collections housed at the Institute are described in detail. Currently the collection consists of 2331 inventoried objects (birds, mammals, alcohol-preserved samples/ objects and some others) and additional approximately 50,000 specimens of insects, stored in 350 drawers. The collection is now well preserved in modern cabinets, in a separate room. Inventory continues, but still larger parts of the collection remain uninventorised in detail. The collection is frequently used for teaching purposes, but also for comparison and determination, documentation and research.

Keywords Technische Universität Dresden • Tharandt • Forest sciences • Zoology • History

23.1 Historical Background

The Forest Academy of Tharandt was founded in the year 1816. Since this year different aspects of zoology became a basic and central part of the curriculum. The exact founding dates of the first Tharandt collections are not available, but in a time

---

I. Brunk (✉) • C. Scheibner • J. Zschille • M. Roth
Fakultät Umweltwissenschaften, Fachrichtung Forstwissenschaften, Institut für Forstbotanik und Forstzoologie, Technische Universität, Dresden, Pienner Straße 7 (Cotta-Bau), 01737 Tharandt, Germany
e-mail: Ingo.Brunk@forst.tu-dresden.de; http://tu-dresden.de/die_tu_dresden/fakultaeten/fakultaet_forst_geo_und_hydrowissenschaften/fachrichtung_forstwissenschaften/institute/forstbotanik/zoologie

L.A. Beck (ed.), *Zoological Collections of Germany*, Natural History Collections,
DOI 10.1007/978-3-319-44321-8_23

of general establishing of systematic collections at universities (e.g. Tübingen, Kiel, Graz, München, Jena, Frankfurt) very soon after founding, the systematic collection of several zoological and other academic objects started. According to Geiler (1968) a first budget of 600 Taler—one and a half times the yearly salary of a university professor (Schuster 2013)—was mobilised for a first acquisition of academic objects, including zoological collections. In 1829 C. L. Krutzsch (1772–1852) established the general insect collection, and in 1830 E. A. Roßmäßler (1806–1867) took over the zoological courses. Somewhat later large amounts of money were spent for a systematic enlargement of the zoological collections (1840–1842: 400 Taler/year, 1843–1845 500 Taler/year (Schober 1866, or up to the year 1850 according to Geiler 1968) and in 1843 1000 additional Taler, Schuster 2013). In April 1849 the collection moved to a new building, and in 1850 it consisted mainly of a mammalian (34 species, 49 objects) and an avian collection (224 bird species, 380 bantlings, additionally eggs and nests) and 34 drawers of insects (2000 species, approx. 4000 pinned specimens). Several additional vertebrate skulls and skeletons, some molluscs and other objects (fishes, amphibians, reptiles, etc.) preserved in alcohol, and a few relevant botanical objects were part of the collections (Geiler 1968). In 1856 M. Willkomm (1821–1895) added his private collection of molluscs as a gift and established during his years of teaching a special collection of insect pests, including larval and pupal stages as well as illustrations of their biology. In 1866 the collection further increased, especially by additional spiders, crustaceans and some few worms (Nitsche 1881). During these years the collection suffered heavy damages due to pest beetles, resulting in the complete destruction of many objects. In 1876 Dr. H. Nitsche (1845–1902) established the Zoological Institute and during this time (up to 1902) much effort was put on conservational issues, including the discarding of many of the destroyed objects (mainly insects and mammals) and the reorganisation of the collections. Much emphasis was placed to limit the collection to well-labelled objects of European origin (Nitsche 1881). Consequently, with some few exceptions tropical species and marine animals (especially fishes and whales) were excluded.

Also historically important objects, like specimens from the avian collection of Christian Ludwig Brehm suffered from pest beetles. The emphasis was on preserving as many birds as possible. Today five objects from the Brehm collection are still part of the collection (Table 23.1).

Under the guidance of Nitsche also, the collection of skulls of birds and mammals was extended. Nitsche was also interested in subfossils. In 1883 he found some prehistoric fragments of aurochs (*Bos primigenius*) and sheep (*Ovis aries*) in a batcave. This intention further benefited from a donation of Mr. (most likely Ethbin) Schollmayer (Ljubljana) comprising several valuable prehistoric mammalian skull fragments (e.g. Brown Bear (*Ursus arctos)* from Franconian Switzerland (Gailenreuther Cave) and Ljubljana).

In 1880 the complete collection of fishes shown at the "Internationale Fischereiausstellung" in Berlin and a unique collection of freshwater pearl mussels (*Margaritifera margaritifera*, four drawers and many other specimens preserved in spirit) from upper Elster region were donated to Tharandt collection.

In 1888 T. Beling donated his important collection of Elateridae larvae (mounted and microscopic slides) to the institute. This collection contained much more species, than his late collection, donated to the Herzogliche Naturhistorische Museum zu Braunschweig.

A small building was used to rear fishes (more than 13 species) and a separate room—in Germany well known as "Karzer"—was afterwards used to rear beetles.

Due to their bad preservation status, the old collection of arthropods (mainly insects) housed in 80 drawers was divided by Nitsche into a general collection of European material used for study (taxonomic research) purposes only and a second one used for teaching purposes. This division is still in use, and every year objects collected by students are added for the latter one.

In 1886 a conservator and later a scientific technician were employed to work in the zoological collections. Especially the first scientific technician deserves to be named here: William Baer (1867–1934). The numerous small drawers made by Baer gave systematic overviews of taxonomic groups (e.g. Coleoptera, Hymenoptera, Diptera, Microlepidoptera), visualised modifications as well as the biology and development (egg, larval and pupal stages) of many specimens. All of Baer's stunning drawers are still in a perfect state and are shown year by year in current teaching courses. Till his retirement in 1930, the work of Baer was especially supported by K. Escherich (1871–1951) and later by H. Prell (1888–1962). The latter one dedicated a gall midge species (*Cecidomyia baeri*) to him and wrote an empathetic necrology (Prell 1934).

During the time of W. Baer, the important insect collections of H. Wiesner, C.F. Sommer, F.W. Fritzsche (see Table 23.1) and some others were integrated (Gisela Förster, pers. comm.), building up the historical core of the collection. This part of the collection—complemented by several more objects and smaller collections of other persons—is still preserved.

After the death of Nitsche, the institute had several directors, especially in the first two decades of the twentieth century. Karl Escherich's directorship lasted only 6 years (1907–1914), but was judged by himself as the most eventful and fruitful years of his life (Geiler 1968).

H. Prell became director of the institute in 1923. During his time research on fur-bearing animals was intensified, but he also published more than 300 papers on entomological, theoretical and applied ecological topics.

During World War II Tharandt was not affected by war destruction, but only few professors remained and the institute was meant to close down (Lochmann unpubl.).

Table 23.1 Objects from collection of C.L. Brehm preserved in Tharandt collection

Common name	Latin name	Object	Locality
Atlantic puffin	*Fratercula arctica*	Bantling	Bremerhaven
Eurasian three-toed woodpecker	*Picoides tridactylus*	Stuffed, winter coat	Kärnten
Ortolan bunting	*Emberiza hortulana*	Stuffed, spring coat	Görlitz
Meadow pipit	*Anthus pratensis*	Stuffed, spring coat	
Mistle thrush	*Turdus viscivorus*	Stuffed, winter coat	

The institutes were protected from plundering by soviet commandants (Schuster 2013).

In 1946 the faculty of forest science in Tharandt was reopened as a part of Technische Universität Dresden and as one of the only three faculties.

In 1957 H. Prell retired (at least he was the acting head of the institute up to August 1958), and some difficult years begun. In 1963 the faculty of forest sciences in Eberswalde was exterminated. In this year some staff members of Eberswalde (e.g. H. Gäbler, F. Kost, W. Bassus) as well as a few collections were moved from Eberswalde to Tharandt, but they were not included into the collection of the institute (Förster, pers.comm.). During this time a separate collection was established at the Institute of Silviculture and Forest Protection. Due to this restructuring of the faculty, resulting in a lack of storing space, some larger parts of the collection, mainly birds and Nitsche's collection of deer (Cervidae) were given away to Waldmuseum im Schloss Grillenburg. Other parts of the collection, for example, some less well-preserved bird bantlings, were simply discarded.

In the years under the directorship of H. Jahnel (1958–1961), H. Geiler (1961–1975) and W. Bassus (1975–1994) the collection was preserved, but not substantially extended. Thanks to the immense work of Gisela Förster of 25 years in caring, preserving, organising and inventorying, we still have the historical core of the collection in a very fine preserved state.

In 2002, the large flooding of the river "Wilde Weißeritz" was a big census for the institutes' work, resulting not only in a loss of minor parts of collection materials (skulls, nests, but especially subfossil fragments of several large animals), which were deposited in a completely flooded separate building, but led also to the reorganisation of the collection. With the help of financial support by the Federal Ministry of Education and Research, new equipment and especially modern cabinets were purchased, and since 2013 most parts of the collection conjoin in a single collection room. But caused by future restrictions in the availability of storage space, another consequence of the inundation was the transfer of some parts of the historical collection (butterflies and moths, eggs, birds and antlers, see Table 23.2) to Senckenberg Museum für Tierkunde Dresden. These parts included the important collections of C.F. Sommer and F.W. Fritsche, as well as some antlers collected during the time of Heinrich Nitsche. The oldest object was an antler of white-tailed deer (*Odocoileus virginianus*), shot in Ecuador in the year 1871 (Bauer and Stefen 2004).

The objects from Waldmuseum im Schloss Grillenburg (e.g. large stuffed objects and antlers) were donated in 2004 to LANU (Sächsische Landesstiftung für Natur und Umwelt), but this museum was closed in 2009. The objects returned first to Tharandt and since 2016 they are deposited in the Naturkundemuseum Potsdam.

Table 23.2 Important historical and actual parts of the collection

Collection	Patria	Objects	Drawers	Remarks
Historical collections	Europe	3500 bird eggs	20	Now part of main collection of SMTD
	Europe	343 mounted birds		Now part of main collection of SMTD
	Europe, Africa, Asia, South America	39 antlers		Now part of main collection of SMTD
	Europe	Antlers and large objects		Now LANU (Sächsische Landesstiftung für Natur und Umwelt)
	Europe	882 mammals (stuffed objects, bantlings, skeletons, skulls)		Incl. a polecat from 1877, main collection
	Europe	570 birds (stuffed objects, bantlings, skeletons, skulls, eggs, nests)		The oldest is a stuffed golden oriole (*O. oriolus*) from 1847
	Europe	Insects (all orders)	198	Main collection, including 34 drawers with specimens only for determination and lectures
	World	Mollusca		16000 objects
	World	697 objects of different animal orders (mainly alcohol-preserved objects and other skulls)		One of the oldest is a common pipistrelle (*P. pipistrellus*) from 1877
	Europe	43 mammal furs		
Beeling, T.	Europe	Elateridae: larva	2	Mounted Elateridae larvae and microscopic slides
Wiesner, H.	Saxony	Coleoptera Saxonia	36	Separate collection
Sommer, C.F. (1857–1899)	Saxony	9041 specimens of Lepidoptera	104	Now part of main collection of SMTD
Fritsche, F.W. (1811–1892)	Europe	3822 specimens of Lepidoptera	20	Now part of main collection of SMTD
Baer, W. (1867–1934)	Mainly Europe	Taxonomy, systematic and biology of different insects orders	116	Separate collections/ drawers

23.2 Inventory and Curation

A manual of inventory was written for some parts of the collection since 1860. Other parts of the collection were inventoried since, and inventorying continues. The listing is now completed for vertebrates and many other taxonomic groups, including all objects stored in alcohol. The inventory of insects is based on a raw estimation of the contents of the drawers, but it will still take some years to finish a full and detailed inventory. Actually emphasis is put on the inventory of the Wiesner collection.

Currently the collection consists of 2331 inventoried objects (birds, mammals, alcoholic objects and some others), and additional approximately 50,000 specimens of insects, stored in 350 drawers (Table 23.1, Figs. 23.1–23.3, 23.4 and 23.5, 23.6 and 23.7, 23.8–23.10, and 23.11–23.14) and other materials.

All bantlings and stuffed objects are frozen at −40 °C at regular intervals to prevent the infestation of parasites and pests. Also all new objects are deep frozen at least 2 weeks prior to being included in the main collection.

At the moment we get an annual small budget from custody of collections of Technische Universität Dresden, mainly used for the digitalisation of Wiesner's beetle collection.

All other work is done along with normal institute work, without any additional staff members.

Fig. 23.1–23.3 View into the collection room (*left*), some bird specimens belonging to collection of C.L. Brehm (*right top*), some alcohol-preserved objects (*right bottom*)

Fig. 23.4 and 23.5 Views into the cabinet: larger objects and alcohol-preserved materials (*left*), part of mammal collection, and some few large birds (*right*)

Fig. 23.6 and 23.7 Views into the cabinet: part of insect collection (*left*), part of collection of skulls and skeletons (*right*)

Fig. 23.8–23.10 Collection of Saxonian beetles from H. Wiesner (*left*), drawer of Wiesner collection (*right top*), drawer of historical insect collection (*right bottom*)

Fig. 23.11–23.14 Details of mammal skulls (*left top*), egg collection (*left bottom*), bird skeletons (*right top*), and old collection manuals (*right bottom*)

23.3 Research

There are few but regular requests on data and specimens for revision, especially for faunistic and taxonomic work (e.g. *Hybomitra*, *Meloe*). Some parts of the collection as well as single species have been revised by specialists. Material from the collection was revised for numerous contributions to "Insektenfauna der DDR" and later publications, e.g. Dr. L. Diekmann (Curculionidae), H. Nüßler, (Cerambycidae, *Carabus*), F. Hieke (*Amara*), Buprestidae (T. Kwast), some Scaraboidea (E. Rössner), several small families of Coleoptera (B. Klausnitzer) and several more.

At the moment there is no overview on whether primary or secondary type specimens are included in the collection, but it seems to be likely, as some publications of staff members were also related to new discovered species. But as part of the collection of C.F. Sommer (now part of Senckenberg Museum für Tierkunde Dresden) several primary types of Microlepidoptera are still waiting to be designated (M. Nuss, personal communication). In addition some of the historical specimens bear red labels, without further information, and need to be analysed in the future.

Skulls, bones and skeletons as well as avian eggs are regularly used for diet analysis of predators. For that, fragments found in scats, pellets or stomach contents were compared with these objects in order to identify prey species. Furthermore larger series of skulls were measured for studies on epigenetic variability of some Mammalian species.

Currently also furs are on a loan for a research project for genetical identification of furs (Projekt: "Molekularbiologisches Echtheitszertifikat für Pelzprodukte"— Forschungsinstitut für Leder und Kunststoffbahnen, Freiberg).

23.4 Current Use and Objectives

At least since the year 1850 (Bericht über die Forstakademie von 1850, Geiler 1968), it has been reported that especially mammalian skulls and skeletons were regularly used in several courses.

Due to the simple fact of having a well-sorted systematic collection with many useful objects available, the collection plays an important role in several academic courses, especially in zoological exercises. Starting taxonomically with worms (Nemathelminthes, Platyhelminthes, Annelida), several exercises focus on the specific characteristics of taxonomic groups of the reign of animals, using objects of the zoological collection. Due to the teaching profile of the Chair of Forest Zoology, much emphasis is put on the determination of arthropods (1 course), specialising later in determination of insect orders (1 course), Hymenoptera and Diptera families (1 course), Lepidoptera families (1 course), Coleoptera families (1 course) and Carabidae up to species level (1 course), followed by the determination of amphibs and reptiles (1 course), mammal skulls and stuffed objects (2 courses) as well as birds (2 courses).

From time to time, the collection room opens its door for a guided tour. Once a year kindergarten groups and school classes have the possibility to see objects and learn something about biology, ecology and taxonomy of selected animal groups. For school classes this can be part of a thematical week, where pupils study a specific biological group or topic.

A few of our objects are currently on a loan to extern exhibitions. A temporary exhibition connecting zoological objects with art is currently curated by the custody of the University and will be shown in a few weeks at Altana-Galery Dresden. Some freshwater pearl mussels are on loan at Museum Adorf, an important historical centre for production of pearl and mother of pearl objects. A drawer of W. Baer showing the biology of nun moth (*Lymantria monacha*) is loaned to SMAC Chemnitz for 5 years.

Acknowledgements We would like to thank the custody of TU Dresden for some financial support and Dr. A. Dwaronat (Bad Freienwalde), J. Faelt (Dresden), G. Förster (Tharandt), E. & K. Lochmann (Tharandt), Dr. A. Milnik (Eberswalde), Prof. Dr. U. Schulz (Eberswalde) and E. Schuster (Tharandt) for comments and informations.

References

Bauer F, Stefen C (2004) Übernahme einer Geweih- und Gehörnsammlung vom ehemaligen Forstzoologischen Institut der Technischen Universität Tharandt. Faunistische Abhandlungen 25:151–153

Geiler H (1968) Die Geschichte des zoologischen Unterrichtes in Tharandt seit dem Jahr 1816. Wissenschaftliche Zeitschrift der Technischen Universität Dresden 17:1727–1740

Lochmann E (unpubl.) Die Geschichte der Tharandter Lehr- und Forschungsstätte. unpubl. Manuscript, 6 pp

Nitsche H (1881) Mittheilungen aus dem zoologischen Institute der königl. Sächsischen Forstakademie zu Tharand. 1. Der zoologische Unterricht und die zoologische Sammlung an der Akademie Tharandt. Tharandter forstliches Jahrbuch 31:158–190

Prell H (1934) William Baer. Tharandter forstliches Jahrbuch:626–632

Schober H (1866) Zur Geschichte der Akademie für Forst- und Landwirthe zu Tharand. Tharander Jahrbuch 17. Festschrift zum 50. Jubiläum der Akademie 1866, 1 Abt:3–236

Schuster E (2013) Chronik der Tharandter forstlichen Lehr-und Forschungsstätte: 1811–2011. Selbstverl. der Fachrichtung Forstwissenschaften der TU Dresden, 398 pp

Chapter 24
DÜSSELDORF: Aquazoo Löbbecke Museum Düsseldorf

Sandra Honigs, Silke Stoll, and Elmar Finke

Abstract The Aquazoo Löbbecke Museum Düsseldorf is a combination of a natural history museum and a zoo. The two main roots of the institution reach back into the nineteenth century. The Aquazoo is mainly engaged in education, research, and documentation concerning conservation, species protection, ethology, reproductive biology, and evolution, as well as being a popular visitor attraction. In the exhibition, both traditional aspects of the institute are entirely interlaced. Evolution is the central overarching topic that dominates the 25 thematic exhibition rooms. These are equipped with interactive learning elements, lucid graphics, and concise texts explaining the live animals and museal objects. The live animal collection with about 450 species focuses on aquatic and terrarium animals. Furthermore, the exhibition includes approximately 1450 natural history exhibits. About 900,000 other objects are hidden in the museum archive. The collections of the house mainly include objects of zoology, botany, geology, human history, and underwater technology. The collection focus lies on malacology and entomology.

Keywords Malacology • Entomology • Geology • Aquarium • Terrarium • Zoo • Conservation • Research • Education

24.1 Historical Background

The Aquazoo Löbbecke Museum is a remarkable place combining a natural history museum with a zoo. Now inseparable, the two main roots of the institution reach back into the nineteenth century. A pharmacist by the name of Carl Heinrich Wilhelm Theodor Löbbecke (Fig. 24.1) founded a private museum focusing on malacology in the nineteenth century. His collection was handed over to the city of Düsseldorf by his widow in 1901, and as a result, the municipal Löbbecke Museum was built on the banks of the Rhine and opened to the public in 1904. Since then, the white marble bust of Theodor Löbbecke, carved by the sculptor Hammerstein, presides over the exhibition (Fig. 24.2). On the other hand, a zoological garden,

S. Honigs (✉) • S. Stoll • E. Finke
Aquazoo Löbbecke Museum Düsseldorf, Kaiserswerther Str. 380, 40474 Düsseldorf, Germany
e-mail: sandra.honigs@duesseldorf.de; http://www.duesseldorf.de/aquazoo

L.A. Beck (ed.), *Zoological Collections of Germany*, Natural History Collections,
DOI 10.1007/978-3-319-44321-8_24

Fig. 24.1 Theodor Löbbecke, pharmacist and founder of the museum (with permission from: Bildarchiv Aquazoo Löbbecke Museum)

Fig. 24.2 Marble bust of Theodor Löbbecke manufactured by the sculptor Hammerstein (with permission from: Bildarchiv Aquazoo Löbbecke Museum)

founded by a local animal welfare association, had existed since 1876 in the district of Düsselthal. In this place, the Löbbecke Museum was merged with the zoological garden in 1930, and it quickly became a popular and thriving institution—that is, until the Second World War broke out. The facilities sustained massive damage in the last years of the war and were abandoned.

Shortly after the war, in 1946, the "Löbbecke Museum & Aquarium" was reopened in a ground-level bunker at a place called Brehmplatz, where the remains of the zoo and what could be saved from the Löbbecke collections were put on display. Then, after about 40 years in the provisional "Zoobunker", the new building set in the public gardens of Nordpark was handed over to the public in 1987. In this place, both traditional aspects were entirely interlaced following a revolutionary concept of the director at that time, Prof. Dr. Manfred Zahn.

As one of the most popular cultural institutions of the North Rhine-Westphalia capital, it was a turning point to be closed to the public in 2013 for necessary renovation. After all, the building had been in operation for 25 years and needed technical improvement as well as a refurbishment of the animal facilities according to recent scientific standards. After 2 years of planning, the construction work started in 2014. The renovation shall be completed in 2017.

24.2 Recent Structure and Special Features

With its double name, the Aquazoo Löbbecke Museum will continue its tradition to be engaged in education, research, and documentation concerning conservation, species protection, ethology, reproductive biology, and evolution, as well as being a popular visitor attraction.

Residing in the middle of the Nordpark, the building was originally designed by the architects Dansard, Kalenborn und Partner as the center point in the middle of four ponds. Their design was based on the architectural models of the baroque ages as the park itself. In the center of the building under a dome made of glass, the tropical hall with crocodiles, tropical plants, turtles, and free-flying butterflies gives an impression of being in a tropical forest far away. Besides, typical tropical crops like coffee, cocoa, banana, pepper, guava, and vanilla are cultivated, thus emphasizing the importance of tropical areas for our daily life. Around this central dome, four wings of building segments radiate outward, representing marine, freshwater, and terrestrial habitats, respectively, supplemented by an exhibition area with arthropods and a display called "Seas and People." Each building segment has a central big enclosure or basin, like the huge Anton Lendle Reef with sharks, rays, and corals (Fig. 24.3) or the African penguin enclosure. In a gentle spiral, the visitors are routed through the impressive and perspicuous exhibition with 6800 m^2. Evolution is the central overarching topic that dominates the 25 thematic exhibition rooms. Each room is equipped with interactive learning elements, lucid graphics, and concise texts explaining the live animals and museal objects. By walking from room to room, visitors follow the history of life on earth (Fig. 24.4), starting in the seas, moving into freshwater, and finally into different terrestrial habitats. The most exciting is the presentation of the hurdles that had to be taken for life on earth: the formation and radiation of living beings, the colonization of freshwater habitats, and the capture of dry land. Young visitors are accompanied by mudskipper "Fred" and his friend "Theodor" (Löbbecke) with trick films, inspiring tips, and questions.

Fig. 24.3 The Copperband butterflyfish (*Chelmon rostratus*) is a typical coral reef dweller (with permission from: Bildarchiv Aquazoo Löbbecke Museum)

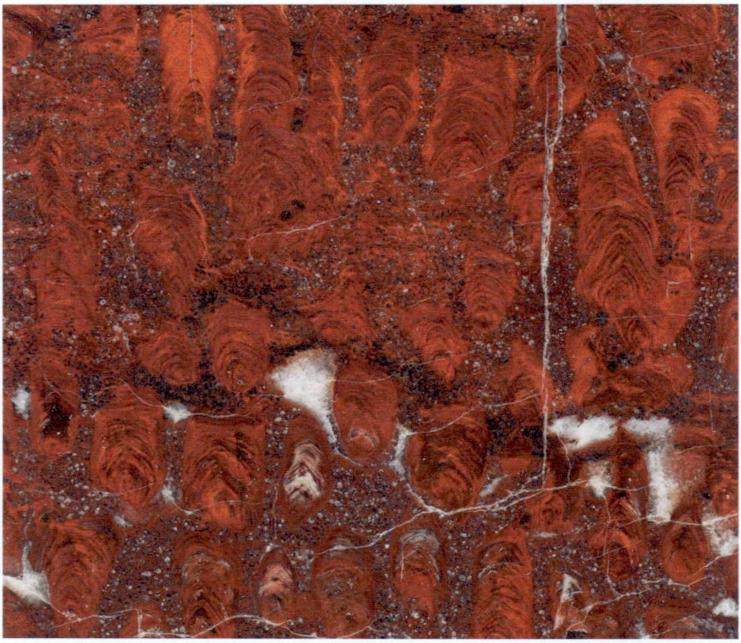

Fig. 24.4 One of the early pieces of evidence for life on earth: a stromatolite from Minnesota (with permission from: Bildarchiv Aquazoo Löbbecke Museum)

Furthermore, an audio guide system for adults (or in plain language) is available to get another kind of access to the exhibits.

The live animal collection focuses on aquatic and terrarium animals. More than 450 species, among them various invertebrates, fish, amphibians, reptiles, birds, and mammals, are listed in the files. According to the endeavor in species protection, rare and threatened species are kept and bred. Since 2008, the "Düsseldorf

Fig. 24.5 This Vietnamese mossy frog (*Theloderma corticale*) offspring is perfectly camouflaged on the moss cushion (with permission from: Bildarchiv Aquazoo Löbbecke Museum)

Amphibian Breeding and Conservation Center" funded by the municipal environmental office is settled in the Aquazoo Löbbecke Museum. Besides the breeding of rare amphibian species, the center is pursuing basic research on the general biology and especially on the reproductive behavior of amphibians (Fig. 24.5). The center aims also to generate positive publicity for this largely underrated group of animals. And, this initiative is accompanied by educational efforts with the aim of promoting their biodiversity.

24.3 Research

In addition, research projects are conducted and supported at the Aquazoo Löbbecke Museum in the fields of geology, entomology, malacology, veterinary medicine, and general zoology. The "Working Group for the Geology of the Rhine Region" and the "Entomological Society Düsseldorf est. 1886 e. V." are affiliated with and regularly meet at the institute. As a member of the European Association of Zoos and Aquaria (EAZA) and the World Association of Zoos and Aquariums (WAZA), the Aquazoo participates in numerous coordinated breeding programs and in the international scientific zoological exchange of experiences. In addition to species protection efforts including breeding programs and research, the Aquazoo is involved in in situ conservation projects as well. Twice a year, landscape protection activities are carried out for the benefit of the Mosel Apollo (*Parnassius apollo*), a rare domestic butterfly. Another project for the protection of native species is the EU Life+ project for the Allis shad (*Alosa alosa*), once a very frequent fish in the Rhine and a popular catch for local fishermen. The reintroduction project is coordinated by an external zoologist residing at the institute.

24.4 Educational Work

As an extracurricular place of learning, teaching content is offered for all grades and types of schools. Furthermore, the educational activities include information desks, animal presentations, guided tours, holiday programs for children, children's birthday parties, special events, outreach programs, teacher trainings, supervision of student research projects, and working groups in partner schools. Objects from the zoological collection of the education department are regularly loaned to schools in the Düsseldorf area for natural history education. Educational activities focus on the observation of animals, particularly in respect of ecological and functional morphological aspects (Fig. 24.6). Materials from the educational collection and live animals are used to support learning and promote a positive connection to nature.

The exhibition of the Aquazoo Löbbecke Museum includes approximately 1450 natural history exhibits (Fig. 24.7). Among them in the lobby, an impressive 17-meter-long skeleton of a Sperm whale (the skull alone weighing about 480 kg), a large ichthyosaur-fossil and the skeleton of an elephant seal are presented.

Fig. 24.6 Who would not be impressed by the observation of a Clearfin lionfish (*Pterois radiata*)? (with permission from: Bildarchiv Aquazoo Löbbecke Museum)

Fig. 24.7 Permian labyrinthodont *Eryops megacephalus* skeleton (with permission from: Bildarchiv Aquazoo Löbbecke Museum)

24.5 The Collections

About 900,000 other objects are hidden in the museum archive. Their documentation is not always complete, since the archive, and thus the documentation suffered losses during the Second World War. In some cases even both, the museum objects and the historical inventory books were destroyed. However, some objects are still being identified and "rediscovered" decades later on the basis of collection labels and catalogs. Losses and improper protective storage during the war, several moves, and little staff are hindering the assignments of several parts of the collections until nowadays. But, since the advent of computers, the two most important and thus the main collections of the museum—the malacological and entomological—had been digitally inventoried.

The collections of the house mainly include objects of zoology, botany, geology, human history, and underwater technology. In addition, there are extensive archive materials, such as photographs, sketchbooks, deeds, notes on the history of the museum and the zoo, their employees, collectors, researchers, sponsors, etc. Furthermore, a library with (partly precious) historical books, once the property of the museum's founder, is maintained on site (Fig. 24.8). The different parts of the collections are clustered by collectors or systematics, depending on the assignment possibility, extent, and place.

Fig. 24.8 Folios of the Ulisse Aldrovandi's *Historia Naturalis* (with permission from: Bildarchiv Aquazoo Löbbecke Museum)

The geological collection mainly contains minerals in high quality, as well as rocks and a huge number of fossils (Fig. 24.9). The mineralogical collection exceeds 3000 specimens, almost all of them generously donated by Franz Hönekopp (1904–1985). The collection is structured systematically according to Strunz and contains minerals of large, hand, and micro size in various compositions (Fig. 24.10), precious or semiprecious stones, and meteorites from all over the world. Rock and fossil samples were collected by major collectors like Joseph Boscheinen, Karl Heinersdorff, Franz Hönekopp, Karl-Heinz Josten, and Wilhelm Josten.

Fossils mediate between geology and biology. The fossil collection contains body fossils, trace fossils, inclusions, casts, and molds (Fig. 24.11). They are arranged according to geological periods, localities, and biological systematics. There are still many unprocessed materials; however, for each of these different collections, catalogs exist, which will form the basis for prospective data processing.

Botanical specimens were collected in typical herbaria, containing vascular plants, lichens, algae, as well as fungi. The collection is dominated by the

Fig. 24.9 Permian palm *Tietea singularis* trunk cross section, Brazil (with permission from: Bildarchiv Aquazoo Löbbecke Museum)

Fig. 24.10 Specimens presented in the mineral exhibition (with permission from: Bildarchiv Aquazoo Löbbecke Museum)

Fig. 24.11 Cretaceous belemnoid *Actinocamax mammillatus* rostrums, Ingaberga, Sweden (with permission from: Bildarchiv Aquazoo Löbbecke Museum)

Fig. 24.12 The teeth of a Tiger shark (*Galeocerdo cuvier*) are specialized to slice through flesh, bone, and other tough substances such as turtle shells (with permission from: Bildarchiv Aquazoo Löbbecke Museum)

herbarium of L. Rabenhorst and the photographic herbarium of Hans Hoeppner. The work on the herbaria is currently suspended.

The zoological collections are very extensive and diverse: there are bones, skeletons, skulls, dermoplastics, furs, skins, eggs, feathers, insect boxes (including eggs, larvae, molts, etc.), shells of mollusks, teeth, antlers, etc. (Fig. 24.12). The remains of the animals can be preserved as dry, frozen, or wet mounts. However, the heart of the museum is the malacological collection gathered by the founder of the museum

Fig. 24.13 Historical specimen containers from the malacological collection made of pillboxes (with permission from: Bildarchiv Aquazoo Löbbecke Museum)

Theodor Löbbecke (1821–1901). Important parts of this collection go back on Peter Wilhelm Ludwig Döring, Wilhelm Dunker, Erich Christian Ludwig Gruner, and Karl Emil Lischke. The database includes more than 300,000 specimens, including extremely valuable-type specimens (Fig. 24.13). The collection with its regional and global specimens is constantly being expanded by ongoing collecting activities.

The second important collection is the entomological compilation. It includes more than 500,000 specimens and is largely digitally recorded. This collection is constantly being expanded, processed, and inventoried (e.g., by the "Entomological Society Düsseldorf est. 1886 e. V."). The collection focuses on the old "Prussian Rhine Province" (on permanent loan from the "Association of Rhinish-Westphalian Lepidopterists") and the Palaearctic. Major collectors were Theodor Hildebrandt, Erich von Metzen, Karl Stamm, Ernst Lueg, Josef Schiffer, and Siegfried Löser. The collection is supplemented by some exotic insects and an extensive wet collection of arachnids.

The vertebrate collection includes trophies, skeletons and skulls, many parts of the legacy of Karl Heinrich Löbbecke (d. 1856), and the animal sculptor Josef Pallenberg (1882–1946). This collection is also being constantly updated and expanded.

The egg collection is primarily made up by the collections of the "Düsseldorf Natural History Society," Karl Heinrich Löbbecke, and Pierre Ghilain. The collection includes about 8000 eggs. This collection is resting but is well documented in various catalogs.

The coral collection originates from self-bred colonies, inherited collections, and customs' confiscations which have been entrusted to the museum. The customs occasionally hand over other quirky confiscations too—mainly from undeclared and illegal duty items, which have been attempted to be imported into Germany by clueless or unscrupulous passengers—for example, crocodile leather bags, ivory artworks, or entirely dried animals. They form part of the "special collection" which are objects that have been adopted but which are outside the main collection focus. This specially separated collection includes as well casts and sculptures from the heritage of the animal sculptor Josef Pallenberg; objects of human history, like stone tools (Fig. 24.14); ethnological objects; or glass slides.

Even a small technology collection is part of the archive. These objects have been purchased from the diving pioneers Hans Hass (1919–2013) and Kurt Schaefer (* 1922) and include underwater cameras (Fig. 24.15), fins, and breathing apparatus.

Fig. 24.14 Pleistocene hand ax (with permission from: Bildarchiv Aquazoo Löbbecke Museum)

Fig. 24.15 Historical diving camera case (with permission and photo credit: Franz Rothbrust)

Chapter 25
EBERSWALDE: Zoological Collections of Eberswalde: Like Phoenix from the Ashes?

U. Schulz, A. Linde, and J. Möller

Abstract Eberswalde was once famous for the entomological collection which was based on the acquisition of specimen from Ratzeburg (1801–1871) and Altum (1849–1900). But these collections were either lost, destroyed, or dispersed at the end of Second World War and in 1963 when the forest academy in northeastern Germany was closed down for political reasons. After the reunification of Germany, a new university was founded in Eberswalde, and since then, zoological collections have been established again: new collections in old rooms—rising like Phoenix from the ashes.

Keywords Entomological collection • Ratzeburg • Altum • Faculty of Forestry • Damages in World War II • Breakup in GDR • Reunification • New collections

25.1 The Famous Historical Zoological Collections of Eberswalde

Until 1830, foresters in Prussia were educated at the University in Berlin, until it was recognized that there was a "lack of a suitable forest nearby" (Dickel 1910). The Director of the Royal Forest Academy, Wilhelm Pfeil (1783–1859), was successful in relocating the Academy to the city of Eberswalde, 45 kilometers northeast of Berlin. In his approach, he was allegedly supported by his benefactor, Alexander von Humboldt (Schirm and Witzenhausen 1959; Schwerdtfeger 1983). The diversely forested area around Eberswalde was much more suitable for the study of forestry and the education of foresters; however, unlike in Berlin, there were no zoological collections for the practical education of students (Pfeil, cited in Dickel 1910). A few specimens for the newly established collection in Eberswalde

U. Schulz (✉) • J. Möller
Faculty of Landscape Management and Nature Conservation, Eberswalde University for Sustainable Development—HNEE, Schicklerstr. 5, 16225 Eberswalde, Germany
e-mail: Ulrich.Schulz@hnee.de; http://www.hnee.de

A. Linde
Faculty of Forest and Environment, Eberswalde University for Sustainable Development—HNEE, Alfred Möller Straße 1, 16225 Eberswalde, Germany

L.A. Beck (ed.), *Zoological Collections of Germany*, Natural History Collections,
DOI 10.1007/978-3-319-44321-8_25

Fig. 25.1 The founder of applied entomology J.T.C. Ratzeburg (*left*) and the Dean of the Eberswalde Faculty B. Danckelmann with a European Owl (*Bubo bubo*) from the historical zoological collection of Eberswalde in 1869 (historical fundus of the library of Eberswalde, HNEE)

were obtained from Berlin, and more were soon acquired by shooting local birds and mammals (Eckstein 1931). Most of the specimens, however, were derived from a zoological collection from Halle: In 1838, over 400 bird samples and nearly 500 bird eggs were purchased from the inheritance of the zoologist Nitzsch. Due to their fragility, these samples were transported to Eberswalde by ship, via the rivers Saale, Elbe, Havel, and the Finow Canal (Eckstein 1931). This activity was initiated by J.T.C. Ratzeburg (1801–1871; Fig. 25.1), who taught natural sciences at the Forest Academy in Eberswalde beginning in 1830.

From today's point of view, Ratzeburg can be considered to be the founder of applied entomology and inventor of forest entomology (Endtmann 2006). His collection of forest pest insects ("Waldverderber") and their natural enemies, among them newly described species of parasitic wasps (Ichneumonidae) and other Hymenoptera (Pteromalidae, Braconidae), established his scientific legacy. This collection contained many type species that generated several publications on taxonomy and can be regarded the basis for his forest zoology text books. Ratzeburg sold this collection in 1842 for a more symbolic price of 2100 mark to the Forest Academy in Eberswalde, where it remained until 1945.

An 1846 inventory in Eberswalde lists 806 bird and 60 mammal samples (Anonymus 1846). The continuous expansion of the collection was paramount for Ratzeburg and his successor, B. Altum (1824–1900; Fig. 25.2). Frequently, taxidermists from Berlin were called to Eberswalde to prepare new mammalian

Fig. 25.2 Ornithologist B. Altum and parts of the historical zoological collection (historical fundus of the library of Eberswalde, HNEE)

samples and to instruct students in dissection and conservation techniques even during the holiday season (Eckstein 1931). Altum donated many of his samples (birds, beetles, butterflies) which he had collected and prepared prior to his time in Eberswalde to the Forest Academy (Milnik and Rohlfien 2000). His collection of butterflies alone comprised over 7000 specimens (Milnik 2006a). According to Eckstein (1931), Altum was honored with the "red eagle medal" for the generous donation of his collection of bird eggs to the Forest Academy. Over a period of 10 years, Altum greatly enlarged this collection. According to Danckelmann (1880), the collection now consisting of 38,738 samples was transferred to the "New Forest Academy" building in 1876, where it required three rooms, especially a two-story "gallery room" (Fig. 25.3).

Three different collections were described: (1) "wildlife zoology" with antlers/horns, feet, and droppings of wildlife and so-called curiosities (in total around 300 samples; (2) "biological collection" with samples referring to the ecology of animals: precious specimen of plants with typical damages caused by mammals, birds, and insects in glass display cases, an entomological herbarium (in total around 800 samples, many dating back to the Ratzeburg collection); and (3) "systematic anatomical collection" of animals and animal parts (e.g., all native mammals with different fur colorations and age classes, all collected and preserved by Altum), many native birds (again varying in color and age; approx. 1200 samples), and around 2100 bird eggs. This was supplemented by a vast number of insects of

Fig. 25.3 The historical zoological collection of Eberswalde in 1930 (Hilf and Schubert 1930)

all orders (mostly from Ratzeburg's, Flotow's, Meyer's, and Altum's collection; around 7000 specimens) and a collection of other invertebrates designed to illustrate the systematics of the animal kingdom (around 780 samples). These numbers, as listed by Danckelmann (1880), are confirmed by handwritten inventories that exist in the historical collection of the HNEE library (Fig. 25.4). The samples were mainly cared for by Karl Eckstein (1859–1939), who was Altum's assistant since 1886 and his successor after 1900. Altum had instructed Eckstein to revise the inventory of the collection which he did over a period of 14 years (Eckstein 1931; Milnik 2006b). Eckstein, however, also added a lot of new samples to the collection—he personally collected new specimens, prepared the samples, and updated the registry. For example, in 1912, he purchased a collection of 494 species of Tenthredinidae from W. Horn. The zoological collection in Eberswalde grew continually, and Eckstein was able to offer samples to other forest academies in Hachenburg, Steinbusch, and Groß Schönebeck and to the universities in Berlin and Königsberg (Eckstein 1931). In 1936, while the National Socialists (NSDAP) were already in power, Eckstein left Eberswalde. A possible reason for his departure may have been alleged defamations by a colleague (M. Wolff) that he was of Jewish ancestry and that he was responsible for falsification of the institute's accounts (Milnik and Rohlfien 1999). Max Wolff was a professor of zoology in Eberswalde from 1914 to 1941. He did not seem to pay much attention to the collections; Schwerdtfeger (1985) reports that Wolff took over the collections from Eckstein in 1932 and managed to degrade it to a "lamentable condition" within a decade. Fritz

Fig. 25.4 Historical inventory lists of the former zoological collections of Eberswalde (© with U. Schulz)

Schwerdtfeger (1905–1986), with the help of assistants, restored the collections beginning in 1942; however, shortly thereafter, the Second World War eventually reached Eberswalde.

25.2 "The Ashes": The Decline of the Zoological Collection in Eberswalde (1945–1963)

In addition to the usual losses in zoological collections due to damages by insects and overzealous students, several dramatic incidents lead to the decay of the collection in Eberswalde. These were mainly direct or indirect damages by war and the closure of the Faculty of Forestry in 1963.

25.2.1 Damage by War: 1945

An aerial bombing raid on the nights of April 24 and 25, 1945, destroyed a part of the Institute for Forest Zoology and devastated the entomological collection (Gäbler 1964). Just prior to the raid, an attempt was made to save at least the famous Ratzeburg insect collection. In those last days of the Second World War, this collection was loaded onto a train in Eberswalde for transport to a presumed safe region in the south of Lower Saxonia. However, the wagon with the collection was destroyed during a bombing raid in Magdeburg (Königsmann 1964; Schwerdtfeger 1985). The fate of the Ratzeburg collection remained unclear. Schedl (1974) reports that he was contacted by persons from Eberswalde after the war, asking whether the collection was with him in Lienz (Tyrolia). As a matter of fact, a part of the Ratzeburg collection by chance remained intact. During the loading of the train in Eberswalde, some drawers, which were stored separately, were obviously overlooked and remained in Eberswalde. They were identified by Königsmann (1964) and donated to the German Entomological Institute (at that time in Berlin—Friedrichshagen; today in Müncheberg) by Gäbler in 1961 (Rohlfien 1979). Five drawers with hymenopteran specimens, among them type species of Ratzeburg, are now safe in Müncheberg.

The total extent of the losses to the collection in Second World War II is unknown. Besides the bombing and fighting in Eberswalde, parts of the collections were ransacked by Russian soldiers or transported to Russia (Ciesla and Joachim 2015). Schwerdtfeger (1985) reports that, immediately after the conquest of Eberswalde by the Red Army, the collections were used by Russian soldiers for "amusement and indulgence."

25.2.2 Flooding: 1946

On February 10–11 1946, just some months after the end of the Second World War, a serious flood in Eberswalde caused further damage to the collection. The small river Schwärze partially submerged the city of Eberswalde, including parts of the university library and the zoological collection (Augustiny 1995). This resulted in a further decimation of the number of zoological samples (Kost 2003).

25.2.3 Closure of the Faculty of Forestry and Breakup of the Zoological Collection: 1963

After the damage and destruction caused by war and flooding, the zoologists Walter Kruel and Hellmuth Gäbler began to pursue the reconstruction of the heavily decimated collection. They purchased the collection of Coleoptera of Mr. Fehse from Thale/Harz, a collection of Macrolepidoptera of Mr. Rieck from Hann. Münden, and a collection of Hymenoptera of Mr. Mertens from Naumburg, to name just a few examples (Gäbler 1964). In addition, the collection was enlarged by personal collections in the field. For this purpose, the zoological institute installed the first biological station in a protected area in the German Democratic Republic on the shore of lake Müritz (Schirm and Witzenhausen 1959). The collection of

Fig. 25.5 Albinotic birds (*Dendrocopos major and Pica pica*) in the historical zoological collection of Eberswalde (Schieferdecker, in Gäbler 1963) lost after 1963

birds (e.g., albinos; see Fig. 25.5) and mammals in Eberswalde had survived the war in much better shape and was constantly enlarged (Gäbler 1964).

All these activities to supplement and renew the collections came to a sudden end in 1963 when the University in Eberswalde was closed down for political reasons. In the early 1960s, students and faculty members were politically active, and partly in opposition to the socialistic government, and several scientists, lecturers, and students were arrested (Ciesla and Joachim 2015). The University in Eberswalde was a thorn in the side of the rulers (SED) of the German Democratic Republic, because "socialistic education of the students" was not included in the curriculum (Joachim 1993). The dean was dismissed because he had failed to "educate the professors, assistants, and students in the socialistic way of thinking" (Milnik 1998), and other scientists in Eberswalde were sentenced to several years in prison (Ciesla and Joachim 2015). There were, however, also economic reasons for the elimination of one of the two Faculties for Forestry in the German Democratic Republic (one in Eberswalde, one in Tharandt close to Dresden).

With the closure of the University in Eberswalde, the zoological collections were redistributed among other institutions, however without any proper records (Kost 2003). It is, therefore, almost impossible to trace the fate of the collections. According to Gäbler (1964), all bird samples originating in Brandenburg were transferred to the Director of the zoological garden in Berlin (H. Dathe). Several mammal samples and the complete collection of bird eggs were transported to the remaining Faculty of Forestry in Tharandt/Saxonia (Kost 2003); other samples were given to the Museum of Natural History in Berlin, to various schools, or to private persons. Only a few samples remained in Eberswalde, among them the famous antlers of the red deer named "Caesar of Rominten." According to Bergmann (pers. comm.), this precious trophy was secretly hidden before the rest of the collection was shipped to Tharandt and ultimately displayed several years later. Many entomological samples were donated to research institutions, for example, the State Institute for Forestry and the German Entomological Institute (today in Müncheberg), where (according to Milnik and Rohlfien 1999) most of the Eckstein collection is located.

Consequently, the history of 133 years of zoological collections in Eberswalde ended with the closure of the University in 1963, resulting in a major loss for Eberswalde (Milnik and Rohlfien 1999).

25.3 Recent Zoological Collections in Eberswalde

After the demise of the German Democratic Republic and the reunification of the two parts of Germany in 1990, a new University for Applied Sciences (HNEE) was established in Eberswalde in 1992. This University resumed the tradition of studying forestry sciences, but additionally, new faculties were established. The Faculty for Landscape Management and Nature Conservation, for example, inhabited some of the old University buildings and also occupied the so-called two-story "gallery room" in 2012 (Figs. 25.3 and 25.6), which originally housed the zoological collections of Ratzeburg, Altum, and Eckstein. After 50 years of various uses and reconstruction following the closure of Eberswalde University in 1963, today, a new zoological collection can be found in this room (Fig. 25.6). The new collection contains over 40,000 animal samples; however, there are only very few "historical" specimens, mostly from the collections of W. Kruel.

Fig. 25.6 The new zoological collection (2015) in the same room like Fig. 25.3: the original zoological collections were destroyed, dispersed, or lost in the years 1945 and 1963 (© with U. Schulz)

In addition to this larger zoological collection, which is now located in the historic academy building on the so-called "downtown campus" of HNEE (where the historical collections were originally housed), there are additional smaller but nevertheless important collections on the "forest campus" of HNEE at the Faculty for Forest and Environment. This includes a wildlife zoology collection with approx. 300 animal skulls (mostly antlers) and 120 mounted and preserved preparations of birds and mammals, all related to hunting activities and used for the education of forestry students and hunters. A forest entomological collection in glass display cases exhibits the most important forest pest insects and feeding traces (e.g., plant parts or wood with feeding traces), and a collection of Carabidae (ground beetles) and bird eggs (collection of Lockow) is used for education in forest ecology. Another forest entomological collection, containing some historical samples, can be found in the Eberswalde Forestry State Center of Excellence (Department of Forest Protection, K. Möller), an institution located on the "forest campus" of HNEE.

25.4 Inventory and Curation

Currently, there is no consistent inventory of the new zoological collections that exist in Eberswalde. The collections of the Faculty for Landscape Management and Nature Conservation (Fig. 25.6) contain over 29,000 insect specimens in drawers, among them 8000 Lepidoptera and over 5000 Coleoptera, but also 178 bird samples (113 species) and 132 mammal samples (69 species) in glass cabinets. As most samples are used in practical exercises for teaching purposes, the inevitable damage to and loss of samples require continuous maintenance and replacement, which is mostly accomplished by teaching personnel and student assistants—there is no curator for the collections. Figure 25.7 shows typical insect display cases and drawers (containing Carabidae, Lepidoptera, and Odonata), which are used by students during practical exercises. Furthermore, smaller collections of specific indicator species are constantly created for master or bachelor thesis and other scientific projects (see Figs. 25.8 and 25.9). Such collections exist for wild bee species (Apidae, 195 species), Carabidae (276 species), and Lepidoptera (576 species). The most common bird and mammal species were collected and prepared for use in lectures and practical exercises. Credit for the continued expansion of these collections can be attributed to the efforts of J. Oehlke, J. Möller, and U. Schulz (Eberswalde), supported by the work of many assistant scientists (among them F. Burger, O. Brauner, I. Brunk, H. Gruß, T. Kolling, M. Sommer, and B. v. Broen).

Fig. 25.7 Insect display cases for student courses in 2015 (© with R. Schlepphorst)

25.5 Current Use and Research

Currently, the collections are used mainly for teaching purposes in zoology, entomology, wildlife biology, and ecology classes in bachelor and master study courses: Forestry, Landscape Management and Nature Conservation, Organic

Fig. 25.8 Part of today's collection of mollusks (© with R. Schlepphorst)

Fig. 25.9 Part of today's collection of Odonata (© with R. Schlepphorst)

Fig. 25.10 White-tailed sea eagle (*Haliaeetus albicilla*) as example for a species relevant for nature conservation in the new zoological collection of Eberswalde (© with U. Schulz)

Fig. 25.11 Great cormorant (*Phalacrocorax carbo*) as example for a species in the new zoological collection of Eberswalde which is controversially discussed in conflicts of various landscape users (© with U. Schulz)

Farming and Marketing, Global Change Management, and Regional Development and Nature Conservation.

Larger specimens like European otter, beaver, or various birds of prey are often provided for use at exhibitions and by NGO's for environmental education in the field of nature conservation. The collections are rarely used for research purposes; however, the entomological collections are used for the determination of indicator species for the evaluation of biotopes in scientific projects. In contrast to the historical collections, there are no type species included in the current collections in Eberswalde. The collections are, however, important for instruction in the various study courses. The vertebrate collection on the "forest campus" is focused on species related to hunting, while the collections on the "downtown campus" emphasize protected species (e.g., see Fig. 25.10) and species which are controversially discussed in conflicts of various landscape users (e.g., see Fig. 25.11).

In summary, the new zoological collections in Eberswalde are adequate for teaching purposes. There is, however, no "Phoenix" among the samples, and today's collections cannot be compared with the legendary and destroyed historical collections of Ratzeburg, Altum, and Eckstein.

Acknowledgment We would like to thank Dr. A. Milnik (Eberswalde), Dr. I. Brunk/Prof. Dr. M. Roth (Tharandt), and E. Schubert (Müncheberg) for their valuable information. In particular, we thank C. Adler, C. Beutel, R. Genthof, and M. Stöhr (Eberswalde) for their help in the library of UESD/HNE Eberswalde.

Sources

Anonymus (1846) Systematisches Verzeichnis der in der Kgl. höhern Forstlehranstalt zu Neustadt-Eberswalde aufgestellten Sammlung ausgestopfter Thiere. Müller-Verlag, Eberswalde

Augustiny H (1995) Die forstliche Bibliothek in Eberswalde, 1. April 1945–31. August 1963. Handschriftliche Aufzeichnung in der Bibliothek der Hochschule für nachhaltige Entwicklung Eberswalde; unpubl., 52 p

Ciesla B, Joachim HJ (2015) Protokoll einer Schließung—das Ende der Forstwirtschaftlichen Fakultät in Eberswalde 1963. be.bra-Verlag, Berlin, in prep

Danckelmann B (1880) Die Forstakademie von 1830 bis 1880. Julius Springer Verlag, Berlin

Dickel K (1910) Über die mit der Universität Berlin verbunden gewesene Königliche Forstakademie (1821 bis 1830) und den Lehrstuhl für Forstwissenschaft an der Universität Berlin (1830 bis 1848). Franz Vahlen-Verlag, Berlin, 123 p

Eckstein K (1931) Die Zoologie in Eberswalde–ein Rückblick auf 100 Jahre. Zeitschrift für Angewandte Entomologie 17(3):631–658

Endtmann KJ (2006) Julius Theodor Christian Ratzeburg. In: Milnik A, Brandenburgische Historische Kommission e.V. (Hrsg) Im Dienst am Wald–Lebenswege und Leistungen brandenburgischer Forstleute. Kessel-Verlag, Remagen, pp 136–137

Gäbler H (1963) Albinos verschiedener Vogelarten in der Zoologischen Sammlung der Forstwissenschaftlichen Fakultät der Humboldt-Universität Berlin in Eberswalde. Beiträge zur Vogelkunde 9(1/2):97–101

Gäbler H (1964) Entwicklung und Tätigkeit des Instituts für Forstzoologie der Humboldt-Universität Berlin in Eberswalde in den Jahren 1953–1963. Zeitschrift für Angewandte Entomologie 55(1–4):295–300

Hilf H, Schubert J (1930) Festschrift zur Hundert-Jahr-Feier der Forstlichen Hochschule Eberswalde–1830–1930. Im Auftrage des Senats der Forstlichen Hochschule herausgegeben. Fritz Lindner Verlag, Düsseldorf, 69 p

Joachim HF (1993) Zur Auflösung der Forstwirtschaftlichen Fakultät Eberswalde. Beitr Forstwi u Landschaftsök 27(4):145–152

Königsmann E (1964) Braconidae aus den Resten der Ratzeburg-Sammlung. Beitr Ent 14:631–661

Kost F (2003) Begleitschreiben zum Inventarium der Zoologischen Sammlung; schr. Mitt. an Archivar Wudowenz über Prof. C. Majunke; Archiv der Bibliothek der Hochschule für Nachhaltige Entwicklung, Eberswalde

Milnik A (1998) In Verantwortung für den Wald—die Geschichte der Forstwirtschaft in der Sowjetischen Besatzungszone und der DDR. MELF, Potsdam

Milnik A, Rohlfien K (1999) Professor Karl Eckstein–Hochschullehrer, Zoologe, Förderer der Heimatkunde. In: Verein für Heimatkunde zu Eberswalde e. V. (ed) Eberswalder Jahrbuch Heimat-, Kultur- und Naturgeschichte, pp 209–223

Milnik A, Rohlfien K (2000) Professor der Zoologie Dr. Bernard Altum (1824–1900). In: Verein für Heimatkunde zu Eberswalde e. V. (ed) Eberswalder Jahrbuch Heimat-, Kultur- und Naturgeschichte; pp 239–248

Milnik A (2006a) Bernard Altum. In: Milnik A, Brandenburgische Historische Kommission e.V. (ed) Im Dienst am Wald–Lebenswege und Leistungen brandenburgischer Forstleute. Kessel-Verlag, Remagen, pp 242–243

Milnik A (2006b) Karl Eckstein. In: Milnik A, Brandenburgische Historische Kommission e.V. (ed) Im Dienst am Wald–Lebenswege und Leistungen brandenburgischer Forstleute. Kessel-Verlag, Remagen, pp 244–245

Rohlfien K (1979) Aus der Geschichte der entomologischen Sammlungen des ehemaligen Deutschen Entomologischen Instituts. Beitr Ent 29(1):415–438

Schedl KE (1974) Zur Synonymie der Borkenkäfer XXIV. Ent Arb Mus Frey 25:333–342

Schirm W, Witzenhausen W (1959) Führer durch die Eberswalder forstlichen Lehr- und Forschungsstätten. Scholz, Eberswalde, 74 p

Schwerdtfeger F (1983) Julius Theodor Christian Ratzeburg (1801–1871). Monographien zur angewandten Entomologie 24. Verlag Paul Parey, 103 p

Schwerdtfeger F (1985) Fazit. Unpublished, 113 S.; Library of Eberswalde, HNEE Eberswalde

Chapter 26
ERFURT: The Naturkundemuseum Erfurt

Matthias Hartmann, Herbert Grimm, and Ulrich Scheidt

Abstract The Naturkundemuseum Erfurt was founded in 1922 through the cooperation of citizen scientists. From the beginning the exhibition style was new for Germany, because the museum presented the plants and animals in their natural habitats and in connection with their environment. In 1968, the museum was closed, and the collections were stored in several buildings in the city centre. After 1990, the modern Naturkundemuseum Erfurt became a reality in a complex in the historical city centre. The style of exhibition was also new for Central Europe. From this time the collections increased rapidly. Today, the zoological collections in total contain more than 1.25 million specimens. The taxonomic focus of the zoological collections is entomology (mainly Coleoptera, Lepidoptera, Hymenoptera, Orthoptera) with 2000 type species (1010 holotypes), ornithology, herpetology and mammalogy. Geographical focal points are Central and Southern Europe, Asia (Southeast and High Asia) and particularly Africa. For Himalayan research in biodiversity, the museum has become a centre of competence in the last decade. Three annual journals and one periodical book series were published by the museum. The museum's special library is of vital importance and the biggest special nonuniversity library in Thuringia. The Naturkundemuseum Erfurt is an important institution for the scientific community, combining science, research, education and preservation of nature and culture.

Keywords Aves • Mammalia • Amphibia • Reptilia • Mollusca • Insecta • Palaearctic • Oriental • Central Europe • Himalaya • Southeast Asia • Thuringia

The Naturkundemuseum Erfurt was founded in 1922 through the cooperation of citizen scientists, as a people's museum. On 29 October 1922, it opened in the historic building named "Haus zum Stockfisch". Most of the collections at this time came from founder members, and only a small proportion is older than 1900. The most important contributor for this time was the entomologist Otto Rapp (1890–1953), who was the head of the museum for a long time. The exhibition

M. Hartmann (✉) • H. Grimm • U. Scheidt
Große Arche 14, 99084 Erfurt, Germany
e-mail: matthias.hartmann@erfurt.de; naturkundemuseum@erfurt.de; http://www.naturkundemuseum-erfurt.de

L.A. Beck (ed.), *Zoological Collections of Germany*, Natural History Collections,
DOI 10.1007/978-3-319-44321-8_26

Fig. 26.1 Ponds on open landscape, view in the exhibition floor 2 (with permission from: Archive NME; photo: F. Behr)

style was new for Germany, because the museum presented the plants and animals in their natural habitats and in connection with their environment. For the first years, the museum was operated by unpaid volunteers. From 1938 to 1940, the first director of the museum was the malacologist Walter Wächtler. From 1941 to 1945, the museum was almost completely closed, and the collections were preserved in several buildings in villages around Erfurt. In this way, the main parts of the collection survived World War II, with only the herpetological collection being destroyed by an aerial mine in 1944 (Figs. 26.1, 26.2, 26.3, 26.4, 26.5, 26.6, 26.7, 26.8, 26.9, 26.10, and 26.11).

In 1945, the Naturkundemuseum Erfurt was the first museum in Thuringia to reopen its doors to visitors. After the death of Otto Rapp, the development of the museum slowed down. In 1968, it was closed, and the collections were stored in several buildings in the city centre, without a real place for a working museum. The damage to the collections during this time was worse than in WWII. The growing environmental movement in the 1970s gave the museum new life, and by the end of the 1980s, there were four scientists, two taxidermists and two technical/economical employees working on a plan for a new and modern museum. After the political changes in Germany in 1990, the modern Naturkundemuseum Erfurt became a reality in the complex of Große Arche 14 in the historical city centre. The collections were now in modern housing and from this time increased rapidly.

Fig. 26.2 View on exhibition part "Noah's Arc—The Conservation of Diversity" (with permission from: Archive NME; photo: F. Behr)

Today, the zoological collections in total contain more than one million specimens, the biggest part being the entomological collection, though the vertebrates are also significant.

The taxonomic focus of the zoological collections is entomology (mainly Coleoptera, Lepidoptera, Hymenoptera, Orthoptera), ornithology, herpetology and mammalogy. Geographical focal points are Central and Southern Europe, Asia (Southeast and High Asia) and particularly Africa.

The zoological collections of the Naturkundemuseum Erfurt are biggest in Thuringia state, and the museum is a municipal one. At present three scientists (including the director), two taxidermists, one librarian, one educational officer and three technical employees are responsible for all duties, and the team is supported by 16 voluntary scientific contributors.

The museum has an exhibition area of 1200 m^2, three labs for taxidermy and one special room for each collection. Because the resources of the city are very limited,

Fig. 26.3 The museum's annual journal *VERNATE*

the museum requires financial support from outside sources, such as the Thuringian government, several foundations and its furtherance society.

Since the 1990s the museum concentrated its research on several taxonomic and geographical points of interest, usually in combination. Besides the fauna of Thuringia and Middle Europe, the main focus is biodiversity research of the Himalayas and Southeast Asia. For Himalayan research in biodiversity in particular, the museum has become a centre of competence in the last decade. Special symposiums (5), publications and 14 expeditions to the Himalayas advanced the development of the collections and international contacts. Through these achievements, the museum is now an important partner in the international network of

Fig. 26.4 The black morph of common hamster (*Cricetus cricetus* L., 1758) is limited in Central Europe at the Thuringian Basin, but here the morph is often encountered (with permission from: Archive NME)

Fig. 26.5 Laotian rock rat *Laonastes aenigmamus*. First described for science in 2005 and is very rarely represented in collections. Photo: S. Brandt (with permission from: Archive NME)

biodiversity research. Results include five edited volumes in the book series *Biodiversity and Natural Heritages of the Himalaya*, and more than 370 special publications about the museums' research in this field have been published in many international journals.

So far, two expeditions to Thailand were mainly focused on detection and biological classification of amphibian and reptile species occurring there. At the same time, ecological and morphological data were collected. The evaluation of the material was carried out in a close cooperation with the museums in Berlin, Bern and Bonn.

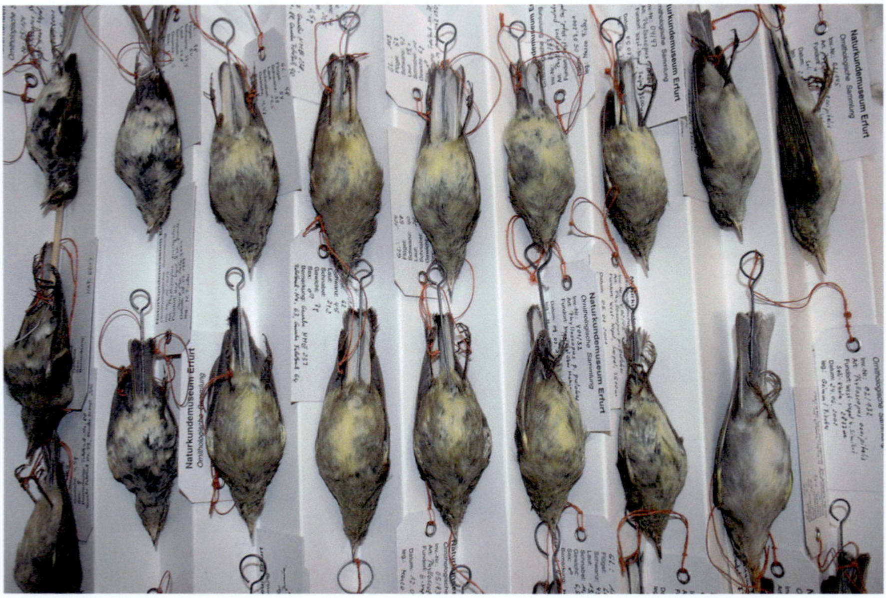

Fig. 26.6 Some warbler skins (*Phylloscopus*) in the bird collection (with permission from: Archive NME)

Fig. 26.7 View in the collection of lark skins (with permission from: Archive NME)

Fig. 26.8 Bearded tit (*Panurus biarmicus*) as full mount in the bird collection (with permission from: Archive NME; photo: F. Behr)

Fig. 26.9 Galapagos land iguana (*Conolophus subcristatus* Gray, 1831), leg. Juan Foerster, 1959; estate of the firm Schlüter (with permission from: Archive NME; photo: F. Behr)

Different research projects are pursued by the curators in addition to their long-term work on relevant collections. These include, for example, eco-faunistic investigations of different organism groups (insects, amphibians and reptiles, birds), such as studies on the ecology, biology and larval development of the common midwife toad *Alytes obstetricans*, studies on the ecological requirements and diet of insectivorous birds and others.

Furthermore, the museum is an Information and Coordination Centre for faunistic and floristic recording activity of volunteers. This has resulted in valuable material in the collections and on the other hand a large, mainly faunistic, data pool. The Coleoptera database alone contains more than 600,000 data sets. Currently four

Fig. 26.10 View in the insect collection (with permission from: Archive NME)

working groups and 16 volunteer staff work as specialists on different organism groups at the museum.

The Naturkundemuseum Erfurt and its furtherance society publish three periodicals: *Veröffentlichungen des Naturkundemuseums Erfurt* (1982–2008), continued from 2009 as *VERNATE*, *Thüringer Faunistische Abhandlungen* (since 1994) and

Fig. 26.11 Detail of the collection part "Coleoptera" (with permission from: Archive NME)

Schriftenschau für den Feldherpetologen (1986 to 2004 as print version, afterwards as an Internet offer, http://www.amphibienschutz.de/literatur/index.html).

The museum is famous in Germany for its educational work, especially its modern exhibitions. Since its reopening in 1995, the museum has followed new teaching models and has provided a new standard in design and arrangement of showcases. The feedback in Germany was extremely positive. For the new exhibitions, as well as the unity of collections, research and educational work, the museum has gained two national awards: in 2004, the museum's award of the Savings Banks of the Hessian-Thuringia Cultural Trust and, in 2007, the special award of the Heinz Sielmann Foundation.

Cooperation between the museum, schools and universities has led to several projects for education and scientific work. For instance, in cooperation with the Applied University of Bernburg (Saxony-Anhalt), the museum commenced a teaching project for curators and taxidermists from Bangladesh. For such cooperation, but also for its own research, the museum's special library is of vital importance. It is the biggest special library in Thuringia for biodiversity research aside from the University Library of Jena.

In all, the Naturkundemuseum Erfurt is an important institution for the scientific community, combining science, research, education and preservation of nature and culture.

26.1 The Collection of Mammals

From the beginning of the museum in 1922, it was a special aim to present a mounted specimen of each mammal species found in Thuringia. Most of these specimens were bought from the taxidermist firms of Dr. Schlüter and Dr. Mass (Halle/Saale). The first small series of skins was prepared in 1923/1924 by the young Andreas Feininger (1906–1999), who later became world famous as a photographer. From this time, dead animals were prepared and preserved by paid employed taxidermists. After World War II, the museum had no taxidermist, so the collecting of mammals almost ceased. Some small collection of skins and mounts came as gifts from Carl August Schmöger (1890–1967), such as animals from Brazil (Espirito Santo and Minais Gerais), which were collected during the 1920s. Also an important collection of hunting trophies from the former German colonies in East Africa was given to the museum after WWII.

Only with a new taxidermist employed from 1977 did the collection begin to increase again, slowly but continuously. After the political change in the 1990s, small mammal series were collected or deposited as a result of ecological studies. These skins, skulls and skeletons are the main part of the mammalogy collection and will be used for faunistic and other scientific research. Animals sent by the public or the authorities serve the same purpose.

Through contact with several zoos in Germany and abroad, we received many dead animals of interesting species. In 2007 the famous Schlüter taxidermy firm (founded in 1853 in Halle, closed in 2007 in Winnenden) sent us their whole stock of animal material. This included very rare species from South America and New Guinea.

Since 2008 all prepared specimens of the scientific collection have tissue samples preserved. The collection is completely digitally registered. Currently 2515 specimens in appr. 300 species are listed (30.vi.2016) (Table 26.1).

Table 26.1 Number of specimens of the mammal collection (on 30.viii.2016)

Mounts	366
Skins	328
Coats	8
Ethanol preserved	4
Horns and antlers	262
Skulls and skeletons	1766
Tissue samples	113

26.2 The Ornithological Collection

In connection with the history of the museum, the ornithological collection is not based on a historical store. The oldest specimens date back to the last quarter of the nineteenth century. The oldest mounted specimen is a rock sparrow from Thuringia from 1879. Founded as a municipal museum, education was in the foreground, first as a show museum aiming to display collections to the people. Therefore, above all, mounted birds came into the collection. Unique pieces from private collectors came as donations or part of collection purchases, but the acquisition of ornithological objects was always limited to a few specimens or very small collections. The most important ones—160 bird skins (Palaearctic)—come from Reinhold Fenk (1881–1953); Max Timpel (1865–1934), 207 bird's egg clutches from the surroundings of Erfurt as well as some part skeletons; and Carl August Schmöger (1890–1967), 70 birds, eggs and skeletons from South America (Brazil).

Some documents were acquired of important faunistic records from Thuringia: cream-coloured courser *Cursorius cursor*, white-backed woodpecker *Dendrocopos leucotos*, booted eagle *Aquila pennata*, little bustard *Tetrax tetrax*, Pallas's sandgrouse *Syrrhaptes paradoxus*, rock sparrow *Petronia petronia* and Leach's petrel *Oceanodroma leucorhoa*, among others. A specimen of a passenger pigeon *Ectopistes migratorius* was also acquired.

Collection growth was connected above all with the availability of zoological taxidermists in the museum. The most important source of collection growth was the collective activity of the employees of the museum and its circle of friends. Therefore, Palaearctic birds—above all from Central Germany—dominate.

In connection with the research activity of the museum in the Himalayas, a total of 206 skins and mounted birds (mainly Passeriformes) from this region were added to the collection. In spite of the main focuses of collecting, bird specimens from all continents, including Antarctica, are in the collection.

At present, birds that are found dead are incorporated to the collection. They are preserved as skins, mounted specimens, skeletons and part skeletons, eggs, loose feathers and tissue samples. Although still many new taxidermies are full mounts (in the museum's permanent exhibition, there are currently 182 mounted birds), it is a current drive to try to increase the collection of skins. Tissue samples of all newly prepared birds are kept.

Present collection:

	Specimens	Species
Mounted birds	1540	584
Skins	1379	126
Skeletons and part skeletons	992	424
Clutches (eggs)	545 (1603)	208
Loose feathers (different completeness)	720	73
Tissue samples	810	252

The collection is properly housed and completely digitised (access file). The status of the collection up to 2012 is published in a catalogue by Grimm (2012).

26.3 The Herpetological Collection

The original small collection of ethanol-preserved herpetological specimens from the museum's inception was destroyed by an aerial mine on 11 November 1944. After WWII no herpetologists worked at the museum for a long time.

Only through donations of a collection from Carl August Schmöger (1890–1967) did specimens of amphibians and reptiles come to the museum's collection. These are remarkable records from the now-destroyed Atlantic rainforests in Brazil.

The increase of specimens and species of the herpetological collection began in the 1980s, the most important part after the 1990s. At present, the herpetological collection consists of around 16,000 amphibians and more than 1500 reptiles of more than 600 species. Ninety percent of the collection is digitally registered. Most of the specimens are preserved in pure ethanol. In addition to this material, more than 30 tortoise shells, some mounts, skin preparations, skulls and skeletons are also held.

The main part of the collection was made in Thuringia as a result of the study of nature reserves or ecological research. With this material, we can document the changes and the development of the landscape in Thuringia. Other local records come from the public and the authorities. This material includes curious records, such as exotic reptiles found living around German cities.

A significant part of the herpetological collection comprises material from the northern Mediterranean fauna. Among the collectors, the samples from Andreas Nöllert (Jena) are the most remarkable. Large series of green toads from Northern Africa, Asia Minor and Central Asia were given by PD Dr. habil. Matthias Stöck (Berlin). Among these materials are species with tri- or tetraploid chromosome numbers—very rare hybrids in nature—and also one paratype of *Bufo [Bufotes] siculus* (Stöck, Sicilia, Belfiore, Buckley, Lo Brutto, Lo Valvo and Arculeo 2008).

A second focus of the collection is South and Southeast Asia, Thailand, Laos, Nepal and Malaysia, with more than 2000 specimens. Most of the material came as donations from volunteers. The main part was presented by Thomas Ihle (Pak Chong/Thailand).

This material also includes the first records of some Asian countries or regions (Tillack et al. 2005, 2006; Manthey and Denzer 2012), as well as small samples of amphibians and reptiles from different regions of New Guinea and Wallacea, sent by Dr. Dmitry Telnov (Dzidrinas/Lettland). A special collection catalogue of Gekkota was published in 2013 (Rösler and Scheidt 2013).

26.4 The Malacological Collection

Several shell collections were conveyed by Erfurt citizens to the precursor of the museum, the scientific study group. Generally, they consisted of attractive mostly marine snails and shells, without data. They still exist even today in the museum, but their scientific value is small.

The snail collection of Werner Boeckel (1909–1941) purchased in 1942 established the basis of the current collection. It comprises around 40,000 specimens of Central European snails as well as some freshwater molluscs. The well-preserved specimens are accommodated in 5700 glass tubes and are listed in a three-part, handwritten catalogue. In the first half of the twentieth century, Boeckel was the most famous malacologist of Central Germany (Bößneck 2009).

A smaller collection of specimens from Thuringia (200 tubes) from Oscar Schmidt (1848–1908) came from the Weimar museum to the collection of the Naturkundemuseum Erfurt. The collection of Walter Wächtler (1901–1943), which remained in the museum, is also of modest scope. Wächtler was a curator between 1929 and 1939 and from 1938 until his death was the leader of the museum (Bößneck and Neumann 2001).

The modern collection of snails and freshwater molluscs is mainly obtained through collecting in connection with environmental planning and survey work. About more than 1460 recorded specimens were spent in the collection last years.

In 2012 and 2013, a collection of land molluscs and freshwater molluscs from Southeast Asia and New Guinea was acquired. Among these animals, mostly preserved in ethanol, are five holotypes and 13 paratypes (Greke 2011). The collector U. Bößneck left four additional paratypes of a snail and a slug species from the Nepal Himalayas to the museum (Wicktor and Bößneck 2004; Gerber and Bößneck 2009). About 60 % of the collection is digitised.

26.5 The Entomological Collection

Among the zoological collections of the museum, the entomological collection is the biggest: more than 30,000 species with around one million specimens in all. All specimens of the main curated collection are labelled with full locality data. An unlisted and unnumbered supplementary collection is used for exhibitions and educational work.

The basis of the entomological collection was laid down by the well-known entomologist Otto Rapp (1878–1953), one of the founders of the museum. In this older part of the collection are important faunistic records from famous Thuringian entomologists like Wilhelm Hubenthal, Arthur Petry, Franz Maaß and Emil Lotze. Also important for regional faunistics are the collections of Peter Heymes (Coleoptera), Rolf-Peter Rommel and Wolfgang Apfel (Lepidoptera). These are all integrated in the general collection of their specific order. The general collection of Lepidoptera is of considerable importance for national science.

The collections of Coleoptera, Orthoptera and Hymenoptera are of international importance because of their holdings of types and the large number of species represented.

At present, the entomological collection has around 2000 species represented by type material (3500 specimens). These include 1010 holotypes. Species catalogues exist for several insect families and orders.

As a result of the main research areas of the museum, the insects of the collection come from the Palaearctic and Oriental regions but also a number from the Afrotropics. Only a very small part was collected in the Neotropics (Brazilian states of Espirito Santo, Minas Gerais and Rio Grande do Sul) between 1907 and 1924 by the private scientist Carl August Schmöger (1890–1967). Besides Middle Europe the most valuable material was collected in the Himalaya region and Southeast Asia. The oriental part was built up from the beginning of the 1990s, when the museum started its intensive field work in this region.

Besides the insect collection, there are small collections of other arthropods, mainly scorpions and spiders, and a historical collection of beetle biology from Middle Europe.

The biggest part of the collection is a result of the scientific work of the museum staff or the associated group of volunteer staff around the museum.

Work in and with the collection was supported by volunteer staff with a high level of scientific competence. Since 2002, Wolfgang Apfel has built up the new collection of "Coleoptera Palaearctic/Oriental region". The collection of Lepidoptera is the responsibility of Andreas Heuer, and the Hymenoptera of Frank Creutzburg.

Through intensive international contacts with many specialists on nearly all taxonomic groups, the general collections are in a modern, revised systematic order.

International research by the museums and its associates increases the number of species, specimens and types every year. So, the museum possesses the most important entomological collection in the Thuringia. In cooperation with scientists worldwide, the material is loaned for many research projects on systematic and phylogenetic questions.

During the last 15 years, more than 550 different papers have been published on material from our collection, showing its great and increasing scientific importance (Table 26.2).

Table 26.2 Number of species and specimens of the entomological collection (on 30.viii.2015) (selected taxa)

Taxa	~Species	~Specimens
Auchenorrhyncha	240	1500
Coleoptera	19,800	840,000
Dermaptera	100	1200
Diptera	1000	28,000
Heteroptera	560	3500
Hymenoptera	2500	38,000
Lepidoptera	3800	37,000
Orthopteroidea	180	1500
Others	2000	3500
Types	1700	2900
Holotypes	1010	

References

Bößneck U (2009) Werner Böckel (1909-1941)—zum 100. Geburtstag eines Erfurter Malakozoologen. VERNATE 28:5–9

Bößneck U, Neumann K (2001) Zum 100. Geburtstag von Walter Wächtler (1901-1943) Malakozoologe und ehemaliger Leiter des Naturkundemuseums Erfurt. Veröffentlichungen des Naturkundemuseums Erfurt 20:5–9

Gerber J, Bößneck U (2009) The genus *Vallonia* in Nepal (Gastropoda, Pulmonata: Valloniidae). Archiv für Molluskenkunde 138:43–52

Greke K (2011) Species of *Ditropopsis* E.A. Smith, 1897 (Architaenioglossa: Cyclophoridae) from the Papuan region. In: Telnov D (ed) Biodiversity, biogeography and nature conservation in Wallacea and New Guinea. Entomological Society of Latvia, Riga

Grimm H (2012) Die Ornithologische Sammlung des Naturkundemuseums Erfurt. VERNATE 31: 139–211

Manthey U, Denzer W (2012) Bemerkungen zur Verbreitung von *Japalura yunnanensis* Anderson, 1878 (Squamata: Agamidae: Draconinae). Sauria 34(2):35–40

Rösler H, Scheidt U (2013) Die herpetologische Sammlung des Naturkundemuseums Erfurt. 1. Teil: Verzeichnis der Geckos (Reptilia: Squamata: Gekkota). VERNATE 32:235–256

Tillack F, Scheid U, Fischer M (2005) Erstnachweis der Himalaja-Katzennatter, *Boiga multifasciata* (Blyth, 1860) für West-Nepal. Sauria 27(3):41–45

Tillack F, Scheidt U, Ihle T (2006) First record of Blakeway's mountain snake, *Plagiopholis blakewayi* Boulenger, 1893, from Thailand, with remarks on the distribution of *Plagiopholis nuchalis* (Boulenger, 1893) (Reptilia, Squamata, Colubridae, Pseudoxenodontinae). Veröffentlichungen des Naturkundemuseums Erfurt 25:181–186

Wicktor A, Bößneck U (2004) *Limax (Limax) seticus* n. sp. from high mountains in Nepal (Gastropoda: Pulmonata, Limacidae). Folia Malacologica 12:183–187

Chapter 27
FLENSBURG: Naturwissenschaftliches Museum Flensburg: Natural History Museum Flensburg

Werner Barkemeyer

Abstract The Natural History Museum Flensburg focuses on North Germany and southern Denmark. The museum, therefore, deals with the fauna, flora and geology of this region as do the collections. The museum has a good bird and old egg collection from the area of Flensburg. The same applies to the collections of Coleoptera, Diptera, Hymenoptera Aculeata and Lepidoptera. The collection data will be available on the website from early 2017 on. Education has always been a key goal of the museum, and there has been a strong ongoing cooperation with the University of Flensburg since 1946.

Keywords Flensburg • North Germany • Denmark • Baltic amber • Schleswig-Holstein

The history of the museum goes back to 1913 when the municipality of Flensburg bought a geological collection from Hans Philippsen (1866–1926). He was a teacher who in his spare time worked as the curator of the collection (Photo. 27.1).

More collections of beetles and molluscs were later handed over, and finally, in 1925 enough material and enthusiasm had been gathered that Flensburg's mayor at the time opened the Naturwissenschaftliches Heimatmuseum. In 2001 the name changed to Naturwissenschaftliches Museum.

The museum is still being run by the municipality. On the permanent staff, there is only one scientist who also acts as the director, collection manager and head of the education and exhibition unit. He is, however, supported by the staff of the municipal museum of arts and culture of Flensburg ("Museumsberg"), the association of friends and sponsors of the Natural History Museum ("Freunde und Förderer des Naturwissenschaftlichen Museums Flensburg e.V."), several volunteers and freelance educators, most of them with an academic background.

W. Barkemeyer (✉)
Naturwissenschaftliches Museum, Museumsberg 1, 24937 Flensburg, Germany
e-mail: Barkemeyer.Werner@flensburg.de; http://www.Naturwissenschaftliches-Museum.Flensburg.de

L.A. Beck (ed.), *Zoological Collections of Germany*, Natural History Collections,
DOI 10.1007/978-3-319-44321-8_27

Photo 27.1 Hans Philippsen (1866–1926), Museum of Natural History Flensburg

Since its beginnings the museum's main focus has been on education. The permanent exhibition deals with the region of North Germany and southern Denmark and is on display on the ground floor in the Heinrich-Sauermann-House (Photo. 27.2) on the Museumsberg which is a part of the museum of arts and culture. The characteristics of the fauna, flora and geology and their interaction are illustrated here by several examples. Temporary exhibitions on various subjects are shown for two or three months a year.

In all six rooms of the permanent exhibition, small chambers called pavilions are equipped with old glass cabinets in which objects, such as birds, mammals and insects, are displayed as was the custom in the old days, i.e. with many objects on display and very few explanations (Photo. 27.3). Posters inform about the life of the collectors whose collections are now a part of the museum. This backdrop spotlights the history of the museum. Outside these chambers between modern glass cabinets and open showcases, visitors can make use of the microscopes, computers and many other modern facilities to make their visit thoroughly enjoyable (Photo. 27.4).

The Eiszeit-Haus (Ice Age House) is located in a public park near the Hans-Sauermann-House. This old building was originally used as stables but is currently a museum warehouse which is open to the public twice a week which is again made possible by volunteers helping out as honorary guards and guides. The major share of the geological and archaeological collection of the museum is stored here. The Eiszeit-Haus is also an education centre offering guided walks, lectures and courses on behalf of the museum.

Photo 27.2 Heinrich-Sauermann-House, Museum of Natural History Flensburg

Photo 27.3 Inside a pavilion in the permanent exhibition, Museum of Natural History Flensburg

The museum cooperates with the University of Flensburg, among others with the Department of Biology and its teaching methods. It focuses on families with children as visitors and offers special programmes for families, schools and nurseries.

Photo 27.4 Permanent exhibition, Museum of Natural History Flensburg

The biological collections are stored in the Heinrich-Sauermann-House. Most of the specimens were collected by private individuals in North Germany, Denmark and Sweden. All the directors of the museum, with the exception of only one, were interested in faunistic studies in North Germany and donated their material and documents to the museum collection. There are no holo- or syntypes in the whole zoological collection.

Insects make up the largest segment of the invertebrates. The collections of Hymenoptera Aculeata, Lepidoptera, Diptera, Coleoptera and Odonata are fairly comprehensive. Most of these specimens are sorted up to species level and stored in more than 500 drawers. The museum also owns a small collection of Diptera Brachycera in Baltic amber. Although most of the details on the insect labels have been recorded, the information has not yet been completely transferred on to an electronic database.

The major part of the collection of Mollusca is made up of shells (snails and marine and freshwater shells) collected and labelled by Hans Philippsen and Otto Plamböck in the first decades of the twentieth century. Approximately 1000 birds are registered in the vertebrate collection. Most of the specimens were bought for the exhibition many years ago. In many cases, however, the collected data of this old material have not been kept in the files or are incomplete.

The most interesting bird specimen is probably the brown noddy *Anous stolidus pileatus* Linnaeus, 1758, which was shot on the west coast of Schleswig-Holstein in 1912. For many decades this one bird was the only record of this species in Europe (Photo. 27.5). The museum owns an egg collection with about 2500 specimens collected or bought by Paulus Paulsen in between 1890 and 1940 with many samples of rare birds from Schleswig-Holstein.

Photo 27.5 Brown noddy, *Anous stolidus pileatus*, from the museum's collection, Museum of Natural History Flensburg

While the museum's prime focus has been on education, an integral part of its policy is to adequately maintain and enlarge the regional collection as this is the scientific backbone of the museum. It will begin putting the data of the biological and geological specimens on the website with links between the exhibition and collections in early 2017.

Reference

Barkemeyer W (1912) Das Naturwissenschaftliche Museum Flensburg. In: Natur- und Landeskunde. Zeitschrift für Schleswig-Holstein, Hamburg und Mecklenburg 119:65–76

Chapter 28

FRANKFURT, DRESDEN, GÖRLITZ, MÜNCHEBERG: Senckenberg: Its Zoological Collections and Their Histories

Michael Türkay, Uwe Fritz, Thomas Schmitt, Willi Xylander, Raffael Ernst, Uwe Kallweit, Klaus-Dieter Klass, Matthias Nuss, Martin Päckert, Christian Schmidt, André Reimann, Katrin Schniebs, Clara Stefen, Andreas Weck-Heimann, and Axel Zarske

Abstract Today, Senckenberg's collections, comprising some 38 million specimens, support a vast research enterprise. Although it is sometimes forgotten, these collections are crucial for day-to-day scientific work and facilitate Senckenberg's mission to better understand biodiversity, evolution and mechanisms within the Earth system. The enormous collection infrastructure encompasses important herbaria, huge geological and paleontological collections and finally the zoological collections, which are introduced in the Senckenberg chapter of this book. The growth of the collections in any particular direction is often contingent on the research projects, special interests and/or current activities of scientists. Yet the collections are cumulative, and many specimens have their roots in the early history of the institution. Therefore, an overview of these collections should also encompass their origin. Facing the current and future challenges of biodiversity research, several German natural history institutes merged in 2009 to form a new

---

Michael Türkay was deceased at the time of publication.

U. Fritz (✉) • R. Ernst • U. Kallweit • K.-D. Klass • M. Nuss • M. Päckert • C. Schmidt • A. Reimann • K. Schniebs • C. Stefen • A. Weck-Heimann • A. Zarske
Senckenberg Naturhistorische Sammlungen Dresden (SNSD), Museum für Tierkunde, Königsbrücker Landstraße, 159 01109 Dresden, Germany
e-mail: Uwe.Fritz@senckenberg.de

T. Schmitt
Senckenberg Deutsches Entomologisches Institut (SDEI), Eberswalder Str. 90, 15374 Müncheberg, Germany
e-mail: Thomas.Schmitt@senckenberg.de; sdei@muencheberg.de

W. Xylander
Senckenberg Museum für Naturkunde Görlitz, Am Museum 1, 02826 Görlitz, Germany
e-mail: Willi.Xylander@senckenberg.de

L.A. Beck (ed.), *Zoological Collections of Germany*, Natural History Collections,
DOI 10.1007/978-3-319-44321-8_28

Senckenberg. As each of these institutes of course has its own tradition, the individual institutes introduce themselves. Each has an impressive history and important collections in which every single specimen preserves information that, once unlocked, fosters a keen knowledge of nature. By the way, reading the history of the Senckenberg Institutes seems like a rendezvous with the who's who of natural history research.

Keywords Museum für Tierkunde Dresden • Staatliche Naturhistorische Sammlungen Dresden • Senckenberg Naturhistorische Sammlungen • Entomology • Malacology • Ichthyology • Herpetology • Ornithology • Mammology • History

28.1 Introduction

Zoological collections are archives, which preserve the primary data for the occurrence of individual species in a particular place at a particular time. Even if the taxonomic classification of a specimen was changed, archives always allow integration into the new system. Within the taxonomic classifications, only the original specimens of a species are invariable, which means they are not replaceable through new collections (Fig. 28.1).

Digitisation does not replace the access to the specimens themselves. Therefore, requests for specific specimens by many scientists from all over the world continue unabated, especially if new characters for the reclassification of a given zoological group are needed.

All this confirms that collections are necessary as an unchangeable reference system for the classification of recent and fossil organisms. They must be maintained for use in perpetuity. Nevertheless, digitisation is an important tool for the access to information, for digital searches of course are much quicker than searching the physical collections themselves. However, in the end, only the access to the collections can provide researchers with the important characteristics of the original specimens.

28.1.1 Structures in the Senckenberg Collections

Senckenberg is composed of six institutes with stations at ten locations. Of these institutes and locations, four have zoological collections:

- Senckenberg Research Institute and Natural History Museum Frankfurt (Headquarters)
- Senckenberg Natural History Collections Dresden
- Senckenberg Museum of Natural History Görlitz
- Senckenberg German Entomological Institute Müncheberg

Despite of the diversity of the individual collections, all institutes follow common principles. Yet due to the long, independent history of these institutes, each has its own character and traditions. Hence, the Senckenberg chapter in this book is

Fig. 28.1 Senckenberg Headquarter with the Natural History Museum Frankfurt on Main

divided into sequenzes that describe the distinctive features of the institutes and their collections. In the Frankfurt Institute, the zoological collections are the second largest after the paleontological (which, like the botanical collections, will not be considered here). In Dresden and Görlitz, the zoological collections are the largest, while Müncheberg is specialised in entomology, so its entire collection is necessarily zoological. The significant collection of marine organisms is a distinctive feature of Frankfurt, while in Görlitz there is a special focus on soil organisms.

For digitising the Senckenberg collection, the management system SeSam is used (www.senckenberg.de/sesam). Because SeSam is connected by a wrapper system with GBIF (Global Biodiversity Information System), all catalogued items also appear on GBIF (www.gbif.org). Presently, most requests for specimen information come via this channel rather than directly. This underscores the importance of such international portals and shows that cooperation makes sense.

28.2 The Collections at the Senckenberg Research Institute, Frankfurt on Main

By Michael Türkay

28.2.1 History

The zoological collections of Senckenberg in Frankfurt are as old as the Senckenberg Natural History Society (Senckenberg Gesellschaft für

Naturforschung). One of the motives for founding this society was the construction of a museum as an appropriate place to house the existing natural history collections in the city. These were private collections of wealthy citizens consisting of animals, plants, fossils and minerals. Even J. W. von Goethe had critically reported on the situation of natural history studies in Frankfurt (Goethe 1816). Following the founding of the society in 1817 and the opening of the museum in 1821, the zoological collections were divided into the following categories: vertebrates (mammals, birds), molluscs, insects and parasitic worms. Skilfully prepared anatomical objects were considered specialties.

The majority of the specimens originated from the private collections of founding members, although one significant early exotic collection came from Brazil. Georg Wilhelm Freyreiss (1789–1825), a naturalist from Frankfurt who lived in Rio de Janeiro, collected many specimens on extended research trips (Freyreiss 1968) and sent them to Berlin and Frankfurt.

Nevertheless, without Eduard Rüppell (1794–1884), all these early collections would probably have remained an interesting episode and Senckenberg perhaps a merely local museum. He was the one who brought together important specimens of natural history from the Middle East, including marine organisms from the Red Sea, so that the young Senckenberg Museum was soon catapulted into the league of world-class museums.

Eduard Rüppell undertook two major collecting trips to Egypt and from there to Eritrea. The first expedition (1822–1827) took him through a vast region of Northern Africa as well as to the opposite side. The places where he has discovered significant collections were Upper Egypt (Fayum), Nubia, Kordofan, Sinai, Gulf of Aqaba, Jeddah and Massaua. The second expedition (1831–1833) took him mainly to inland Abyssinia. The extensive material collected on both trips was sent to Frankfurt and formed a new core for all zoological collections of Senckenberg (Mertens 1949). This was the starting point for the Senckenberg expertise in zoology. Whereat through the exchange of duplicates additional valuable material could be acquired from other institutions which was a further enrichment.

In 1870, a number of Frankfurt naturalists and members of the Senckenberg Society established a research fund, which—in recognition of his extraordinary efforts—was named after Rüppell. The aim of this "Rüppell Stiftung" was to finance further expeditions and thus to promote the increase of the Senckenberg collections (Knoblauch 1920). From that date until 1917, twelve expeditions were undertaken.

28.2.1.1 The Entomological Collections in Frankfurt

As in many other museums, the entomological collections are the largest of Frankfurt's zoological holdings. They comprise circa 3.7 million specimens, of which the majority belong to the following three orders:

- Coleoptera
- Lepidoptera
- Hymenoptera

but also containing nearly all other insect orders.

The oldest collections are from Brazil (collected by Freyreiss), followed by those from Northern Africa and the Middle East (Rüppell). Another significant early collection was that of Carl von Heyden (1793–1866), a founding member of the Senckenberg Society, and his son Lucas von Heyden. Their emphasis was Palaearctic microlepidoptera (the father) and beetles (the son), but the collections also contained many other orders of insects. Of great importance were also the collections dispatched from China by Johann Tobias Ronnefeldt, a merchant from Frankfurt. Max Saalmüller (1832–1890) worked on a significant collection of moths and butterflies from Madagascar at the Senckenberg. A very important addition was the Lepidoptera collection of Adalbert Seitz, editor of the spectacular monograph series "The Macrolepidoptera of the World" (*Die Großschmetterlinge der Erde*). Seitz worked as curator of Lepidoptera from 1919 to 1938 at the Senckenberg Institute.

Subsequently, further collections of insects came from Mexico, El Salvador, Bolivia, Peru, Argentina, Chile, Northern Africa, Madagascar, Nigeria, Cameroon, Tanganyika, Inner Africa, South Africa, the Philippines, the Moluccas, the Faroe Islands and Kei Islands, Java, Micronesia, Australia and South Georgia. Along with these objects from overseas, further material came from Europe, including the Canary Islands and Madeira (Franz 1967). More recently, the Coleoptera collections from Venezuela and Nepal, bombycoid moths from Mexico as well as further material from different parts of Germany and Europe were important additions.

Together with an old collection of Georg Semper (1837–1909), the huge Philippine Lepidoptera collection of Colin G. Treadaway F.R.E.S. (appointed honorary research associate at the Senckenberg Entomology II division) is of great importance. Together they form one of the worldwide largest collections from that region and contain a large number of primary and secondary types. For the Hymenoptera the main emphasis of the collection is the western Palaearctic. Through recent scientific work, the Symphyta are well represented and documented.

Beside the large orders already mentioned above, there is a significant collection of thrips (order Thysanoptera) in Frankfurt, which was curated by Richard zur Strassen (1926–2013). He worked on the taxonomy and zoogeography of these tiny insects and was considered one of the very few specialists of the world. Therefore, zur Strassen often identified the material of other scientists. The collection, mostly accumulated personally, includes about 50 % of the 5300 species known to date (Figs. 28.2, 28.3, 28.4, 28.5, 28.6, and 28.7).

Fig. 28.2 *Smerithus ocellata* (Sphingidae) (with permission from: Senckenberg Forschungsinstitut und Naturmuseum; photo: Sven Tränkner)

Fig. 28.3 Tropic Nymphalide butterflies (with permission from: Senckenberg Forschungsinstitut und Naturmuseum; photo: Sven Tränkner)

28.2.2 The Invertebrates in Frankfurt

28.2.2.1 The Mollusc Collections

The second largest zoological collection is that of Mollusca with roughly 700,000 lots, including about 30,000 type lots. The collection started in 1823 with a small number of marine molluscs. However, it grew rapidly and gained importance with the specimens of Eduard Rüppell, chiefly from the Red Sea and the Abyssinian highlands, and by exchange with foreign collectors and museums. Due to the activities and influence of Wilhelm Kobelt (1840–1916) and the curation through Fritz Maas (1886–1969) and Adolf Zilch (1911–2006), the proportion and significance of marine molluscs has diminished since 1869 in favour of terrestrial and limnic molluscs. Today, these form one of the world's largest and scientifically

Fig. 28.4 *Above*: *Lycaena sp*. possibly *Lycaena hippothoe*; *below*: *Polyommatus sp*. possibly *Polyommatus bellargus* (with permission from: Senckenberg Forschungsinstitut und Naturmuseum; photo: Sven Tränkner)

Fig. 28.5 Dr. Wolfgang Nässig, curator of the Senckenberg Entomology II (Lepidoptera) (with permission from: Senckenberg Forschungsinstitut und Naturmuseum; photo: Sven Tränkner)

Fig. 28.6 Drawers from the beetle collection (with permission from: Senckenberg Forschungsinstitut und Naturmuseum; photo: Sven Tränkner)

Fig. 28.7 Dr. Damir Kovac, curator of the Senckenberg Entomology I (Coleoptera, Thysanoptera, Strepsiptera) (with permission from: Senckenberg Forschungsinstitut und Naturmuseum; photo: Sven Tränkner)

most valuable collections with particular empasis on the Palaearctic, Southeast Asia, China and the Philippines. Amog others it contains the collections of E.A. Rossmaessler (1806–1867). O. Boettger (1844–1910), P. Ehrmann (1868–1937), O.F. von Moellendorff (1848–1903) and H. Schütt (1923-2009). Of special systematic significance are the door snails (Clausiliidae), of which Senckenberg has the worldwide largest collection with an unparalleled number of types. Other collections of special significance are those of the Cyclophoroidea, the freshwater Rissooidea, Helicoidea and the freshwater mussels (Unionida).

At present the marine part of the Frankfurt mollusc collection covers 40 % of the roughly 33,000 species contained in that collection. A large proportion comes from the seas around the Arabian Peninsula. However, an impressive number of the marine species also comes from Hainan and Colombia. The molluscs from the deep sea and from hydrothermal vents, which were collected through the expeditions with the German research vessels Meteor, Sonne and Valdivia are of similar great significance as the big stocks from European seas and the Eastern Atlantic (Zilch 1967; Janssen 1992, 2011; Senckenberg homepage). (Figs. 28.8, 28.9, and 28.10).

28.2.2.2 Spiders and Their Relatives

Currently, the arachnid collection comprises some 500,000 specimens in 100,000 lots including around 20,000 species and 11,000 type series. On average 20 % of all species worldwide are included. The earliest historical specimens came predominantly from Central Europe. However, further specimens from tropical regions and from overseas were acquired early on from the Middle East and North Africa via Eduard Rüppell, from the Moluccas via Willy Georg Kükenthal (1861–1922), from the Faroe Islands and Kei Islands via Hugo Merton (1879–1940) and from Micronesia and Melanesia via Senckenberg curator Eugen Wolf. Considerable growth

Fig. 28.8 Specimens from Senckenberg's important collection of Phillipine land snails (with permission from: Senckenberg Forschungsinstitut und Naturmuseum; photo: Sven Tränkner)

Fig. 28.9 Views from Clausiliidae Clausiliiden (with permission from: Senckenberg Forschungsinstitut und Naturmuseum; photo: Sven Tränkner)

Fig. 28.10 Preserved material from hydrothermal vents (with permission from: Senckenberg Forschungsinstitut und Naturmuseum)

was achieved by the acquisition of the collections of Carl-Friedrich Roewer and Hermann Wiehle and later also of collections from El Salvador (A. Zilch) and Peru (H. W. Koepcke). In recent years, Jochen Martens (honorary research associate) has integrated his collections from the Himalayas, while the current curator Peter Jäger permanently enriches the collections with many new species of sometimes extraordinary spiders from Laos and other countries in Southeast Asia.

Beside spiders and camel spiders at Senckenberg in Frankfurt, the collection of harvestmen is probably the largest collection in the world with respect to the number of species. It contains more than 3600 species with a high percentage of type material. The arachnid collection also contains a large number of mites (Acari), especially water mites (Hydrachnidia and Halacaridae). In total there are 17,000 catalogued specimens identified to species level, of which about 1700 are holotypes and 1400 paratypes. The large collection of freshwater mites from Karl and Karl-Otto Viets (father and son) is noteworthy.

Historically the provenance of the myriapod (millipede) collection was the same as that of the arachnid collection. Currently, there are some 6700 specimen lots identified to species level, including about 450 holotypes and 880 paratype series. Otto Kraus curated the collection from 1954 to 1969. Later, guest researchers like Golovatch, Stoev and Wesener continued the work in the collections (Fig. 28.11).

28.2.2.3 The Crustacean Collections

The crustacean collection is also one of the oldest in the museum. In 1824, Senckenberg member Carl von Rothschild of Naples contributed specimens from the Mediterranean. However, it was through the agency of Eduard Rüppell that the crustacean collection experienced its first major expansion. Rüppell sampled four regions around the Red Sea intensively: the Gulf of Suez (At Tur) and the Gulf of Aqaba, the region around Jeddah and the southernmost region of Massaua (Mertens 1949). A major exchange of specimens with the museums of Leiden (1829) and Paris (1830) led to further increase in species representation. The first handwritten catalogue, compiled by Adolf Reuss in 1832, lists 259 items, of which 108 were from Rüppell. During the following years, a number of collections from many regions of the world came in: Europe, Brazil, North and East Africa, Madagascar, Moluccas, Melanesia and Micronesia, South Georgia and finally the deep sea. The last resulted from the German deep sea expedition of the Research Vessel Valdivia (1898 to 1899).

After 1948, under the lead of Richard Bott (1902–1974), the research focus in the crustacean division shifted to freshwater crabs; and the new subject has caused a further increase. By now it is one of the most important globally and is very rich in type material. In addition, as with the molluscs above, extensive deep sea decapod collections have been added since 1967 through the expeditions by the German research vessels Meteor, Sonne and Valdivia.

In recent times, a joint project with King Abdul Aziz University in Jeddah expanded the Red Sea collections considerably (www.redseabiodiversity.org). New focal areas were also developed: the North Sea, the marine waters around the Arabian Peninsula and East Asia, especially the Sea of Japan. Currently, the crustacean collection comprises around 62,000 lots with about 650,000 individuals (Bott 1967; Türkay 1981, 1992, 2015) (Figs. 28.12 and 28.13).

Fig. 28.11 *Above*: Dr. Peter Jäger, curator of the arachnid collection, please shift; *below*: *Steginoporella buskii* Harmer, 1900 from the north coast of Socotra, Yemen. Belongs to bryozoans (with permission from: Senckenberg Forschungsinstitut und Naturmuseum; photo: Sven Tränkner)

Fig. 28.12 *Saron sp.* from the Red Sea. Belongs to the crustacean collections (with permission from: Senckenberg Forschungsinstitut und Naturmuseum; photo: Sven Tränkner)

Fig. 28.13 *Uca tetragongon* from the Red Sea. Belongs to the crustacean collections (with permission from: Senckenberg Forschungsinstitut und Naturmuseum; photo: Sven Tränkner)

28.2.2.4 Other Invertebrates

The remaining, smaller, mostly marine invertebrate groups of the collections comprise around 75,000 specimens. The majority of the material on the whole represents four groups consisting of: Porifera with about 11,000 lots, Cnidaria with about 16,000 lots, Polychaeta with about 19,000 lots and Bryozoa with about 20,500 lots. Currently specialists are studying Porifera, Polychaeta and Bryozoa intensively; Cnidaria (more specifically Octocorallia) were also studied until recently.

As with the larger collections treated before, the influence of scientific studies on size and increase of collections is obvious. The early historical expeditions are also the base for the collections of these groups. In addition, the intensive sampling in the North Sea and in the deep sea in recent years has increased the numbers of marine invertebrates considerably. Because the Porifera are an important animal group for research in the Antarctic, their number in the Frankfurt collections has increased considerably through the intensive recent work by the scientists.

Important parts of the bryozoan collection are the mostly fossil specimens from the legacy of Eberhard Voigt (1905–2004) and the donation of recent Bryozoa from Heinrich Ristedt (University of Bonn). The photos with bryozoans refer to this part.

Fig. 28.14 Steginoporella buskii (2) (with permission from: Senckenberg Forschungsinstitut und Naturmuseum; photo: Sven Tränkner)

The Voigt collection represents the globally most comprehensive sample of Cretaceous bryozoans. It consists of 160 types, 2200 published originals and about 300,000 additional specimen series from Ordovician to Recent. The Ristedt collection comprises about 15,000 specimen boxes with material predominantly from the Mediterranean, the Antarctic and from many localities of the Indo-Pacific.

For the Polychaeta collection and those of other invertebrate groups, a detailed list of regions and collectors is available on the Senckenberg homepage (Kraus 1967; Senckenberg homepage) (Figs. 28.14, 28.15, and 28.16).

28.2.2.5 The Fish Collections

The collection of fish consists of about 54,000 lots containing about 138,000 individuals. About half of the collection is of marine origin, the rest freshwater. As with many other Senckenberg collections, these began with Eduard Rüppell. The fish that he collected on his two expeditions to the Red Sea and the Nile were sent to Frankfurt. The skeletons of the fish that he prepared became signature pieces of the collection and played a great role in anatomical descriptions. Skeleton preservation was continued until modern X-ray machines replaced this laborious work. Subsequently, fish from many parts of the world have found their way into the collection. Nevertheless, as in early days, marine species from the Red Sea and the northwest part of the Indian Ocean remained the focus.

Thus, many specimens from marine waters around the Arabian Peninsula, including the Red Sea, the Gulf of Aden, the Gulf of Oman and the Persian/Arabian Gulf, are represented in this globally most extensive collection. Besides individual collecting trips, this is also a result of large expeditions with participation of

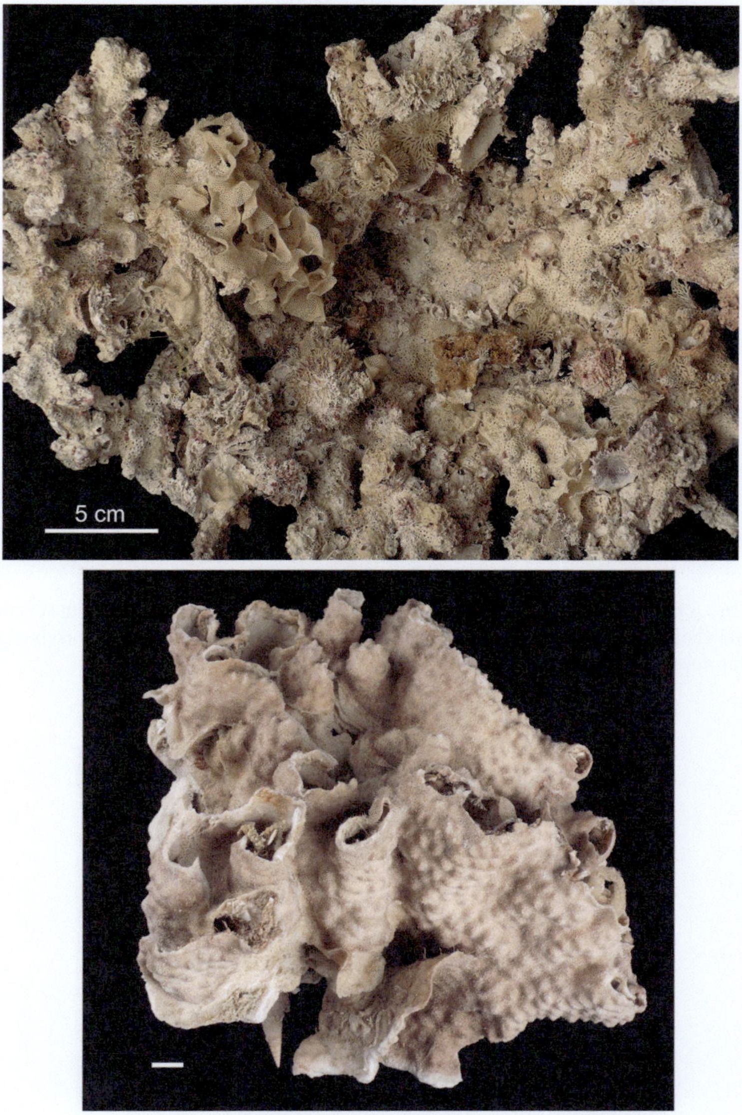

Fig. 28.15 *Above*: dead coral with a diverse bryozoan settlement, found near Abd al-Kuri, Socotra west coast, Yemen. *Below*: *Celleporaria agglutinans* (Hutton, 1873), Otago shelf, New Zealand (with permission from: Senckenberg Forschungsinstitut und Naturmuseum; photo: Sven Tränkner)

Senckenberg scientists: the Xarifa Expedition (1957/1958), Meteor cruise 1 (1964/1965), Sonne cruise 2 (1977), Valdivia cruises 22 (1979) and 29 (1981) and Meteor cruise 5 (1987). Most of these sea expeditions focussed on the deep sea of this area, which has contributed considerable material to the collection. Additional sources of material were the work of Senckenberg scientists on a marine wildlife sanctuary in

Fig. 28.16 A bryoherm—*Hippomenella vellicata* (Hutton, 1873)—overgrown with several cyclostome and cheilostome bryozoan species (with permission from: Senckenberg Forschungsinstitut und Naturmuseum; photo: Sven Tränkner)

Fig. 28.17 *Harpadon erythraeus* (with permission from: Senckenberg Forschungsinstitut und Naturmuseum)

the Persian/Arabian Gulf as response to the oil spill of the Gulf War in 1991 and the Red Sea Biodiversity Project mentioned earlier under the crustacean collections. While the focus has been on marine fish, the extensive freshwater collection from the Middle East is one of the globally most comprehensive ones (Figs. 28.17, 28.18, and 28.19).

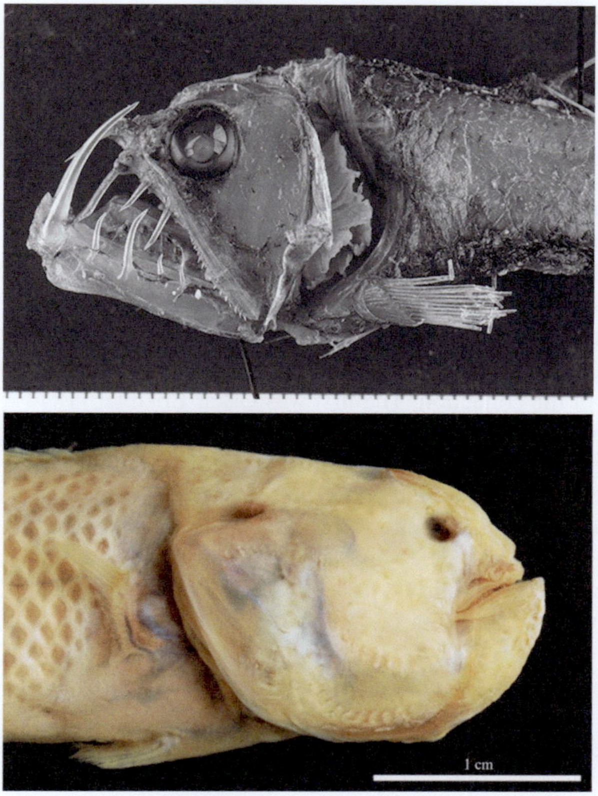

Fig. 28.18 *Above*: *Chauliodus pammelas*; *below*: *Trypauchen raha*. Both refer to the "Fish Collections" (with permission from: Senckenberg Forschungsinstitut und Naturmuseum)

Fig. 28.19 Dr. Eduard Rüppell (with permission from: Senckenberg Forschungsinstitut und Naturmuseum)

28.2.3 The Terrestrial Vertebrates in Frankfurt

28.2.3.1 The Herpetological Collections

The core of the collections of terrestrial vertebrate, just like those of aquatic animals, was formed by the samples sent by Freyreiss and Rüppell. This material also contributed to the growth of the Senckenberg biodiversity archives, because several specimens of the same species were exchanged for material from other museums and private collectors. Thus, these duplicates helped to multiply in short order the number of species in the Frankfurt collections (Klausewitz 1967; Klausewitz et al. 1992; Senckenberg homepage).

In the herpetological division, two persons were principally responsible for the rapid growth of the collection. The first was Oskar Boettger (1844–1910), head of the division from 1875 until the year of his death. During his time as curator, many specimens from a multitude of localities found their way to Frankfurt and became the basis for the 230 species and varieties that Boettger described as new. Some of these were part of larger collections from Madagascar, the Moluccas, New Guinea and from Australia; colleagues and friends from many countries, however, sent material. Boettger also published three catalogues of the herpetological collection, which document the holdings of the herpetological collection at the end of the nineteenth century.

The second person of high significance in this respect was Robert Mertens (1894–1975), curator of herpetology from 1920 to 1960, but active until the year of his death. He collected much material during the course of numerous expeditions and enriched the collection considerably. Besides Europe, these trips brought him to North Africa, Cameroon, Kenya, Namibia, South Africa, Pakistan, Indonesia, Australia and many countries in Latin America.

In recent years, the scientific activities by the current curator Gunther Köhler and his coworkers are concentrated on the Neotropics. Through his activities, the Central and South American parts of the herpetological collections have grown considerably and about 100 species of amphibians and reptiles, especially in the genus *Anolis*, have been described. The majority of this type material is part of the Senckenberg collections. Presently, the collection comprises about 110,000 specimens (Mertens 1967; Klemmer 1992; Senckenberg homepage) (Figs. 28.20 and 28.21).

28.2.3.2 The Ornithological Collections

The ornithological collection is one of the largest amongst the vertebrate collections. It includes about 100,000 specimens representing at least 75 % of the species known to date. Three collectors were especially important in the history of acquisitions: Eduard Rüppell, Carlo von Erlanger, and Hans von Berlepsch. The last, in particular, provided the greatest number of bird species. Geographically, the focus

Fig. 28.20 PD Dr. Gunter Köhler, curator of the Senckenberg herpetological collections Frankfurt (with permission from: Senckenberg Forschungsinstitut und Naturmuseum; photo: Sven Tränkner)

of the collection is in Europe, North Africa and South America. Most of the historical material consists of skins with feathers. Of special significance are the specimens of extinct bird species: 30 such species are represented by material in the collection. Besides the skins, there are also collections of eggs and nests (Steinbacher 1967; Peters 1992; Senckenberg homepage).

Skeletons were present even in the early collections, such as those of Eduard Rüppell. But while skeletons and fluid-preserved whole specimens are of importance for anatomical and phylogenetic studies, they are underrepresented in the collection. During Dieter Stefan Peters' tenure as curator of the ornithology division, the scientific interest shifted to the fossil birds. Comparison of fossils with the recent fauna was only possible through an expansion of the osteological and anatomical collection, which was pursued after 1976. The current curator, Gerald Mayr, has continued and expanded the work on fossil birds and phylogeny since 1997. The osteological collection has now grown into one of the largest in Europe. It includes more than 20,000 specimens representing 2070 species (Mayr 2011) (Figs. 28.22 and 28.23).

28.2.3.3 The Mammal Collections

In the beginning the mammal collection was one of the smaller collections. Eduard Rüppell contributed a number of large mammals as well as smaller specimens like bats, rodents and insectivores. Besides Rüppell's collections from northeast Africa,

Fig. 28.21 *Above left*: Dr. Robert Mertens (1894–1975), curator of the herpetological collections from 1920 to 1960. *Above right*: Oskar Boettger (1844–1910), head of the division from 1875 to 1910. The fish belongs to "The Fish Collections". Credit: *Labeo coubie* Rüppell, paratypus (with permission from: Senckenberg Forschungsinstitut und Naturmuseum)

material from South Africa, southern Europe, South America and North America could also be acquired, and even in later years these regions remained the main source for collected material and research. Additional material came from Australia as well as from East and Southeast Asia. One of the most prominent curators was Fritz Römer (1866–1909), Senckenberg director from 1907 to 1909 (Gutmann 1967). However, the main increase of the mammal collections happened after 1946.

Heinz Felten, curator of the mammal division from 1955 to 1987, changed the focus of the collections towards small mammals like rodents, insectivores and bats. Material of these mammal orders came in regularly from many regions of the world, especially tropical America, Southeast Asia, Europe and the Mediterranean region. During his curatorship the division increased and Felten was joined by the

Fig. 28.22 Dr. Gerald Mayr, curator of the Senckenberg ornithological collections Frankfurt (with permission from: Senckenberg Forschungsinstitut und Naturmuseum; photo: Sven Tränkner)

Fig. 28.23 *Middle*: Dr. Gerald Mayr with the tiny skeleton of a hummingbird; *left*: drawer with skeletons of kingfishers. Both refer to "The Ornithological Collections", Frankfurt. *Right*: *Centropyx paulensis*. Belongs to "The Herpetological Collections", Frankfurt (with permission from: Senckenberg Forschungsinstitut und Naturmuseum; photo: Sven Tränkner)

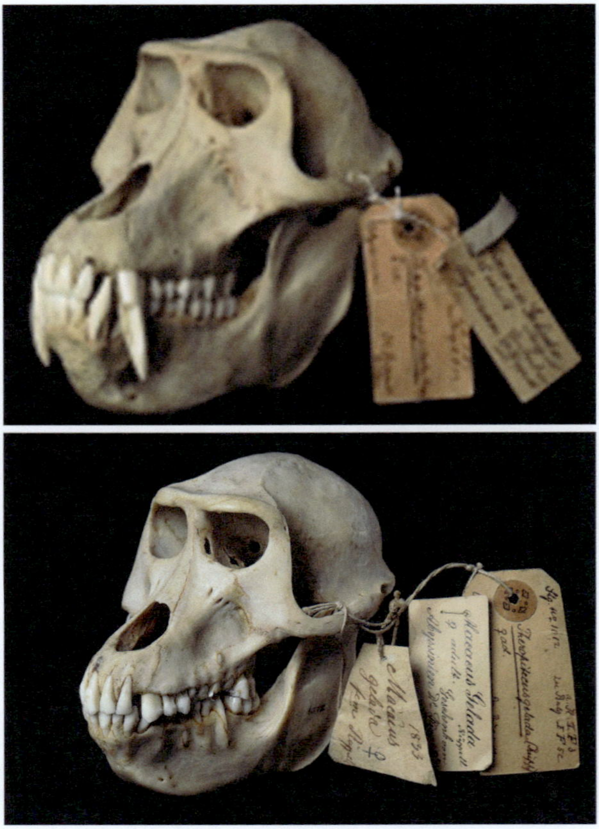

Fig. 28.24 Both pictures show *Theropithecus gelada*. Both specimens were presents from Eduard Rüppell. Refer to the Senckenberg Mammal Collections, Frankfurt (with permission from: Senckenberg Forschungsinstitut und Naturmuseum)

co-curators Dieter Kock and Gerhard Storch, each of them focused on other animal groups.

Through these endeavours, the holdings grew to 95,000 recent, 66,000 fossil, about 7000 anatomical specimens and 1000 histological slide series. The bat collection is one of the most important worldwide and contains 90 % of all known genera. In addition, the primate collection is one of the most important in Germany (Felten 1967; Storch 1992; Senckenberg homepage).

Today, the mammalogy division is also in charge of the comparative anatomy collection. This collection is very old, dating to the middle of the nineteenth century. One of the most prominent curators was Fritz Römer (1866–1909), Senckenberg director from 1907 to 1909 (Gutmann 1967). A further curator was appointed in 1967, Wolfgang Friedrich Gutmann (1935–1997), who enriched the collection considerably with new anatomical objects and histological material. After his death several collections (Starck, Pilleri, Kuhn, Klima, Edinger) were added.

During the following years, the collections were supervised by Thomas Martin and Ottmar Kullmer, supported by Virginie Volpato. Currently Irina Ruf is the curator in charge (Figs. 28.24, 28.25, 28.26, and 28.27).

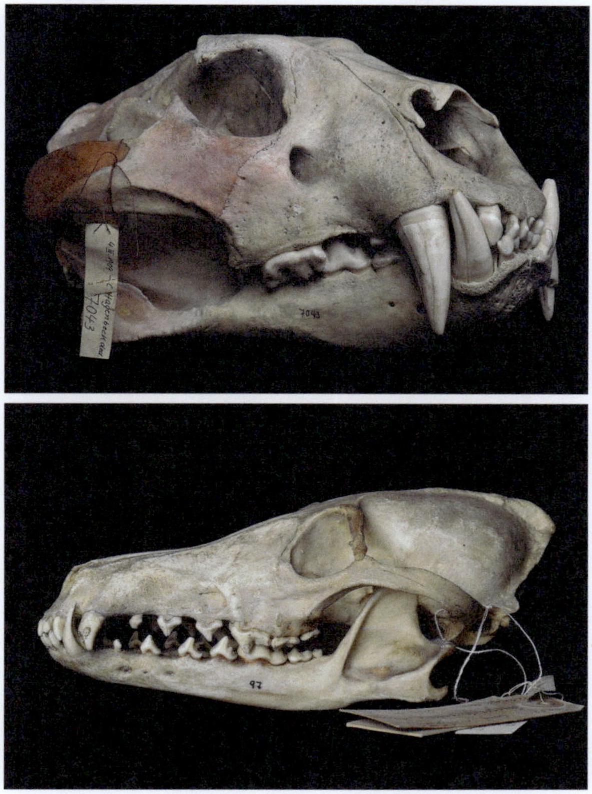

Fig. 28.25 *Above*: *Panthera leo tigris*; *below*: *Canis simensis*. Both pictures refer to the Mammal Collection Frankfurt (with permission from: Senckenberg Forschungsinstitut und Naturmuseum)

Fig. 28.26 *Equus quagga quagga*. An extinct species. The Senckenberg Natural History Museum shows one of the yet 24 worldwide existing specimens (with permission from: Senckenberg Forschungsinstitut und Naturmuseum)

Fig. 28.27 The photo on *top* in the *middle* shows a staff member in the Frankfurt mammal collection. The one *left* under it shows *Pagurus bernhardus* from the Dogger Bank in the North Sea. This belongs to the crustacean collections (with permission from: Senckenberg Forschungsinstitut und Naturmuseum; photo: Sven Tränkner)

28.2.3.4 The Senckenberg Natural History Collections Dresden

By Raffael Ernst, Uwe Kallweit, Klaus-Dieter Klass, Matthias Nuss, Martin Päckert, Christian Schmidt, André Reimann, Katrin Schniebs, Clara Stefen, Andreas Weck-Heimann, Axel Zarske and Uwe Fritz

28.2.4 The Ups and Downs of an Impressing History

In 1560, Saxon Elector August I (1528–1586) founded his "Chamber of Arts" (Kunstkammer). Amongst many impressing artefacts, there were also representative specimens of natural history. An important step in the history of this collection was their separation from the objects of art through the well-known Saxon Elector and Polish King August II, called August the Strong (1670–1733). In 1728, he moved the natural history cabinet to the "Palais de Sciences", the well-known Dresden Zwinger (Fig. 28.28). Therefore, the year 1728 is understood as the year of origin of the Dresden Natural History Museum.

In 1820, thus almost a whole century later, Heinrich Gottlieb Ludwig Reichenbach (1793–1879) took over the position as inspector of the Natural History Collections (Fig. 28.29, left). This was the start of a crucial phase because Reichenbach soon has opened the collections to the public. Besides, he also has focused his research in particular on the fauna of Saxony. Finally, in 1857, he has separated together with his colleague Hanns Bruno Geinitz (1814–1900) the geological, mineralogical and paleontological collections from the botanical and zoological museum. This was an important decision, because it led to the today's "Museum für Mineralogie und Geologie" which is now also part of Senckenberg and houses significant mineralogical, petrological and paleontological collections.

During the Dresden May Uprising of 1849, large parts of the Zwinger fell a victim to fire, and most of the collections were destroyed. Only few specimens were rescued, mainly birds. After the destruction, Reichenbach had to overcome strong opponents who voted against a rebuilding of a zoological museum. But, thanks to Reichenbach's worldwide connections and because of his friendship with Saxon King Friedrich August II (1797–1854), he succeeded in restoring the holdings within an admirably short time.

In 1874, Adolf Bernhard Meyer (1840–1911) followed Reichenbach as director (Fig. 28.29, right). Meyer has sold many zoological, anthropological and ethnographical specimens to the museum, which he had obtained during expeditions to Sulawesi, the Philippines and New Guinea (1870–1873). Meyer was mainly interested in ornithology, mammalogy, anthropology, ethnography and evolution and zoogeography. For instance, he has translated several books by A. R. Wallace into German and has published a German synthesis of Wallace and Darwin's theories on evolution.

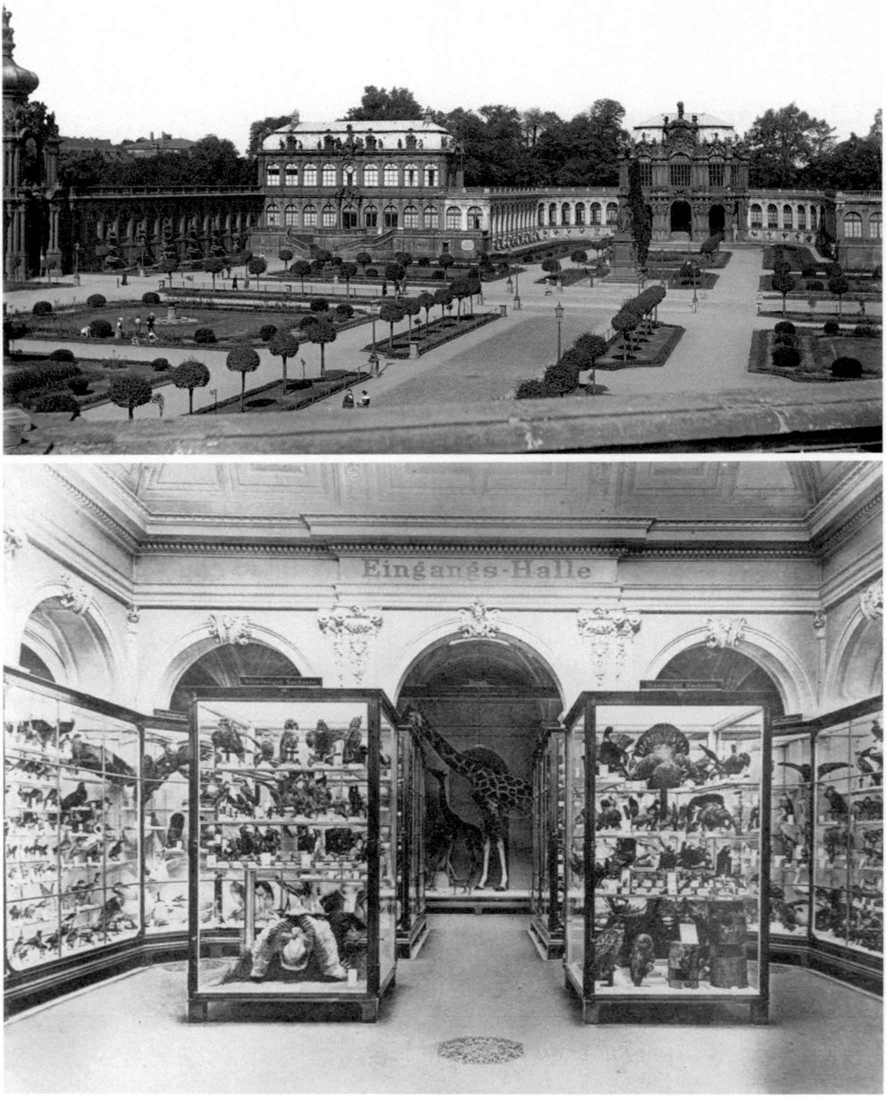

Fig. 28.28 The historical Palais des Sciences, the Dresden Natural History Museum in the Zwinger, before the Second World War

Fig. 28.29 (*left*) Heinrich Gottlieb Ludwig Reichenbach (1793–1879), long-term director of the Royal Natural History Museum at Dresden; (*right*) Adolf Bernhard Meyer (1840–1911), influential director of the Dresden Museum in the late nineteenth century (with permission from: Senckenberg Naturhistorische Sammlungen Dresden (SNSD))

Under Meyer's directorship, the museum was entirely reorganised and thus gained a modern profile. The botanical collection was transferred to the Technical University at Dresden in 1875, and a new division for anthropology and ethnography was founded. Then the name of the museum has been changed from "Königliches Naturhistorisches Museum" in "Königliches Zoologisch-Anthropologisch-Ethnographisches Museum Dresden" (Royal Zoological-Anthropological-Ethnographical Museum Dresden). Besides, Meyer has separated the exhibitions from the scientific collections according to the system of the British Museum in London. In addition, he has acquired many extinct animals for the collections, e.g. an almost complete skeleton of a Steller's sea cow (*Hydrodamalis gigas*).

Meyer's work was so successful that Steward Culin, an American ethnographer, in 1898 mentioned the museum in the American newspaper Philadelphia Post as "the model museum of the world" and Meyer "as one of the foremost museum administrators in Europe". One aspect of this success story was that special iron collection cabinets were developed by Meyer together with the Dresden company August Kühnscherf & Söhne. These "Kühnscherf cabinets" were later bought by many other museums in Europe and overseas, for example, in Berlin, Prague, St. Petersburg, Vienna and Washington (Hertel 1978).

However, Meyer, who originated from a Jewish-German family in Hamburg, became a victim of anti-Semitic intrigues and was discharged in 1906. With him, the Dresden Museum undoubtedly lost one of its most productive, influential and important directors.

It seems as a dark piece of irony that the once by the Jewish-German Meyer established new Darwinist profile of the museum with its anthropological-ethnic aspects later was misused by the Nazis for their purposes and even has contributed to one of the darkest chapters in German history. During this shameful period, the museum's name was "Staatliches Museum für Tier-, Völker- und Rassenkunde" (State Museum for Zoology, Ethnology and Racial Sciences).

After Meyer, for a longer time none of the directors has played an important role, perhaps except for Arnold Jacobi (1870–1948), who considerably expanded the entomological collections. To enumerate the other directors would be beyond the scope of this contribution. General aspects of the further history can be found in Kummerlöwe (1939), Reichert (1954, 1956a, b) and Hertel (1978).

In the 1930s, the museum moved from the Zwinger to another building in Dresden, which provided more space for the exhibition. Since 1940 and during World War II, fortunately large parts of the museum collections and the library could be evacuated and stored in sixteen castles and manors in central and eastern Saxony. However, due to orders of the Nazi Government, the exhibition had to remain open for the public. Unfortunately not all scientific collections could be moved. The alcoholic collections, amongst others the ichthyological, herpetological and marine biological specimens, were deposited in the basement of Dresden castle.

During the first bombing of Dresden, on October 7th, 1944, parts of the museum building were destroyed, and with it, many specimens of the entomological research collection. A few months later, in the night of February 13th to 14th, 1945, the rest of the exhibition, as well as the alcoholic collections in the castle, shared the fate of the historic city of Dresden and were destroyed nearly entirely (Reichert 1954, 1956a; Obst 1977).

Following World War II, many specimens of the surviving collections were confiscated by the Red Army and transported to the natural history museums in Moscow and Leningrad. There they remained for years and were given back only in 1978 to celebrate the 250th anniversary of the natural history museum in Dresden. After the war, the rests of the collections and the library step by step were brought back to the destroyed city of Dresden. This was a long process, which for a major part of the collections has lasted more than 10 years. However, the last batch of the specimens from a Saxon castle arrived only in 1999! Robert Reichert (1897–1959), director from 1947 to 1957, has reported in two articles on the hard postwar times, by the way, without mentioning any confiscated specimen (Reichert 1954, 1956a).

After Reichert, the directorship was taken over by Wilhelm Götz who had moved for political reasons from Stuttgart to the German Democratic Republic (1957–1966). Götz had planned to alter the museum into an entomological research institute. For that purpose, he wanted to donate the vertebrate and non-entomological collections to other institutes and to merge the museum with the Deutsches Entomologisches Institut (German Entomological Institute). The entomologists should get the space that would have become free by other collection parts. This plan had not found any political support. Except a few specimens that had already been transferred to the National Museum in Budapest, which was completely destroyed in 1956 during the Hungarian revolution, the Dresden collections remained where they were. Götz was discharged as director, but remained in the museum as guest researcher. Interestingly, he was during this time officially employed at the Zoologisches Museum (Natural History Museum) Berlin.

A further essential change in postwar years was the separation of the zoological collections from the anthropological-ethnological collections which constituted from then onwards the Dresden Museum of Ethonology (Museum für Völkerkunde).

At first, the complete zoological museum was housed in a more or less undestroyed part of the Zwinger in postwar times. Later, only the small exhibition remained there while the research divisions as well as the scientific collections with after all still about 6.5 million specimens were moved into the ruined former Saxon parliament, the "Ständehaus". Due to the bad logistic and financial situation during Socialist times, they remained there until 1998.

Under the directorship of Fritz Jürgen Obst, the collections finally found a safe home in the two new research buildings in Dresden-Klotzsche (Fig. 28.30). One building honours Adolf Bernhard Meyer; the other received its name in remembrance of Robert Reichert. In these buildings, to which later a third office building was added, the collections now are housed under excellent conditions (Fig. 28.31) and thus provide formidable prerequisites for modern biodiversity research. This could be boosted by the acquisition of modern laboratories for genetic research under the current director Uwe Fritz.

28.2.5 The Realm of Collections

28.2.5.1 The Invertebrates in Dresden

The collection of lower invertebrates, including what was formerly often referred to as "Vermes", bears a small but significant historical part of the collections of the former Zoological Museum of the University of Leipzig. It also comprises materials of Georg Friedrich Rudolf Leuckart (1822–1898), the German founder of the research branch of parasitology, who led the Institute of Zoology since 1870. Further material originates from Hinrich Nitsche (1845–1902) and Carl Chun (1852–1914), the latter a disciple of Leuckart, and head scientist of the German Deep Sea Expedition named after the ship "Valdivia".

28.2.5.2 The Non-insect Invertebrates

In the collection of higher non-insect invertebrates are housed about 20,400 items with 94,800 specimens of arachnids, crustaceans, myriapods, echinoderms and tunicates. Because the original type-rich collection was largely destroyed in 1945, more than half of the material was acquired after the World War II. A considerable acquisition for these and other collections was the majority of the holdings of the Zoological Institute Leipzig in 1968. It brought material of a wide geographical

Fig. 28.30 A. B. Meyer Building, one of the new research and collection facilities of the Senckenberg Natural History Collections Dresden (with permission from: Senckenberg Naturhistorische Sammlungen Dresden (SNSD))

Fig. 28.31 Collection room in the A. B. Meyer Building

range, but mostly from Europe and South America and to some extent Africa and Australasia. The Arachnida collection is by far the biggest with about 15,000 items and at least 80,000 specimens.

Another major part constitutes the collection of Heinz Hiebsch, a Saxon arachnologist, which was bought in 1976. In addition, the holdings comprise material from Southeastern Europe, Africa, Australia and South America. The crustacean collection contains 3500 items with approximately 10,000 specimens. It includes 123 type specimens and material mainly from Europe, South America, Australia and Southeast Asia. The remaining collections are smaller and comprise together 1900 items with 4800 specimens. Historically important invertebrate material originates from W. Gueinzius (South Africa), F. T. Doflein (Japan), C. Felsche (Indonesia), B. Dybowski (Lake Baikal) and L. W. Schaufuss (Indo-Pacific). Additionally, there are specimens from the Albatross Expedition, the German South Pole Expedition and the German Deep Sea Expedition.

With well over one million specimens, the mollusc collection is the second largest zoological collection in Dresden. Historically important are the collections from H. E. Anton and Th. & P. Reibisch containing type material, specimens collected by E. F. Poeppig in the early nineteenth century in South America and by W. Gueinzius in 1843 in South Africa as well as local collections from A. Vohland, K. Büttner and A. Schlechter for Saxony, from K. Regius for Saxony-Anhalt and from R. Haldemann for Brandenburg. The latter are important documentations of the mollusc fauna of these federal states. Furthermore, the collection contains unique samples of Saxon freshwater pearl mussels (Fig. 28.32).

28.2.5.3 The Entomological Collections in Dresden

The entomological collections in Dresden comprise by far the largest amount of specimens, with more than five million lots. The majority of insects are kept as dry specimens, but especially dipterans are also alcohol preserved, and specimens of many small species are mounted for microscopic purposes. While virtually all

Fig. 28.32 Part of the rich collection of Saxon freshwater pearl mussels (*Margaritifera margaritifera*)

insect groups are present in Dresden, amongst them the recently discovered species-poor order Mantophasmatodea with comparatively many specimens, only such collection parts will be mentioned below which are represented by significant individual numbers.

Generally, the entomological collections in Dresden are very type rich with name-bearing types of more than 14,000 insect taxa. In particular, the collection of beetles (Coleoptera) is renowned worldwide. It comprises more than two million specimens that represent about 80,000 species, with primary types for circa 11,000 taxa. Significant persons in the early period of collection development were curators T. F. W. Kirsch and K. M. Heller, who between 1875 and 1927 recruited several valuable private collections, such as the one of J. K. E. Faust with its type material for ca 2500 weevil species. The Curculionoidea are indeed the most important systematic focus of the collection today, represented by ca 20,000 species and 7500 taxa with primary types. The Dresden Curculionoidea collection is amongst the most important ones worldwide. Further taxa of particular relevance—also measured based on their representation by primary types—are the species-rich Scarabaeidae (chafers and relatives, mainly by the C. Felsche collection with ca 11,000 species) and the less speciose Mordellidae (mainly K. F. Ermisch collection), Lycidae (with the collection of R. Kleine) and Byrrhidae (mainly based on recent collecting by O. Jäger). Focal geographic regions are (beside Saxony), with regard to earlier periods, South America, Southeast Asia (mainly the Philippines: collections of W. C. M. Schultze in the 1920s) and the Himalayas (circa 70,000 specimens) in terms of more recent collections.

During the nineteenth century, Dresden became a metropolis of Lepidopterology. For that reason, Dr. Otto Staudinger (1830–1900) has established there his globally active insect trading company in 1859. In 1862, the Entomological Society "Entomologischer Verein Iris zu Dresden" was founded and published the famous journal "Deutsche Entomologische Zeitschrift Iris". Its membership comprised well-known lepidopterologists from all over the world. Because of this heyday, many butterfly and moth collections were given to the Museum of Zoology Dresden. The preparator Johannes Dreaseke (1892–1970), who for many decades has made an excellent service in the collections, has successfully saved large parts of the collection from the destruction of World War II. Today, the Lepidoptera are deposited in more than 6000 drawers with over 800,000 butterfly and moth specimens of 23,500 species from all over the world (Fig. 28.33). The number of primary types is at about 1000. Some of the most important contributions are those of European and Southeast Asian Lycaenidae by Carl Ribbe (1860–1934), the Lepidoptera from Tibet by Walter Stötzner (1882–1965) and the Zygaenidae by Manfred Koch (1901–1972). On a regional scale, the Dresden collection is a comprehensive archive of documents of the last 150 years that attest extinctions, fluctuations and the first arrivals of Lepidoptera in Saxony. Amongst the specimens are ones from Ernst Möbius (1869–1945), who published the fauna of "Macrolepidoptera" from the kingdom of Saxony in 1905 as well as the fauna of "Microlepidoptera" of Dresden and its surroundings in 1936. Today, the research is on a worldwide scale, with a focus on the biodiversity, phylogeny and systematics of the snout moths, a group comprising about 15,500 species (Nuss et al. 2003–2015).

The Diptera collection consists of more than 600,000 pinned specimens (10 %), microscopic slides (20 %) and alcohol-preserved specimens (70 %). Since 1997, a specific collection of fossil fungus gnat-like flies (Sciaroidea) is established, mainly amber specimens. A special treasure is the collection of 18,000 slides of the family Cecidomyiidae and 3000 affiliated herbs. There are type specimens of approximately 2000 taxa, mainly name-bearing types, with a focus on South America.

The Hemiptera collection comprises 370,000 specimens, amongst them type specimens of circa 770 species. It consists mainly of Auchenorrhyncha and Heteroptera. Most of the older non-Palaearctic material had been acquired via the insect trading company Staudinger & Bang-Haas and were studied by Arnold Jacobi (who was director of the Museum from 1906 to 1936). Hans Schiemenz (1964–1990) and Rainer Emmrich (1970–2004), who both curated the Hemiptera, mainly collected the European material. The collections from Hans Joachim Müller include also the greater part of the collection of Herman Haupt. Since 2009, also the Auchenorrhyncha and Heteroptera collection of Reinhard Remane (Professor of Zoology at the Marburg University) is kept in Dresden. These specimens count ca. 230,000 and were collected from 1944 to 2008, mainly in the western Mediterranean region, the Canary Islands and Germany. Very interesting material comes from two stays in Iraq (1957–1958) and Sudan (1961–1962).

The Orthoptera and "smaller insect orders" consist of ca. 65,000 specimens. Types represent about 170 species. An important part of the collection is material gathered by Wilhelm Götz in Southeast Europe.

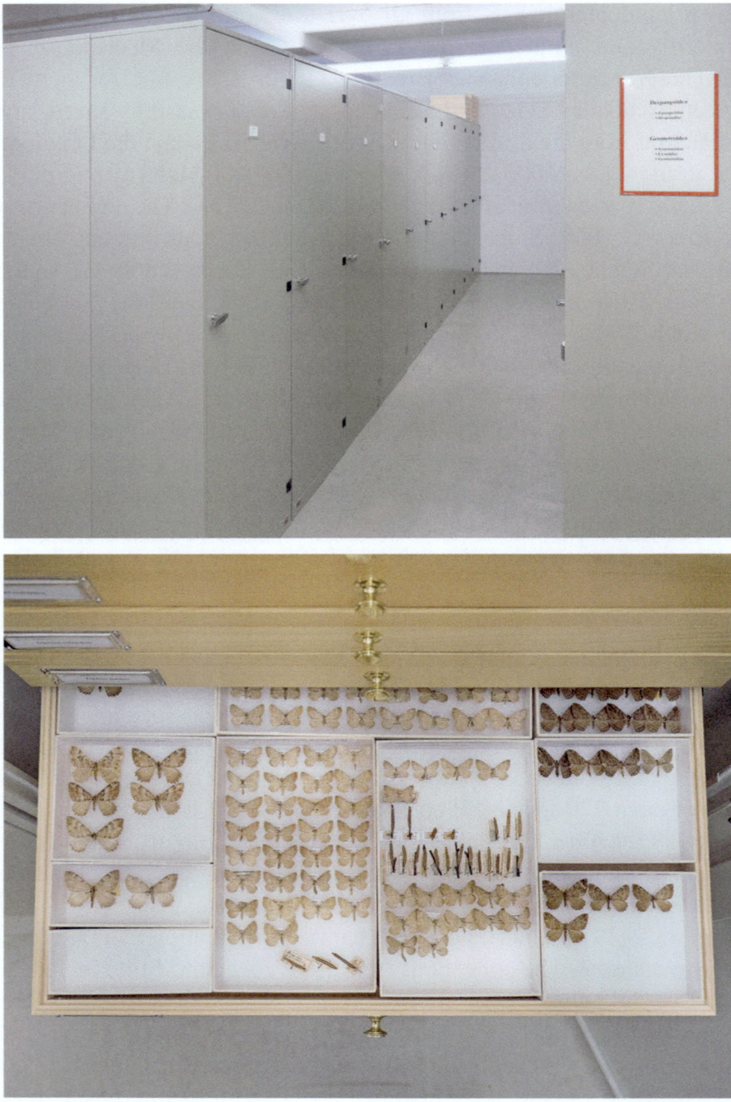

Fig. 28.33 Part of the Dresden butterfly collection

28.2.5.4 The Vertebrates in Dresden

The development of the vertebrate collections has been documented in detail by a catalogue starting with the inauguration of Adolf Bernhard Meyer in 1874, even though there have been originally many older specimens, many of them later destroyed during the 1849 revolution. Like most Dresden collections, also the objects of mammals cover a worldwide scope with valuable specimens of extinct species. Furthermore, the monkey and ape collections are species rich and remarkable. The exact tally of mammal specimens is difficult to estimate, as the losses during the war have not been subtracted. At the end of 2014, 27,477 mammals were registered. Important collectors were Menden, Schulz-Kampfhenkel, Hinsche, Schadenberg and Schierbrand und Kleinschmidt. Material collected in South America by Eduard Poeppig came to the museum in the 1970s when the University in Leipzig was restructured and the Zoological Museum there dissolved. Additional valuable mammal and bird material originates from the former museum of Julius Riemer, which was integrated virtually completely in the Dresden collections. The mammal types of Dresden were summarised in the paper by Feiler (1999). For a long time, curators were not assigned to special collections, but worked on several taxonomic groups. Dr. Alfred Feiler, at the museum from 1970 to 2000, was the first curator only responsible for mammals. Dr. Thomas Ziegler succeeded him, and since 2002, Dr. Clara Stefen is the curator of the mammalogical collection.

The bird collection in Dresden comprises approximately 92,000 items and has a worldwide coverage with a strong geographical focus on Europe, East Asia and Papua New Guinea. Up to now, holo-, syn- and lectotypes of 364 taxa have been catalogued as well as numerous paratypes (Eck and Quaisser 2004; Quaisser and Eck 2006). Like the mammal collection, the ornithological collection houses many representatives of extinct species (Fig. 28.34). Due to the historical losses in 1849 and 1945, the oldest specimen to date is from 1810 (Roselaar 2003). From 1874 onwards, Adolf Bernhard Meyer has catalogued the historical collections and materials of his expeditions to Southeast Asia and New Guinea. With the beginning of the twentieth century, the Dresden Museum received valuable material from two East Asia expeditions led by Walther Stötzner. Hugo Weigold, who afterwards went on in central and northern China until 1919, collected the bird skins from his

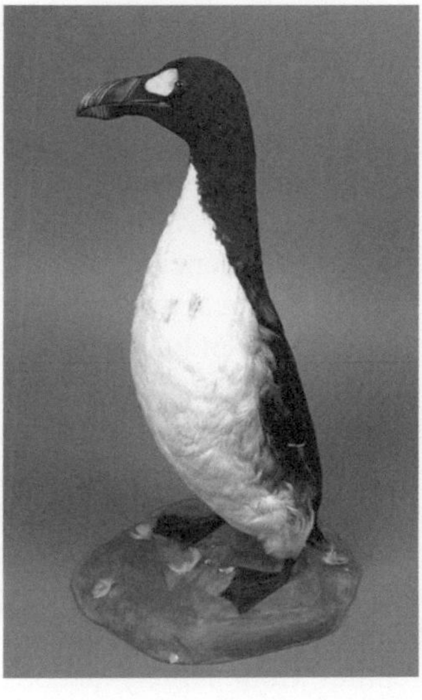

Fig. 28.34 Great Auk (*Pinguinus impennis*), one of the many representatives of extinct vertebrate species in the Dresden collection

first expedition to Tibet in 1913–1915. The material from Stötzner's expedition to Manchuria in 1927/1928 was contributed from several collectors.

At the time of the German Democratic Republic, curator Siegfried Eck dedicated most of his research to the comparative morphology and systematics of Eurasian birds. He regularly initiated the acquisition of skin collections like the regional Saxon collections of Ernst Bährmann (Eck 1984) and the large second collection of Otto Kleinschmidt (Eck 2001). Eck also acquired the egg collection by Wolfgang Makatsch that, with about 7000 individual clutches, is one of the largest in Germany (Roselaar 2003). The historical bird collection from Eduard Poeppig's expedition to the Americas in the years 1827–1837 comprises a few hundred bird specimens, in particular from Chile, Peru and Brazil. The historical bird collection from Eduard Poeppig's expedition to the Americas in the years 1827–1837 comprises a few hundred bird specimens, in particular from Chile, Peru, and Brazil.

Already before the German reunification, Eck has maintained an intense East-West German cooperation with Jochen Martens of the University of Mainz. Martens has regularly donated material from the Himalayas and China to the Dresden Museum.

When the Dresden Museum was incorporated into Senckenberg in 2009, the current curator Martin Päckert extended the integrative taxonomic research focus to a comparative genetic, bioacoustics and morphological approach. The recent acquisitions are two large feather collections of nearly 7000 individual feather mounts.

A distinct herpetological collection in Dresden was systematically started in 1874. The present holdings are an excellent and dunning example of a collection that has seen dramatic changes and periods of turmoil (reviewed by Fritz 2002 and literature cited therein). During the bombing of Dresden at the end of World War II, the collections were struck twice by fire, which resulted in most severe losses. The alcohol collections, including the herpetological holdings, were hit severely, which has reduced the latter from 6704 to only 98 specimens.

In an effort to re-establish the collection in subsequent years, the museum received material from various sources, including previous university collections. Amongst the specimens received, those from the collection of the former Zoological Museum of the University of Leipzig were probably one of the most diverse additions. It contains holdings collected and catalogued by Eduard Friedrich Poeppig (1798–1868) and later revised by Willy Hennig (1913–1976). Fritz Jürgen Obst, the former curator of the Dresden collection, has extensively commented this (1977). His great merit is the rebuilding of an important herpetological collection in Dresden. Recently, valuable but incompletely documented holdings, including type material, collected by Karl Georg Friedrich Rudolf Leuckart (1822–1898), were rediscovered amongst the very same material.

In 2015, the herpetological collection contains approximately 50,000 specimens, amongst them 449 type specimens (28 holotypes, one syntype, 419 paratypes, one lectotype and one paralectotype) representing seventy-four taxa (Fig. 28.35). Amongst the most strongly represented taxa are Caudata, Testudines and Ophidia. However, important specimens of other taxonomic groups, such as amphibians and reptiles from largely underrepresented biogeographic regions (e.g. the Guiana Shield Region of South America, Central African Rainforests of Angola and the

Fig. 28.35 Type specimens (turtles and tortoises) from the herpetological collection

Southeast Asian countries like Laos and Vietnam), have complemented the collection in recent years. Today, also an exceptional osteological collection of turtles belongs to the vertebrate specimens. This is unique in terms of available taxa and scope, i.e. specimens mainly designated for comparative morphological and phylogenetic research of fossil and extant taxa (Fig. 28.35).

The ichthyological collection in Dresden was founded in 1880. A curator, however, was hired only in 1961. Until then, colleagues from other museums mainly curated the collection and conducted scientific research. Therefore, Peters (Berlin), Fischer (Hamburg), Steindachner (Vienna) and Günther (London) were working on material, which was collected by A. B. Meyer in the Indo-Australian Archipelago. This collection consisted of about 2000 specimens in 546 species. Nineteen of this species were new to science at that time, and twelve of these species are valid today. The types and paratypes of these taxa were deposited unfortunately in other institutes and museums and not in Dresden. Steindachner determined specimens, which were collected by A. Stübel (1835–1905) in South America during his collection trip (1868–1877). This material consists of 121 specimens in 55 taxa. Five species new to science were found in this material. Only Benno Wandolleck (1864–ca. 1930) as a scientist from Dresden described a species collected by Stübel in 1916. Large parts of the old collection were destroyed during the World War II. Nevertheless, still valuable samples of faunistically important species now extinct in Saxony (*Acipenser sturio, Alburnoides bipunctatus, Coregonus oxyrhynchus*) from the River Elbe around Dresden dating from 1880 as well as type specimens from that time are preserved.

Corresponding to the current research focus, the collection comprises now many Characiformes and Siluriformes of South America. It consists today of 35,000 items; 33,000 of them are digitised. Eighty-eight holotypes, 1029 paratypes from 147 taxa, 58 syntypes from 10 taxa and 7 paralectotypes from 3 taxa are present.

28.3 The Senckenberg Museum of Natural History Görlitz

By Willi Xylander

28.3.1 A Glance at the History

The Natural History Museum in Görlitz originated from the Natural History Society Görlitz, founded in 1811 as an Ornithological Society. However, already before 1830 the society had broadened its scope and then included zoology, botany, geology as well as geography, medicine and science history. Since 1827, the society regularly published its own journal.

At the start, the collections mainly had been private and belonged to citizens from Görlitz and its surrounding. Thanks to Georg von Möllendorff (1811–1861), a former explorer and diplomat and then president of the society, it had become possible to unite all these collections in a new building in 1860. He initiated a museum building, financed by prominent citizens from Görlitz (Fig. 28.36). For the first time, enough space was provided for a reasonable development of the constantly growing collections.

Von Möllendorff appointed pharmacist Reinhard Peck (1823–1895) to be the first curator of the collections and later director of the museum. Peck soon opened the collections that at first had been accessible exclusively for the society members to the public. Peck was—what we today would call—an excellent networker. He had connections with many important representatives of the city and so got the money for further interesting objects enlarging the collections.

After Reinhard Peck had died, Hugo von Rabenau (1845–1921), a botanist, succeeded him in office the next 20 years.

Fig. 28.36 View at the Senckenberg Naturmuseum Görlitz (main building) (with permission from: Senckenberg Museum für Naturkunde Görlitz)

In 1934, the society acquired the former trade clubhouse “Gewerbevereinshaus” for the further grown collections. Now, it was also possible to start giving public lectures in an extra hall.

In 1949, the biology teacher Dr. Traugott Schulze (1907–1973) became the director. In consequence of World War II and the political circumstances, at that time the museum belonged to the city of Görlitz. Nevertheless, Schulze succeeded in bringing the museum on the list of the research museums of the State Secretariat for Higher Education and Applied Science (GDR) (Staatssekretariat für das Hoch- und Fachschulwesen). Thus, in 1953 the museum became a research institute (Hammerschmidt 2011; Xylander and Düker 2009).

After Schulze has left Görlitz, Dr. Karl-Heinz Großer, a well-known forest botanist, followed him, but quit this position in 1959 to focus on his research work at the Institute of Regional Science and Conservation in Halle.

Finally, Dr. Wolfram Dunger, a soil biologist from Leipzig, became the director in December 1959. His idea was to investigate soil biodiversity by using collections as taxonomic reference. Dunger has clearly aligned the scientific profile of the museum. Over the years, he appointed a dozen of scientists who either already were or—under his supervision—became experts for nearly all relevant groups of soil animals. Until 1990, he increased the number of scientists from two up to more than fifteen (Hammerschmidt 2011; Voigtländer 2002; Xylander and Düker 2009; Xylander 2017).

Soon, after he had taken over the director's position, Dunger started research projects on soil animal communities of lignite mines dumps. This, because he regarded mining and dumping of soil as example for impacts on soil animal communities on the landscape level (Wanner et al. 1999; Xylander 1999, 2000, 2015a, b). The lignite mine Berzdorf, which was very close to Görlitz, offered ideal conditions for such research.

Dunger built up contacts with other researchers in the East and in the West. Therefore, the reputation of the museum grew, and its high scientific level was internationally acknowledged. Due to this international reputation, the museum became, after the political change and a comprehensive evaluation process, a State Museum of the Free State of Saxony.

In October 1995, PD Dr. Willi Xylander followed Dunger as director and carried on the work keeping the research focus. Between 1996 and 2003, the Museum building and the Reinhard-Peck-Haus were modernised, the latter houses the botanical collections and belongs to the Museum since 1977. A new library (1995) and a building for the soil zoology research were acquired (1999 Wolfram-Dunger-Haus). In 2012, additionally about 1000 sqm could be gained in a building at the Sonnenplan. At present, the soil-zoological collections and collections of lichens, mosses and fungi are housed there.

On January 1st, 2009, the Museum in Görlitz merged with the Senckenberg Gesellschaft für Naturforschung in Frankfurt on Main, which led to a change in the structures. Presently, there are three departments (soil zoology with seven curators, zoology with four curators plus one PostDoc and Botany with four curators). Chairs of the departments are Professor Dr. Xylander (soil zoology), Professor Dr. Ansorge (zoology) and Professor Dr. Wesche (botany).

So, the zoological collections belong to two departments. This structure reflects the museum's history:

The collections of vertebrates, molluscs and winged insects from the Division of Zoology and Paleobiology, which comprise also many historic objects from the nineteenth and early twentieth century.

The collection of soil animals, which were mostly acquired during Dunger's time as director of the museum.

28.3.1.1 The Collections of the Zoology Division

The Zoology Department is divided into three divisions working with recent taxa (vertebrates, molluscs and pterygota insects) plus a division for palaeozoology and geology.

The Zoological curators are the following:

- Prof. Dr. Dr. h.c. Hermann Ansorge (mammalogy, head of the department)
- Dr. Bernhard Seifert (entomology, Pterygota)
- Dr. Heike Reise (malacology)

Fig. 28.37 Specimen from Görlitz bird collection (with permission from: Senckenberg Museum für Naturkunde Görlitz; photo: Ekkehard Mättig)

To the collections of the Mammalia Division, curated by Professor Ansorge, belong a large recent scientific collection of mammals and birds and the historical collection of the Natural Science Society of Görlitz. The latter includes fish, amphibians and reptiles, e.g. von Möllendorff's material from the Philippines, Sumatra, Madagascar and Southwest Africa (see Dunger 1986). The also existing collections of Anamnia and reptiles, however, are not part of the recent research profile.

The largest part of the historical collection is built by the ornithological collection which comprises more than 6000 habitus preparations of 2539 species representing nearly all orders of birds, in particular from Europe and Southeast Asia but also from South America, Africa, Australia and Antarctica. The collection also provides extinct species such as *Ara tricolor*, *Emeus crassus* and *Heteralocha acutirostris* (Fig. 28.37).

Remarkable, with regard to the recent collection profile, are the specimens of rare species from Upper Lusatia, with recorded data from the early nineteenth century until today, like the collections of J. von Zittwitz, Von Möllendorff and von Loebenstein. Due to the concept of Professor Ansorge and after a period of stagnation, now, the ornithological collection grows fast. At present, it counts about 12,000 specimens and objects, of which the most are accessible via the Internet (Ansorge 1986, 2008; see a list of important objects in Ansorge 2008).

The collections of Mammalia originate from the early years of the Naturforschende Gesellschaft and are featured by mostly exotic specimens from Africa (Steudner) and Southeast Asia (von Möllendorff), while specimens of the local fauna were rarely collected at that time. However, with Hermann Ansorge, the number of specimens has increased significantly. Currently, there are over 34,600, mainly represented by a skull collection. Nevertheless, it also comprises complete skeletons, furs and tissue samples. The taxonomic focus lies on carnivores, espe-

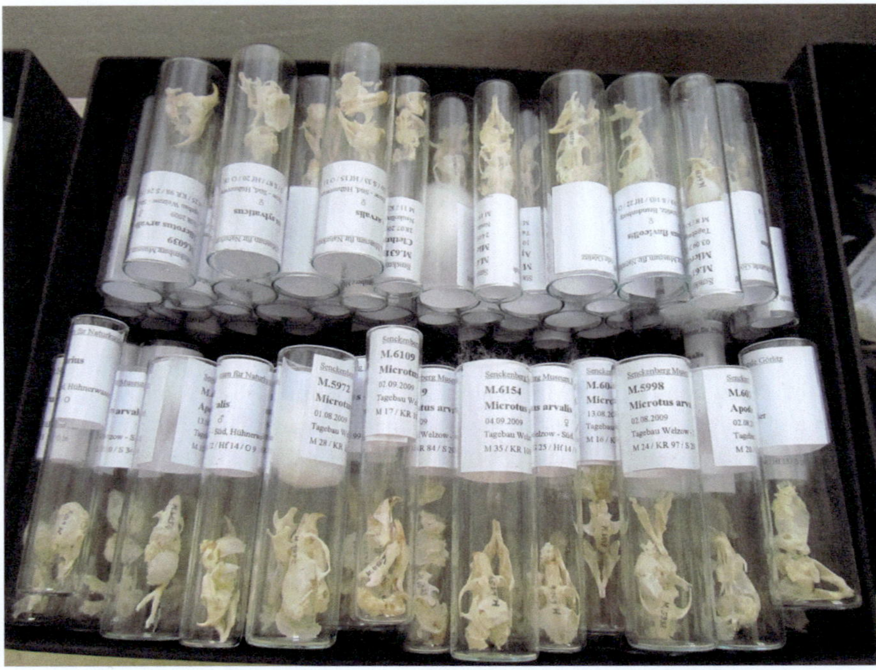

Fig. 28.38 Skull collection of small rodents (with permission from: Senckenberg Museum für Naturkunde Görlitz)

cially from Europe, including a large collection of wolves from the recent German population, as well as from other regions of the Palaearctic, not to forget the Central Asian equids (wild ass, Przewalski's horse) (see Dunger 1986; Ansorge et al. 2012).

The focus of research in the Mammalia Division lies in population biology, in feeding ecology and in conservation of various carnivores and equid mammals. The spectrum of applied methods ranges from the use of morphology, e.g. non-metric character analyses, population ecology up to molecular methods. In addition, since the 1990s, tissue and organ samples from birds and mammals are collected and kept deep-frozen for further research. So, the Mammalia Division in Görlitz presents one of the oldest museum collections of tissue samples from vertebrates in Germany (Fig. 28.38).

28.3.1.2 The Entomological Collections in Görlitz

Worth being mentioned is also the entomology collection with its historical roots reaching back to the year 1827. Today, Dr. Bernhard Seifert curates about 1.3

million counting specimens. Many of these were collected in Upper Lusatia and Silesia (for details, see Franke 1982). However, also further parts of the world are included.

Interesting pieces of this collection are those from August Kelch (purchased in 1859 by the Naturforschende Gesellschaft) and of Küsell (donated 1882/1883). They count 10,000 specimens of 2800 species. Most important "with regard to the number and scientific relevance" is the collection of Schwarz (Liegnitz). Purchased in 1899, it comprises circa 8650 species of West-Palaearctic beetles (Franke 1982; see also Dunger 1986). The entomological collection also has been enriched through taking over important material from the museum in Bautzen (1970–1974).

In the historical collection, lepidopterans are many in number with about one thousand species from Europe and hundreds from Africa, America (e.g. USA, Venezuela, Argentina and Brazil) and Asia (Franke 1982). Also historical and of importance is the collection of microlepidopterans from Sommer. And the Hymenoptera from Kelch as well as from K.H.C. Jordan with the ants and further taxa, associated with ants, need to be mentioned. Not less important and as well from Bautzen is the collection of Diptera with type material (by Heinrich Kramer). Furthermore, Engelmann has donated a collection of doublets of Heteroptera from Central Europe from the important bug specialist K.H.C. Jordan (Dunger 1986; Franke 1982).

The most type specimens of the Entomological Division are ants with types of one hundred two species, fifty-seven primary types (holo-, lecto- and neotypes) and more than 1500 paratypes (from descriptions of new species mainly by Seifert). Furthermore, there is type material from Auchenorrhyncha with forty specimens, Sternorrhyncha with ten, Diptera also with about forty and some other taxa with ten in total.

Recent research in the division of entomology nearly exclusively focuses on ants (Formicidae) of the Palaearctic region. Seifert and coworkers investigate the taxonomy of these animals as well as the biology, evolution, population biology and behaviour. For this purpose, Seifert uses complex morphometric measurements and data for species description and separation as well as for the detection of hybrids. After the takeover of the very important ethanol collection by Buschinger in 2015, the number of specimens of ants in the collection reaches about 600,000, which means two-thirds of the determined specimens of the entire entomological collection. At present, the collection is in the Wolfram-Dunger-Haus (Fig. 28.39).

Fig. 28.39 By Dr. Bernhard Seifert (with permission from: Senckenberg Forschungsinstitut und Naturmuseum; photo: Sven Tränkner)

28.3.1.3 The Malacological Collection in Görlitz

The malacology collection, currently curated by Heike Reise, comprises circa 6000 species and about 25,000 series of specimens (Reise, pers. comm.) with a biogeographical focus on the western Palaearctic and a taxonomic focus on terrestrial Gastropoda.

The collection, first mentioned in 1827, has grown continuously throughout the nineteenth century. Indeed of importance are the historical parts, collected by von Möllendorff, Borcherding and Lepsius and some North American Unionids, probably from Lea (Dunger 1986, Reise pers. comm.).

Most of the material consists of gastropods, but also includes bivalves and a few other molluscs. The historical and most other recent collection consist of shells, whereas the recent research collection from Dr. Reise and coworkers shows a high ratio of soft bodies of snails and slugs in alcohol. At present, this collection counts circa 22,000 slug specimens. It is stored at the Wolfram-Dunger-Haus.

The recent research focus in the Malacology Division lies on terrestrial slugs, especially on the genera *Deroceras* and *Arion* as well as on a number of agricultural and horticultural pests with a more or less worldwide anthropogenic distribution.

Research projects deal with species-specific mating behaviour, species differentiation by combining behavioural, anatomical and molecular character sets and questions of allopatric and parapatric speciation (Fig. 28.40).

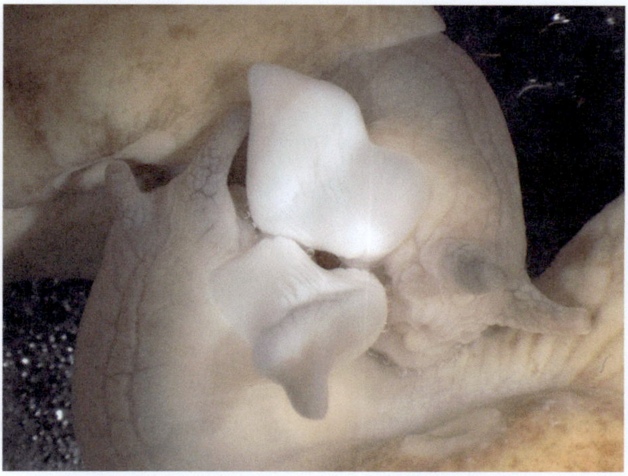

Fig. 28.40 Caught in act: a mating pair of slugs. Sexual behaviour may be used for species differentiation (with permission from: Senckenberg Museum für Naturkunde Görlitz; photo: J.M.C. Hutchinson)

28.3.1.4 The Collections of the Soil Zoology Department in Görlitz

The collection of soil animals comprises about 4.5 million specimens representing nearly all groups of terrestrial soil animals. This collection that has been founded by Professor Wolfram Dunger reflects his research focus on soil ecology on the species level and the taxonomic research on soil animals (Xylander 1999, 2000, 2015a, b, 2017).

A large part of the collection is available via a comprehensive data warehouse: www.edpahobase.org (see Burkhardt et al. 2014).

The following specialists curate the several parts of the collections:

- Myriapoda (Dr. Karin Voigtländer, also responsible for Lumbricidae and terrestrial Isopoda)
- Collembola (Dr. Hans-Jürgen Schulz, also responsible for other taxa of Apterygota, e.g. Archaeognatha, Zygentoma)
- Gamasina (Dr. Axel Christian, also responsible for other taxa of Arachnida)
- Actinedida (Dr. David J. Russell, also coordinates the mesofauna ecological research)
- Oribatida (Dr. Ricarda Lehmitz)
- Nematoda and Tardigrada (Dr. Karin Hohberg)
- Platyhelminthes (Prof. Dr. Willi Xylander)

The largest collection is that of Collembola with about 1.4 million determined specimens. It was founded by Dunger (see Dunger 1986) and is curated by H.-J. Schulz (Fig. 28.41). It consists of more than 800 species and includes over 1000 type specimens (holo- and paratypes) from various specialists such as Gerhard Bretfeld, Wolfram Dunger, Mikhail Potapov and Hans-Jürgen Schulz.

It comprises collectibles, mainly from Europe, and from the eastern Mediterranean (especially Crete), Russia and Mongolia. Much of the material is preserved in ethanol, but there are also more than 100,000 specimens on microscope slides,

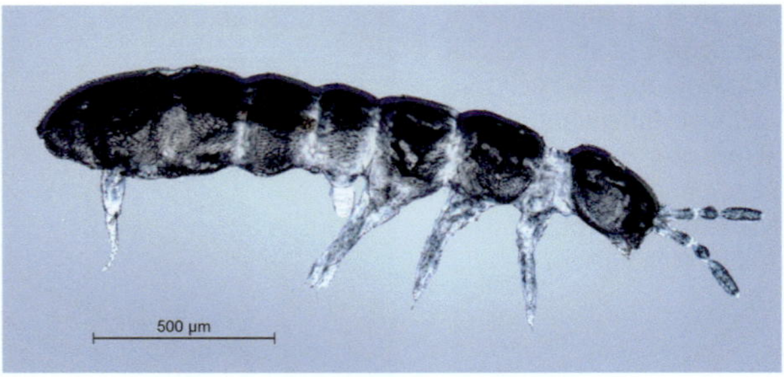

Fig. 28.41 A spring tail (Collembola): *Folsomia sexoculata* (with permission from: Senckenberg Museum für Naturkunde Görlitz; photo: H.-J. Schulz)

partly as whole mounts like in the collections of Schulz and Dunger, or separated (head, antenna, furca, thorax and abdomen) as in the Bretfeld collection.

With several hundreds of specimens, Protura, with material from Dunger, Nosek and Balkenhol, and Diplura, Archaeognatha and Zygentoma (Helmut Sturm) are part of the Apterygota collection.

Another large collection of soil animals is that of the division of Arachnida with more than 300,000 specimens (Fig. 28.42), of which about 180,000 are determined to species level (Christian, in litt.). The curator is A. Christian who has his taxonomic expertise in Gamasina and Ixodida (Fig. 28.43).

These collections, which mainly originate from Europe, include also material from Asia, America and Africa, with a variety of type material of about 1600 type specimens from more than 390 species of Uropodina (M. Hutu and J. Starý), Gamasina (A. Christian, W. Karg and G. Dosse) as well as recent and fossil spiders from the Wunderlich collection, partly from amber. Highlights are the unique holotype specimens of the worldwide only two known species of the primitive mite group Opilioacarida from Dominican amber (*Paracarus pristinus*, J. Dunlop, Wunderlich and Poinar 2004). Additionally the holotypes of two ricinulei species from amber belong to this collection.

The Myriapoda collection, founded by Dunger and now curated by K. Voigtländer, comprises nearly 100,000 specimens, especially of Diplopoda and Chilopoda, but also of Symphyla from the Palaearctic including numerous items from Central Europe and the Iberian Peninsula. Prominent are collections from Italy by Buchner because they were determined by Karl Wilhelm Verhoeff (Dunger 1986), one of the most important Myriapoda taxonomists of all times. They are a reference to his view on a species. These collectibles mean blessing in disguise, because many of those from Verhoeff had been deposited in Dresden and got lost during World War II. That is why these remaining parts are of particular relevance.

A recent focus of research in these divisions lies in the determination characters of developmental stages (especially of *Lithobius* spec.). Of special interest is also the biodiversity of Myriapoda in Central Europe and the circum Mediterranean countries.

New collections of Actinedida, Platyhelminthes, Nematoda and Tardigrada were founded between 2001 and 2009. Up to now, the collections of determined

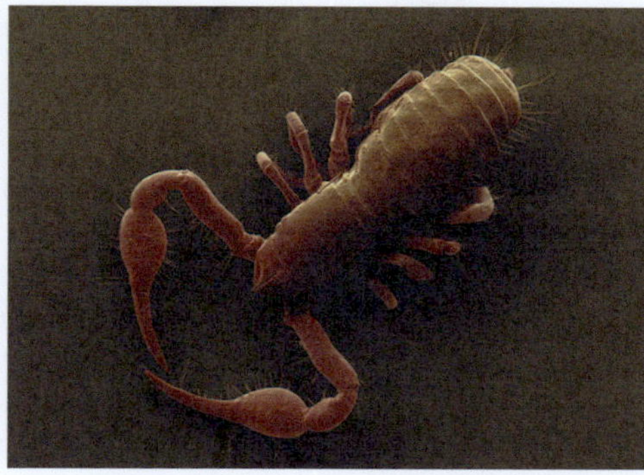

Fig. 28.42 Pseudoscorpion (with permission from: Senckenberg Museum für Naturkunde Görlitz; photo: Volker Hampe)

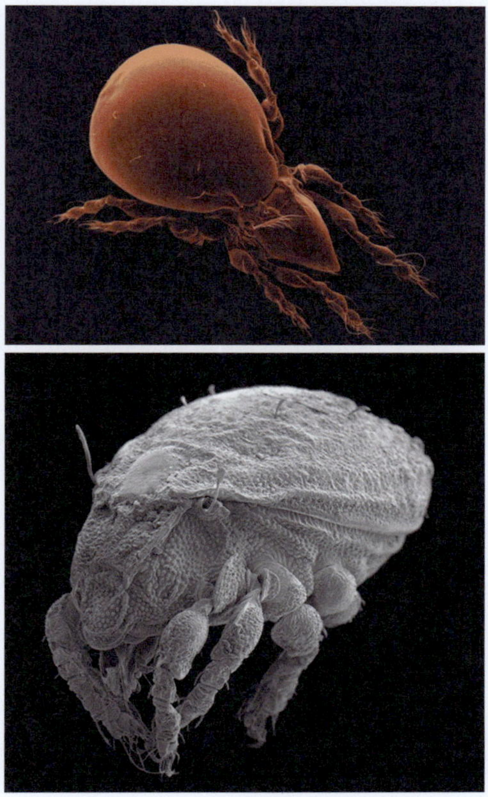

Fig. 28.43 A moss mite (Oribatida): *Scutovertex* (with permission from: Senckenberg Museum für Naturkunde Görlitz; photo: Volker Hampe and Birgit Lang)

specimens only count in thousands. However, due to active research projects, they are growing fast.

Among the recent divisions, only the Oribatida collection has a longer tradition (Fig. 28.43). At present, there are more than 100,000 specimens from about 1000 species, mostly in Ethanol and glycerol including about 50 type specimens. Former curators were Hans-Dieter Engelmann and Thomas Schwalbe (see Voigtländer

2002). Now, after A. Christian temporarily was responsible for the collection until 2013, R. Lehmitz curates it.

28.3.1.5 The Natural History Exhibitions

Beside the work in research and collection, Senckenberg Museum of Natural History Görlitz shows the biological and geological exhibitions on about 1200 Sqm. These permanent exhibitions deal with the Upper Lusatia region to which Görlitz belongs. Subjects are the Earth History as well as the animals and plants of the region. In addition, also animals from the savanna and from the tropical rain forests can be seen. Moreover, two of the halls present touring exhibitions. A vivarium shows living fishes, reptiles, amphibians and mammals from tropical rain forests and of course from the Upper Lusatia.

As a speciality, the museum regularly produces international touring exhibitions, which pick up on research subjects.

With the special emphasis on soil zoology, the following have been created:

- “Life in the soil”, presented 1995–2002
- “Beneath our feet”, 2004–2012
- “The thin skin of earth”, started in autumn 2015, ongoing
- Furthermore, “Wolves” (start in 2007, recently updated, ongoing)
- “Research in Mongolia” (2012, ongoing)

These exhibitions are always presented for many years. In total, they attracted more than 2 million visitors in Germany and other countries (Ansorge et al. 2012; Hohberg and Xylander 2004; Hohberg et al. 2011; Meiners and Xylander 2012; Leinfelder and Xylander 2012; Xylander 2006, 2011a, 2012, 2015a, b, Xylander et al. 2013).

28.4 The Senckenberg German Entomological Institute Müncheberg

By Thomas Schmitt

28.4.1 A Changeful History

The Senckenberg German Entomological Institute (German: Senckenberg Deutsches Entomologisches Institut, SDEI) in Müncheberg (Fig. 28.44) is the only research institution in Germany exclusively dedicated to entomological research. It was founded as Deutsches Entomologisches Nationalmuseum (DENM) in 1886 in Berlin by Dr. Gustav Kraatz who also was its first director.

Fig. 28.44 View of the SDEI building in Müncheberg (with permission from: Senckenberg Deutsches Entomologisches Institut (SDEI); photo: Christian Kutzscher)

Under the next director, Dr. Walther Horn, leading the institute until 1939, the institution's name was changed to Deutsches Entomologisches Museum (DEM) in 1911 and to Deutsches Entomologisches Institut (DEI) in 1920.

His successor, Prof. Dr. Hans Sachtleben, organised the evacuation to Schloß Blücherhof in Mecklenburg in 1943 due to World War II and the retransfer to Berlin in 1950, but to Friedrichshagen in the eastern part of the city and not the western Berlin-Dahlem where the institute was located earlier. Hence, all the important collections safely survived the turbulences of the war and the postwar time.

In 1951, the DEI became part of the newly established German Academy of Agricultural Sciences (German: Deutsche Akademie der Landwirtschaftswissenschaften).

The next director, Prof. Dr. Heinz Fankhänel, leading the DEI from 1962 to 1970, organised its move to Eberswalde in 1963. In 1970, the DEI lost its independence and its internationally well-known name. In the following year, the institute became Division Taxonomy of Insects (formerly DEI) (German: Abteilung Taxonomie der Insekten (ehemals DEI)) of the newly established Institute of Plant Protection Sciences Kleinmachnow, Division Eberswalde (German: Institut für Pflanzenschutzforschung Kleinmachnow, Bereich Eberswalde). Head of the division was Prof. Dr. Günter Morge (1971–1984) followed by Dr. Günther Petersen (1984–1989), Dr. Reinhard Gaedike (1989–1990) and Prof. Dr. Werner Ebert (1990).

After the German reunification, the name German Entomological Institute was reinstalled. When Prof. Dr. Joachim Oehlke was director (1990–1993), the institute became Project Group in the KAI[1] from 1992 to 1993. Under the directorship of Prof. Dr. Holger Dathe from 1993 to 2010, the DEI first became Project Group Entomology at the University for Applied Sciences Eberswalde (German: Projektgruppe Entomologie an der Fachhochschule Eberswalde) in 1994, then registered association German Entomological Institute (German: DEI e.V.) in 1997 and, in the year 2000, changed to the Centre for Agricultural Sciences (Zentrum für Agrarlandschaftsforschung, ZALF), becoming a department of ZALF with the old name DEI. In 2009, the DEI merged with Senckenberg, since then being named Senckenberg German Entomological Institute (SDEI). Currently, its director is Prof. Dr. Thomas Schmitt.

All over this changeful history, many entomologists have brought together one of the best entomological collections of Germany with some special parts being amongst the best ones worldwide. This was highly influenced by the specific taxonomic focus of some of its heads to specific insects groups as, for example, to staphylinid beetles or to tiger beetles. Due to the concept of bringing together collections and the intensive work on them, an entomological library has been developed at the SDEI, which is internationally renowned for its completeness.

Today, the insect collection of the SDEI contains more than 3 million pinned insects and an unknown number of specimens preserved in alcohol. The collection represents about 275,000 species. Due to the about 25,000 type specimens, the SDEI collection is one of the most important insect collections of Europe. The fundaments of the DEI collection were the donations by G. Kraatz, but also by L. von Heyden, K. Letzner, G. Metzler, W.H. Rolph and A. & J. Stern. Thereafter, the collection was enlarged, e.g. by the following:

- Collections by the members of the institute (e.g. in course of expeditions of the DEI)
- Exchanges with institutions and specialists (e.g. museums in London, Paris and Stettin)
- Receipt of doublets as reward for determination of material by the scientists of the institute
- Donations or inheritances of determined material, sometimes including type material, from institutions and private entomologists (e.g. collections of H. Sauter from Formosa/Taiwan)
- Buying from commercial insect traders and collectors in the past (e.g. O. Staudinger & A. Bang-Haas, E. Heyne, A. Heyne, H. Rolle, A. Kricheldorff)

[1]KAI = Coordination and Processing Institution of institutions of the former Academy of Science of the GDR (German: Koordinierungs- und Abwicklungsstelle für Institutionen der ehemaligen Akademie der Wissenschaften der DDR).

Over the past 130 years, more than one hundred collections were integrated into the main collection of the SDEI, some of these being of high relevance.

28.4.1.1 The Sauter Collection

One of these collections of outstanding international relevance is the Sauter collection. Hans Sauter (1871–1948), a German entomologist, emigrated to Japan and Formosa (Taiwan), where he extensively collected insects. Due to a specific agreement, the major part of Sauter's material has become part of the SDEI collection. With 2258 type specimens, this collection is one of the internationally most attractive collections of the institute even today and one of the reasons for the ongoing cooperation with institutions in Taiwan and mainland China.

28.4.1.2 Busy Bees on Expedition

The collections of the SDEI also gained much of their significance by the collecting activities of its staff. In this respect, the expedition to Albania in 1961 was one important highlight, during which about 42,000 specimens were collected. Another more recent expedition was to the Far East of Russia. Numerous additional collection trips of members of the SDEI to various countries of Europe and to North America, but also Africa and New Zealand, have contributed to the rich collection of today.

28.4.1.3 How It Is Organised

The entire collection is structured into five divisions, each one headed by a scientist assisted by a technician, sometimes complemented by third-party funded scientists.

These divisions are:

- Coleoptera
- Diptera
- Hemimetabola (i.e. Heteroptera, Cicadina, Saltatoria, Odonata)
- Hymenoptera, Strepsiptera and Siphonaptera
- Lepidoptera, Trichoptera and Neuropteroidea

The Coleoptera collection forms the largest part of the SDEI collection embracing more than 40 % of all specimens (1.7 million pinned specimens representing about 160,000 species). Due to the particular interest of the founder of the institute, G. Kraatz, and other scientists working here, like L. Dieckmann, W. Horn, K.-H. Mohr and S. Schenkling, this collection is not only one of the largest of Europe but also particularly rich in type material (ca. 13,800 types).

Outstanding material is particularly available for the families Cicindelidae, Chrysomelidae, Cleridae, Curculionidae and Staphylinidae. Type catalogues for most of the families have been published, and lists of species or at least genera have been worked out and are partly available from the web pages of the SDEI.

The Diptera collection includes more than 850,000 specimens and about 7500 types, most of which have been published in type catalogues. Furthermore, comprehensive overviews of the species with the number of specimens are available at the web pages of the SDEI. Due to the taxonomic focus on Sciaroidea and the active sampling activities of the members of this division, the collections on this group are of particular importance in the Diptera collection.

The collection of hemimetabolous insects contains about 146,000 specimens in 14,000 species. The particular value of this collection is due to its about 1700 types, many of which have been described by G. Breddin for bugs, plant hoppers and cicadas. Most of the type material has been published in various type catalogues. Lists of species including the respective numbers of specimens are partly available from the web pages of the SDEI.

The Hymenoptera collection harbours about 200,000 specimens in about 15,000 species. This collection is particularly valuable due to the integration of the collections of K. Bleyl, K.F. Ermisch, F.W. Konow, K.F. Lange and W. Mink as well as the active collection activities of the division's members. The types of this collection were published in several type catalogues; an updated list can be obtained from the web pages of the SDEI.

The focus of the Lepidoptera collection is on micro-moths (Microlepidoptera). In particular, the collection of the Tineidae is remarkable due to the work of G. Petersen and R. Gaedike. As distinguished specialists have revised the collection, a large number of type materials by E. Meyrick, E. Stand, L.B. Prout and M.J. Bastelberger are available. The register of this type material can be downloaded from the web pages of the SDEI. The entire collection is available online as digital photographs of the insect boxes.

References Frankfurt Collections

Bott R (1967) Crustaceen-Sektion. Senckenbergiana biologica 48(Sonderheft B):75–78

Felten H (1967) Geschichte der Säugetier-Sektion. Senckenbergiana biologica 48(Sonderheft B):11–17

Franz E (1967) Geschichte der Entomologischen Sektion. Senckenbergiana biologica 48 (Sonderheft B):55–72

Freyreiss GW (1968) Reisen in Brasilien. Monograph Series Etnografiska Museet, 13, 99 pp, Stockholm [Published posthumously on the basis of a manuscript of Freyreiss]

Goethe JV von (1816) Über Kunst und Altertum in den Rhein- und Main-Gegenden. Stuttgart, Heft 1, 196 pp, (Cotta)

Gutmann WF (1967) Zur Neueinrichtung der vergleichend-anatomischen Sektion. Natur und Museum 97(8):333–334

Janssen R (1992) Sektion Malakologie. In: Türkay M, Ziegler W (eds) 175 Jahre Senckenbergische Naturforschende Gesellschaft. Jubiläumsband II. Geschichte der wissenschaftlichen Abteilungen und Sektionen. Senckenberg-Buch 68:78–86, Frankfurt a. M. (W. Kramer)

Klausewitz W (1967) Geschichte der Ichthyologischen Sektion. Senckenbergiana biologica 48 (Sonderheft B):41–54

Klausewitz W, Krupp F, Köhler C (1992) Sektion Ichthyologie I. In: Türkay M, Ziegler W (eds) 175 Jahre Senckenbergische Naturforschende Gesellschaft. Jubiläumsband II. Geschichte der wissenschaftlichen Abteilungen und Sektionen. Senckenberg-Buch 68:42–50, Frankfurt a. M. (W. Kramer)

Klemmer K (1992) Sektion Herpetologie. In: Türkay M, Ziegler W (eds) 175 Jahre Senckenbergische Naturforschende Gesellschaft. Jubiläumsband II. Geschichte der wissenschaftlichen Abteilungen und Sektionen. Senckenberg-Buch 68:39–41, Frankfurt a. M. (W. Kramer)

Knoblauch A (1920) Die Preis- und Stipendien –Stiftungen der Senckenbergischen Naturforschenden Gesellschaft. Berichte der Senckenbergischen Naturforschenden Gesellschaft 50:9–93, pls 1–5

Kraus O (1967) Geschichte der Evertebraten-Sektionen. Senckenbergiana biologica 48 (Sonderheft B):73–90

Mayr G (2011) Senckenbergs Vogelskelette. Eine junge Sammlung als Grundlage aktueller Forschung. Natur Forschung Museum 141(3/4):82–89

Mertens R (1949) Eduard Rüppell: Leben und Werk eines Forschungsreisenden, 388 pp., Frankfurt a. M. (W. Kramer)

Mertens R (1967) Die herpetologische Sektion des Natur-Museums und Forschungs-Institutes Senckenberg in Frankfurt a. M. nebst einem Verzeichnis ihrer Typen. Senckenbergiana biologica 48 (Sonderheft A):1–106

Peters DS (1992) Sektion Ornithologie. In: Türkay M, Ziegler W (eds) 175 Jahre Senckenbergische Naturforschende Gesellschaft. Jubiläumsband II. Geschichte der wissenschaftlichen Sektionen 1967-1992. Senckenberg-Buch 68:32–38, Frankfurt a. M. (W. Kramer)

Senckenberg Homepage: www.senckenberg.de

Steinbacher J (1967) Geschichte der ornithologischen Sektion. Senckenbergiana biologica 48 (Sonderheft B):19–40

Storch G (1992) Sektion Mammalogie I. In: Türkay M, Ziegler W (eds) 175 Jahre Senckenbergische Naturforschende Gesellschaft. Jubiläumsband II. Geschichte der wissenschaftlichen Abteilungen und Sektionen. Senckenberg-Buch 68:15–19, Frankfurt a. M. (W. Kramer)

Türkay M (1981) 150 Jahre senckenbergische Krebstierforschung. Natur und Museum 111 (5):151–157

Türkay M (1992) Sektion Crustaceen. In: Türkay M, Ziegler W (eds) 175 Jahre Senckenbergische Naturforschende Gesellschaft. Jubiläumsband II. Geschichte der wissenschaftlichen Sektionen 1967-1992. Senckenberg-Buch 68:116–121, Frankfurt a. M. (W. Kramer)

Türkay M (2015) Die Nordsee-Sammlung. Natur, Forschung, Museum 145(7/8):212–213, Abb. 1–3

Zilch A (1967) Geschichte der malakologischen Sektion. Archiv für Molluskenkunde 97 (1/6):7–43

References Dresden Collections

Eck S (1984) Katalog der ornithologischen Sammlung Dr. Udo Bährmanns (3. Fortsetzung). Zoologische Abhandlungen Staatliches Museum für Tierkunde Dresden 39:71–98

Eck S (2001) Otto Kleinschmidts zweite ornithologische Sammlung im Museum für Tierkunde Dresden. Zoologische Abhandlungen Staatliches Museum für Tierkunde Dresden 51:119–132

Eck S, Quaisser C (2004) Verzeichnis der Typen der Vogelsammlung des Museums für Tierkunde in den Staatlichen Naturhistorischen Sammlungen Dresden. Zoologische Abhandlungen Staatliches Museum für Tierkunde Dresden 54:233–316

Feiler A (1999) Ausgestorbene Säugetiere, Typusexemplare und bemerkenswerte Lokalserien von Säugetieren aus der Sammlung des Staatlichen Museums für Tierkunde Dresden (Mammalia). Zoologische Abhandlungen Staatliches Museum für Tierkunde Dresden 50:401–414

Fritz U (2002) Herpetology and herpetological type specimens at the Museum für Tierkunde Dresden with a bibliography of herpetological contributions by Fritz Jürgen Obst (Amphibia, Reptilia). Faunistische Abhandlungen Staatliches Museum für Tierkunde Dresden 23:3–34

Hertel R (1978) Das Staatliche Museum für Tierkunde Dresden und seine Publikationen. Zoologische Abhandlungen Staatliches Museum für Tierkunde Dresden 35:V–VIII

Kummerlöwe H (1939) Rede zur Wiedereröffnung des Museums für Tierkunde 4. Oktober 1937. Abhandlungen und Berichte Museum für Tierkunde und Völkerkunde Dresden 20:1–15

Nuss M, Landry B, Vegliante F, Tränkner A, Bauer F, Mally R, Hayden J, Segerer A, Li H, Schouten R, Solis MA, Trofimova T, De Prins J, Speidel W (2003–2015) Global Information System on Pyraloidea. http://www.pyraloidea.org/

Obst FJ (1977) Die Herpetologische Sammlung des Staatlichen Museums für Tierkunde Dresden und ihre Typusexemplare. Zoologische Abhandlungen Staatliches Museum für Tierkunde Dresden 34:171–186

Quaisser C, Eck S (2006) Korrekturen und Ergänzungen zum Verzeichnis der Typen der Vogelsammlung des Museums für Tierkunde in den Staatlichen Naturhistorischen Sammlungen Dresden. Zoologische Abhandlungen Staatliches Museum für Tierkunde Dresden 55:129–138

Reichert R (1954) Ein Stück Museumsgeschichte. Abhandlungen und Berichte Museum für Tierkunde Dresden 22:1–11

Reichert R (1956a) Über 200 Jahre Museumsgeschichte. Dresdner Wissenschaftliche Museen. Dresden und Leipzig (Steinkopff)

Reichert R (1956b) Persönlichkeiten, von denen unsere Sammlungen erzählen. Dresdner Wissenschaftliche Museen Beiträge zur 750-Jahr-Feier unserer Stadt. Dresden und Leipzig (Steinkopff)

Roselaar C (2003) An inventory of major European bird collections. Bulletin of the British Ornithologists' Club 123A:253–337

References Görlitz Collections

Ansorge H (1986) Die Vogelsammlung des Staatlichen Museums für Naturkunde Görlitz—Belege zur Ornis der Oberlausitz. Abh. Ber. Naturkundemuseum Görlitz 60(5):1–12

Ansorge H (2008) 3. Wir über uns—Die Ornithologische Sammlung des Naturkundemuseums. In: Jahresbericht 2005-2007. Abh. Ber. Naturkundemuseum Görlitz, pp 290–298

Ansorge H, Hanelt M, Schmidt-Corsitto K, Selenge S, Treiber J, Wesche K, Xylander W (2012) Im Land der Gräser und wilden Pferde—Biologische Forschungen in der Mongolei. Eine Ausstellung aus Anlass des 50-jährigen Jubiläums der Zusammenarbeit deutscher und mongolischer Biologen 1962-2012. Senckenberg Museum für Naturkunde Görlitz (Hrsg.). Eigenverlag, S. 1–72

Burkhardt U, Russell DJ, Buryn R, Decker P, Döhler M, Höfer H, Lesch S, Rick S, Römbke J, Trog C, Vorwald J, Wurst E, Xylander WER (2014) The Edaphobase Project of GBIF-Germany—a new online soil-zoological data warehouse. Appl Soil Biol 83:3–12

Dunger W (ed) (1986) Die Sammlungen des Staatlichen Museums für Naturkunde Görlitz—Forschungsstelle—Anläßlich des 175. Jahrestages der Gründung der Naturforschenden Gesellschaft zu Görlitz. Abh. Ber. Naturkundemuseum Görlitz 59 (Suppl):1–72

Franke R (1982) Zur Geschichte der entomologischen Sammlungen des Staatlichen Museums für Naturkunde Görlitz. Abh. Ber. Naturkundemuseum Görlitz 56(3):1–21

Hammerschmidt J (2011) 200 Jahre Naturforschende Gesellschaft und Museum für Naturkunde Görlitz. Peckiana 7:1–146

Hohberg K, Xylander W (2004) Unter unseren Füßen—Lebensraum Boden. Ein Begleiter durch die Wanderausstellung. SMNG, Eigenverlag, S. 1–40

Hohberg K, Zumkowski-Xylander H, Xylander W (2011) Wölfe. Begleitheft zur gleichnamigen Ausstellung des Senckenberg Museums für Naturkunde Görlitz. 2. Aktualisierte und überarbeitete Aufl. Eigenverlag, SMNG, Görlitz. S. 1–40

Leinfelder R, Xylander WER (2012) Naturwissenschaftliche Museen und Museen für Naturkunde. In: Museen zwischen Qualität und Relevanz—Denkschrift zur Lage der Museen (Hrsg. B. Graf & V. Rodekamp). Berliner Schriften zur Museumsforschung Bd. 30:357–366

Meiners U, Xylander WER (2012) Sammlungen—Grundlage der Museen. In: Museen zwischen Qualität und Relevanz—Denkschrift zur Lage der Museen (Hrsg. B. Graf & V. Rodekamp). Berliner Schriften zur Museumsforschung Bd 30: 73–88

Voigtländer K (2002) Wir über uns—Die Abteilung Bodenzoologie stellt sich vor. In: Jahresbericht 1999-2001. Abh. Ber. Naturkundemuseum Görlitz

Wanner M, Dunger W, Xylander WER (1999) Fünfzig Jahre Bodenzoologie—der Beitrag des Staatlichen Museums für Naturkunde Görlitz. Informationen des Sächsischen Museumsbundes 18:68–71

Xylander W (1999) Prof. Dr. Wolfram Dunger. Abh. Ber. Naturkundemuseum Görlitz 71(2): 415–416

Xylander W (2000) Prof. Dr. Wolfram Dunger—Zum 70. Geburtstag und zur Ernennung zum Ehrenbürger der Kreisfreien Stadt Görlitz. Neues Lausitzisches Magazin—Zeitschrift der Oberlausitzischen Gesellschaft der Wissenschaften zu Görlitz. Neue Folge 3:10–12

Xylander WER (2006) In die Ferne schweifen—Zur Konzeption und Realisierung von naturkundlichen Wanderausstellungen. Museumskunde 71(1):69–75

Xylander WER (2011a) Präsentation von Wirbellosen Tieren in naturkundlichen Ausstellungen. Tagungsband der Jahrestagung der Böhmisch-Bayerisch-Sächsischen Museen in Ceska Lipa "Natur im Museum". Museum Bulletin 14:67–76

Xylander W (2012) Wenn Wissen auf Reisen geht—Wie Senckenberg seine Forschungsergebnisse vermittelt. Natur Forschung Museen (Sonderheft "Museen—Wissenschaft als Erlebnis") 142(5/6):198–205

Xylander W (2015a) Die Bodenzoologie Görlitz—Forum für den internationalen Wissenstransfer. Natur Forschung Museum 145(3/4):78–83

Xylander WER (2015b) Zwischenstelle Museum—Bodenbiologie, Biodiversität, Bodenschutz und Transfer. In: Wessolek G (ed) Von ganz unten—Warum wir unsere Böden besser schützen müssen. Oekom-Verlag, München, pp 97–104

Xylander WER (2017) Prof. Dr. Wolfram Dunger – Bodenzoologe und Museumsmann. Natur Forschung Museum 147(5/6)

Xylander W, Düker C (2009) Das Senckenberg Museum für Naturkunde Görlitz. Natur und Museum 139(1/2):30–39

Xylander W, Walker C, Herkner B (2013) Exhibitions of the Senckenberg museums: Windows into research. In: V. Mosbrugger (Hrsg.) Senckenberg 2011-2012, pp 148–163

Chapter 29
GIEẞEN: University Collections: Justus Liebig University Gießen

Eva Diehl, Birgit Jauker, Christian Albrecht, Thomas Wilke, and Volkmar Wolters

Abstract The collection of the Department of Animal Ecology and Systematics at the Justus Liebig University Gießen comprises a broad range of invertebrates and vertebrates, skulls, eggs, bird nests, and genetic resources. In total, more than half a million specimens of organismic samples and more than 20,000 specimens of genetic samples are stocked. The collection is complemented by anatomic maps and models. As the Justus Liebig University strongly emphasizes organismic aspects of academic training in biology, a large part of the collection is regularly used for teaching Germany's native fauna. For example, approximately 5000 invertebrates out of 90 taxa are used for mandatory courses on taxonomic identification. The collection also contains voucher specimens of scientific studies and research projects of the past decades. A collection's highlight is the material sampled for the faunistic inventory of the Hoher Vogelsberg area in Hesse, which is part of the largest volcanic region in Europe. Another highlight is one of the largest DNA reference collections for worldwide freshwater mollusks. It is on of the most complete wet collections of macrozoobenthic taxa of worldwide ancient lakes, i.e. extant lakes orders of magnitudes older than most lakes on earth. This part of the collection is also a valuable DNA reference collection for some poorly studied regions such as the Tibetan Plateau.

Keywords UGSB • Teaching collection • Vertebrates • Invertebrates • Freshwater mollusks • DNA • Ancient lakes • Vogelsberg

29.1 General Information

The collection at the Justus Liebig University Gießen consists of two major parts: the Justus Liebig University (JLU) Gießen Zoological Collection of the Animal Ecology Group (Prof. Dr. Volkmar Wolters) and the University of Gießen

---

E. Diehl • B. Jauker (✉) • C. Albrecht • T. Wilke • V. Wolters
Department of Animal Ecology and Systematics, Research Centre for Biosystems, Land Use and Nutrition, Heinrich-Buff-Ring 26-32, 35392 Gießen, Germany
e-mail: Birgit.Jauker@allzool.bio.uni-giessen.de; Christian.Albrecht@allzool.bio.uni-giessen.de

L.A. Beck (ed.), *Zoological Collections of Germany*, Natural History Collections,
DOI 10.1007/978-3-319-44321-8_29

Systematics and Biodiversity (UGSB) Collection of the Systematics and Biodiversity Lab (Prof. Dr. Thomas Wilke, Fig. 29.2). According to sampling location, documentation, and purpose, the JLU Zoological Collection comprises (a) the Teaching Collection for educational purposes mainly comprising Germany's native fauna, especially birds and arthropods, the latter being stored without reference data; (b) the Voucher Collection of scientific projects, mainly arthropods, sampled nationally and internationally, encompassing detailed reference data; and (c) the Regional Collection of animals from diverse habitats of the Hoher Vogelsberg region in Hesse, Germany (Fig. 29.1, left). The UGSB Collection is primarily a research collection that consists of (a) the DNA prep Collection, (b) the Wet Collection, and (c) the Dry Collection. The main purpose of the collection is the

Fig. 29.1 Sample regions for the collections of the Justus Liebig University Gießen: nature reserve Hoher Vogelsberg (*left*, source: V. Mader) and the ancient Lake Malawi (*right*, source: C. Albrecht)

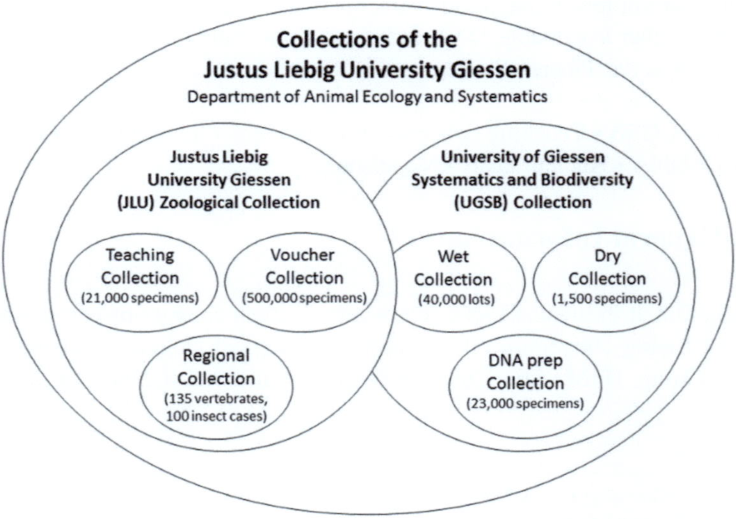

Fig. 29.2 Collections of the Justus Liebig University Gießen, Department of Animal Ecology and Systematics: the Justus Liebig University Gießen Zoological Collection and the University of Gießen Systematics and Biodiversity Collection with estimates for parts

long-term maintenance of DNA resources for evolutionary biology and biogeography studies with a taxonomic focus on aquatic invertebrates, particularly mollusks. The extensive DNA prep Collection also serves as voucher depository for externally funded research. A regional focus is on Africa, Asia, and SE Europe (Fig. 29.1, right).

The JLU Zoological Collection (except the Voucher Collection) is mostly based on the intensive sampling and collecting of Germany's native fauna during the 1960s, 1970s, and 1980s initiated by Prof. Dr. Heinz Scherf (*1928–†2004). Currently, the Teaching Collection comprises approximately 1250 vertebrates with 500 mammals including 450 skulls, 600 birds, 126 fishes, 20 amphibians, and 12 reptiles (Fig. 29.3). Mammals and birds were mounted as taxidermy mounts or round mounts by the former staff taxidermist Herwig Püschel. The Teaching Collection further comprises 120 clutches and several nests of birds. Although primarily focusing on today's fauna of Germany, several historic and exotic objects add to the collection, e.g., a bluethroat (*Luscinia svecica*) of 1907 (Fig. 29.3) or a southern rockhopper penguin (*Eudyptes chrysocome*) from the Berlin Zoo in 1904. Concerning invertebrates, snails, seashells, crustaceans, centipedes, millipedes, spiders, and a wide range of insects are included in the Teaching Collection. Parts of the insect collection have been handed over to the Senckenberg Nature Research Society, Frankfurt, in the mid-1990s. The roughly 20,000 invertebrates remaining are used as course material for academic education on zoological taxonomy and systematics. They were mostly collected for this purpose and stored without reference data. Damaged material is regularly substituted. The Regional Collection was founded in 1963, when the zoological field research station Künanzhaus was established with the aim of an extensive faunistic inventory of the Hoher Vogelsberg region. Since then regular surveys have been conducted. This collection includes 135 vertebrates (round mounts and skulls of small mammals and birds) and a reference collection consisting of 100 insect cases, e.g., with Coleoptera (36 cases) and Lepidoptera (19 cases). The Voucher Collection currently comprises roughly 500,000 arthropod individuals stored in insect cases or alcohol. Voucher specimens include pollinators (e.g., hover flies and wild bees), predatory

Fig. 29.3 Exhibits from the Justus Liebig University Gießen Zoological Collection (from *left* to *right*): a bluethroat (*Luscinia svecica*) of the year 1907, a Eurasian pygmy shrew (*Sorex minutus*), and a rook's clutch (*Corvus frugilegus*, source: Justus Liebig University Gießen)

arthropods (e.g., carabid beetles and spiders), soil fauna (e.g., collembolans and mites), and limnic taxa. These specimens are kept for a minimum of 10 years.

The UGSB Collection was founded in 2004 by Prof. Dr. Thomas Wilke and Dr. Christian Albrecht. The collection includes highly valuable materials from regions rarely surveyed and from taxa prone to extinction or already extinct. The collection strongly builds on regular field trips and expeditions but also on donations. Currently, the DNA prep Collection comprises 23,000 specimens, the Wet Collection comprises more than 40,000 lots, and the Dry Collection adds 1500 specimens. The UGSB Collection is growing by approximately 3000–5000 lots per year. The largest part of the UGSB Collection consists of freshwater macrobenthos, particularly species from ancient lakes mostly belonging to the Gastropoda and Bivalvia, but also includes crustaceans, insects, leeches, and other taxa. Samples in the collection cover all ancient lakes with particular emphasis on Ohrid, Baikal, Malawi, Tanganyika, Victoria, Titicaca, and the Malili lakes (e.g., Albrecht and Wilke 2008; Schultheiß et al. 2009; Stelbrink et al. 2015). Many of these lakes are in grave danger of losing their (endemic) diversity. Therefore, the UGSB Collection is an extremely valuable archive for future reference. Furthermore, rarely studied regions are well represented in the UGSB Collection such as the Tibetan Plateau, the South American Altiplano, Wallacea, or African freshwaters. Besides invertebrates, the UGSB Collection contains reasonable numbers of blood and feather samples from the white stork *Ciconia ciconia* and various bird species of Australia. During the last years, important additions of marine taxa, mostly fishes and corals, have been registered related to research conducted in the framework of the Centre of Excellence in Marine Sciences (CEMarin). All samples in the UGSB Collection are geo-referenced and databased. The digital information covers sampling, locality, and storage details. Taxonomic workup is usually to the family and genus level and very often to the species level. Additionally, environmental samples (such as the whole bottom samples from lake dredging campaigns) are maintained. To date, access to the UGSB database is provided via the internal university network; a potential external web access is being discussed.

The major part of the collections is maintained at the Department of Animal Ecology and Systematics and in adjacent university buildings in Gießen (except the Regional Collection). The arthropods of the Teaching Collection and the Voucher Collection are kept in a 4 °C climate chamber. The taxidermy mounts are stored in glass cabinets visible to the public (Fig. 29.4, left). Other vertebrate objects are stored in drawers underneath and in cupboards (Fig. 29.4, right). At least once a year, several of the mounted specimens are used for short-term exhibitions on current zoological or ecological topics. The UGSB Collection is located in a dedicated underground collection room equipped with explosion-proof freezers for the Wet Collection (−20 and −80 °C), ultralow freezers for DNA preps (−80 °C), and cabinets for the Dry Collection (Fig. 29.5, middle). In order to assure that the wet samples remain suitable for DNA extractions over extended periods of time, samples are kept both at −20 °C and in 80 % ethanol in specialized cryotubes (Fig. 29.5, right) or polycarbonate/polypropylene containers (Fig. 29.5, left). Furthermore, the underground collection room features stereomicroscopes and a

Fig. 29.4 Exhibition and storage of the Teaching Collection: taxidermy mounts in glass cabinets accessible to the public (*left*) and vertebrate skulls in drawers underneath (*right*, source: Justus Liebig University Gießen)

Fig. 29.5 Storage and handling of the University of Gießen Systematics and Biodiversity Collection: specimens from the Wet Collection stored permanently in 80 % ethanol in polycarbonate/polypropylene containers (*left*) in explosion-proof freezers in the underground collection room (*middle*) and DNA preps in cryotubes (*right*, source: C. Albrecht)

computer desk for on-site data access. Within the department, two laboratories with five to six stereoscopes and one to two microscopes each are available for the determination of species by scientific staff and students. The university further provides several rooms with up to 200 workplaces equipped with stereoscopes and microscopes that are used in determination classes. Moreover, high-resolution pictures from a Keyence microscope can be projected via a digital projector.

The Regional Collection is stored at the field station Künanzhaus located on the mountain Hoherodskopf (764 m) in the nature reserve Hoher Vogelsberg approximately 50 km east of Gießen. The field station offers 10 beds and 12 workplaces equipped with stereoscopes and is used for long-term taxonomic courses that involve fieldwork such as arthropod surveys or bird monitoring.

Responsible for the JLU Zoological Collection is Prof. Dr. Volkmar Wolters, head of the department and chair of the Animal Ecology Group. Permanent educational staff member Dr. Birgit Jauker is taking care of the Teaching Collection, i.e., inventory, documentation, and loans. She is teaching the determination classes for undergraduates and has several years working experience in a museum. Replacing damaged arthropod individuals for determination classes requires huge efforts. The continuous resampling is financed by the Animal Ecology Group. Maintenance tasks such as pest control within the collection and repairing damaged objects are either done by members of the department or given to external personnel on a contract basis. Responsible for the UGSB Collection is Prof. Dr. Thomas Wilke, head of the Systematics and Biodiversity Lab and leader of the Evolutionary Biology Group. He has long-term experiences in collection work, among others from positions held at the Academy of Natural Sciences in Philadelphia, USA. Dr. Christian Albrecht (assistant professor), leader of the Aquatic Biodiversity and Biogeography Group, is curating the collection and is taking care of, e.g., databasing, loan, and documentation. He has long-term experiences in collection work in international museums in the USA, Europe, and Africa. Curatorial support comes from a permanent half-time technical assistant staff member of the Systematics and Biodiversity Lab. So far, there has been no external funding for collection maintenance.

29.2 Research

Within the JLU Zoological Collection, national and international research projects and scientific studies including PhD, master, and bachelor theses of the Animal Ecology Group built up the Voucher Collection. Specimens have mainly been collected to investigate the impact of land use on functionally important arthropods in agricultural landscapes. For example, the effect of management intensity on predator diversity and pest control potential was studied by collecting web-building spiders and their prey (Diehl et al. 2013; BIOPLEX project). Jauker et al. (2009) investigated the impact of landscape structure on pollinators including wild bees and hover flies (DBU-funded PhD thesis). Most studies were conducted in Hesse, Germany. Other specimens were sampled at the Biodiversity Exploratories' sites all over Germany (e.g., Birkhofer et al. 2015) or at the International Rice Research Institute, Philippines (John et al. 2015; DFG research unit ICON). Collection-based research uses existing voucher specimens to answer specific research questions, e.g., on the effect of agricultural management on phylogenetic distances in ground beetle communities or phenotypic traits such as body size characteristics in wild bees (Warzecha et al. 2016; Fig. 29.6).

Fig. 29.6 Voucher specimens from scientific studies: wild bees from flowering fields (*left*, source: D. Warzecha), ground beetle *Poecilus cupreus* (*middle*, source: V. Mader), and selected endemic mollusk species from the ancient Lake Prespa, Balkans (*right*, source: C. Albrecht)

Scientists from the Animal Ecology Group collaborate with international universities (e.g., Lund, Sweden or Moscow, Russia), the Senckenberg Nature Research Society, the Academy of Nature Conservation of the State of Hesse (Naturschutz-Akademie Hessen), and regional conservation projects (e.g., Naturschutzgroßprojekt Vogelsberg). The JLU Zoological Collection is listed in the information system on collections and museums at German universities (http://www.universitaetssammlungen.de/) established by the Humboldt University of Berlin. Parts of the voucher specimens can be found in a web-based database that currently includes 2.5 million records and was established to investigate the impact of human land use on species diversity (Newbold et al. 2015; http://www.predicts.org.uk/). Several scientists of the group participate in an online platform designed to compile spatially explicit records of bird species in Germany (www.ornitho.de) or share their findings using social media (e.g., Twitter). Concerning the Regional Collection, the group cooperates with taxonomic experts by lending specimens from different taxonomic groups for taxonomic review or the composition of red lists. Numerous results from the faunistic inventory of the nature reserve Hoher Vogelsberg have been published in a journal pertaining explicitly to the research station Künanzhaus.

All research projects of the Systematics and Biodiversity Lab (Prof. Dr. Thomas Wilke) involve collection-based research. Major contributions to the UGSB Collection came from long-term projects funded by the DFG such as the priority programs "Radiations—Genesis of Biological Diversity" and Tibetan Plateau: Formation-Climate-Ecosystems (TiP) or the international scientific deep-drilling project Scientific Collaboration on Past Speciation Conditions in Lake Ohrid (SCOPSCO). Moreover, a project within the DFG research unit "RiftLink" has been contributing significant amounts of new materials from many parts of sub-Saharan Africa. Due to new projects and intensive collaboration with African universities from South Africa, Uganda, Tanzania, and the Democratic Republic of Congo, African samples are constantly added to the collection. The CEMarin, initiated by the University of Gießen, is an international consortium based on a long-term collaboration of research institutions in Columbia and Germany. Marine samples (e.g., corals and sponges) are increasingly stored in the UGSB Collection, most often as long-term loan according to CBD rules. New materials, mostly macrozoobenthos, also come from the Pontocaspian region as part of the recently funded Marie Curie Actions Initial Training Network (ITN) Pontocaspian Rise and Demise (PRIDE).

The Systematics and Biodiversity Lab has developed a wide network of both professional and laymen researchers donating materials to the UGSB Collection. Cooperations exist, among others, with the Humboldt-Universität zu Berlin and Freie Universität Berlin; the Limnological Institute of the Russian Academy of Sciences, Irkutsk; the Hydrobiological Institute, Ohrid, Macedonia; the Mbarara University of Science and Technology (MUST), Uganda; and the Australian National Wildlife Collection, Canberra. The collection is regularly consulted in international freshwater mollusk assessments carried out by the IUCN. Dr. Albrecht is an appointed assessor of the IUCN SSC Mollusc Specialist Group. For example, collections from the Balkans (e.g., Fig. 29.6, right) based on more than a decade of intensive sampling were used for the most recent Red List of nonmarine mollusks of Europe (Cuttelod et al. 2011). The UGSB Collection also has international importance due to the fact that some current species invasions in Europe are well documented by time series collections. A prominent example is the notoriously invasive quagga mussel *Dreissena rostriformis* (Heiler et al. 2013). Publications basing on UGSB material can be found at http://www.uni-giessen.de/cms/faculties/f08/departments/tsz-en/wilke/publications.

29.3 Educational Work

During the academic training of biologists, the large Teaching Collection from the JLU Zoological Collection and parts of the UGSB Collection are used. Vertebrate objects are presented in general zoology courses as well as in specific ornithology or mammalogy courses. The invertebrates are used in large numbers as practicing and exam material in animal determination classes for bachelor students and teacher trainees. Local schools also borrow material for class projects. In advanced classes, students learn about collection-based research, insect preparation techniques, and handling animal collections, e.g., starting their own collection and designing an exhibition on a current zoological or ecological topic by using the available collection material. In several field excursions, interested students can study the terrestrial, limnic, and marine biodiversity, e.g., at the field station in the nature reserve Hoher Vogelsberg, the North Sea, and the Red Sea or in Namibia. During an international course on soil animal taxonomic identification, selected students from over ten European countries receive practical knowledge from worldwide recognized experts for numerous invertebrate groups. This course is organized by members of the Animal Ecology Group in cooperation with the Russian Academy of Sciences and the Swedish Agricultural University. Bachelor and master theses are often supervised by the department's taxonomic experts as they involve the determination of specific animal groups, e.g., analyzing invertebrate diversity within different habitats. Ultimately, graduates from the Department of Animal Ecology and Systematics are well prepared for a profession that requires taxonomic expertise.

References

Albrecht C, Wilke T (2008) Lake Ohrid: biodiversity and evolution. Hydrobiologia 615:103–140

Birkhofer K, Diekötter T, Meub C, Stötzel K, Wolters V (2015) Optimizing arthropod predator conservation in permanent grasslands by considering diversity components beyond species richness. Agric Ecosyst Environ 211:65–72

Cuttelod A, Seddon M, Neubert E (2011) European red list of non-marine Molluscs. Publications Office of the European Union, Luxembourg

Diehl E, Mader VL, Wolters V, Birkhofer K (2013) Management intensity and vegetation complexity affect web-building spiders and their prey. Oecologia 173:579–589

Heiler KCM, Bij de Vaate A, Ekschmitt K, von Oheimb PV, Albrecht C, Wilke T (2013) Reconstruction of the early invasion history of the quagga mussel (*Dreissena rostriformis bugensis*) in Western Europe. Aquat Invasions 8:53–57

Jauker F, Diekötter T, Schwarzbach F, Wolters V (2009) Pollinator dispersal in an agricultural matrix: opposing responses of wild bees and hoverflies to landscape structure and distance from main habitat. Landscape Ecol 24:547–555

John K, Jauker F, Marxsen J, Zaitsev AS, Wolters V (2015) Earthworm bioturbation stabilizes carbon in non-flooded paddy soil at the risk of increasing methane emissions during the wet season. Soil Biol Biochem 91:127–132

Newbold T, Hudson LN, Hill SLL, Contu S, Lysenko I et al (2015) Global effects of land use on local terrestrial biodiversity. Nature 520:45–50

Schultheiß R, Van Bocxlaer B, Wilke T, Albrecht C (2009) Old fossils–young species: evolutionary history of an endemic gastropod assemblage in Lake Malawi. Proc R Soc London Ser B 276:2837–2846

Stelbrink B, Shirokaya AA, Clewing C, Sitnikova TY, Prozorova LA, Albrecht C (2015) Conquest of the deep, old and cold: an exceptional limpet radiation in Lake Baikal. Biol Lett 11

Warzecha D, Diekötter T, Wolters V, Jauker F (2016) Intraspecific body size increases with habitat fragmentation in wild bee pollinators. Landscape Ecol 31:1449–1455

Chapter 30
GÖTTINGEN: The Zoological Museum of Göttingen University

Rainer Willmann

Abstract The Zoological Museum of the University of Göttingen dates back to the collections of Johann Friedrich Blumenbach and the "Königliche Academische Museum" that was founded in 1773. In the nineteenth century, it housed one of the most famous zoological collections worldwide with particularly valuable herpetological material. However, due to a shift from collection-based work to more general aspects of biology and a neglect of classical biodiversity research, parts of the collections left Göttingen or were lost during the twentieth century.

Major improvements took place only during the last 20 years. The collections have been relocated to acceptable rooms (ca. 350 m^2 plus ca. 60 m^2 for a considerable collection for teaching purposes). Several collections were acquired, and research on the systematics of particular taxa contributes to extending the collections. According to current estimates, the museum contains about 150,000 objects and units. A database is nearing completion. The exhibition halls cover about 330 m^2. However, as the number of staff and the museum's budget are limited, it is difficult to offer changing exhibitions focusing on varying topics.

The University of Göttingen plans to assemble many of its numerous smaller collections in the historical building that contains the Zoological Museum, while several departments not associated with museum-based work will be relocated. Further improvement for the zoological collections can be expected from this development.

Keywords Göttingen • Blumenbach, Johann Friedrich • Royal Academic Museum • Königliches Academisches Museum • Natural curiosities • Mammals, amphibians, birds, snakes, Testacea • Schaalthiere • Conchylien • Berthold, Arnold Adolph • Keferstein, Wilhelm Moritz • Ehlers, Ernst • Naples • World War I • World War II • Heberer, Gerhard • Ax, Peter • Museum Alexander Koenig • Seba, Albertus • d'Argenville, Dezaillier • Linnaeus • AnimalBase • Bonn • Monitors • Lizards • Crustaceans • Fish • Thylacinus • Tasmanian Tiger • Insects • Mantodea • Phasmatodea • Sperm whale • Huia • *Heteralocha acutirostris* • Norfolk Island Kaka • *Alca impennis* • Great Auk • *Drepanis coccinea* • Cook, James

R. Willmann (✉)
Zoologisches Museum der Georg-August-Universität Göttingen, Berliner Str. 28, 37073 Göttingen, Germany
e-mail: gtroest@gwdg.de; rwillma1@gwdg.de

L.A. Beck (ed.), *Zoological Collections of Germany*, Natural History Collections,
DOI 10.1007/978-3-319-44321-8_30

In 1773, the Royal Academic Museum (Königliches Academisches Museum) was founded in Göttingen in which several collections of "natural curiosities" and many other objects, such as physical instruments, coins and libraries, were assembled. The first curator of the museum, and hence also of the zoological collections, was the famous German anatomist and co-founder of biological anthropology, Johann Friedrich Blumenbach (1752–1840). In 1778, he listed more than 12,000 objects, among them 307 devoted to mammals; 73 to birds; 197 to amphibians, snakes and other reptiles; and 1476 to "Testacea—Schaalthiere—Conchylien" (Nawa 2010). Due to the large amounts of specimens later acquired by Blumenbach, sections of the collections were transferred to the botanical garden (this applies to the herbarium) and to buildings that were newly bought by the Göttingen University from 1792 onwards. In 1840, after Blumenbach's death, the university enlarged the treasures of the academic museum by buying Blumenbach's private collections.

From 1840 to 1861, the zoological collections were curated by the herpetologist Arnold Adolph Berthold (1803–1861). Berthold was followed by Wilhelm Moritz Keferstein (1833–1870), and in 1870 Ernst Ehlers (1835–1925) became the head of the collections. Keferstein pointed out that the building used in his times was unsuitable for the collection and initiated the idea of a new building for the natural history collections. However, he did not live to see the new building, which was ready to receive the collections in 1878.

In the late nineteenth century and at the beginning of the twentieth century, the collections were among the most famous zoological collections worldwide. In 1877 one was successful in obtaining thousands of birds belonging to Major Kirchhoff from Nuremberg. A year later, the Rosenhauer insect collection arrived in Göttingen. Australian mammals, amphibians, birds and fish; marine animals from Naples; specimens from Borneo, Mexico and elsewhere; and further insect collections followed (Ehlers 1901; Willmann 2001). Capable preparators were employed who contributed to the reputation of the museum.

However, the collections belonged to a university where the directors did not always appreciate the value of one of the grandest zoological collections in Germany. After World War I, research shifted more and more from collection-based work to more general aspects of biology. Again, increasing numbers of students had to be taught in a building which was originally not devoted for these purposes, and, in addition, geology, palaeontology and anthropology resided in the same place, while new space was hardly ever provided. Exhibition halls were turned into lecturing rooms, parts of the collections left Göttingen, and due to inappropriate storage, some of the old insect collections even decayed.

From 1944 on, Göttingen was threatened by World War II bombing. This meant an end to the exhibitions because hundreds of objects had to be stored in a safer cellar. The bird collection was transferred to the church of the nearby village of Reinhausen. On April 7 and 8 of 1945, parts of the museum building were destroyed by bombs.

While most of the collections survived the event, it was almost impossible to provide a suitable storage for them. In 1952 one could read in the newspapers that the "great Zoological Museum" still does not provide access for the public. However, in 1956 parts of the mammal and bird collections were on display (Piepho

1956), and the anthropologist Gerhard Heberer was responsible for a modern exhibition on the evolution of man. In 1969, the opening of a modern zoological exhibition took place in a new building which was added to the old one as the new director, Peter Ax, was lucky to get funds for that purpose. The new halls extended over 250 m^2. One topic at that time was colours in the animal kingdom and mimicry and, another one, of course, animal diversity. One sacrificed, however, exhibition halls in the centre of the traditional buildings, among them the bird's hall, which was about 180 m^2 large.

For most of the collections, there was almost no room left, and the majority of specimens were not meant to go on display. Therefore, it was decided to dispose many objects instead of applying for personal and permanent financial support for the research collections. Their fate may be illustrated by using the amphibians as an example. Until 1974, the herpetological collection of the museum Alexander Koenig in Bonn was of minor importance. Ten years later, however, that collection had become a respectable assemblage of amphibians, snakes, monitors, lizards, etc., the reason being that it had received the herpetological collection from Göttingen in 1977 with thousands of jars, including specimens from the Blumenbach collection (Böhme and Bischoff 1984). Other parts of the Göttingen zoological collections went to Frankfurt (crustaceans), Hamburg (fish) and Hannover (most of the Kirchhoff collection of birds). Older scientists who once worked in Göttingen remember that a remarkable assemblage of bones was thrown into the backyard of the Institute of Zoology. Palaeontologists were successful in saving, among others, the skull of a *Thylacinus* (Tasmanian tiger). However, almost all of this material was destroyed. This led to the discouraging wording, to be read in *Die Zeit* in 1993 (Weymayr 1993), that the zoological collections in Göttingen have more or less ceased to exist. Fortunately, it was not quite so bad. For example, a large collection of sea cucumbers from the middle of the nineteenth century, most of them holotypes that were regarded as lost, was recovered and thus returned to the scientific community (Willmann 2001).

Staff and finances. As the head of the Institute of Zoology was later also (after the foundation of several zoological departments) the head of the Department of Animal Morphology and Systematics and the museum's director, it largely depended on the respective keeper of this position how much support the zoological collections gained. For a long time, even portions of the budget that were meant for the museum were used for other purposes. Only in 2005 did the museum get a budget of its own. A little later, the University of Göttingen became aware of the uniqueness and importance of its 30 collections that were scattered all over the campus, many of which were almost forgotten and had no one who looked after them. In order to avoid situations as the aforementioned and due to the fact that at other universities collections were lost or purposely destroyed (because they require financial support), the position of a top curator was created in Göttingen ("Zentralkustodie") who shares the responsibility with the heads of departments, the directors and the collections' curators.

The present and the future. Until 2005, the museum building of 1878 contained various zoological departments, part of the Department of Anthropology and small temporal working groups of other faculties. Currently, it mainly houses the

Departments of Animal Ecology and, of course, the Zoological Museum and the Department of Animal Morphology and Systematics. In the beginning of the 1990s, with Gert Tröster as the new curator of the Zoological Museum and the author of this article as its new director, the main tasks were to save what could be saved and to check what was left. Large parts of the collections not meant for display were housed in corridors and unsuitable rooms throughout the building and even in an old building of the university library. Stepwise, the collections have been relocated to rooms with acceptable conditions, although these are still not ideal.

Several insect and other collections were acquired in recent years, including collections that formerly were used in schools. The latter fact is unfortunate because this implies that preserved animals and plants are no longer often used in schools. A collection of insects preserved in alcohol is being set up. As a result of third-party-funded research, large collections of Mantodea (currently on loan with its main student in Bad Dürkheim), Phasmatodea and other insects came into existence. Recently, single larger mammals, such as a hippopotamus and a giraffe, were dissected in order to put their skeletons on display, and the museum now houses a large collection of horse skulls. A database covering the entire collection is nearing completion. As mentioned above, the museum now has its own budget and additional personnel especially for publicity work, for example, for teaching school children (pupils from most schools in the area regularly visit the museum).

A large collection devoted to teaching is kept in separate rooms. It is comprised of units mainly assembled before 1900 and contains many valuable specimens. Numerous historical zoological plates (prints as well as hand paintings) that were used for teaching in former times are also kept in the museum. (Recent plates are still used during practical courses.)

However, the situation in the Zoological Museum is far from satisfactory. Current tasks include the extension of the collections, the preservation of the specimens according to latest developments in the respective techniques, the adaptation of the collections to the needs of modern research, etc. However, the number of staff is, to put it mildly, limited, and the budget is not sufficient to allow collecting trips where necessary, for example, in order to support research on certain taxa or to add specimens to the museum that might be important for comparison purposes.

Exhibitions. In the exhibition halls of the Zoological Museum of Göttingen, insights into the fauna of Central Europe are combined with an exhibition on animal diversity. One focus is on the evolution and diversity of whales and their fate under human influence. This part of the exhibition resulted from the dissection in January 1998 by the museum staff and members and students of the Department of Morphology and Systematics of a 17-m-long sperm whale that had been stranded at the North Sea coast of Schleswig-Holstein. It took the museum staff 3 years of preparation work to produce casts of organs and to put the skeleton on display (Frenz and Paulat 2001). In the "black hall", skeletons of various species of mammals and sauropsids are shown in order to provide visitors with insights into comparative skeletal anatomy. One case is devoted to extinct birds, presenting, for example, a pair of Huias (*Heteralocha acutirostris*, extinct since 1907), a Norfolk Island Kaka (extinct since 1851) and a skeleton of *Alca impennis*, the Great Auk (combined from remnants of several specimens, extinct since 1844). The museum furthermore possesses vouchers of species that are now extremely rare. Again, a specimen of the Hawaiian *Drepanis coccinea*, part of the Forster-James Cook collections donated on behalf of the king of England to the university in 1782, belongs to the museum's more interesting bird specimens (on display in the Institute of Ethnology of Göttingen University).

The position of the Zoological Museum of Göttingen in the university's framework. The Zoological Museum gained renewed importance when, in 1999, the foundation of the Centre of Biodiversity Research and Ecology was initiated at Göttingen University by the author, with a study programme of its own. One focus of the study programme is the multitude of animals and plants, their evolution and ecology. The zoological collections became an integral part of bachelor, master and doctoral theses, e.g. on various insect groups. Biohistorical museum-based work has opened new fields in the Institute of Zoology and Anthropology. For example, the museum and the department participated in a new edition of the once famous four-volume work of Albertus Seba, a vast set of books on natural history from the middle of the eighteenth century (Willmann and Rust 2001; Willmann et al. 2001), and in the "Conchyliologie" of Dezallier d'Argenville (Willmann 2009; Willmann and Willmann 2009). Both of these are sources of many of Linnaeus' descriptions. Another achievement is "AnimalBase", which provides online access to information on species described in the few decades after the publication of Linnaeus' 1758 edition of the *Systema Naturae*. AnimalBase has links to almost all zoological taxonomic literature of that period of time and therewith to the original descriptions, which are often difficult to find otherwise (open access). This was achieved during a project in cooperation with the university library of Göttingen that started in 2000 (Early Zoological Literature online).

Imminent changes. The building housing the zoological collections opened as a natural history museum in 1878 and is protected according to laws for historical monuments. Under the name "Forum Wissen", the University of Göttingen attempts to assemble most of its smaller collections and objects of particular public interest in the building, while all departments including the Department of Animal Morphology and Systematics are being transferred to other places in the university. An improvement for the collections can be expected in the course of a restructuring of the building and major renovations, while the zoological exhibitions are likely to remain largely as they are.

Addendum

Staff

One preparator, one technical support (teaching school children) (50 % position), one curator (Akademischer Oberrat, teaching obligation [university courses and lectures]) and one director (Prof. C 4; also head of the Department of Animal Morphology and Systematics, teaching obligation).

Space

Exhibition halls: 330 m^2
Collection rooms (distributed over four levels in the same part of the building): 350 m^2
Study collection: 63 m^2
Room for teaching school children: 23 m^3 (level 2)
Taxidermist's laboratory: 49 + 42 m^3 (in two buildings)

Collections

According to a recent estimation, the research collections in the Zoological Museum of Göttingen University contain more than 150,000 specimens:

Insecta 46,000, most of them pinned
Madagascar collection (alcohol preservation) about 40,000 specimens
Molluscs 14,400 units (about 500,000 specimens)
Vertebrates (excluding material preserved in fluid and excluding birds) 750
Aves > 4800 (including an egg collection of about 1000 exemplars)

Alcohol Collection (excluding insects):

a. Various taxa (from vertebrates to arthropods) ca. 2500 units (about 7500 specimens)
b. Polarstern collection/Pantopoda 7500 specimens

Study Collection

About 330 vertebrates (mainly skeletons and skeletal parts, such as skulls) and 800 invertebrate animals, most of the latter preserved in alcohol.

Literature

Birukow GAx P (1967) Aus der Geschichte der Zoologie in Göttingen. Verh Deutsch zool. Ges 1966 (= Zool Anz Suppl 30):48–53
Böhme W, Bischoff W (1984) Die Wirbeltiersammlungen des Museums Alexander Koenig. III. Amphibien und Reptilien. Bonner zool Monogr 19:151–213
Chronik des Zoologischen Museums seit 1909 (geführt bis 1946). Unpublished
Ehlers E (1901) Göttinger Zoologen. In: Festschrift zur Feier des hundertfünfzigjährigen Bestehens der Königlichen Gesellschaft der Wissenschaften zu Göttingen, Berlin, S. 391–494
Frenz L, Paulat M (2001) Die Verwandlung des Moby Dick. Geo 6:100–122

Nawa C (2010) Sammeln für die Wissenschaft? Das Academische Museum Göttingen (1773-1840), 159 p., Göttingen
Piepho H (1956) Das Göttinger Zoologische Museum in der Nachkriegszeit. Beiträge zur Naturkunde Niedersachsens. Festschr. zum 70. Geburtstag von Dr. H. Weigold, 72–74
Wagenitz G (1987) Die Geschichte des Fachbereichs Biologie. Georgia Augusta 47:29–35
Weymayr C (1993) Eingemachte Typen. Die Zeit 7:37
Willmann R (2001) Das Zoologische Museum der Universität Göttingen. In: Hoffmann D, Maack-Rheinländer K (eds) "Ganz für das Studium angelegt": Die Museen, Sammlungen und Gärten der Universität Göttingen. Wallstein Verlag, Göttingen, pp 249–259
Willmann R, Rust J (2001) Der Thesaurus Albertus Sebas als zoologisch-botanisches Werk. In: Seba A (ed) Das Naturalienkabinett. Vollständige Ausgabe der kolorierten Tafeln. Taschen Verlag, Köln, S. 26–39
Willmann R, Rust J, Wissemann V, Bierwirth K-D, Bradler S, Engmann, S-E, Hehn N, Hörnschemeyer T, Klug R, Kutzke H, Mielke W, Vogel G, Wedmann S, Welter-Schultes F, Willmann S (2001) Die Bestimmung der Arten im Thesaurus von Albertus Seba. Kommentare zum Thesaurus Albertus Sebas and Appendix. In: Seba A (ed) Das Naturalienkabinett. Vollständige Ausgabe der kolorierten Tafeln. Taschen Verlag, Köln, pp 543–582
Willmann R (2009) The scientific importance of Dezallier d'Argenville's Conchyliologie. In: Dezallier D'Argenville A-J (ed) La Conchyliologie Shells/Muscheln/Coquillages. Taschen Verlag, Köln, pp 38–39
Willmann R, Willmann S (2009) Plates with identification of species. In: Dezallier d'Argenville A.-J (ed) La Conchyliologie. Shells/Muscheln/Coquillages. Taschen Verlag, Köln, pp 45–206

Chapter 31
GÖTTINGEN: Collections in the Departments of "Wildlife Sciences" and "Forest Zoology and Forest Conservation"

Bernhard Weißbecker, Gerrit Holighaus, and Niko Balkenhol

Abstract Both departments are parts of the *Faculty of Forest Sciences and Forest Ecology* within the *Georg-August-University of Göttingen*. Their fields of interest partly overlap which is reflected also in the composition of their scientific collections. The *Department of Forest Zoology and Forest Conservation* holds a comprehensive collection of the birds of Lower Saxony with about 500 specimens and the historical collection of W.G. Glimmann that comprises birds of prey, owls and wildfowl. Also the insect collection has a considerable historic value: some of the 35,000 specimens were collected by the famous entomologists J.T.C. Ratzeburg and A. Förster. According to the original focus of the *Department of Wildlife Sciences*, its collection comprises a huge selection of trophies, especially antlers or horns of typical game species. Of historic value are the obsolete game tag register of Germany and some relicts of the royal Hanoverian hunting court, which was dissolved in 1868. Currently, the collection is being rearranged to highlight the important roles that wild animals play in maintaining the functionality and resilience of ecosystems.

Keywords Birds • Insects • Wildlife • Trophies • Glimmann • Ratzeburg

---

B. Weißbecker (✉) • G. Holighaus
Department of Forest Zoology and Forest Conservation, Georg-August-University Göttingen, Büsgenweg 3, 37077 Göttingen, Germany
e-mail: bweissb@gwdg.de

N. Balkenhol
Department of Wildlife Sciences, Georg-August-University Göttingen, Büsgenweg 3, 37077 Göttingen, Germany

L.A. Beck (ed.), *Zoological Collections of Germany*, Natural History Collections,
DOI 10.1007/978-3-319-44321-8_31

31.1 Introduction

Both departments and their collections have their origin in the Royal Forest Academy that was founded 1868 in Hannoversch Münden (now officially named "Hann. Münden"). In 1939 the Forest Academy was incorporated into the University of Göttingen, and finally in 1970 the departments moved into their new buildings in Göttingen. Over the decades, the research fields of both departments varied depending on prevailing silvicultural subjects and preferences of their respective department heads. Research in the *Department of Forest Zoology and Forest Conservation* was always focused on insects—especially in their role as forest pests. The idea that also endangered insect species deserve protection followed in later years. The *Department of Wildlife Sciences* in its beginnings had a strong focus on hunting but today concentrates on the management and conservation of wildlife and their habitats and ecosystem functions. Birds, mice and other small mammals are of some interest for both departments, since some of them are counted among the forest pests that constitute the "forest conservation" subject. Zoological collections were established in both departments to support research and teaching activities and have grown considerably over the years.

31.2 Department of Forest Zoology and Forest Conservation

Besides a rather unsorted conglomerate of various items used as teaching aids for lectures in basic zoology, the *Department of Forest Zoology and Forest Conservation* holds comprehensive collections specialised on two classes of animals: birds and insects.

The bird collection with about 500 specimens covers the major part of the native bird species of Lower Saxony. After a reorganisation in 2010, the birds are shown in display cabinets dedicated to different habitats in order to inform students about the typical avifauna of biotopes like coniferous forests, deciduous forests or coastal areas (Fig. 31.1). A special highlight within this assemblage is a historical collection of birds of prey, owls and wildfowl compiled by Wilhelm Georg Glimmann (1802–1876). After the death of Glimmann, who was a private collector, his excellently preserved exhibits were purchased by the department. Due to a misspelling in an article of 1908, the collection became known as the "Gliemannsche Vogelsammlung". Originally and typical for its time, Glimmann's birds were presented in small display cases arranged with a modelled environment containing, e.g. nests or eggs. However, the historical wooden cases were not sealed very well and often were too small to present the exhibits adequately. To prevent attacks by insect pests of a very different nature, such as museum beetles, around 1990 a decision had to be taken: either to protect the bird specimens in the best possible way or to preserve the "historic arrangement" of the exhibits. Finally, the former

Fig. 31.1 Display cabinet showing the avifauna of a coniferous forest (Photo: © with Richard Schütz)

option was selected—as might be expected from a department that is committed to insect control and to scientific rather than historic principles. However, a few display cases were kept in the original state that still gives a feeling of the "historic spirit". The bird collection is regularly in use for teaching purposes but will also be opened for interested visitors.

The collection of insects with a total of ca. 35,000 specimens also has a long history. From the beginning, it was a scientific collection. Several types of specimens exemplify the definition of newly described species in the nineteenth century. Parts of the collection can be traced back to the important entomologists Julius Theodor Christian Ratzeburg (1801–1871) and Arnold Förster (1810–1884). A specimen presumably labelled by Ratzeburg himself is shown in Fig. 31.2. As

Fig. 31.2 Parasitic wasp (Ichneumonidae) collected by J.T.C. Ratzeburg (Photo: © with Gerrit Holighaus)

author of the fundamental textbook *Die Forst-Insecten*, published in three volumes from 1837 to 1844, Ratzeburg is regarded as the father of forest entomology. The insect collection is not open to the public; however, we welcome scientists from other departments, cities or states to perform studies with our material.

31.3 Department of Wildlife Sciences

The *Department of Wildlife Sciences* was originally founded as a department for hunting studies, with the sole aim to provide a "scientific justification" for trophy hunting and the associated exhibitions of harvested trophies. Consequently, the large majority of items in the current collection are either antlers or horns of typical game species. These trophies mostly stem from species native to Germany and Central Europe, especially roe deer (*Capreolus capreolus*) and red deer (*Cervus elaphus*), but also include many specimens from Asia and Africa (e.g. sambar deer, *Rusa unicolor*; greater kudu, *Tragelaphus strepsiceros*), as well as a few exhibits from North America (e.g. moose, *Alces alces*; wapiti, *Cervus canadensis*).

Trophies, especially antlers, are often regarded as indicators of animal vitality and quality and are commonly used in wildlife management to identify animals that can legally be harvested. To illustrate the variability in trophies as well as their development, the collection entails several time series of deer antlers of single individuals throughout their life. The collection also includes various forms of "abnormalities" that can occur due to biotic and abiotic disturbances.

Another large portion of the collection consists of lower jaws of roe and red deer that were originally used for age estimation of harvested individuals throughout Germany. In addition, the collection harbours the obsolete game tag register of Germany, which was started in 1904 and was used for several decades to gather information about harvest numbers for certain big game species. For this, hunters were required to send special postcards to the department that listed the species, age class and kill site for every harvested animal.

Fig. 31.3 Raccoons (*Procyon lotor*) (Photo: © with Gerrit Holighaus)

To illustrate developments in hunting techniques and hunting ethics, the collection also includes various types of relevant equipment, such as historic weapons and different kinds of traps, several of which are now outlawed. Some of these historic items originally belonged to the royal Hanoverian hunting court, which was dissolved in 1868.

Under new leadership, the scope of the department changed substantially in 1972. The focus now lay much more on the conservation of animals in human-dominated environments. Consequently, during the next 33 years, the departmental collection expanded its focus from game species and hunting trophies towards non-game species, including rare and endangered species such as the great bustard (*Otis tarda*) or invasive species such as raccoons (*Procyon lotor*, Fig. 31.3). Many of the exotic trophies entailed in the collection were also contributed during this time and were either collected during expeditions or donated from various sources. The collection was also expanded with items displaying various animal signs, including scats, feathers and tracks.

After a hiatus from 2005 to 2011, a new department was formed that now focuses on the holistic and complex interactions between humans, wildlife and ecosystems. Consequently, the collection is currently undergoing another transition, with the aim to use it more frequently for teaching. Specifically, the goal is to rearrange the collection so that it illustrates the dynamics of human-wildlife interactions through history and to highlight the important roles that wild animals play in maintaining the functionality and resilience of ecosystems. The renovated collection will also include workspace for students to make hands-on experiences in species identification and age estimation techniques possible. Increased efforts are currently undertaken to use the vast materials available in the collection for scientific research, for example, with respect to genetic analyses based on the collections of trophies and lower jaws.

Chapter 32
GREIFSWALD: The Zoological Museum of the University Greifswald: Past, Present, and Future

Peter Michalik

Abstract The Zoological Museum forms a unit with the Zoological Institute of the University Greifswald. It was established by the university in 1836 with a focus on the Pomeranian fauna. In the course of the last centuries, the museum grew quickly mainly based on numerous international expeditions, and nowadays it houses approximately 3.5 million specimens including representatives of most animal groups. The collections are used for research, teaching, and public outreach. Current research focuses on taxonomy, systematics, and evolutionary morphology of arthropods especially insect and arachnids. In 2015, the museum moved into a renovated building with state-of-the-art facilities. The personnel consist of a curator, a taxidermist, and a technician.

Keywords Pomerania • Entomology • GDR • Academic reform • micro-CT • Types

32.1 Brief Overview of the History of the Zoological Museum Greifswald

The history of the Zoological Museum Greifswald goes back to 23 November 1836 when a small collection of biological objects were moved to "new" rooms in an administration building of the university located in the J.-S.-Bach-Strasse 12 (formerly known as Büchstrasse 12) (Fig. 32.1). This negligible collection was overseen by Prof. Dr. Christian Friedrich Hornschuch (1793–1850), but especially the conservator Wilhelm Schilling (1790–1874) and the curator Dr. Friedrich Christian Heinrich Creplin (1788–1863) ensured a rapid growth of the zoological collections (Fig. 32.2). In these early years, activities clearly concentrated on local fauna. Schilling was a passionate hunter and accumulated a large amount of local vertebrates resulting in the oldest and largest collection of birds from Western Pomerania known to date.

P. Michalik (✉)
Zoologisches Institut und Museum, Loitzer Str. 26, 17489 Greifswald, Germany
e-mail: michalik@uni-greifswald.de; https://zoologie.uni-greifswald.de

L.A. Beck (ed.), *Zoological Collections of Germany*, Natural History Collections,
DOI 10.1007/978-3-319-44321-8_32

The attractiveness of such collection was documented already during this period as many citizens of Greifswald visited the Zoological Museum regularly. After Hornschuch passed away, Prof. Dr. Julius Münter (1815–1885) became the director of the Zoological Museum and Botanical Garden. The latter was certainly more in Münter's interest, and thus it was of importance that he contracted Dr. Reinhold Buchholz (1837–1876) in 1864 as curator of the zoological collection (Fig. 32.2). Buchholz was very active and attended several expeditions to Arctic regions and Western Africa documented by a large amount of material housed in the museum nowadays. However, these expeditions were often dramatic as, e.g., the expedition to the North Pole in 1869 got shipwrecked near the coast of Greenland and Buchholz and members of the crew persevered for 200 days on an ice floe before rescued. In 1876, Buchholz became the first full professor for zoology in Greifswald but died only three months later after he experienced a severe pneumonia. His successor was Prof. Dr. Adolf Gerstaecker (1828–1895, director 1876–1895), who had a broad interest on zoological topics (e.g., anatomy of mammals) but is mostly known for his entomological contributions (Fig. 32.2). For example, he described numerous new insect species from Africa based on the material collected by Buchholz. The insect collection of Gerstaecker is housed in 240 drawers and contains more than 500 types (Fig. 32.3). Gerstaecker further increased the zoological collections by using his own material but also by purchasing several large entomological collections (e.g., collection of tropical butterflies of Carl Friedrich Pogge III).

Besides his scientific activities, Gerstaecker tried to improve the situation within the building which was obviously an urgent matter. For example, the roof was so broken that it rained on his desk. Even though he had a severe verbal exchange with the university administration, the building was not renovated before 1923–1926, which marked the founding period of the Zoological Institute and Museum (see below). It was not before Gerstaecker's successor Prof. Dr. Christian Gustav Wilhelm Müller (1857–1940, director 1895–1923) that this critical situation began to change. Müller was able to convince the university to purchase an adjacent building which relaxed the situation in the museum (Fig. 32.2). He reorganized the collections and started to include museum material in teaching activities—a practice which outlasted until today. Moreover, he modernized the teaching facilities resulting in an increased attractiveness for prospective students (e.g., only in 1913 156 students were registered for zoology). Besides his merits for the infrastructure and teaching program, he was scientifically very active resulting in an enormous amount of material collected in Europe but also South America with a focus on Brazil, where his brother, the famous natural scientist Dr. h.c. Johann Friedrich Theodor (Fritz) Müller (1822–1897), lived and worked. His favorite group was certainly seed shrimps (Ostracoda)—documented by his extensive collection with numerous types. It can be concluded that the directorship of Müller marks a turning point as he integrated the Zoological Museum with its extensive collections into an academic context leading to the formation of the nowadays known "Zoological Institute and Museum Greifswald" (ZIMG). His successor Prof. Dr. Paul Buchner (1886–1978, director 1923–1926) was finally able to undertake a substantial renovation and extension of the building, and from 1926 the institution carried its official name. Since then the institution was constantly growing,

Fig. 32.1 The old building of the ZIMG in the J.-S.-Bach-Str. 11/12 (*left*) and the new facilities in the Loitzer Str. 26 (*right*)

in personnel but also in space, and currently the ZIMG consists of five full professorships dedicated to a variety of disciplines including, e.g., physiology, ecology, behavior, and systematics. The Zoological Museum is an integrative part of the department of General and Systematic Zoology and overseen by a curator and technical personnel. Even though the history of the museum is tightly linked to the institute, I will only focus on the zoological collections in the following. For a detailed review of the history of the ZIMG, see Kämpfe and Michalik (2011).

The successor of Buchner was Dr. Ernst Matthes (1889–1958, director 1927–1935) who initiated and overseen the complete reorganization of the main exhibition used for teaching but also for public outreach ("Schausammlung"). During World War II, the ZIMG suffered from a decreasing number of students and a high fluctuation of personnel. In 1946, the institution was maintained by two people only (Dr. Heinrich Hertweck, 1906–1985, and Ilselotte Groth, 1923–2012, Fig. 32.2), and 22 rooms were used as living/storage rooms which had to be reclaimed. Especially, the commitment of Ilselotte Groth resulted in a nearly untouched survival of the zoological collections during these difficult times. It was not before 1953, with the appointment of Prof. Dr. Rolf Keilbach (1908–2001, director 1953–1973) as a director, that the ZIMG prospered again (Fig. 32.2). With regard to the Zoological Museum, entomology was certainly the focus since then, and the entomological collections grew constantly. This was intensified with the appointment of Prof. Dr. Gerd Müller-Motzfeld (1941–2009) as curator in 1978 and as full professor in 1992 after the German reunification (Fig. 32.2). His activities focused on faunistic and taxonomic studies of Palearctic ground beetles (Carabidae) resulting in an extensive collection (>200 insect drawers and >500 vials, >4400 species, type specimens of 98 carabid species) but also an enormous amount of sorted material from more than 20 years of faunistic and ecological projects in Western Pomerania (>20,000 vials/year) (Michalik et al. 2010).

By far the largest impact on the Zoological Museum was caused by the third academic reform in the GDR in 1968. The strict government-imposed research focus in Greifswald was microbiology, and all other disciplines were downsized rigorously with dramatic consequences for the zoological collections. Since all

Fig. 32.2 *Upper row*, C.F. Hornschuch, W. Schilling, R. Buchholz, and A. Gerstaecker; *middle row*, G.W. Müller, R. Keilbach, I. Groth, and G. Müller-Motzfeld; and *lower row*, the collection of skeletons including marine mammals before the third academic reform in 1968 (*left*) and the insect collection after this reform packed in a little room by destroying parts of the old cabinets (*right*)

Fig. 32.3 *Upper row*: The wet collection (*left*) and dry collection (*right*) in the attic of the old building under poor conditions. *Middle row*: The new public and teaching collection housed in illuminated showcases. *Lower row*: The new insect collection with the restored historic cabinets (left) and an impression of the highly valuable and well-preserved collection of Gerstaecker

groups working on zoology were concentrated ("packed") in the main building, parts of the collections were "moved" to the attic and other hidden/remote places (Fig. 32.3). For example, the historic cabinets of the insect collections were concentrated in a small room by piling them up on top of each other (Fig. 32.2). The valuable collection of primate skulls from Gerstaecker was nearly destroyed, and especially the collection of birds and mammals was now housed in poor condition. The apparent rush in this emergency situation by moving this part of the collection in the non-insulated attic is best illustrated by an episode experienced by the present taxidermist Hans-Knut Weidemann. After his appointment in 2000, he found a nearly complete skeleton of the extinct Tasmanian tiger on the attic and could also recover some small parts of it by sifting the dirt from the floor. Moreover, the skeletons of several whales, which were the main attraction of the exhibition on marine vertebrates in ZIMG (Fig. 32.2), were "donated" to the Marine Museum in Stralsund as, e.g., the skeleton of a fin whale (*Balaenoptera physalus*), which became one of the most famous objects of this museum later on. Ironically, the Zoological Museum also grew as a result of this drastic reform. During a visit in Rostock in 1969, the former curator Ilselotte Groth rescued several insect collections (e.g., Clasen collection which is one of the oldest and most valuable beetle collections of Mecklenburg-Western Pomerania) of the Zoological Institute Rostock which were already placed outside beside the trash bins. All these examples clearly illustrate that this dictated reform was certainly a serious danger for university collections during GDR times.

32.2 Current Situation of the Zoological Museum

32.2.1 Infrastructure

The current situation is rather pleasant as the Zoological Institute and Museum will be relocated to new or completely renovated buildings finishing in 2017. During 2014–2015 the Zoological Museum moved into a renovated building (located in Loitzer Str. 26, Fig. 32.1) finally providing adequate facilities for a long-term storage of the collections. For this purpose new cabinets, illuminated display cases (Fig. 32.3), and mobile shelving systems were installed. Moreover, the historic insect cabinets were restored in order to store part of the valuable collections of pinned insects in its original trays (Fig. 32.3). In total, the museum houses approximately 3.5 million specimens and more than 800 types in the following collection units (for a detailed list of the collections, see https://zoologie.uni-greifswald.de):

- Wet collection (representatives of nearly all animal groups, extensive material from faunistic studies)
- Dry collection (including mollusk shells, corals, sponges, and wax models)
- Insect collection (contains only pinned insects)

- Collection of stuffed birds and mammals (contains mostly mounted specimen)
- Collection of skeletons including representatives of all vertebrate groups
- Collection for teaching and public outreach (including representatives of all animal groups presented in illuminated display cases)
- Collection of representatives of the local fauna (housed nearby the practical rooms in the new lab building of the ZIMG)

32.2.2 Research

As outlined above, the collections of the ZIMG having a long history reflected in the diversity of the represented animal groups. However, the focal point was and still is certainly on arthropods especially on insects and arachnids resulting in constantly growing collections. For example, numerous expeditions to Middle and Central Asia organized by Müller-Motzfeld during the 1990s and 2000s generated a large collection of ground beetles of these regions, and recent expeditions to South America and Australia proliferated the collection of spiders and other arachnids.

Current research focuses mainly on taxonomy, systematics, ecology, and evolutionary morphology of arachnids and insects using a variety of methodologies including invasive and noninvasive techniques. Especially the application of noninvasive methods as, e.g., microcomputed X-ray tomography (μCT) became very popular in recent years and offers a high potential for collection-based research (Faulwetter et al. 2013). Since the biology section of our university owns an Xradia XCT200 imaging system, we could successfully include μCT in taxonomic research on extant spiders for the first time (e.g., Michalik et al. 2013). Based on this initial study and in cooperation with Dr. Martín Ramírez from the Museo Argentino de Ciencias Naturales "Bernardino Rivadavia" in Buenos Aires (Argentina), we developed a research program addressing the taxonomy, systematics, and evolutionary morphology of haplogyne spiders with a focus on Neotropical and Australasian taxa including μCT analyses. Further current projects using this technique are focused on ground beetles and spiders in Baltic and Burmese amber (e.g., Schmidt et al. 2016).

32.2.3 Teaching

Since the Zoological Museum is an integrative part of the ZIMG, the personnel (curator, taxidermist, and technician) and the collections are included in the teaching of several B.Sc. and M.Sc. programs (e.g., Biodiversity and Ecology). Besides the practical training on, e.g., animal determination and morphological methods, specimens are also included into lectures on systematic zoology. Furthermore, our facilities for taxidermy allow to include methods of conservation and preparation of animals in teaching activities.

32.2.4 *Public Outreach*

For the public the collections can only be accessed by guided tours—a possibility which is, e.g., frequently used by schools. Moreover, the museum is regularly involved in outreach activities of the university and the city of Greifswald.

32.3 Future of the Zoological Museum

The most challenging task in the near future is certainly the digitization of all collections. So far, only a very tiny fraction of the collections, i.e., the collection of birds of Pomerania, is digitized and publicly accessible (http://www.wissenschaftliche-sammlungen.uni-greifswald.de). Furthermore, we were part of a larger collaborative research project funded by the German Science Foundation (DFG) aiming to digitize Palearctic ground beetle collections of museums located in Northern Germany. In the course of this project, we introduced the software Specify® as general management database. Besides the entry of specimen data, we also plan to include high-resolution images of at least all type specimen. For this purpose the BK PLUS Lab System (Dun, Inc.) was purchased establishing an imaging platform allowing to cover the whole range of specimen sizes. All data will be publicly accessible on relevant online resources. Another aim of our digitization efforts is the interlinkage with teaching activities by interactively using high-resolution images for training in animal determination.

References

Faulwetter S, Vasileiadou A, Kouratoras M, Dailianis T, Arvanitidis C (2013) Micro-computed tomography: Introducing new dimensions to taxonomy. ZooKeys 263:1–45

Kämpfe L, Michalik P (2011) 175 Jahre Zoologie in Greifswald—ein Rückblick. Greifswalder Universitätsreden 143:7–27

Michalik P, Fusser M, Sonnenberg D, Schmidt J (2010) Catalogue of the types of Gerd MÜLLER-MOTZFELD's Carabidae collection, and notes on the insect collections at the Zoological Institute and Museum (ZIMG) of the University of Greifswald (Germany). Entomologische Blätter 106:203–232

Michalik P, Piacentini L, Lipke E, Ramirez MJ (2013) The enigmatic Otway odd-clawed spider (*Progradungula otwayensis* Milledge, 1997, Gradungulidae, Araneae): Natural history, first description of the female and micro-computed tomography of the male palpal organ. ZooKeys 335:101–112

Schmidt A, Belousov I, Michalik P (2016) X-ray microscopy reveals endophallic structures in a new species of the ground beetle genus *Trechus* Clairville, 1806 from Baltic amber (Coleoptera: Carabidae: Trechini). ZooKeys 614:113–127

Chapter 33
HALBERSTADT: Museum Heineanum

Bernd Nicolai

Abstract Building on its nearly 200-year history and its important collections, the Heineanum has devoted itself in particular to ornithology. The basis is the bird collection of Ferdinand Heine senior (○1809 †1894): from 12,367 specimens (7,313 mounted specimens and 5,054 skins) at that time, still about 11,600 specimens remain today. The Heine collection is noteworthy and has more than 310 type specimens of over 200 taxa. Most of them were described by Jean Cabanis and/or Ferdinand Heine junior in their collection catalogue 'Museum Heineanum' (published 1850–1863).

Since 1909 public exhibitions have existed, and a complete natural museum was built up after World War II: collections (currently about 35,000 pieces: among others skins/skeletons of birds and mammals, feather collections, clutches/eggs, two skeletons of dinosaurs), permanent and special exhibitions, library (more than 22,500 volumes), research and publication (edit. of two journals). The Heineanum gets a significant support from the museum society ('Förderkreis für Vogelkunde und Naturschutz am Museum Heineanum e.V.').

Keywords Museum Heineanum • Ornithological collections • Type specimens • Type catalogue

33.1 Collection History

Building on its nearly 200-year history and its important collections, the Heineanum has devoted itself in particular to ornithology. The name 'Museum Heineanum' is derived from the landowner and public official Ferdinand Heine senior (1809–1894). His name appears on the title page of the first volume of a catalogue of the bird collection (Fig. 33.1) that was compiled by Jean Louis Cabanis (Cabanis 1850/51). At that time the collection already contained several thousand specimens. In his foreword to the above catalogue, Cabanis supplies concrete

---

B. Nicolai (✉)
Museum Heineanum, Domplatz 36, 38820 Halberstadt, Germany
e-mail: nicolai@halberstadt.de; http://www.heineanum.de

L.A. Beck (ed.), *Zoological Collections of Germany*, Natural History Collections,
DOI 10.1007/978-3-319-44321-8_33

Table 33.1 Overview and chronological categorization of the authorities responsible for the Museum Heineanum collection

Time period	Ownership status	Administrative offices **Directors** Curators	Administrative offices Taxidermists	Remarks
(*ca.* 1830) to 1894	Private collection 'Museum Heineanum'	Ferdinand **Heine** senior	P.L. Martin Carl Müller F. Tiemann ??	Scientific treatment Jean Cabanis catalogue (1850–63)
1994–1907 (1909)		Ferdinand **Heine** junior		
(1907) 1909–1936	(Private collection) administered by the town of Halberstadt	Ferdinand Heine Ernst Heine Otto Heine **August Hemprich**	?? Bernhard Gebser (since 1926 self-employed taxidermist in Halberstadt)	Contract between F. Heine and Halberstadt for 30 years
1936–1945 and 1949	Charitable trust 'Ferdinand Heine'sche Vogelsammlung'	Stiftungsvorstand **August Hemprich**		Charitable trust contract between O. Heine and Halberstadt
(1945) 1949–1956	Non-independent charitable trust (town Halberstadt)	**Carl Michaelis** (1946–1953) Rudolf Busch		Resolution by Halberstadt town council
1956–1964	Department of the Municipal Museum	**Gerhard Ruhe** (1953–1964) Kuno Handtke (1956–1964)	H.J. Hrncirik (1962–1985) Heidrun Scheidt (1966–2006) Wolfgang Sucker (1977–1993) Detlef Becker (from 1987)	Assessment of the collection: Wilhelm Meise in 1948, Hans von Boetticher in 1955/56
1964–1979	Halberstadt municipal museum	**Kuno Handtke** (1964–1979)		Resolution by Halberstadt town council Since 1992 supported by the Friends of the Museum Heineanum
From 1979		**Helmut König** (1979–1991) Bernd Nicolai (1981–1990) **Bernd Nicolai** (1991–2016) Rüdiger Becker (from 2016)		

evidence of the start of the collection. This would place the beginning of the collection in the year 1833 (Table 33.1 and Figs. 33.1, 33.2, 33.3, 33.4, and 33.5).

Already by 1843 the collection had grown to a substantial size, and in that year, it was *founded in its scientific form* (Heine and Reichenow 1890). The first definite numbers concerning the holdings can be read in the report on the 14th Meeting of the German Ornithologists' Society in Halberstadt and Braunschweig from 29 September to 02 October 1862. At that time '*the entire collection* [comprised] *about 10,500 specimens*'. The rapid expansion of the bird collection in the 1840s

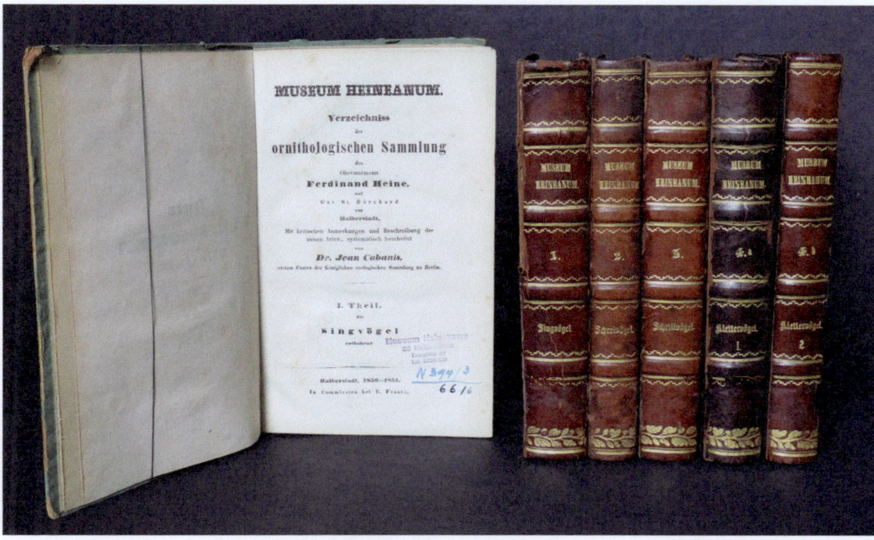

Fig. 33.1 Volumes of the collection catalogue 'Museum Heineanum' and the front page of the first volume of Cabanis (1851)

and into the 1860s was paralleled by its systematic ordering and scientific arrangement, which was by this time in the hands of Ferdinand Heine junior (1840–1920), under the direction of Cabanis. On the occasion of the annual meeting of the German Ornithologists' Society in 1862, Heine junior gave a guided tour of the collection.

In the following decades, however, the collection grew only modestly. At least, in the 'nomenclator' of Heine and Reichenow (1890), there were 11,968 specimens listed in printed form.

On the basis of a contract with the town of Halberstadt, the collection moved from Heine's St. Burchard Estate to the Cathedral Square (Domplatz) in a side building of Spiegel's curia, immediately adjacent to the cathedral. The opening as a permanent exhibition on 23 September 1909 attracted considerable public attention. A 'Guide to the Museum Heineanum' compiled by Hemprich (1910) contained a 'table with details of the number of species, prepared specimens, skins, and total number of items as of 23rd Sept. 1909'. According to this publication, the entire Heine collection consisted of 12,367 specimens (7,313 mounted birds and 5,054 skins). Although in subsequent years relatively little new material was acquired, some of it was very important, e.g. 200 birds from Sumatra from the former inhabitant of Halberstadt Runge (donation, 1914), 1,250 hummingbirds from Willy Schlüter (purchase, 1928) and over 400 hummingbirds from Rudolf H. Jung (purchase, 1936).

But the most significant episode in the more than 180-year history of the collection was without doubt the destruction of Halberstadt on 8 April 1945 and the following hard post-war years. While the skin collection remained almost unharmed, there was substantial damage in the exhibition building, badly affecting

Fig. 33.2 Cover of the type catalogue of the Museum Heineanum with the type specimens of bird collection (Quaisser and Nicolai 2006)

Fig. 33.3 View into a collection cupboard of the Museum Heineanum with bird specimens

the cabinets and cases containing the mounted specimens. Heine's egg collection was more or less completely destroyed (Busch 1957). Following this, the specimens sustained further damage because of inadequate housing. Some badly affected specimens were destroyed by the taxidermist B. Gebser without consideration of their scientific value or documenting exactly what was lost. Some of the type specimens that are catalogued today as having disappeared were probably among them.

An exhibition was soon opened again, thanks to expert advice and support from Berlin and by Wilhelm Meise and Hans von Boetticher (Coburg) (Busch 1957). In 1956 Kuno Handtke was appointed as full-time curator, and he dedicated himself to the creation of a general natural history museum. Accordingly, alongside caring for Heine's bird collection, he made great efforts in expansion and in the acquisition of new material. The main sources were random finds of dead animals, a very limited amount of officially sanctioned active collecting, and the incorporation of items from private collectors or other museums.

From the beginning of the 1970s, some natural history museums were encouraged to raise their profile, among them the Heineanum in the field of ornithology. Non-ornithological material was sent to other museums, and in return the Heineanum received ornithological specimens from, e.g. Haldensleben, Magdeburg, Oschersleben and Stendal. Special mention here must be made of the parts of the collection of Gottlieb von Nathusius (1939). The largest complete collections to come were in October 1994, with the acquisition of the Kummer (1993) oological

Fig. 33.4 View into the original cupboard of the egg collection 'Kummer'

collection (already purchased in 1979) with almost 3000 clutches, as well as the collections of Ernst Hesse, Gustav Schulz and Viktor Stimming.

33.2 Collection Characteristics

The Heine collection is characterised above all by its breadth and variety. For instance, Heine and Reichenow (1890) listed 5187 species and Hemprich (1910) even 5431. Among them are ten specimens of seven birds that are now extinct: Labrador Duck, *Camptorhynchus labradorius*; Passenger Pigeon, *Ectopistes*

Fig. 33.5 Front of exhibition building 'Museum Heineanum' (fully developed for the presentation of Heine's collection from 1909 up to now)

migratorius; Carolina Parakeet, *Conuropsis carolinensis*; Norfolk Kaka, *Nestor productus*; Ivory-billed Woodpecker, *Campephilus principalis*; Imperial Woodpecker, *Campephilus imperialis*; and Huia, *Heterolocha acutirostris*. Just how valuable this material is can be judged by the fact that on one of the huias, an also extinct feather louse *Huiacola extinctus* (Mey 1990) was discovered for the first time.

Some families and groups of species are represented more extensively surely first and foremost because of the interests of both the Heine's and include important specimens, e.g. hummingbirds and cuckoos. Recently above all, materials of Black Redstart *Phoenicurus ochruros* (250 specimens), Red Kite *Milvus milvus* (60 specimens) and Wryneck *Jynx torquilla* (more than 1200 eggs) were collected.

The sources of Heine's specimens were mostly natural history dealers and professional field collectors. Connections to other museums also brought new material to Halberstadt, principally the close contact to the Zoological Museum in Berlin. In this way bird skins for the Heine collection were obtained from, e.g. W. F. Hemprich, C.G. Ehrenberg and Th. von Heuglin from Northeast Africa; L. Krebs from Southern Africa; G.A. Fischer from East Africa and Zanzibar; C.C. Platen and A.R. Wallace from Celebes, Borneo and the surrounding South Sea Islands; and R. Schomburgk from British Guiana. Alongside dealers and museums, the purchase and exchange of items, including some type specimens, from other well-known private collections were Ferdinand Heine's third important source of bird material. A list compiled by Busch (1966) contains more than 80 well-known sources. Two of these, among many others, were the collections of Christian Ludwig Brehm and Ferdinand von Homeyer.

Especially noteworthy are finally the numerous type specimens of the Heine bird collection published by Quaisser and Nicolai (2006) (Fig. 33.2): in total 328 specimens (94 holotypes, 228 syntypes, 6 lectotypes) of 210 taxa. Most were described by Jean Cabanis and/or Ferdinand Heine junior in their collection catalogue 'Museum Heineanum' (published 1850–1863). Furthermore, types described by C.L. Brehm and A.E. Brehm (7), E.F. von Homeyer (3), A. Reichenow (3), W.F. Hemprich and C.G. Ehrenberg (2), L. Reichenbach (2), W. Schlüter (2), P.L. Sclater (2), H. Lichtenstein (1) and H. Schlegel (1) are listed. Twenty type specimens of 17 taxa were either destroyed during World War II (or shortly afterwards) or got lost in another way. Since finishing the works on the type catalogue (2005), further type specimens could be identified (five syntypes out of three taxa; Nicolai in prep.).

Finally further large collections can be found in the Heineanum besides the bird preparations (see Table 33.2, Figs. 33.3 and 33.4).

33.3 Research

With the development of an ordinary natural history museum and the profiling towards a special museum of ornithology, the research of the native avifauna was promoted. Thus the editorship of 'Mitteilungen der Interessengemeinschaft Avifaunistik DDR der Biologischen Gesellschaft der DDR (1968–1975)' was up to Helmut König in the Heineanum. This work got its climax in the organisation of the breeding bird mapping of the GDR (1978–1982/83) with the headquarter in Halberstadt (König et al. 1978) and the later editing of 'Atlas der Brutvögel Ostdeutschlands' (Nicolai 1993). The avifauna investigations in the Harz region were and are furthermore supported (e.g. Hellmann et al. 1998; Hellmann and Wadewitz 2000; Nicolai 2002b; Nicolai and Wadewitz 2003; Wadewitz 2012).

At the same time, studies of morphology, biology and ecology got increasing attention.

Exemplarily the following species and articles are mentioned: Eurasian Bullfinch, *Pyrrhula pyrrhula* (Handtke 1975, 1977); Middle-spotted Woodpecker, *Picus medius* (Günther 1992); Common Swift, *Apus apus* (Günther et al. 2005); Spotted Flycatcher, *Muscicapa striata* (Holz 1993); Little Owl, *Athene noctua* (Nicolai 1994); Black Redstart, *Phoenicurus ochruros* (Nicolai et al. 1996; Nicolai 2002a); Red Kite, *Milvus milvus* (Nicolai et al. 2009a, b; Mammen et al. 2014); and Wryneck, *Jynx torquilla* (Becker et al. 2014). The food and feeding of birds are of special interest among others (e.g. Nicolai 1992, 2006, 2010).

Finally two scientific periodicals are well established currently: since 1976 the 'Ornithologische Jahresberichte des Museum Heineanum' (hitherto 31 volumes) and since 1990 the 'Abhandlungen und Berichte aus dem Museum Heineanum' (hitherto ten volumes and ten special ones).

Table 33.2 Overview and short characterisation of zoological collections in the Museum Heineanum [Stand: 2015]

Collection sort	Number of pieces	Comments
Birds (A1): skins, mounted specimens	18,850	Heine's collection (11,589 specimens), preserved local species (>3500 specimens) and purchases, gifts, taking deliveries after 1909
Birds (A11): feather collections	2,000	Collected feathers (e.g. rests of prey of predators), plucked feathers (e.g. at skeleton preparation)
Birds (A12): skeletons/ part-skeletons	2,900	Preserved local species, mainly (partial) skeletons after preparation of skins (from the middle of 1970s)
Birds (A14, K14): clutches	7,200 [>35,000 eggs]	[ca. 530 species] especially collections of Richard Schmidt (1865–1936), Max Hübner (1864–1939), Johannes Kummer (1914–2003)
Birds (A1): nests	120	Especially native examples (up to now not specifically collected)
Mammals (M1): skins	1,750	Faunistic examples, especially bats and other small mammals; collection is continued?
Mammals (M12): skulls/ (partial) skeletons	1,850	
Skeletons of dinosaur	2	*Plateosaurus quenstedti*, *Eurycleidus arcuatus*
Pictures of birds	700	Small-sized, coloured original drawings of H.G.L. Reichenbach (1793–1879)

33.4 Exhibitions

In the permanent topical exhibitions of the Heineanum (Fig. 33.5), extensive information about the class of birds is given: 'Birds of the World' (286 different species of birds from hummingbird to ostrich, systematically classified, demonstrate shape and colour variety), 'World of Birds" (17 thematically arranged charts about biology of birds, from the feather to cultural history), 'Birds of Harz Mountains and the Foreland' (160 local species in their characteristic living space, their voices callable), 'Fascinating world of birds—AGENDA Systematics 2000' (evolution of species richness and endangering of bird life), 'Red Kite—Roter Drachen—Rotmilan' (extensive presentation of biology, ecology, cultural history and endangering of the species) and 'Halberstädter Dinosaur' (*Plateosaurus* and *Plesiosaurus* found in the urban area of Halberstadt).

These exhibitions help in intensifying museum educational service of all ages. Environmental education and protection of species and nature are the aim. Numerous materials (worksheets, cut-out sheets, colouring booklets, games, etc.) developed in the museum are offered and can be used.

References

Becker D, Tolkmitt D, Nicolai B (2014) Zur Brutbiologie des Wendehalses *Jynx torquilla* im nordöstlichen Harzvorland: IV Brutgröße und Fortpflanzungsziffer. Ornithol Jber Mus Heineanum 32:43–57

Busch R (1957) Zur Chronik des "Museum Heineanum" in Halberstadt. Falke 4:45–52

Busch R (1966) Über die Bezugsquellen der Vogelbälge im Museum Heineanum zu Halberstadt. Unveröff. Mskr

Cabanis J (1851) Museum Heineanum. Verzeichnis der ornithologischen Sammlung des Oberamtmann Ferdinand Heine auf Gut St. Burchard vor Halberstadt. I. Theil, die Singvögel enthaltend. (R. Frantz) Halberstadt [1850-1851] VIII, 230 pp

Günther E (1992) Untersuchungen zum Brutbestand, zur Bestandsentwicklung und zum Habitat des Mittelspechtes (Dendrocopos medius) im nordöstlichen Harz (Sachsen-Anhalt). Ornithol Jber Mus Heineanum 10:31–53

Günther E, Hellmann M, Nicolai B (2005) Baumbrütende Mauersegler *Apus apus*—Relikte uralter Waldqualitäten. Vogelwelt 125:309–318

Handtke K (1975) Zur Rassenzugehörigkeit der Gimpel (*Pyrrhula pyrrhula*) aus dem Harz und seinem nördlichen Vorland. Naturkdl Jber Mus Heineanum 10:33–41

Handtke K (1977) Biometrische Untersuchungen an Skeletten des Gimpels, Pyrrhula pyrrhula. Ornithol Jber Mus Heineanum 2:23–46

Heine F, Reichenow A (1890) Nomenclator Musei Heineani Ornithologici. Verzeichniss der Vogel-Sammlung des Königlichen Oberamtmanns Ferdinand Heine. (R. Friedländer) Berlin [1882-1890], VI, 373 pp

Hellmann M, Wadewitz M (2000) Die Vögel der Brockenkuppe. Ornithol Jber Mus Heineanum 18:1–49

Hellmann M, Günther E, Nicolai B (1998) Die Vögel des Brockenurwaldes: Vorkommen, Siedlungsdichte, Avizönose. Ornithol Jber Mus Heineanum 16:103–136

Hemprich A (1910) Führer durch das Museum Heineanum (Heine'sche Vogelsammlung) in Halberstadt—Domplatz 36. (Museum Heineanum) Halberstadt, 80 pp

Holz R (1993) Zur Brutperiode und zum Neststand des Grauschnäppers, Muscicapa striata, in Sachsen-Anhalt. Ornithol Jber Mus Heineanum 11:75–90

König H, Dornbusch M, Saemann D (1978) Meßtischblatt-Kartierung der Brutvögel der DDR. Falke 25:102–103

Kummer J (1993) Geschichte der oologischen Sammlung Kummer. Ornithol Jber Mus Heineanum 11:47–52

Mammen U, Nicolai B, Böhner J, Mammen K, Wehrmann J, Fischer S, Dornbusch G (2014) Artenhilfsprogramm Rotmilan des Landes Sachsen-Anhalt. Ber. d. Landesamtes f. Umweltschutz Sachsen-Anhalt H.5

Mey E (1990) Eine neue ausgestorbene Vogel-Ischnocere von Neuseeland, *Huiacola extinctus* (Insecta, Phthiraptera). Zool Anz 224:49–73

Nicolai B (1992) Untersuchungen zur Nahrung und zum Nahrungserwerb des Hausrotschwanzes (*Phoenicurus ochruros*). Ornithol Jber Mus Heineanum 10:75–105

Nicolai B (1994) Steinkauz *Athene noctua*—Artenhilfsprogramm des Landes Sachsen-Anhalt. Ministerium f. Umwelt u. Naturschutz Sachsen-Anhalt, Magdeburg (32. S)

Nicolai B (2002a) Ökologie und Brutbiologie des Hausrotschwanzes *Phoenicurus ochruros gibraltariensis* (S.G. Gmelin 1774) in Halberstadt. Ornithol Jber Mus Heineanum 20:3–55

Nicolai B (2002b) Die naturkundlichen Sammlungen des Museums Heineanum. In: Görgner E, Heidecke D, Klaus D, Nicolai B, Schneider K (eds) Kulturerbe Natur: Naturkundliche Museen und Sammlungen in Sachsen-Anhalt. Mitteldeutscher Verl, Halle (S.), pp 44–55

Nicolai B (2006) Nahrungsangebot und selektive Nahrungswahl des Steinkauzes *Athene noctua*. Populationsökologie Greifvogel- u. Eulenarten 5:557–566

Nicolai B (2010) Wanzen (Heteroptera) als Hauptbeute beim Hausrotschwanz *Phoenicurus ochruros* im Spätsommer 2010. Ornithol Jber Mus Heineanum 28:75–82

Nicolai B (Hrsg.) (1993) Atlas der Brutvögel Ostdeutschlands. Fischer-Verlag, Jena, 314 S.
Nicolai B, Wadewitz M (2003) Die Brutvögel von Halberstadt—Ergebnisse einer Brutvogelkartierung 1998 bis 2002. Abh Ber Mus Heineanum 6, Sonderh, 188 S.
Nicolai B, Schmidt C, Schmidt F-U (1996) Gefiedermerkmale, Maße und Alterskennzeichen des Hausrotschwanzes *Phoenicurus ochruros*. Limicola 10:1–41
Nicolai B, Hellmann M, Günther E (2009a) Artenschutz beim Rotmilan—Zur aktuellen Situation in seinem Welt-Verbreitungszentrum Deutschland/Sachsen-Anhalt (Grundlagen, Probleme, Aussichten). Nat u Landschaftsplanung 41(3):69–77
Nicolai B, Neuhaus R, Holz R (2009b) Museum Heineanum: Geschichte und Bedeutung. Abh Ber Mus Heineanum 8, Sonderh. Halberstadt, 183 pp
Quaisser C, Nicolai B (2006) Typusexemplare der Vogelsammlung im Museum Heineanum Halberstadt. Abh Ber Mus Heineanum 7(2 Sonderheft):105 Seiten
von Nathusius G (1939) Meine Vogelsammlung mit Beobachtungen aus den Jahren 1904-1939. Beitr Avifauna Mitteldt 3(Sonderh):1–26
Wadewitz M (2012) Brutvögel auf einem Trockenrasen am Huy 1995–2012: Auf und Ab mit Büschen. Ornithol Jber Mus Heineanum 30:63–80

Chapter 34
HALLE-WITTENBERG: The Zoological Collection of the Martin Luther University in Halle-Wittenberg

Karla Schneider and Frank Steinheimer

Abstract The Zoological Collection, founded in 1769 as a natural history collection with 2500 zoological and mineralogical objects, is now an integral part of the *Zentralmagazin Naturwissenschaftlicher Sammlungen* (ZNS) of the Martin Luther University Halle-Wittenberg. The ZNS conducts research and teaching with these collections and uses them for the transfer of knowledge outside of a university context. The scientific value of the Zoological Collection is based on its large number of type specimens and exhibits of very rare specimens as well as the long series with animals from South America, Cuba, Mongolia and Central Europe. In 2012, two parts from the collection were included in Germany's national list of valuable cultural assets. Today it is a centre for biodiversity research in Saxony-Anhalt. It is used within the framework of national and international cooperation to clarify systematic, phylogenetic, ecological and history of science-related questions as well as in the creation andup editing of red data books, checklists and species and habitat protection programmes, along with art and film projects, dissemination of knowledge, university public and press relations and building a sense of collective identity. The caption summarises the most important parts of the collection, provides a historical overview of collection's development and describes its current, wide-ranging use.

Keywords Halle University collection • Goldhagen • Nitzsch • Burmeister • Schönwetter • Mongolia • Type specimens

34.1 Introduction

The Elector Frederick the Wise founded one of the oldest German universities, the Wittenberg "Leucorea", as early as 1502. Famous scholars such as Martin Luther (1483–1546) and Philipp Melanchthon (1497–1560) helped make Wittenberg into

K. Schneider (✉) • F. Steinheimer
Zentralmagazin Naturwissenschaftlicher Sammlungen, Martin-Luther-Universität Halle-Wittenberg, Domplatz 4, 06108 Halle (Saale), Germany
e-mail: karla.schneider@zns.uni-halle.de; http://www.naturkundemuseum.uni-halle.de

L.A. Beck (ed.), *Zoological Collections of Germany*, Natural History Collections,
DOI 10.1007/978-3-319-44321-8_34

the intellectual centre of the Reformation. Nearly 200 years later, in 1694, Friedrich III inaugurated the University of Halle, "Fridericiana". The jurist Christian Thomasius (1655–1728) and the philosopher Christian Wolff (1679–1754) are considered as spiritual founders of the Halle University. The university was the centre of the early German Enlightenment and, thanks to the activity of August Herrmann Franke (1663–1727), the centre of pietism. As part of the territorial reorganisation after the Napoleonic wars, in 1817 the Prussian politicians united both universities to become the Frederick University, which was renamed the Martin Luther University in 1933.

34.2 History and Scope of the Collection

The Zoological Collection can also look back on a long and successful tradition of biological research until well into the eighteenth century. Initially founded as a university museum and a part-time public museum, it was expanded mainly as a teaching and research collection in the last century. Today it is an integral part of the *Zentralmagazin Naturwissenschaftlicher Sammlungen* (ZNS) of the Martin Luther University and one of the major research collections in Germany. It is the oldest and most extensive of its kind in Saxony-Anhalt. It is firmly integrated into the national and international biodiversity and phylogeny research, national archaeozoology, regional nature conservation work, university teaching and extra-curricular education as well as public and press relations and the university's preservation of traditions. It is considered as an archive for the global knowledge of the beauty, variety, biodiversity and distribution of wildlife on our planet. The Zoological Collection is also an essential reference collection for zoological classification (taxonomy) and the associated naming (nomenclature). Its approximately 2.5 million objects cover all animal groups, from single-celled organisms, sponges and corals to flat and roundworms through to molluscs, crustaceans, spiders, insects and vertebrates. The extensive and well-documented material comes mostly from South America, the Caribbean, Central Asia and Central Europe. The high percentage of greatly endangered or already extinct species, such as eggs of the Syrian Ostrich *Struthio camelus syriacus* W. Rothschild, 1919; the only preserved specimen worldwide outside Central Asia of the beaver subspecies *Castor fiber birulai* Serebrennikov, 1929, which lives only in the Mongolian-Chinese border area; an ex-mounted specimen of a male Pink-headed Duck *Rhodonessa caryophyllacea* (Latham, 1790); one of the two worldwide-known specimens of the Rufous-legged Owl from the South American Chiloe Island *Strix rufipes sanborni* Wheeler, 1938 and one of 60 preserved Tasmanian Tigers *Thylacinus cynocephalus* (Harris, 1808), and 2000 type specimens, which are classified at each new description of an animal species, are also preserved (Figs. 34.1, 34.2, 34.3, 34.4, 34.5, 34.6, 34.7, 34.8, 34.9, 34.10, 34.11, 34.12, 34.13, and 34.14).

Historians consider the appointment of Johann Friedrich Gottlieb Goldhagen (1742–1788) as the first professor of natural history in 1769 as the birth of the

Fig. 34.1 Exterior view of the late classicist building at the Domplatz in Halle (Saale), in which the Zoological Collection has been housed since 1886 (Photo: © with Frank Steinheimer 2012)

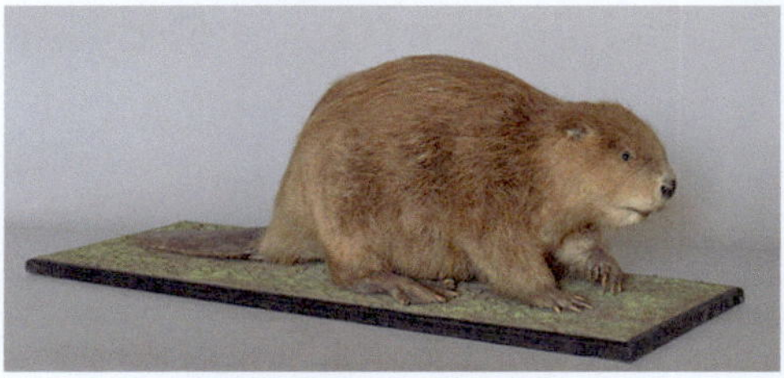

Fig. 34.2 Taxidermy specimen of the beaver subspecies *Castor fiber birulai* Serebrennikov, 1929, from Central Asia. This preserved specimen is probably the only one worldwide to be found in a collection outside the region (Photo: © with Markus Scholz 2015)

Zoological Collection. Goldhagen acquired Gründler's[1] natural history collection in 1775 for the purpose of demonstration in his lectures and developed it, together with his own collection, into a well-ordered natural history collection boasting nearly 2500 exhibits. The university purchased this natural history collection in

[1]Gottfried August Gründler (1710–1775) is known for having redesigned the art and natural history room of the Francke Foundations and as a copper engraver of zoological works, including those of Jacob Theodor Klein (*Historia piscium naturalis*).

Fig. 34.3 Several exhibits of the collection of corals from the late eighteenth century, which most likely found their way into the collection via Johann Reinhold Forster. "Invent" stands for the first (old) inventory of 1788 (Photo: © with Markus Scholz 2015)

Fig. 34.4 Bannerfish (*Heniochus* sp.) from the collection of Marcus Élieser Bloch (1723–1799). Bloch was the first scientist to systematically preserve tropical fish. His main collection is located at Berlin's Museum of Natural History and is regarded as the world's oldest still existing fish collection. The specimen originates from the Indian Ocean and came to Halle (Saale) before 1800 (Photo: © with Frank Steinheimer 2012)

1787. It formed the basis for the university's later zoological as well as its mineralogical collection, which was managed separately from 1805 onwards. Johann Reinhold Forster (1729–1798) took over responsibility for the collection

Fig. 34.5 Microscopic slide box with material from the insect order Phthiraptera (chewing lice) of O. Taschenberg (1854–1922); the shown family of Goniodidae is especially rich in types; the collection was reworked by S. Kéler in 1935 (Photo: © with Markus Scholz 2015)

in 1779, shortly after his famous circumnavigation of the globe with James Cook. Several corals and molluscs from the Pacific originate from Forster's collection from this period and are thus most likely from Cook's second circumnavigation. At the same time, historical dry fish specimens found their way to Halle (Saale) by way of Berlin. The Zoological Collection had a very eventful history up to 1815. After the city of Halle was occupied by the Napoleonic troops in 1806, the university was closed for 2 years, and the collection had to be transferred precipitously. Its condition noticeably degenerated over the following years (Heidecke 1994).

The rebirth and development into today's Zoological Collection began with the appointment of Christian Ludwig Nitzsch (1782–1837) as full professor of natural history and director of the collection. By preparing approximately 300 bird skeletons, mounting more than 200 birds and mammals as well as purchasing and intensively exchanging specimens especially with the Berlin Museum, Nitzsch managed to rebuild an extensive collection of more than 1400 vertebrate species as well as a large insect collection until 1823 (Heidecke 1994). Thanks to Prussia's far-sighted policy, an almost complete set of duplicate copies of the latest discoveries of Brazilian birds and mammals was stored in Halle (Saale). Today they sometimes constitute the only evidence of the relevant species descriptions after Berlin's Museum of Natural History was damaged during the Second World War. In addition to ornithology, Nitzsch was interested in external and internal parasites, especially species of Mallophaga, also referred to as hair and feather lice. He wrote down his studies from 1800 to 1837 in five handwritten volumes entitled: "Epizoografische Adversarien". These volumes comprise 1720 pages including sketches, notes about observations, preliminary comments on a natural history of wingless parasitic insects and 101 separate coloured drawings. The collection of preserved insects itself consists of 64 glass tubes with individuals from 50 different species. Among these, there are 30 reliable type series. The insects were embedded in balsam and are located on specimen slides in oval cut-outs of cardboard. Christian Ludwig Nitzsch's Mallophaga collection is scientifically and culturally

Fig. 34.6 Insect box of Carl Hermann Conrad Burmeister's collection from South America, here showing the scarab beetle family (Scarabaeidae). All animals with red labels are so-called types, the first of their kind to have ever been described (Photo: © with Markus Scholz 2015)

unique. His publications and the preserved specimens are of fundamental importance worldwide for the classification of this group of insects. He also described many other species such as the rabbit louse (*Haematopinus eurysternus* Nitzsch 1818) and the bee louse (*Braula coeca* Nitzsch 1818) and introduced a number of new genera. Nitzsch's mallophaga collection was revised by Stefan von Kéler in 1935 and published in 1941 under the title "Systematic list of the collection of Mallophaga of the Zoological Institute of the University of Halle established by Chr. L. Nitzsch and elaborated and enriched by Ch. G. A. Giebel and O. Taschenberg". It is the oldest part of the collections in the field of entomology and forms the basis of global research on hair and feather lice and the basis of their taxonomy. This collection is still a fundamental scientific resource even today and is indispensable for systematic studies on this group of insects. It was included in Germany's national list of valuable cultural assets in 2012. Nitzsch's research initiated a change in the Zoological Collection. He increasingly reorganised the

Fig. 34.7 View into one of the two large halls of the educational and display magazine in the ambience of the late nineteenth century (Photo: © with Markus Scholz 2012)

collection of curiosities into a systematically structured collection that was perfectly suitable for use in research and university teaching.

After his death, Carl Hermann Conrad Burmeister (1807–1892) took over the official duties. He continued with the mounting work, ordered the objects according to taxonomic considerations and provided the collection specimens with coloured labels according to their geographical origin. He gave the university his private insect collection, which included 10,000 specimens of some 5000 species. The collection was significantly expanded through purchases in Paris and London as well as other donations, including the collections of Junghuhn (Java), Zimmermann (North America), Ecklon and Drége (Cape region, South Africa) and Beschke (Brazil). In Burmeister's "List of mammals, birds and amphibians mounted in the Zoological Museum of the University of Halle-Wittenberg", published in 1850, he refers with pride to an inventory of 1803 mounted vertebrate specimens. This inventory was then greatly expanded during his two journeys in South America. There he collected 922 mammals, 5400 birds, 200 eggs, 1146 amphibians and reptiles, about 100,000 insects and 3000 other invertebrates. Of particular importance are the types described by Burmeister, which he identified with an asterisk on the label. These include 9 mammals, 25 birds, 5 amphibians, 6 reptiles and more than 1000 insect types. The family of scarab beetles (Scarabaeidae) is one of the focal points when it comes to insect types. Burmeister moved to Buenos Aires in 1861 and continued the expansion of the Museo Publico in key areas as its director.

Fig. 34.8 Detail from Dietrich von Schlechtendal's gall collection. This exceptionally rare collection shows plant galls and their originator (Photo: © with Markus Scholz 2015)

A special merit of his successor Christoph Gottfried Giebel (1820–1881) is the integration and scientific review of Burmeister's huge South American collection and the purchase of missing taxa in the collection from Banka and Borneo (Deissner), from Illinois (Brendel), Santiago de Chile (Philippi) and Egypt (Reil). Professor of mineralogy and director of the mineralogical collection of Halle University, Ernst Friedrich Germar's (1786–1853) Curculionidae (weevils) collection was acquired along with a few specimens from Say's famous expedition to the Rocky Mountains in the years of 1819–1820. Privy Council Dr. Eduard Suffrian's (1805–1876) extensive collection of beetles, comprising 35,565 specimens, was added as a donation in 1876, as well as Hamburg portrait painter Hermann Steinfurth's (1823–1880) conch collection (shells/snails) in 1880 and Judicial Council Georg Adolf Keferstein's (1793–1884) lepidopteran collection (butterflies, 440 boxes) in 1885.

The Zoological Collection was moved to its final location in a late classicist building (a former medical and surgical clinic) on Domplatz 4 in the centre of Halle's old town district in 1886. It now consists of a display and teaching magazine and scientific collections. Beginning in 1890, the public collection displayed the zoological system on the basis of models as well as wet and dry preserved specimens. Today visitors are presented with a wide variety of about 9000 stuffed animals distributed over 720 m^2 of exhibition space in 189 historic glass cabinets in

Fig. 34.9 Today most mammalian and avian specimens are stored, to save space, as so-called skins in sealable tongue and groove boxes. The photo shows representatives of the Mongolian gerbil species *Meriones meridianus* (Pallas, 1773) (Photo: © with Markus Scholz 2015)

two halls and the corridors of the building, in the original ambience of the late nineteenth century. The exhibition and teaching magazine is not usually accessible to the public, but is opened for tours and as part of special events, and attracts several thousand visitors every year.

The collection inventory was expanded through a variety of activities over the first half of the last century. The native fauna was increasingly included in the collection, and a special provincial collection was built up in order to familiarise the citizens of Halle with the local wildlife. A donation of 163 species from Otto Goldfuss' (1831–1905) well-known mollusc collection was added in 1903. Approximately 500 additional species were also acquired from his estate. The purchase of an extensive collection of beetles from Pastor Theodor Müller (1841–1903) took 2 years to complete. As the university was lacking funds, four Halle citizens finally bought the collection for 3000 Reichsmark and transferred it to the university in 1908. Another very significant addition took place in 1911. The widow of Viktor von Röder (1841–1910) fulfilled her husband's last wish, giving the Zoological Collection a valuable library along with its extensive and multi-type collection of dipterous insects. Von Röder collected primarily dipterous insects in his neighbourhood (Hoym in Saxony-Anhalt) but maintained an intensive international

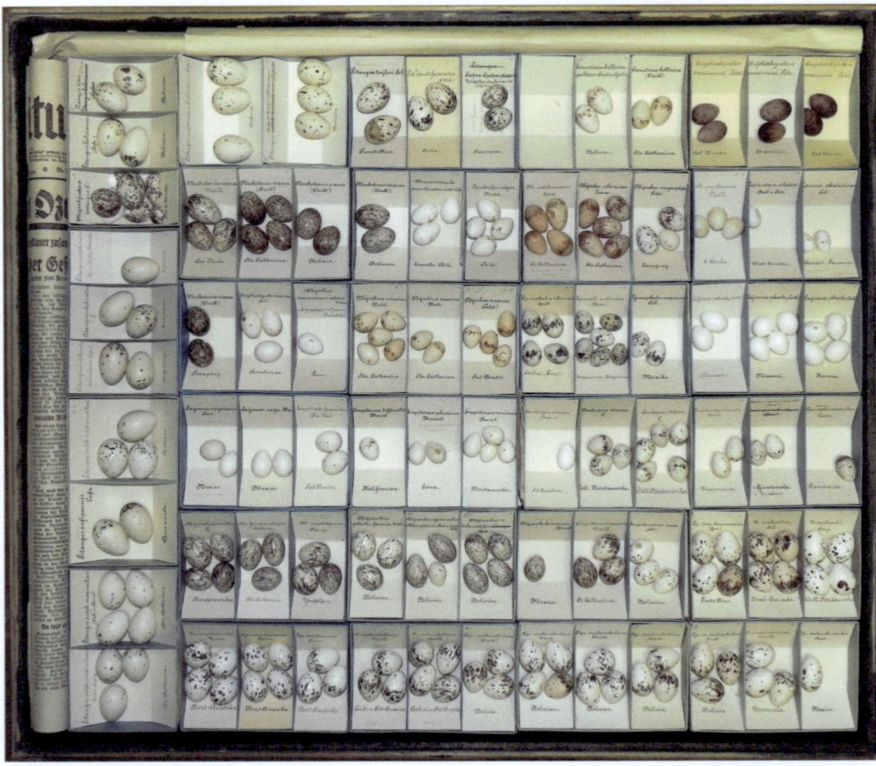

Fig. 34.10 Max Schönwetter's eggshell collection is one of the three most important collections of its kind worldwide. The photo shows a box of blown eggs of various species from the American bird family Tyrannidae (Photo: © with Hans-Jürgen Altner 2011)

exchange. He described more than 81 new taxa (Gattermann and Neumann 2005). It was during this period that Dietrich von Schlechtendal's (1834–1916) gall collection was donated to Halle as well as a large number of invertebrates from the Sunda Expedition by Bernhard Rensch (1900–1990) and Gerhard Heberer (1901–1973). Distinguished taxidermists like Adolf Haug (1887–1961) and Hugo Bleil (1880–1961) contributed to a didactic reworking of the public collection with the respective custodians. Haug, for example, prepared lifelike mammalian taxidermy specimens, including lions, sea bears and a group of African dwarf antelopes, dik-diks, while Bleil produced 15 large mammal models to a scale of 1/6 or 1/25 of their natural size and a very lifelike orang-utan. The lepidopteran collection of Wolfgang Bath (1882–1932) was acquired in 1932 and that of Privy Council Rudolf Heinrich (1859–1939) in 1938. Both collections included animals from the Central European fauna region and supplemented the existing butterfly collection, which included primarily exotic specimens. The collection managed to come through the Second World War and the following years without any damage but also without tenure of a professorship. Thoughtful Soviet army personnel brought a beetle

Fig. 34.11 This historic collection specimen is the only exemplar of the Fisher's Estuarine Moth *Gortyna borelii* (Pierret, 1837) from the city of Naumburg. This butterfly species is on the verge of extinction today (Photo: © with Joachim Händel 2015)

collection (15 cabinets, each with 40 boxes) of landowner Georg Dieck (1847–1925) to Halle in 1946, which they had found in the attic of his Zöschen estate near Merseburg. The collection's very bad condition resulted in large parts of it being sorted out over time.

Rudolf Piechocki (1919–2000) served as curator of the collection from 1959 to 1984. Many new contributions, mainly skin or fur and skeletal specimens from around the world, were the result of his work on many levels. Shorter collecting trips to the Barents Sea (1952), Bulgaria (1953), and Naples (1956) were followed by major expeditions to China (1956: 702 bird skins), Armenia (1959), Cuba (1967–1968: more than 1000 bird skins from 128 species, ever since then the world's third largest Cuban bird collection after Havana and Washington DC) and to Mongolia. In the tradition of scientists Georg Wilhelm Steller (1709–1746) and Peter Simon Pallas (1741–1811), who were associated with Halle, a cooperation agreement was concluded between the universities of Ulaanbaatar and Halle (Saale) in 1967. Rudolf Piechocki and Michael Stubbe (*1939) were its pioneers from the very start and instrumental in its success. Numerous expeditions (from 1962 up to the present), international conferences and more than 250 scientific publications as well as the publication of the journal *Research on the Biological Resources of Mongolia* have led to internationally recognised results, and with more than 4000 bird and mammal skins, it counts as one of the largest

Fig. 34.12 The wealth of material in the training courses at the *Zentralmagazin Naturwissenschaftlicher Sammlungen* (ZNS) is rarely equalled at other universities in Germany. The taxidermy course teaches various taxidermy and preservation techniques for natural history collections (Photo: © with Joachim Händel 2010)

Mongolian vertebrate collections worldwide. Research into endangered species, such as beavers, birds of prey and wild donkeys, as well as the protection of their populations, extensive biodiversity research with due regard to the vertical zonation of flora and fauna, the impact of changes in the natural environment and cultivated land on wildlife as well as complex biological explorations of major Mongolian nature reserves and the development of measures for nature conservation management are just a few aspects that have shaped this cooperation and continue to do so (Stubbe et al. 2005). Some of the applied results were used to preserve the Southeast Asian beaver, which boasted a higher population in Mongolia in the past. In 1962, the first Mongolian-German biological expedition led zoology staff from Halle to the last remaining beaver population in the flood plain of the Bulugun (Bulgan-gol) river, which was not only put under protection in the following years but also stabilised through targeted management.

On the initiative of Piechocki, the GDR's Central Conservation Management designated the Zoological Institute of the Martin Luther University as the collection centre for findings of the remains of endangered animals. Since 1954, 24 species and more than 2100 animals have been processed, their cause of death studied in more detail, and stored, including 190 White-tailed Eagles (*Haliaeetus albicilla*), 100 Eurasian Eagle-owls (*Bubo bubo*), 65 Great Bustards (*Otis tarda*), 62 Little Owls (*Athene noctua*), 770 beavers (*Castor fiber*), 707 Eurasian River Otters (*Lutra*

Fig. 34.13 The collection "Anderwelten" (Other Worlds) by the fashion designer Pia Fischer makes creative use of animal biodiversity. The individual garments tell a story of strength and fragility, individuality and blemishes, in the form of a fable. Fashion photographer Marco Warmuth illustrated the relationship between these costumes and the Zoological Collection (Photos: © with Marco Warmuth 2013)

lutra) and more than 100 wildcats (*Felis silvestris*). Since 2009, all of the wolves (*Canis lupus*) killed in Saxony-Anhalt have been stored in the collection as skeletal and fur specimens. The Zoological Collection benefited from the following important insect collections over the years: the Hymenoptera collection (hymenopterous insects) of school teacher and Leopoldina member Hermann Haupt (1873–1959) in 1959, the Coleoptera collection (beetles) of Hermann Köller (1885–1979) in 1970 and the Aphodius collection (dung beetles) of Igor Sergeevič Grebenšikov (1912–1986), which contained many different species. The scientifically invaluable eggshell collection of Max Schönwetter (1874–1961) was acquired in 1969. It boasts a stock of 19,206 bird eggshells from 3839 taxa and forms the basis of Schönwetter's "Handbuch der Oologie" (1960–1992), the only standard work of reference for the science of birds' eggs in the world. The collection was also included in Germany's national list of valuable cultural assets in 2012. A collection catalogue and an extensive worldwide correspondence about oology complete this collection, which Max Schönwetter developed into the Germany's largest and most species-rich private oological collection in 60 years of collecting, through

Fig. 34.14 View into the room dedicated to the special exhibition *Cicadas—An Electromechanical Sound Installation on Evolution* from the year 2012. In the foreground, real cicada collection objects; in the background, the colonies of artificial cicadas by internationally renowned Argentinean sound artist Edgardo Rudnitzky in a space stage design by Oliver Proske (Photo: © with Markus Scholz 2012)

exchanges, purchases and transfers of individual historical collections. Today it is one of the world's three largest egg collections in terms of the number of species covered (along with the collection at the Western Foundation for Vertebrate Zoology in California and the Natural History Museum of London/Tring).

With the development of modern research disciplines from the 1950s onwards, multiple technical modifications were made. Starting in 1952, Werner Somburg (1907–1978) took over the responsibility for looking after the insect collection. He played a key role in its preservation, enlargement and reorganisation. Reconstruction and moving the collection into the current premises took place in 1956–1957. This was probably also responsible for Max Richter's (1883–1984) large butterfly collection, which encompasses the whole of Europe, coming to Halle. It includes 10,000 butterflies belonging to 780 species from the Thuringian region. The work was then continued by Manfred Dorn (1931–2005), research assistant from 1973, then professor of entomology and responsible for the entomological collections. He focused mainly on applied problems, in particular the pollinating function of wild bees in alfalfa crops. He set up a regional Apoidea collection (bees, wasps and bumblebees), into which he integrated his own collection. More recently, further collection material has found its way to Halle in the form of donations and bequests. This included the 43-box butterfly collection of Konrad Drechsler (1939–1999), former chairman of the Entomological Society in Halle, in 2000, as well as important butterfly material from Central Asia/Mongolia in 2010 collected by Wolf-Dietrich Busching (1954–2010), director of the Naumann Museum in Köthen. The last few years have seen the start of technical reconstructions and renovations of rooms, changes that are required in order to subdivide and rework parts of the scientific collections. A room for scientific mammal skins was set up in 2014. The 70,000 objects included in the mollusc collection were moved from the roof area into a modern collection space. A new archive room for the c. 2000 historic teaching charts is under construction.

The extensive collection material allows the ZNS employees to not only promote research by third parties, from checklists to various animal groups and geographical regions, phylogeny research and evolutionary biology through to very specific environmental issues on a given species, but also contribute to current research by way of research projects and their contacts with 37 national and 53 international institutions (as at 2015). The museum's own international research focuses on beetle and butterfly science, ornithological nomenclature, archaeozoology, in the field of oology (science of birds' eggs) and osteology (the study of bones), studying various taxidermy methods and research on endangered vertebrates, while its regional activity is primarily focused on nature conservation issues. Since 1957, studies carried out on the causes of death of the endangered native birds and mammals have been printed in numerous publications on systematics, morphology, reproductive biology and population ecology. Direct human hunting of the Elbe Beaver combined with increasing habitat destruction led to the extinction of almost all beaver populations in Central Europe. A small residual population survived in the Elbe river basin of Central Germany. Scientists from the Zoological Collection under Piechocki and Dietrich Heidecke (1945–2011) conducted basic research, developed a monitoring programme and managed to have the beavers put under protection. The development of Elbe Beaver population has thus enjoyed complete documentation in the Saxony-Anhalt region since 1970 by ZNS. Saxony-Anhalt, Saxony and Brandenburg are once again home to more than 6650 beavers today (Heidecke 2011). A cooperative research project with the Federal Research Institute for Animal Health, supported by the state veterinary authorities, saw the provision of collection material and expertise in classifying species for the purpose of studying bat rabies. This expertise also determines the cooperation with the Helmholtz Centre for Environmental Research UFZ, e.g. in recording the biodiversity of insects and other animals that visit flowers. Since the 1990s, there has been close cooperation with the State Agency for Environmental Protection in Saxony-Anhalt. Among other things, preparatory studies were undertaken on animal populations in Saxony-Anhalt, on red data books as well as on species and habitat conservation programmes. As part of the acquisition of selected invertebrate species listed in Annex II of the Flora-Fauna-Habitat Directive, the population of the Fisher's Estuarine Moth *Gortyna borelii* (Pierret, 1837)—a rare butterfly—is currently being assessed. Recent additions are studies (that have been cited worldwide) on impact victims of wind turbines, especially on bats whose specimen materials are archived at the ZNS. The ZNS is now considered as the state of Saxony-Anhalt's most important reference collection for the biodiversity. When it comes to the field of entomology, the collection is currently growing most quickly, thanks to donations and assessment activities under nature conservation law. In terms of archaeozoology, the ZNS is involved in researching the respective ecological and cultural-historical contexts and classifies the remains of vertebrate bones and eggshells from various sites in Central Europe from 300,000 years ago (Schöningen site) up to the modern era with Martin Luther's "cesspit". The ZNS itself publishes an average of one to two dozen scientific papers a year, of which about half in peer-reviewed journals, and presents their research in a number of conference papers. The Zoological Collection heads the national expertise in the macroscopic taxidermy of invertebrates with their review of the corresponding standard work (Piechocki and Händel 2007).

34.3 Collection Use

The ZNS is one of the few institutions in Germany involved in object-related teaching and transfer of knowledge about species. Courses on the classification of native animals, such as special entomology and special ornithology (including zoological nomenclature), taxidermy as well as the protection and animal systematics, are offered at the university level. Collection excursions, which focus in detail on a group of animals in the collections and include 2–3 h per course, not only teach the phylogeny of animals based on the exhibits but can also resolve evolutionary questions based on individual objects and provide an understanding of ecological relationships as well the history of collection and science. The ZNS is involved in teacher training and other training courses in urban (Long Night of the Museums, Children's City), university (Children's University, Long Night of the Sciences) and nationwide events (such as International Museum Day, Heritage Day), drawing several thousand visitors a year to the Zoological Collection. Three projects funded by the BMBF (German Federal Ministry of Education and Research) are currently underway in the museum educational programme on the ZNS collections, among others, in the MuseobilBOX programme of the Bundesverband Museumspädagogik e.V. (Federal Museum Education Association) as part of the programme "Culture makes us strong: alliances for education" and in the Zoological Collection, the project "Which animal is right for me".

The ZNS has been cooperating with the Giebichenstein School of Art and freelance artists, local associations, collections and museums for years now. This often brings a very fruitful exchange of viewpoints and new ways of looking at things to the Zoological Collection from the field of art, cultural history, history of science, aesthetics and design theory. In cooperation with the Werkleitz e.V. association, a documentary film about the Zoological Collection was shot by media artist and filmmaker Pim Zwier of Amsterdam and entitled "Atemberaubend—Breathtaking"; as a valuable window of time, it reflects the activity of the curators and taxidermists surrounding the Collection, as well as currently being the only document that successfully captures the ambience, from the characteristic sound to the systematics. The film is very popular at international film festivals, thanks to its very unique blend of curiosity, art film and zoological documentation.

Every year the ZNS reports on its activities and achievements in the form of a bound annual report. It also organises its own special exhibition once a year. The theme for this came from the Zoological Collection in 2012 and was dedicated to evolution and occupation of an ecological niche as manifested by the singing of cicadas. In cooperation with the internationally renowned sound artist, Edgardo Rudnitzky of Buenos Aires, and in a technical implementation by the Colombian Juan Pablo Ayala-Cortés integrated into the interior design of the stage designer Oliver Proske from Berlin, the ZNS presented this exhibition, which was significantly funded by the German Federal Cultural Foundation, in Halle (Saale) and Potsdam. It was the first time in the world that a scientist and a sound artist cooperated to explain the process of evolution with a work of art (Steinheimer 2012).

In the medium term, two more large and globally significant, the domesticated animal and Geisel Valley/Eocene fossil collections and several smaller collections will be moving into the Zoological Collection building on the Domplatz in Halle (Saale) with its total usable space of 3300 m^2, including three additional nationally valuable cultural assets. As a central university institution with its own budget and statutes, reporting directly to the rector of the university, the ZNS will then leverage synergies in custodian and collection management to ensure professional development and supervision of all collections, offer background tours on different topics and use special exhibitions to continue communicating the knowledge of the latest research results to the society at large. In-house research continues its international focus on the science of domesticated animals, Eocene palaeontology, zoological nomenclature and the exploration of the natural resources of Mongolia, while entomology and malacology in particular contribute important contract research when it comes to statutory national obligations, i.e. Natura 2000, the Flora-Fauna-Habitat Directive, Bird Protection Directive, the Bonn Convention on the Conservation of Migratory Species and red data books. The archaeozoological work in the field of oology is currently being established due to the need for expertise here throughout Germany, in addition to what is already Germany's largest archaeozoological and osteological reference collection on mammals and birds at the ZNS. The Zoological Collection is currently curated by two professional taxidermists, a curator, the head of the ZNS as well as a dozen associated scientists, volunteers and project staff.

References

Heidecke D (1994) Die Zoologischen Sammlungen. In: Speler R-T (Hrsg) 300 Jahre Universität Halle: 1694—1994; Schätze aus den Sammlungen und Kabinetten. Verlag Janos Stekovics, Halle, pp 309–322

Heidecke D (2011) Geschätzter Bestand der Biber (*Castor fiber*) in der BRD (Zählungen im Zeitraum 2007–2010). Unveröffentl. MS (Zusammenstellung). 02.05.2011

Gattermann R, Neumann V (2005) Geschichte der Zoologie und der Zoologischen Sammlung an der Martin-Luther-Universität Halle-Wittenberg von 1769 bis 1990. Abhandlungen der Sächsischen Akademie der Wissenschaften zu Leipzig. Mathematisch-naturwissenschaftliche Klasse. 9/3:22–166, Verlag der Sächsischen Akademie der Wissenschaften zu Leipzig

Nitzsch ChL (1818) Die Familien und Gattungen der Thierinsekten (insecta epizoica); als Prodromus einer Naturgeschichte derselben, von Dr. Chr. L. Nitzsch, ordentl. Professor d. Naturgesch. zu Halle. Mag. Ent. 3, Hrsg: Germar EF, Zincken JLTF, J. Chr. Hendel und Sohn, Halle (Saale), p. 305, 315

Piechocki R, Händel J (2007) Makroskopische Präparationstechnik/Wirbellose, Leitfaden für das Sammeln, Präparieren und Konservieren. 5. überarbeitete und aktualisierte Auflage, E. Schweizerbart'sche Verlagsbuchhandlung, Stuttgart, pp 1–346

Schönwetter M (1960–1992) Handbuch der Oologie. Herausgegeben und ergänzt von W. Meise. 4 Bände. Akademie-Verlag, Berlin

Steinheimer F (Hrsg) (2012) Cicadas—Ein elektromechanisches Klangkunstwerk zur Evolution. Katalog zur Sonderausstellung. Zentralmagazin Naturwissenschaftliche Sammlungen der Martin-Luther-Universität Halle-Wittenberg, Halle (Saale), pp 1–76

Stubbe M, Stubbe A, Samjaa R, Wesche K (2005) Vier Jahrzehnte erfolgreiche Wissenschaftskooperation der Universitäten Halle und Ulan-Bator. Erforsch. biol. Ress. Mongolei (Halle/Saale) 9:11–41, Salzland Druck GmbH Co. KG, Staßfurt

Chapter 35
HAMBURG: Back to the Future: The *Centrum für Naturkunde* on Its Way Toward Reestablishing a Natural History Museum in Hamburg

Matthias Glaubrecht

Abstract Comprising today approximately ten million specimens, Hamburg's zoological collection at the newly established *Centrum für Naturkunde* (CeNak) goes back to the *Naturhistorisches Museum*, founded in 1843. It includes several historical collections that originate from the initiative and interest in natural history of civilians, merchandisers, traders, and owners of seagoing vessels. Only in 1891 that these collections were adequately stored and put on display in a then most innovative museum building in the city's center that soon became one of the largest and most important natural history museums, second only to that in Berlin; it was also the one of its kind with most visitors for five decades. The museum and parts of its dry collections—mostly in entomology, malacology and mammalogy, and those in the exhibition—were destroyed during the "Operation Gomorrha" bombing of Hamburg and the subsequent "Feuersturm" in the morning of 30 July 1943, with the ruins being knocked down in 1951. Other parts, essentially the large alcohol collections and those of birds, have been stored elsewhere during WWII. Since then only provisionally housed the museum and its staff became part of the Universität Hamburg in 1969 and moved into a new building in the early to mid-1970s at its current location. As part of a growing mass university, these collections were long neglected and without substantial means to accomplish this goals, with a small display room opening on 2000 qm in 1984 only. With the founding of the CeNak in 2014 plans are underway for establishing again a modern natural history museum in Hamburg.

Keywords Naturhistorisches Museum 1891 • Allied bombing 1943 • Zoological Museum 1969 • CeNak 2014 • Integrated biodiversity research • Evolutionary systematics

---

M. Glaubrecht (✉)
Centrum für Naturkunde—CeNak/Center of Natural History Hamburg, Universität Hamburg—Zoological Museum, Martin-Luther-King-Platz 3, 20146 Hamburg, Germany
e-mail: cenak@uni-hamburg.de; http://www.cenak.uni-hamburg.de

L.A. Beck (ed.), *Zoological Collections of Germany*, Natural History Collections,
DOI 10.1007/978-3-319-44321-8_35

35.1 Year of Foundation and Age of Parts of the Collection

The *Naturhistorisches Museum Hamburg*, or Museum of Natural History (here called NatHistMus), was created when on 17 May 1843 the founding contract was signed by the Hamburg senate and the *Naturwissenschaftlicher Verein*. Itself established in 1837 only, this Hamburg association of natural science agreed on the fusion of its own collection with that of the *Johanneum*, a distinguished academic gymnasium since 1613 (with its roots going back to 1529). Both already possessed considerable collections, going back to the seventeenth century and accumulated through donations and acquisitions, such as, e.g., the curiosity cabinet of the Peter Friedrich Röding in 1849 or the impressive beetle collection of Johannes Amsinck in 1833.

Evidently, individual objects in these collections go back as far as 1684; such as, for example, CeNak's emblematic skull of a female narwhale (*Monodon monoceros*). As the only object that survived the bombing of the Hamburg NatHistMus in July 1943 when the building was destroyed, it mirrors the rich history and fate of its collections. Aptly called "Lisa"—from da Vinci's famous and equally unique painting in the Louvre—this narwhale skull exhibits two long projecting tusks (instead of none, as usual in females, or only one as in males).

In contrast to many other collections started as "Wunderkammer", or curiosity cabinets of regional kings and other dignitaries, the Hamburg collections originated from the initiative and interest in natural history of some of its civilians, merchandisers, traders, and owners of seagoing vessels. Well known among them was, at his time, Peter Friedrich Röding (1767–1846). Being the owner of a private art and natural history museum, he was once declared by the famous German writer Heinrich Heine in the 1820's as one of the seven "Merkwürdigkeiten," or curiosities, of Hamburg that any visitor must have seen (Heine 1887). Through an entry in his travel notebook, we know of another famous writer and poet, Adelbert von Chamisso, who in 1815 visited the Museum Röding in Hamburg. Being on his way to Copenhagen to join a Russian expedition around the world, Chamisso saw in Röding's collection, among other curiosities, also the female narwhale skull with the two tusks then already famous (see Glaubrecht and Bein 2017).

The NatHistMus had been run by the so-called *Museumskommission*, a coalition of natural history association of men and state representatives who operated the museum as a committee of volunteers, each in charge of one taxonomic area within the collection, ruling under a *Kollegialverfassung* (which was substituted only with the appointment of the first director five decades later). Thus, as Nyhart (2009: 207) noted rightly, like so much else in Hamburg's cultural life, the museum was a public–private enterprise. However, over the years, it became increasingly obvious that this arrangement was not promoting the overall interests of the museum. Its collections were long time provisionally housed at the *Johanneum*, a school for higher education, located in the nineteenth century in the St. Johannis-Kloster, a former monastery at the Rathausmarkt near the inner part of the city center's Lake Alster which later became the site of Hamburg's impressive town hall. For most of that century, international trade was the main source of material, resulting in many objects for the NatHistMus that also profited enormously from the sale of the Museum Godeffroy. This private museum, actually one of the largest ever, was founded in 1861 by Johan Cesar VI Godeffroy, who asked his captains, contractors, and traders to collect in particular in the South Sea and Eastern Australia. After the Museum Godeffroy was forced to close after bankruptcy, all zoological objects were sold and given in 1886 to the NatHistMus (see Kranz 2005; Scheps 2005).

As the museum's first professional and full-time paid scientific worker and curator, Johann Georg Pfeffer (1854–1931) was employed in 1880 and the second curator came in 1883, Johann Wilhelm Michaelsen (1860–1837). Its first full-time director from 1882 to 1889 was Heinrich Alexander Pagenstecher (1825–1889), a formerly retired zoology professor and director of the Zoological Institute from Heidelberg. In particular, instrumental in establishing the museum as an institution, however, was Hamburg's seven-time mayor Gustav Heinrich Kirchenpauer (1808–1887). Although a lawyer and journalist by profession, one of Kirchenpauer's avocations was natural history, in particular the taxonomy of marine invertebrates which he first studied in the River Elbe estuary. He described and named eight genera and 77 species of hydroids (of which about half of the species are still recognized as valid), and one new genus and 26 new species-group taxa of bryozoans, later working on a global scale. Unfortunately, most of Kirchenpauer's collection, among them many types, was destroyed in Hamburg during WWII (Calder and Brinckmann-Voss 2011).

Opened on 17 September 1891, after nearly a decade of planning and constructing, the NatHistMus moved to its own building, designed in Italian renaissance style by the architects Manfred Semper and Carl Philipp Krutisch, with 100 m in length, 36 m width, and 32 m height and with open galleries surrounding the major hall (see Figs. 35.1 and 35.2). It was located near Hamburg's later main train station, where it formed part of a museum assemblage together with the Hamburg *Kunsthalle* and the *Museum für Kunst und Gewerbe*. The new museum was inaugurated by the new director since 1889, Karl Matthias Kraepelin (1848–1915), who had become professor at the *Johanneum* two years earlier (after teaching there since 1878). Being a "local talent," Kraepelin was not only recognized for his commitment to reform science education in schools (see Nyhart 2009) but also a distinguished naturalist specialized in the study of scorpions, centipedes, and spiders. He in the subsequent two decades made this museum an internationally renowned research institution. In this context, he explicitly named, next to geographical distribution, zoological systematics as the main task ("die wissenschaftliche Hauptaufgabe") of this museum. Kraepelin was extremely successful in building up the museum's global collections through private donations and in context with the colonial movement. Kraepelin (1899: 10–11; 1901: 132) reported that from 1891 to 1899, within less than a decade, the museum holdings

Fig. 35.1 The *Naturhistorisches Museum*, or Museum of Natural History, in Hamburg designed in Italian renaissance style opened in 1891 (CeNak Archive)

Fig. 35.2 The main exhibition hall of the *Naturhistorisches Museum Hamburg*, with large whale skeletons on display until the 1940's (CeNak Archive)

have been nearly doubled, with an increase of about 15.000 catalogue numbers or c. 60.000 specimens per year. At the same time, the NatHistMus, promoting the modern dual arrangement of separating research collections from exhibitions, was visited by an average of 140.000 people per year. Kraepelin was among those museum scientists who saw it as their role to present evolution to a broad public through the new public exhibits and to conduct research related to the history of life (Nyhart 2009: 26). We need to remember that at that time the Hamburg NatHistMus was, together with the one in Berlin, the leading German research museum, presenting a newly biologized picture of nature and of zoological research, with its holdings being an archive containing the records of natural history and with systematics playing an important role (see Nyhart 2009: 239–240).

In 1914, when also the museum's name changed into *Zoologisches Museum*, the zoologist Hans Lohmann (1863–1934) followed as director, serving until 1933. Lohmann, who had an expertise in polar plankton research, also became the first professor of zoology at the newly founded University of Hamburg in 1919 (to which the museum, however, did not belong until 1969). In 1934 Berthold Klatt (1885–1958) followed as forth director. The NatHistMus was destroyed during the Allied bombing of Hamburg in the early morning of 30 July 1943 ("operation Gomorrha"), only 3 months after the celebration of its centenary (Fig. 35.3). Significant parts of its scientific collections, in particular the alcohol-preserved collections, were removed earlier and stored outside the museum building or even Hamburg, together with some parts of the dry collections, in particular those of birds. Noteworthy, the alcohol collections were not removed for their enormous scientific value or as cultural heritage but to protect Hamburg's citizens from the risk of burning (Hallermann 2007: 22). An estimated fifth of these wet collections have fallen prey to the post-wartime black market. In contrast to today's

a **b**

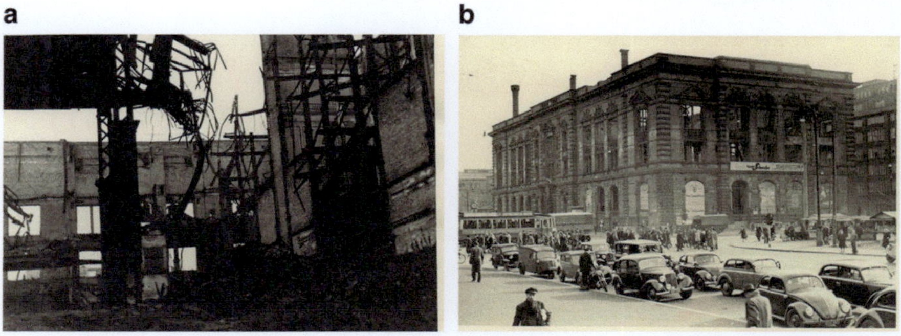

Fig. 35.3 The ruins of the *Naturhistorisches Museum Hamburg*, (**a**) destroyed by the bombing of the city's center on 30 July 1943 and (**b**) shortly before its final removal in 1951 (CeNak Archive)

use of methylated ethanol, the preserving spirit then could be redistilled. This fate of the museum's collection became part of the postwar literature thanks to the famous German writer Siegfried Lenz (1964). While only the iconic narwhale skull with the double tusks was rescued due to a curious event, many other rare objects on display were destroyed in the exhibition hall, among them the most impressive skeletons and partial casts of some larger whales (Fig. 35.2) and even the very rare skeleton of the extinct Steller's sea cow (then called *Rhytina gigas*; see Mohr 1950).

After the destruction of its building, those parts of the NatHistMus collections that have been rescued were provisionally deposited in the Botanical State Institute in Hamburg's city center; in 1953 they moved to a former bunker near today's university campus. With the end of the war, those parts of the collection that have been deposited in the eastern German state of Saxony came under the ruling of the Soviet military administration and returned from there in the late 1950's. Only after many frustrating years Curt Kosswig (1903–1982), who was director from 1955 to 1969, managed to negotiate the building of a new facility for both a museum and the zoological institute (that had once been part of the NatHistMus). Until then having being an independent research institute of the state of Hamburg (*Staatsinstitut*), the natural history collections and its staff were formally integrated into the university in 1969, and then renamed as Zoological Institute and Zoological Museum Hamburg (ZIM). It moved, 1972–1976, into a new building on its current location (see Fig. 35.4). In 1984 a small exhibition, presented on 2.000 qm, was opened. However, without professional staff for exhibitions and education being employed, museum scientists were long left alone with running this exhibit which has not seen much improvement since then (see below).

Fig. 35.4 The Zoological Museum Hamburg as part of the newly founded *Centrum für Naturkunde* (CeNak) in its university building from the early 1970's (CeNak Archive)

Larger collections in particular for the entomological and malacological departments were taken over from the natural history section of the nearby *Museum Altona* when it was dissolved in 1979. Also, the rich fish collection of the Hamburg *Institut für Seefischerei* (ISH) of the *Bundesforschungsanstalt für Fischerei* (BFAFi) was taken over in 1993 (see under the respective collections).

Despite continuous efforts by Hamburg's museum zoologists and the curators of the constituent collections organized in the *Verbund der Naturwissenschaftlichen Sammlungen Hamburgs* (VNSH; see Brandt et al. 2010) that long in vain tried to convince university and federal state authorities of the need for and the establishment of a new natural history museum at a more prominent site, the respective commitment of Hamburg state officials was lacking. Even worse, in 2005 the *Behörde für Wissenschaft und Gesundheit* of the Freie und Hansestadt Hamburg (i.e., the Free and Hanseatic City) intended, after ill-advised ideas on the digitalization of its inventories, to close down the zoological collections. This approach, fortunately, not only resulted in strong media echo in German newspapers but ultimately also in a rather successful—as it turned out—evaluation by the German *Wissenschaftsrat*. In 2009 this board recommended several steps for the improvement of the museum's situation, asking the senate of the city of Hamburg to take action within three years.

Finally, after decades of unsuccessful attempts to improve the situation of the natural history collections, in October 2014 the *Centrum für Naturkunde* (CeNak), or Center of Natural History, was founded by the Universität Hamburg. Albeit virtual to date, the three collections of the Geological-Paleontological Museum, the Mineralogical Museum, and the Zoological Museum (ZMH) now again form an essentially autonomous research institution independent of the faculty, directly answering only to the presidency of the Universität Hamburg (for details on the new structure, see below).

With the exception of these latest developments, the history of the NatHistMus Hamburg has been described in many details in particular by Weidner (1969, 1967, 1993), based on former accounts by Kraepelin (1899, 1901) and in particular by Panning (1955, 1956, 1957, 1958), Ladiges et al. (1968), Hoerschelmann et al. (1993), and Kraus (1994). Additionally, in providing the historical context, many aspects can also be found in Köstering (2003) and Nyhart (2009). Currently, Köstering (in prep.) is writing the first profound history of the NatHistMus and the zoological collections in Hamburg, covering the period from the year of its foundation in 1843 to the late 1970's.

35.2 Number of Species and/or Specimens, Focal Points

Within the newly founded CeNak, the Zoological Museum Hamburg (ZMH) is the largest of its three museums, comprising approximately ten million specimens housed in nine research collections with taxa from the entire animal kingdom. Together with the objects on show in its permanent exhibition, it renders the CeNak

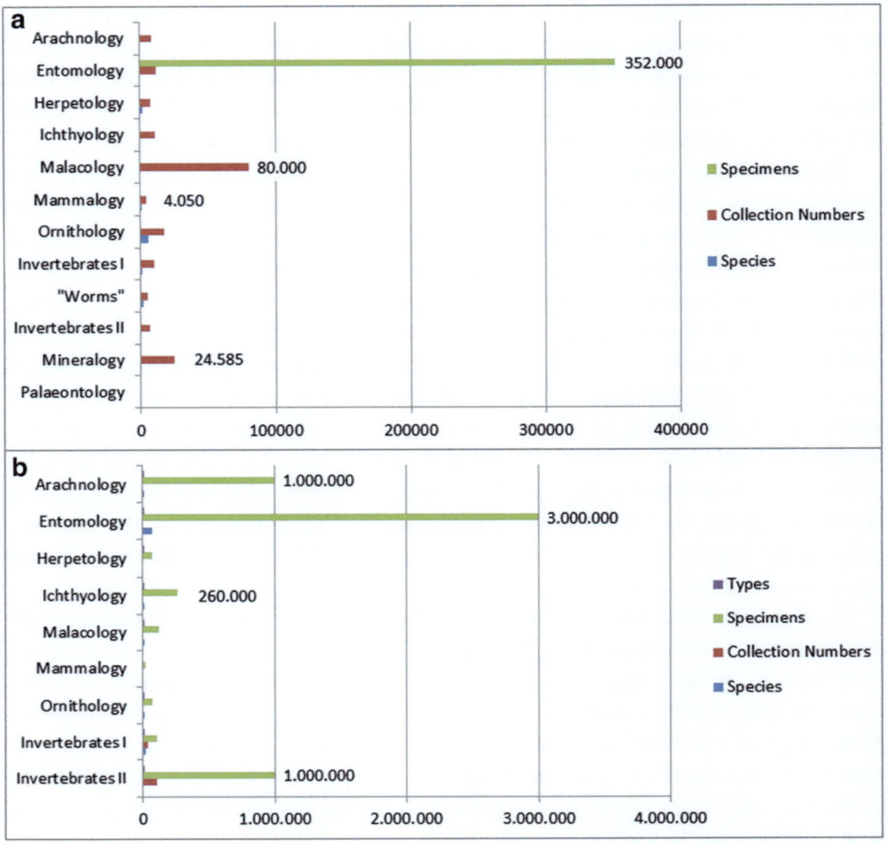

Fig. 35.5 Number of specimens and catalogued entries, respectively, (**a**) in the former *Naturhistorisches Museum* around 1900 and (**b**) as currently estimated for the zoological collections of the *Centrum für Naturkunde*

one of the most important scientific research museums in Germany, despite its losses during WWII (see Fig. 35.5). Certain parts of its collections, as will be detailed here, are among the most important of its kind at a national as well as international scale, in particular those of the same invertebrate groups such as, for example, the Oligochaeta and Polychaeta, the Crustacea, the Acari and Arachnida, as well as the Tunicata, the fishes, and ungulates. When compared with other museums, the number of zoological specimens alone makes the ZMH the fourth or fifth largest, after Frankfurt, Berlin, and Munich, in its size comparable to the museum in Stuttgart and larger than that in Bonn.

However, as it holds true for other natural history collections, figures about zoological inventories are tentative at best. To date, no reliable attempt has been made within Germany (e.g., among those collections represented by the DNSF), not to speak of any international approach (e.g., within the CETAF facilities), to consistently assemble figures on collection size in a comparative fashion. Thus, each museum presents more or less reliable estimates independently. However, these are hardly comparable, as the factual basis for the counts are seldom or not at all mentioned (not to speak of being evaluated). For the Hamburg collections, our estimates are given here as they are retrievable from former reports, in particular from a guidebook for the Zoological Museum (Brandt et al. 2008). Although viewed against a preliminary evaluation of existing data bases, these estimates for the ZMH should be considered as approximate only and thus taken with considerable care, as in all other cases.

35.3 Arachnology: Arthropod Collections

approx. 1 mio. specimens, 8.800 species—3.200 types known (of 2.000 primary types)

This scientific collection, for most of its time closely associated with the department of Entomology in the ZMH (see below), houses a total of 860.000 preserved specimens, not counting the rich Acari collection. It includes arachnids and other arthropods, such as in particular the Pararthropoda, Chelicerata, and Myriapoda. A most numerous and important collection is that of the Acari, with c. 3.600 species represented. Catalogue entries are present for a total of 55.000 specimens, of which 18.000 are stored in ethanol and 37.000 are microscopic slides. Of particular historical interest, for example, is a rich collection of Australian spiders, collected by Amalie Dietrich for the Museum Godeffroy in 1862–1872 in Queensland and worked up by L. Koch and E. von Keyserling, with an estimated more than 800 of the c. 2.000 araneaen species. The collection is internationally renowned in particular for its material in Tardigrada, Arachnida, and heterostigmatic Acari, being consulted frequently by scientists for its many types as well as the historical material. In addition, rich in species numbers and important are the collections of diplopods (1.060 species) and scorpions (452 species).

The curatorial position was long held, from 1959 to 1992, by Gisela Rack, followed by Hieronymus Dastych, who retired in October 2012. For more details, see Dastych in Brandt et al. (2008: 50–51).

35.4 Entomology: Insect Collections

approx. >3 mio. specimens, 65.000 species—c. 10.000 types

Once one of the largest entomological collections in Germany, the Zoological Museum suffered greatly from the loss of essential parts of its collection in 1943 which was assumed to house in this department up to then about 1.5 mio. insects, with an estimated 10.000 types (Fig. 35.5). Parts of these historical collections, going back to the time of Hanseatic merchandisers with their interest in natural history collections, were rescued, while important parts of the collection, such as of holometabolic insects with many types, were destroyed. Today, CeNak's entomological collection holds again up to at least 3 mio. specimens, with 1 mio. specimens pinned and in alcohol each, supplemented by well over 5.000 microscopic samples and 500 modeled objects, and about 10.000 types, with a focus in Acercaria (>4.000), Coleoptera (>3.100), and Lepidoptera (>1.500), rendering it one of the largest and most important among German museums.

The international significance of the collection is reflected by loan requests for some 5.000 specimens to around 20 different countries each year. Collected from all around the world, containing numerous unique specimens, the collection is subdivided into three sections: the scientific-systematic collection, scientific special collections (including insects that damage plants and textiles), and a comprehensive teaching collection. Only a very small fraction of its holdings are catalogued and/or digitalized, now using Specify, transferring data from scanned catalogue cards (65.000) to Excel sheets (50.000 entries), resulting in about 30.000 entries (as of 2015).

The collection is growing on average by 10.000 specimens per year and, thus, continues to document the biodiversity of insects, the largest class of animals on Earth. Edited by H. Strümpel and H. Dastych, the entomological department published its own journal "Entomologische Mitteilungen aus dem Zoologischen Museum Hamburg" for many years (with back issues being available online under https://www.cenak.uni-hamburg.de/forschung/zoologie/entomologie/zeitschrift.html); it will be continued within a relaunched journal *Evolutionary Systematics* published for the entire Zoological Museum in Hamburg, for which currently new plans are about to be implemented.

35.5 Herpetology: Reptilia and Amphibia

65.000 specimens—194 nominal types, with 333 type specimens registered

This department holds the fifth largest herpetological collection in Germany, with about 70 % reptiles and 30 % amphibians, stored essentially in 17.500 jars, supplemented by several hundred dried skeleton and skin specimens and 259 sets of serial histological sections. About half of the collection, with data from c. 33.000 individuals, are now digitized, using FileMaker Pro software (July 2015), among those are 190 types that are available online through GBIF; type catalogues have been published by Hallermann (1998, 2006). As the inventory catalogue was destroyed in 1943 during WWII, the holdings and data of the collections had to be restored from the information preserved on the original labels, which was started by H.-W. Koepcke (1914–2000); see Hallermann (2007, and references therein). The collections hold some important historical specimens, among them, e.g., reptiles collected by Maximilian Prince zu Wied-Neuwied (1782–1867) during his famous expedition to Brazil 1815–1817; described in 1825, these are among the oldest preserved holdings in Germany. The ZMH also comprises the herpetological collections of Johann Gustav Fischer (1819–1889), the most distinguished Hamburg herpetologist at his time. He described 106 reptile and six amphibian taxa, of which only 34 and four, respectively, are still present at the ZMH, while the remainders were lost during WWII. For a more detailed account of the history of the museum's herpetology, see Hallermann (2007).

35.6 Ichthyology: Fish Collection

260.000 specimens, 70.000 series, 8.100 species—1.340 type series with 3.700 type specimens

The 260.000 specimens in this collection, comprising about 70.000 series, represent about 8.100 species from 60 orders and 378 families, which are about 25 % of all known fishes worldwide. Thiel et al. (2009) have given a detailed account on the development of the ZMH fish collection, based on the analysis of its data base. The foundation of the collection goes back to the NatHistMus in Hamburg, with initially low numbers, amounting, e.g., in 1850, only 110 lots, but increased markedly during the second half of the nineteenth century and surpassed the 4400 mark in 1900. Overall, about 6 % of the current collection stems from this period between 1851 and 1900. The first two decades of the twentieth century saw a relative rapid expansion of the collection, continuing to profit from expeditions exploring unknown habitats and territories. An estimated 14 % of the holdings, which are mostly (70 %) stored in alcohol, were destroyed during WWII. The most significant, rendering the ZMH collection now beyond doubt the largest of marine fish in Germany and also one of the largest worldwide, came in 1993 with the official takeover of 23.860 marine fish lots from the former *Institut für Seefischerei* Hamburg (ISH), after the collection had been transferred to the ZMH already in 1978; for a brief history of the ISH fish collection, designed and initiated by Gerhard Krefft (1912–1993), see Stehmann (1997). It resulted in an increase of 75 % of all actual catalogue lots stored now in the ZMH. Today these holdings are partly (ca. 65 %) digitalized, using our own Access database, with about 46.500 entries (as of 2015). These are also to be found via online data search such as FishBase (www.fishbae.org; www.cenak.uni-hamburg.de/sammlungen/zoologie/ichthyologie.html).

The collection maintains a worldwide coverage, containing about two thirds of all fish species, 70 % of all fish genera and 98 % of all fish families known to occur in the Atlantic Ocean, including the Antarctic sector, and has considerable series from the Indian Ocean and large numbers of lots originating from the Pacific Ocean and also from Europe, Asia, Africa, and North and South America. Actinopterygii and Elasmobranchii together comprise more than 93 % of the orders represented in the ZMH fish collection database. For more details, see Thiel et al. (2009).

The research of the ZMH Ichthyology Group today includes modern taxonomic and biogeographic approaches important in present global biodiversity research. One research focus is, next to biosystematics, of selected groups of teleosts and elasmobranchs, the study and modeling of ecology and distribution of diadromous fish, as well as ongoing surveys of the fish fauna in the North and Baltic Seas and their adjacent waters and, in particular, in the River Elbe [see Thiel et al. (2009), Thiel and Thiel (2015)]. It renders the ZMH fish collection as an indispensible tool among international reference centers dealing with biodiversity documentation and conservation of fishes.

35.7 Invertebrates I

>100.000 specimens, 31.000 catalogue entries, c. 20.000 species—c. 3.300 types

Within its diverse collection, about 23 groups of multicellular animals (Metazoa) are represented, with currently more than 31.000 catalogue entries (Fig. 35.6). There are three main parts of international rank. First, among the Oligochaeta, there are 4.200 catalogue entries, with 1.346 types, including the historical collection in particular of Hamburg's longtime curator and oligochaet specialist Wilhelm Michaelsen (1860–1937), who was working from 1887 to 1923. Thus, of the

Fig. 35.6 A view of the collection Invertebrates I of the Centrum für Naturkunde (CeNak Archive)

c. 8.000 species of Oligochaeta known worldwide, 16 % of all named taxa are represented by type material in the CeNak, i.e., every fifth oligochaet in the world is represented in Hamburg by its name-bearing specimens. Second, about 3.700 catalogue entries of Tunicata are present, with 1.343 entries for the Ascidia (with 309 types). Third, there are 1.538 series of Bryozoa with 54 types known. Other parts comprise Echinodermata with 3.900 series and 209 types, Cnidaria (3.260 lots, 156 types), Porifera (1.988 series, 215 types), Hirudinea (979 series, 35 types), and Chaetognatha with 700 series. More than 25.000 entries are currently digitally registered, using FileMaker Pro.

35.8 Invertebrates II

>1 mio. specimens, 100.000 catalogue numbers—11.000 types known

This collection essentially of crustaceans and marine bristle worms houses, with more than 100.000 catalogue numbers, the largest type collection in Germany. It holds 44.888 catalogue numbers for the Crustacea, among them also 6.458 types, and 27.693 catalogue numbers for the Polychaeta, with c. 4.363 types. More than 90 % of the species in this collection are from marine habitats, with 70 % of them coming from the southern hemisphere, in particular the South Polar Ocean. Using the database Sesam, type entries from former catalogue cards were digitalized. The collection comprises 368 families of the Crustacea (with c. 2.100 genera) and 48 families of the Polychaeta (with c. 1.350 genera), with Amphipoda (c. 250.000 specimens) and Isopoda (c. 170.000 specimens) being the most numerous parts within the malacostracan Crustacea, and the Copepoda (c. 260.000) and Ostracoda (25.000) among the “entomostracan” crustaceans. A systematic focus are several groups of Crustacea, among them, e.g., the Peracarida, the Decapoda, and the Copepoda, with a geographical focus on taxa from the Southern Ocean and various deep-sea basins. A special research interest of the current curator is the systematics and phylogeny as well as the documentation and analysis of the biodiversity and biogeography of Peracarida with focus on Isopoda.

35.9 Malacology: Mollusk Collection

120.000 series, 10.000 species—1.787 types

The CeNak Mollusk Collection includes primarily snails, mussels, and cephalopods from every continent and ocean on Earth. Once its collection numbered about 300.000 series, but the whole dry collection was lost as a result of the destruction of the Natural History Museum in 1943 (see Fig. 35.5). Of its more than 120.000 series, currently (as of 2015) more than three quarter (100.700) are digitalized, using Specify, comprising 100 % of its wet and 80 % of its dry collection, of which more than 35 % are georeferenced. In addition there is a tissue collection for genetic studies with more than 5.700 samples.

The alcohol collection of the NatHistMus and the former collection of the *Museum Altona* formed the base of the recent collection, the latter including part of the collections of Otto Semper, with types, e.g., of Karl Theodor Menke and Ludwig Pfeiffer. The collection has especially increased in size by yields of research vessels ("Walther Herwig," "Polarstern") and the accession of private collections, in particular those of Rolf Brandt from Thailand collecting nonmarine mollusks, as well as Eberhard Claus, Peter Glöer, and Wolfgang Fauer. Focuses in the collection are land snails from southeastern Europe, Turkey, the Caucasus region, South America (Columbia, Peru), and eastern Africa (Uganda, Rwanda), freshwater mollusks from Thailand, and marine mollusks from the Antarctic and Patagonian regions. The cephalopod collection with about 3.250 series is the largest in Germany and among the largest worldwide.

Research focuses largely on species diversity, evolution and distribution patterns, as well as the phylogenetic relationships of land snails. A current project, for example, is investigating how environmental factors and historical processes have led to a high level of species diversity in the Caucasus.

35.10 Mammalogy: Mammal Collection

c. 20.000 specimens—number of types not known

The collection comprised about 10.000 skeletons and 3.000 skins, supplemented by 5.000 specimens stored in alcohol, rendering it one of the most important in Germany. In particular, this latter section of the collection is of importance as it comprises historical material, including organs and embryos, while the dry collection was completely destroyed in the former NatHistMus building in 1943 (Fig. 35.5). Using Specify about 16.000 specimens are digitalized, with next to all skins this comprises about 95 % of the skeletons, 80 % of the wet material, and 60 % of the histological sections. The focus of the collection is on hoofed animals, African primates, and marine mammals, with postwar collections of larger mammals, represented by skulls and mostly complete postcranial skeletons, and from taxa that are hardly retrievable today from their natural habitats. The collection comprises material derived from major expeditions to India and Angola as well as other parts of Africa in the 1950's–1960's, e.g., from Manfred Röhrs and Henriette Oboussier, with large and complete series of bovid material.

Most recently, in 2014, the CeNak took over from Günter Bräuer a most valuable and unique (paleo-)anthropological collection that has been assembled over the last four decades. With 250 casts of fossil hominids, it is certainly the largest of its kind in Germany. In addition, a collection of prehistorical anthropology and human osteology was donated, comprising 1500 specimens of medieval skeletons (in 650 boxes), most valuable for morphological-anatomical comparisons.

Research in the collection focuses on functional aspects of mammal teeth and their chewing, using three-dimensional data analyses of dental surface structures that enable the study of diet and life history of (even extinct) species and to reconstruct past and present environments including influences such as of climate change.

35.11 Ornithology: Bird Collection

71.000 specimens, 3.500 species—c. 142 known types

The ornithology collection, for which once a catalogue was printed in 1898, suffered from substantial losses during WWII. It gained c. 9.000 specimens again since 1967 and has yet doubled since then. Today a total of 30.000 specimens of mounted or skinned birds are listed again in the catalogue, supplemented by 20.000 feather samples, 4.000 skeletons, and 2.000 alcohol samples, plus a total of 15.000 specimens in the egg collection. Of the species-rich collection, about 4.000 numbers (with about 25.000 individuals) have been digitalized and types included in the GBIF database. The curatorial position has not been available since 1997, with the collection being provisionally administered by the acting curator for herpetology.

35.12 Today's Conditions of Infrastructure (Staff, Rooms, Laboratories, Exhibitions, Financial Support, Perspectives)

Organizational Structure CeNak's structure today is only to be understood from a historical perspective of our institution which seems to be quite unique among the larger natural history museums in Germany. Integrating its important collections into the Universität Hamburg in 1969, as part of the Zoological Institute and Zoological Museum (ZIM), resulted in the successive loss of the museum's autonomy, with transfer of virtually all decision making to the Faculty of Mathematics, Informatics and Natural Sciences (MIN), but was not necessarily based on better competence. This increasing and unhealthy dominance of university processes and interests, in concert with the rigorous decrease in financial support of building maintenance and collection-related staff members, has created a disastrous situation for the museum. After two major evaluations by external experts, among them most importantly by the German *Wissenschaftsrat*, an independent advisory board in scientific and research questions for the German government, this situation of the ZMH became at least public and most obvious.

Although the Hamburg Senate for more than the last decade remained indolent and inactive, the university in 2012 finally took over responsibility in paving the way for both the establishment of a new director and the foundation of the CeNak.

In this context, a new institutional concept was accepted and implemented in the course of the new appointment. The presidency of the Universität Hamburg decided in May 2014 to found the new Centrum für Naturkunde (CeNak), or Center of Natural History, followed by the inauguration of its new scientific director in October 2014. Effectively, the CeNak now is no longer part of the MIN Faculty, thus independent of other university administrative structures and only responsible to the president of the university. It integrates under its (to date only virtual) roof the three natural history collections of the former Zoological Museum, the Geological-Paleontological Museum, and the Mineralogical Museum, which, though, remain for the time being at their respective localities.

Within the CeNak, the ZMH comprises its former nine departments, each with its own collection and curator (the latter with the exception of ornithology): mammals, birds, reptiles and amphibians, fish, insects, spiders, mollusks, as well as the two invertebrate sections, viz., for Crustacea and Polychaeta, and all remaining, respectively (see Fig. 35.7). In addition, an exhibition on 2.000 qm is immensely popular as a place of learning for school children and university students alike (for more on the exhibition and didactic concept, see below).

Personnel The CeNak currently (as of August 2015) has a total of 99 employees of different status. Among them are 35 employees whose salaries are paid through the university budget, with 12 scientists altogether, i.e., professors and curators (of which ten are in the Zoological Museum), as well as two scientific qualification positions (associated with the newly established professorship for *Biodiversität der Tiere*). In addition, there are 21 employees in science-supporting services, such as

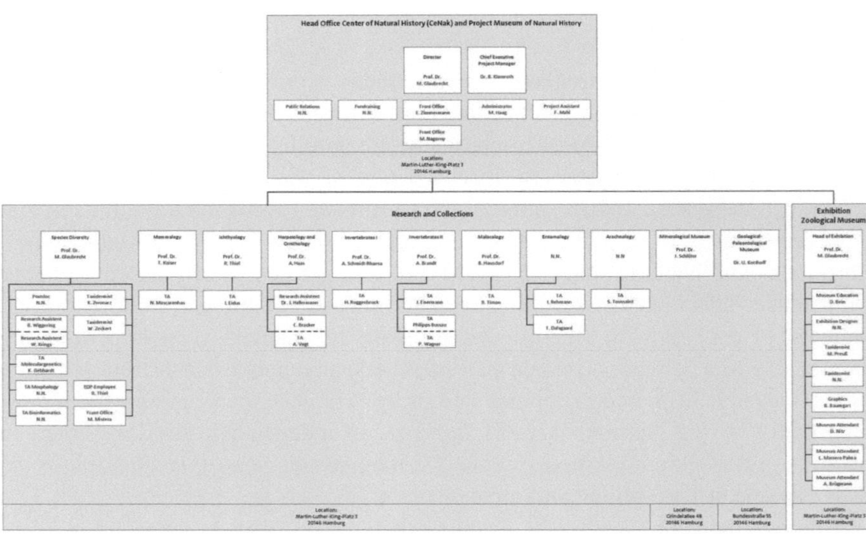

Fig. 35.7 Organigramm, or composition of the positions and personal in the Centrum für Naturkunde Hamburg (CeNak)

collection maintenance, preparators, and other staff in administration, direction, and exhibition. Another total of 31 employees, essentially Ph.D. students and student assistants, are paid through third-party money, i.e., grants from founding bodies such as DFG, BMBF, DAAD, and others. In addition to these total of 66 employees, the CeNak has currently another 33 members, such as bachelor, master, and other students, professor emeriti, volunteers, and guest researchers.

A cooperation contract with the MIN faculty regulates the engagement of CeNak's scientists, professors as well as curators and others, in teaching activities at the Universität Hamburg. In mutual agreement between university, faculty, and CeNak's director, the teaching load was now reduced and fixed, in order to set free more of the working time of scientific staff members for museum and collection affairs.

Rooms and Laboratories Built in the early to mid-1970's, the ZMH facility at Martin-Luther-King-Platz has lacked careful maintenance for decades and is among those university buildings desperately in need for renovation or, alternatively, rebuilding on site or elsewhere. While planning toward this goal was long not visible, the spatial situation has been shown to be close to critical by the evaluation of the *Wissenschaftsrat* in December 2008. Major improvements, i.e., renovating the most necessary on site, were mutually agreed on in the course of the appointment of CeNak's director and are currently implemented, to be finalized in 2016. Plans are discussed to also improve the situation of the exhibition, but are not completely agreed on yet.

Nevertheless, in its museum building, the ZMH possesses currently a well-equipped central molecular lab and several morphology labs, including the scanning electron microscopy facility shared with the Zoological Institute. It is planned in the course of renovating parts of the Zoological Museum building to create central labs for molecular genetics and morphology. To run these, new staff is currently hired, with positions granted by the university for the next 5 years in the course of the inauguration of the new director.

The ZMH also have a *Präparatorium*, including workshops and maceration facility, with two preparators working full time for both the research collections and the exhibition. The CeNak continues to share other facilities, such as technical workshops and services, their library, and their lecture halls with the closely associated Zoological Institute in an adjacent part of the same building.

Exhibition Once among the most attractive and important of all zoological museums in Germany, and second only to the *Museum für Naturkunde* in Berlin, the public exhibition of the NatHistMus in Hamburg suffered greatly from the Allied air raid on 30 July 1943. It took nearly half a century, until 1984, when again the so-called Schausammlung opened its doors to the public (see Fig. 35.8). Unfortunately, this exhibition on only 2.000 qm is well hidden within the MIN campus at its current location. Being the only object once on display in the former NatHistMus that survived, the narwhale skull with two tusks, recently named "Lisa" now (see above), became the new icon. The current exhibition was created in the early 1980's and has not yet seen much of an improvement in terms of design

Fig. 35.8 View of the exhibition of the Centrum für Naturkunde Hamburg (CeNak)

and didactic concept. Its basic idea, essentially exhibited by larger vertebrates, was described in more detail in Brandt et al. (2008). Referring to it here in more detail will be superfluous, as with the new CeNak implemented, key positions within the newly established exhibition department will now be filled, and a new exhibition concept will be created and implemented in the near future.

Nevertheless, the exhibition continues to generate increasing numbers of visitors. With about 20.000 visitors per year in 2004 and 30.000 in 2007, it was even doubled in number until reaching 65.000 in 2014 (see Fig. 35.9). For a comparison with other museums in Hamburg, we note that the *Museum für Völkerkunde* counts 130.000 and the *Kunsthalle* 380.000 visitor per year, however, on much more space and with by far other means. The successful development in the ZMH's exhibit is undoubtedly the results of an extraordinary engagement of the museum didactic services, creating next to regular-guided tours and education in particular for school classes also smaller special exhibits and events.

Financial Support As a result of its new institutional structure, effective October 2014, the CeNak built over the course of the first year of existence its own administration, human resource planning, and financial budget, albeit in close association with the central university administration. At the moment, its budget is comparably low, with approximately around 3 mio Euro per year spent essentially on the salaries of CeNak's personal and for the most basic support of its collection, exhibition, and research.

Perspectives With the implementation of the CeNak, it is planned to (re-)establish a modern natural history museum in Hamburg, taking up after more than half a century the tradition of the former NatHistMus, in particular the most successful and prosperous of its phases when Karl Matthias Kraepelin developed its leading

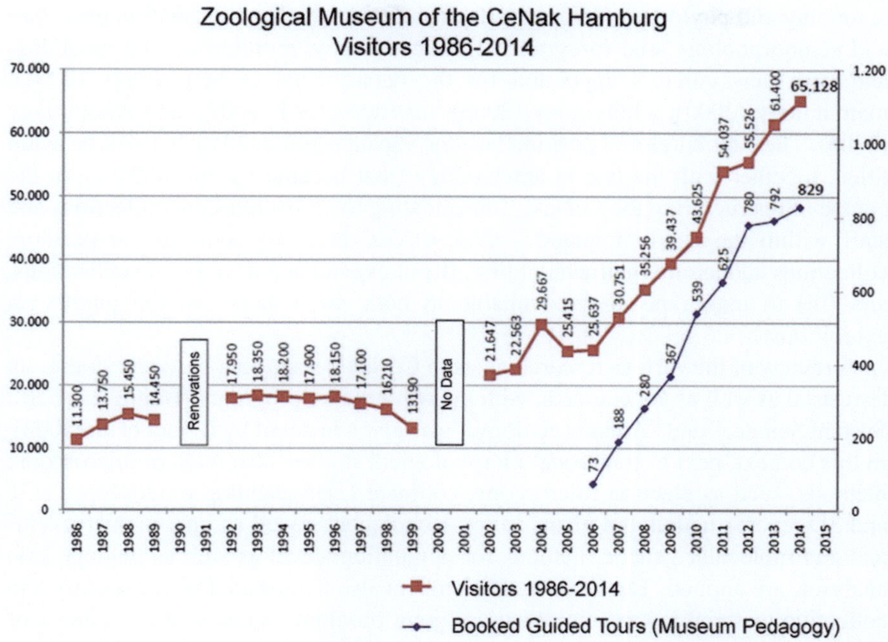

Fig. 35.9 The increase of visitors in the exhibition of the Zoological Museum of the Centrum für Naturkunde Hamburg (CeNak)

role in research, exhibition, and education (see above). Details of these plans are currently discussed and negotiated with the various partners in the university, the city state's administration, and with private donators.

The vision and concept of an *Evolutioneum* as Hamburg's new Natural History Museum for the twenty-first century has been developed by its current director in the course of his appointment. Pending on the political commitment of the Hamburg Senate for the implementation of a new master plan for his museum, including in particular a new building with a museum exhibition as showcase of its research, CeNak's two declared aims are, first, to reestablish the museum as an excellent research institution taking up its former tradition and, second, to bridge the gap between academia and the general public.

35.13 *Research*: Examples of Today's Research, Potencies

The ZMH enjoys considerable national and international recognition in the scientific community for its achievements in collection-based research by its leading scientists and their broad expertise in organismal zoology. The ZMH has seen over the last decade a complete generation turnover in its leading scientific staff, with three research programs guiding new staff acquisition in the past, viz., (program I) taxonomy and phylogeny, (program II) evolutionary biology, speciation processes and ecomorphology, and (program III) biogeography, monitoring, and modeling, and with new curators appointed for the departments of herpetology (2004), mammalogy (2005), ichthyology (2006), invertebrates I (2007), and entomology (2008). The later curatorial position became vacant again in 2010, but will be again filled, together with the one in arachnology (that became vacant in 2012), in the course of restructuring the CeNak. Consolidating the three museum collections and staff within the newly founded CeNak serves three key aims: (i) to preserve collections and assure their availability, (ii) to expand and develop the collections, and (iii) to make specimens available to both researchers and the public via exhibitions.

A review of the various research goals of CeNak's scientists, based on studies on terrestrial as well as aquatic taxa, with the two broader programs "Biological Earth System Science" and "Aquatic ecology," was given in detail by Brandt et al. (2008). In this context, next to traditional morphological studies, also modern microscopic methods, such as electron microscopy, confocal laser scanning microscopy, uCT and 3D reconstruction and visualization techniques as well as immunohistochemical and molecular genetic methods for populations-level as well as phylogenetic analyses, are applied. The ZMH staff members also contributed with expertise and collection material toward the success of joint programs, such as, e.g., "Census of Marine Life," "Tree of Life," "Deep Metazoan Phylogeny," "Biota" and "Diversitas," as well as "Clisap" (see Fig. 35.10).

The ZMH as part of the CeNak is integrated in the major national and international research and museum networks, including CETAF (Consortium of European Taxonomic Facilities), DNSF (Konsortium Deutsche Naturwissenschaftliche Forschungssammlungen), NORe (Museumsverbund der Nord- und Ostsee Region e.V.), and Deutscher Museumsbund. We agreed on a formally legalized

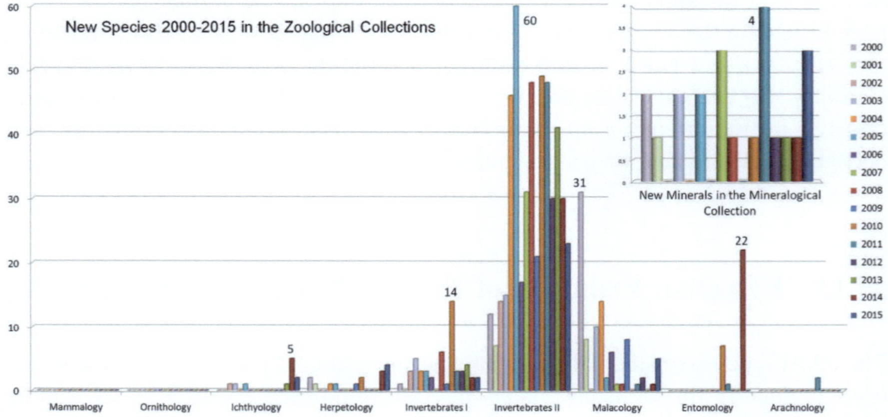

Fig. 35.10 The number of new species described by scientists from the Zoological Museum of the Centrum für Naturkunde Hamburg within the past 15 years (CeNak)

cooperation granting joint research in taxonomic and systematic studies with the *Deutsches Zentrum für Marine Biodiversitätsforschung* (DZMB), founded in 2001 as external department of the Forschungsinstitut Senckenberg in Frankfurt (FIS) and administered through the Senckenberg facilities in Wilhelmshaven. Another legalized cooperation is established between the *Johann Heinrich von Thünen-Institut für Seefischerei* (vTI) in Hamburg, from which the ZMH has taken over a very rich fish collection in 1993. The vTI continues to transfer ichthyological specimens of its research activities.

In the context of restructuring the museum, also its own journal will be relaunched. The former "Mitteilungen aus dem Zoologischen Museum Hamburg" has a long tradition going back to 1883, when it was founded as "Jahresbericht über das Naturhistorische Museum zu Hamburg" and has been continuously been published until 2010 (see online under https://www.cenak.uni-hamburg.de/uebercenak/journale.html). Currently new plans are about to be implemented for a new open access journal *Evolutionary Systematics* to be published for the entire CeNak.

Contributing Toward an Encyclopedia of Life Within a period of the last 15 years, from 2000 to 2014, a total of 500 new taxa on the species level have been described by scientists at the CeNak (see Fig. 35.10), in addition to a total of 34 new minerals by CeNak's mineralogist. Based on its research collections, the scientists in the CeNak will continue with museum-based zoological research in the field of taxonomy and systematics, evolution and ecology, as well as ecosystem function of aquatic and terrestrial taxa. The main focal research areas have been described by the individual scientists of the ZMH for their responsible departments and collections in Brandt et al. (2008). More recent updates are available online on CeNak's website for the respective departments.

In the near future, an integrated biodiversity research program, along the lines of a vision on *Discovering biological diversity—its evolution and future*, will be developed for the CeNak in close association with its researchers. We will continue to combine a special focus on marine habitats, from the local River Elbe estuary to the polar regions, and on selected terrestrial to limnic taxa from various regions in the world, accentuating an evolutionary systematics perspective, i.e., combining taxonomy-based descriptive approaches in systematics, phylogeny, and biogeography with more causal and analytical approaches as described by Glaubrecht (2010). This research program will rest on three methodological bases, viz., morphology, molecular genetics (including genomics), and biogeographical and ecological modeling. This program will be supplemented by a biohistory agenda, looking into the life history of individual objects as well as taking into focus the historical development of collections and museums; for some examples of actual research, see the CeNak websites, for the theoretical reasoning in particular those under the department for "Biodiversität der Tiere."

In this context, the implementation of a working program on the intelligent digitalization of our research collection (i.e., including all researchable metadata) will be crucial. We also plan to establish a third collection strategy, viz., in addition to the dry and wet collection, to build a “cold archive” where extracted DNA and tissue samples will be stored for future molecular genetic research.

35.14 Promotion of Junior Staff

Flexibility in terms of permanent or temporarily available positions to promote junior staff is quite limited in a German university institution nowadays. Next to two scientific qualification positions associated with the newly established department for “Biodiversität der Tiere,” currently a total of 31 employees are paid through third-party money. These are essentially Ph.D. students and student assistants, thus promoted as upcoming scientists, in addition to the supervision of bachelor and master students. CeNak has no formal postgraduate or Ph.D. promotion program on its own, but is part of those programs developed in responsibility of the faculty and/or Universität Hamburg.

35.15 Publications: Written or Other Used New Media

Over the last couple of years, the scientific staff of the CeNak, here counting 12 fully employed scientists and their staff (mainly assistants and Ph.D. students), wrote a total of about 100 original and peer-reviews scientific publications per year, including books and book chapters plus several popular science publications. For example, in 2014, these original publications numbered 96 plus 8 scientific books, i.e., in average 8.7 per then fully employed scientist (note that some staff positions of the CeNak were not filled yet at that time).

We found, in a web-based bibliographic search, that a collection material of the CeNak was mentioned, for example, from the entomology department in 96 publications within the last 8 years (2008–2015), thus resulting in about 12 publications per year where authors used and/or referred to material from the CeNak. We can assume that in other departments, in particular those holding the species-rich invertebrate taxa, would exhibit similar figures, adding up in about 100 publications per year mentioning CeNak material (Fig. 35.11).

35.16 Publications: Done by Lectures, Workshops, Meetings, and Expeditions

The scientific staff is, as elsewhere, engaged in scientific lectures, workshops, and other meetings; however, we haven’t assembled a detailed listing of these activities.

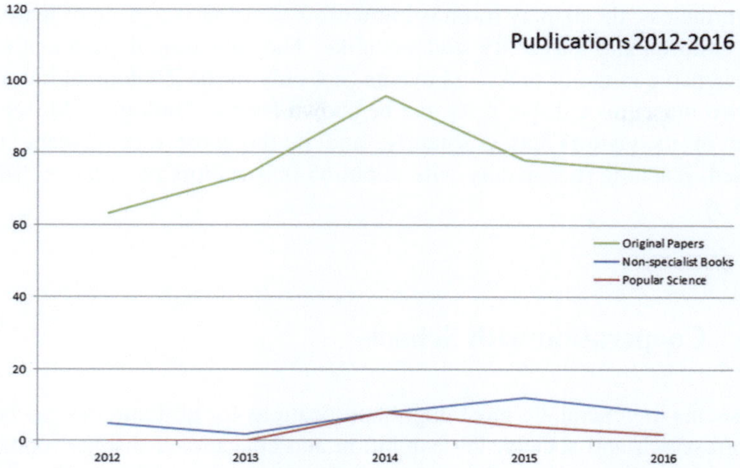

Fig. 35.11 The number of publications by staff members of the Centrum für NaturkundeHamburg (CeNak), for the period 2012–2016 in the three categories as given

Scientists of the CeNak are regularly involved with major marine expeditions in an international context, for example, in expeditions of the research vessel "Sonne," "Meteor," and "Polarstern," as well as others. These are mainly cruising in Antarctica, but also in the Middle Atlantic and the Pacific. More low-key, i.e., individually organized, terrestrial expeditions were regularly going to regions in Southeast Asia, e.g., Thailand and Indonesia, as well as Australia and Africa. Details on this can be found on the individual websites via CeNak's homepage, as well as in the individual publications and reports of its scientists.

35.17 Didactic Conceptions of Exhibitions

Given the integration into the Universität Hamburg since 1969, a small exhibition was only opened in 1984. It was designed as "Schausammlung" or display room, i.e., presenting selected zoological objects of special interest, such as particularly larger mammals and birds, with the idea to allow an overview of this part of global biodiversity, ranging from marine whales to European ungulates. As research in the ZMH was essentially disconnected from its permanent exhibition, no special signature of these activities was visible and accessible to the visitor; also no particular object and their individual history have been presented. Substantial financial support and trained staff was long lacking. Only with the newly established CeNak it will be possible to make up for these deficiencies in the past.

Nevertheless, the display room is immensely popular as a place of learning for school children and university students alike. The education department offers a comprehensive range of tours and events (not only in the Zoological but also the other two museums exhibits). As can be shown for the Zoological Museum, the number of its visitors has constantly, and in the most recent years, rapidly increased, reaching its capacity with about 65.000 visitors per year in 2014 (see Fig. 35.9).

35.18 Cooperation with Schools

As museums in general are most important locations for bridging the gap between science/research and society, the exhibition and educational display room of the ZMH provides an "außerschulischer Lernort" or external school room for educating pupils in natural sciences and biology, respectively. Accordingly, a large fraction of the increasing number of visitors in the ZMH exhibit is from schools and kindergarten. As the university also has special curricula for teachers, there is a close cooperation with the educational department established, making use of a special collection (with about 3.000 catalogued items and/or specimens in addition to those in the scientific research collections). In 2014 a total of 870 didactic programs have been offered and booked, with a third of the visits coming from classes of the Hamburg gymnasiums. There is a focus on evolutionary biology for which the exhibition staff with its two permanent positions and 16 freelance members is particularly trained.

35.19 Use of Social Media (Youtube, Facebook, Twitter, Instagram, and Local Platforms)

Information on CeNak, its exhibitions, and events can be found online on its website as well as on Facebook via its homepage under https://www.facebook.com/pages/Zoologisches-Museum-Hamburg/102096659875733?sk=timeline (Fig. 35.12).

Fig. 35.12 A view of the Malacology collection, Centrum für Naturkunde Hamburg (CeNak)

References

Brandt A, Haas A, Misof B, Thiel R (2008) Zoologisches Museum Hamburg. Eigenverlag Zoologisches Museum, Hamburg

Brandt A, Kotthoff U, Kranz G (2010) Naturwissenschaftliche Museen und Sammlungen Hamburg—VNSH. Zoologisches Museum Hamburg, Hamburg

Calder DR, Brinckmann-Voss A (2011) Gustav Heinrich Kirchenpauer (1808–1887) of the City of Hamburg, and his research on hydroids and bryozoans. Zootaxa 2742:49–59

Glaubrecht M (2010) Evolutionssystematik limnischer Gastropoden. Unpublished Habilitation Thesis, Humboldt Universität, Berlin. Available from https://www.cenak.uni-hamburg.de/forschung/zoologie/biodiversitaet/forschungsschwerpunkte/evolutionssystematik.html

Glaubrecht M, Bein D (2017) From curiosity object to iconic status—the unique "unicorn whale" in the Center of Natural History in Hamburg. Zoosyst Evol 91 (in prep.)

Hallermann J (1998) Annotated catalogue of the type specimens of the herpetological collection in the Zoological Museum of the University of Hamburg. Mitt Hambg Zool Mus Inst 95:197–223

Hallermann J (2006) Additions to the catalogue of the type specimens of the herpetological collection in the Zoological Museum of Hamburg. Mitt Hambg Zool Mus Inst 103:137–147

Hallermann J (2007) Zur Geschichte der herpetologischen Sammlung des Zoologischen Museums Hamburgs, mit besonderer Berücksichtigung von Dr. Johann Gustav Fischer (1819–1889). Sekretär 7(1):20–32

Heine H (1887) Aus den Memoiren des Herren von Schnabelewopski. In: Heine H, Werke S (eds). Verlag H. Dürselen, Leipzig

Hoerschelmann H, Villwock W, Dzwillo M, Hartmann G, Hartmann-Schröder G, Rack G, Dastych H, Wilkens H, Hoerschelmann H, Schliemann H, Münzing J (1993) 150 Jahre Naturhistorisches Museum zu Hamburg. Mitt Hambg Zool Mus Inst 90:7–29

Köstering S (2003) Natur zum Anschauen. Das Naturkundemuseum des deutschen Kaiserreichs 1871–1914. Böhlau, Köln

Köstering S (2017) Haus der Weltnatur. Das Naturhistorische Museum in Hamburg. Abhandlungen des Naturwissenschaftlichen Vereins in Hamburg (in preparation)

Kraepelin K (1899) Das Naturhistorische Museum in Hamburg und seine Ziele. Verh Dtsch Zool Ges 1899:7–18

Kraepelin K (1901) Das Naturhistorische Museum. In: Hamburg in naturwissenschaftlicher und medizinischer Beziehung. Herausgegeben zur 73. Versammlung Deutscher Naturforscher und Ärzte. Verlag von Leopold Voss, Hamburg, pp 124–149

Kranz H (2005) Das Museum Godeffroy, 1861–1881. Naturkunde und Ethnographie der Südsee. Altonaer Museum, Hamburg

Kraus O (1994) Naturwissenschaftlicher Verein und Zoologisches Museum Hamburg—von der Bürgerinitiative zur staatlichen Einrichtung. Verh Naturwiss Ver Hambg (NF) 34:385–397

Ladiges W, Kosswig C, Haffner K (1968) 125 Jahre Naturhistorisches Museum zu Hamburg: 25 Jahre Wiederaufbau. Mitt Hambg Zool Mus Inst 65:5–59

Lenz S (1964) Lehmanns Erzählungen. Aus den Bekenntnissen eines Schwarzhändlers. Hoffmann & Campe, Hamburg

Mohr E (1950) Ein Hautstück der Stellerschen Seekuh, *Rhytina gigas*, Zimm. 1780. Zool Anz 145 (7–8):181–185

Nyhart LK (2009) Modern nature. The rise of the biological perspective in Germany. University of Chicago Press, Chicago, IL

Panning A (1955) Beiträge zur Geschichte des Zoologischen Staatsinstituts und Zoologischen Museums Hamburg. 1. Die Gründung des Naturhistorischen Museums. Mitt Hambg Zool Mus Inst 53:1–23

Panning A (1956) Beiträge zur Geschichte des Zoologischen Staatsinstituts und Zoologischen Museums Hamburg. 2. Teil. Der Aufbau der Archiv-Sammlung. Mitt Hambg Zool Mus Inst 54:1–20

Panning A (1957) Beiträge zur Geschichte des Zoologischen Staatsinstituts und Zoologischen Museums Hamburg. 3. Teil. Die Vermehrung der Sammlung in Verbindung mit Hamburgs Handel und Kaufmannsreederei. Mitt Hambg Zool Mus Inst 55:1–23

Panning A (1958) Beiträge zur Geschichte des Zoologischen Staatsinstituts und Zoologischen Museums Hamburg. 4. Teil. Die Förderung des Museums durch Hamburgs Linienreederei. Mitt Hambg Zool Mus Inst 56:5–59

Scheps B (2005) Das verkaufte Museum. Die Südsee-Unternehmungen des Handelshauses Joh. Ces. Godeffroy & Sohn, Hamburg und die Sammlungen "Museum Godeffroy". Abh Naturwiss Ver Hambg (NF) 40 (Goecke & Evers, Keltern-Weiler)

Stehmann M (1997) Gerhard Krefft (1912–1993) and post-World War II collection building in ichthyology at the Institut für Seefischerei, Hamburg: a melding of applied and basic research. In: Pietsch TW, Anderson WD (eds) Collection building in ichthyology and herpetology, vol 3, American Society of ichthyologists and herpetelogists special publication. Allen Press Inc., Lawrence, KS, pp 121–132

Thiel R, Thiel R (2015) Atlas der Fische und Neunaugen Hamburgs. Arteninventar, Ökologie, Verbreitung, Bestand, Rote Liste, Gefährdung und Schutz. Freie und Hansestadt Hamburg, Behörde für Stadtentwicklung und Umwelt, Amt für Natur- und Ressourcenschutz, Abteilung Naturschutz, Hamburg

Thiel R, Eidus I, Neumann R (2009) The Zoological Museum Hamburg (ZMH) fish collection as a global biodiversity archive for elasmobranchs and actinopterygians as well as other fish taxa. J Appl Ichthyol 24(suppl 1):9–32

Weidner H (1967) Geschichte der Entomologie in Hamburg. Abh Verh Naturwiss Ver Hambg NF 9(Supplement):1–387

Weidner H (1969) 125 Jahre Naturhistorisches Museum zu Hamburg. Abh Verh Naturwiss Ver Hambg NF 13:5–30

Weidner H (1993) Bilder aus der Geschichte des Zoologischen Museums der Universität Hamburg. Die Zoologischen Sammlungen im Naturhistorischen Museum zu Hamburg während seiner Kollegialfassung 1843 bis 1882. Mitt Hambg Zool Mus Inst 90(Suppl): 1–380

Chapter 36
HEIDELBERG: The Zoological Museum of the University of Heidelberg

Volker Storch

Abstract The “Zoologisches Cabinet” of the University of Heidelberg was founded in 1819 by Friedrich Tiedemann. In its modern outfit, it was presented to the general public in the 1970’s. Since 1979/1980, Sunday lectures in each winter semester have attracted thousands of interested people. The museum was closed less than a decade before its 200th anniversary. The Sunday lectures still attract 400 and more visitors each winter Sunday.

Keywords Tiedemann • Bronn • Evolution • Extinction • Thylacinus • Ectopistes

In 1819, Friedrich Tiedemann (1781–1861) founded the “Zoologisches Cabinet”. Tiedemann, as a professor of anatomy and physiology at the University of Heidelberg since 1815, had learned from the famous palaeontologist Georges Cuvier in Paris that a zoological collection could be very useful for teaching students in the field of human and comparative anatomy. He was supported financially by the senate of the university and the ministry and bought a collection of birds and mammals. Tiedemann though being a professor of the faculty of medicine had a deep interest in zoology and published books and articles on the anatomy of the human brain, the heart of fish, the anatomy and biology of the flying dragon *Draco volans* from SE Asia and the biology of amphibians, echinoderms, etc.

Heinrich Georg Bronn (1800–1862) was the first full professor of zoology at the University of Heidelberg. He was one of the most prominent palaeontologists and zoologists in nineteenth-century Germany. In 1833, he became the director of the Zoological Cabinet. In the same year, he published a catalogue of mammals of the “Museum Academicum Heidelbergensum” as he called the “Zoologisches Cabinet”. In 1860, Bronn translated and published Charles Darwin’s *The Origin of Species*. The German version came out only 10 months after Darwin’s opus magnum. Bronn added much to the Zoological Cabinet. His private collection of exotic species was bought by the state, and his private collection of locally collected specimens was donated to the cabinet. Bronn had an extremely broad knowledge of

V. Storch (✉)
Centre for Organismal Studies, Universität Heidelberg, INF 230, 69120 Heidelberg, Germany
e-mail: volker.storch@cos.uni-heidelberg.de

L.A. Beck (ed.), *Zoological Collections of Germany*, Natural History Collections,
DOI 10.1007/978-3-319-44321-8_36

various fields, as his lectures covered zoology and botany, forestry, agriculture and palaeontology. His name is still well known because of the handbook *Bronn's Klassen und Ordnungen des Tierreichs*.

After Bronn's early death, Heinrich Alexander Pagenstecher (1825–1889) became the director of the museum. He, too, had broad interests, published articles on deep-sea animals and edited four volumes on general zoology (1875–1881). He, amongst others, bought a specimen of the Australian thylacine *Thylacinus cynocephalus* from the East India Company. This Australian marsupial was driven to extinction by the European settlers. The last animal died in the Hobart Zoo in 1936 (see below). Pagenstecher left Heidelberg in 1882 and became the director of the newly founded Museum of Natural History in Hamburg.

Pagenstecher's successor was Adam Otto Bütschli (1848–1920), who was a former student of chemistry, mathematics and mineralogy with a deep interest in geology and palaeontology, which was stimulated by Karl Alfred von Zittel (a former student of Heinrich Georg Bronn), who contributed considerably to the establishment of palaeontology at German universities. Bütschli became a research assistant of both the biologist Rudolf Leuckart in Leipzig and the zoologist Carl Möbius in Kiel, who founded the zoological museums in Kiel and in Berlin. Bütschli was a prominent scientist, who published amongst many other important papers, his 2000 page opus eximium "Protozoa". Whilst Bütschli did not add much to the collections, it should be noted that under his influence a new building was erected as an Institute of Zoology. The collections were kept in a makeshift building in the backyard and were mainly used for demonstration purposes for students. Luckily, the collections survived the two world wars without too many losses; however, they were in a bad shape, because nobody had taken professional care of them. In the 1930's, the interest of local zoologists in the collections had declined considerably. Some material was given to museums in Mannheim, Karlsruhe, Hamburg and Halle, where, due to the war, unfortunately many objects were destroyed. Zoologists at the Zoology Institute in Heidelberg were not deeply interested in any kind of collection until Professor Manfred Lüdicke (curator of the museum from 1959 to 1976) took over. During these years, he acquired rare vertebrates, and Dr. Gunter Konrad, once a technician (Präperator) at the museum in Heidelberg, made a valuable donation, amongst others he gave several birds of paradise for display. In 1964, the Zoological Institute and the Zoological Museum moved to the new campus "Im Neuenheimer Feld". In the beginning, the exhibition of the most interesting and most beautiful objects occupied two floors in the new building. However, one floor was given to a new department soon thereafter. In 1979, the remaining parts of the exhibition were presented to the general public in a very good and modern condition (Fig. 36.1). The scientist in charge, Professor Heinz Moeller, initiated at the same time a series of public Sunday lectures which have been accepted by hundreds of visitors each winter semester and which until to date (2016) attract 400–600 people every Sunday in every winter semester (October–February) (Moeller and Sparing, 1985).

According to Möhl (2002), the Zoological Museum was considered one of the most popular institutions of the University of Heidelberg. According to him, this had to do with the Sunday lectures which attracted thousands of interested people

Fig. 36.1 Panorama of the Zoological Museum

over the decades and which were made possible by the financial support received from over 400 members of the "Freundeskreis des Zoologischen Museums". Professor Moeller retired in 2001. He repeatedly expressed his concern that the museum could be closed one day. He felt that an exhibition like this—with over 1200 items on display in over 20 showcases on 450 m^2 plus audiovisual aids—was not as highly esteemed by the new decision makers as compared to those at the beginning of his activities. The position of the only technician to the museum had been regarded as "not necessary" by the university. As a result skilled maintenance of the museum was in danger. After Moeller's retirement as a professor, Siegfried Sparing was the last technician who held a position at the museum. Thanks to his skills, the "Zoological Museum of the University of Heidelberg", as it was called officially, presented itself as a magnet for local people, for school children and high school and university students. The latter deepened their knowledge on biodiversity and evolution. In large showcases, invertebrates and vertebrates were on display (especially for students of biology, who studied for their examination in biodiversity), with other showcases with extinct species created concern (e.g., passenger pigeon (Fig. 36.2) and thylacine); another focus was the evolution of primates including *Homo sapiens*. A very special attraction was a showcase with stuffed birds whose songs could be heard from a cassette activated by the push of a button by the visitor. Since Moeller's special interest was in Australian marsupials, special attention was given to these. Moeller (1997), he published a monograph on *Thylacinus cynocephalus*.

Volker Storch took over parts of Heinz Moeller's duties in 2001: he organized the Sunday lectures and took care of the museum assisted at first by Siegfried Sparing and later by Bernhard Glaß. Over a decade later, in 2012 Professor Moeller's concern became a reality. The museum was closed, items on loan were given back to the owners, and most showcases were destroyed (they had not been manufactured for being moved). The majority of items that had been on display for years or decades were transferred to another building on the university campus, where they await to be used again—at another place and another time. The layout of the former museum was redeveloped to create rooms for seminars and practical training.

Fig. 36.2 Passenger pigeon (last record: 1 September 1914)

References

Moeller HF (1997) Der Beutelwolf. Westarp, Magdeburg, p 195

Moeller HF, Sparing S (1985) Zoologisches Museum der Universität Heidelberg. Westermanns Monatshefte, Braunschweig, p 128

Möhl KH (2002) Heidelberger Museen und zugängliche Sammlungen. Verlag Rhein-Neckar-Zeitung, Heidelberg, p 101

Chapter 37
JENA: The Scientific Zoological Collections at the Phyletisches Museum in Jena: Historical Development and Conservational Challenges

Gunnar Brehm, Dietrich von Knorre, and Martin S. Fischer

Abstract The zoological collections at the Friedrich Schiller University Jena have a complex history that dates back to the eighteenth century. Although material was gained in sum over long periods, storage capacities were always a limiting factor, material was often insufficiently documented, and some of it was lost or destroyed. Here, we report briefly on the history of the zoological collections and their importance and value in the light of the biodiversity crisis and legal impediments. Conservational problems include permanent curation, temperature and humidity fluctuations as well as pests and the use of poisons.

Keywords Scientific collection • Jena • Phyletisches Museum • Conservation • Haeckel

37.1 Early History of the Zoological Collections in Jena

The complex early history of the zoological collections in Jena was described in detail by von Knorre (1983), von Knorre and Beutel (2006) and Fischer et al. (2008). In this overview, we only highlight some of the most important dates and persons. In 1700, Wilhelm Ernst, duke of Sachsen-Weimar, started building up a 'Naturalien- und Kunstkammer' in the castle of Weimar. In 1779, the collection was complemented by Duke Carl August of Sachsen-Weimar with the purchase of an important collection of natural curiosities and fossils sampled by Johann Ernst Immanuel Walch (1725–1778), with many specimens described and illustrated by Walch (1773). The cabinet was set up in the Jena city castle (Jenaer Stadtschloss). The Walch cabinet was thus the nucleus of the first collection for research and teaching of natural objects in Jena. This publicly accessible 'Carl August Museum'

G. Brehm (✉) • D. von Knorre • M.S. Fischer
Institut für Spezielle Zoologie und Evolutionsbiologie mit Phyletischem Museum, Erbertstraße 1, 07743 Jena, Germany
e-mail: gunnar.brehm@uni-jena.de

L.A. Beck (ed.), *Zoological Collections of Germany*, Natural History Collections,
DOI 10.1007/978-3-319-44321-8_37

(later, Großherzogliches Museum) was supervised and curated by various persons, the most famous of them Johann Wolfgang von Goethe (1749–1832); for a detailed list, see von Knorre (1983). The museum cabinets included objects ranging from minerals, fossils, organs of domesticated animals as well as bones, skins and dried plants.

After 1850, the collection was split and its zoological part indicated as 'Zoological Cabinet' and later as 'Zoological Museum'. Still, large parts of the collection have been housed in the Jena city castle until it was knocked down in 1905 to give way to the main university building. Many objects were lost or destroyed in this period either due to limited space for storage or ignorance of the value of the objects.

Ernst Haeckel was director of the 'Großherzogliches Zoologisches Museum' from 1862 onwards, and he became a regular professor for zoology in Jena in 1865. During many years of his early career, working space was very limited, and a separate building for the Institute of Zoology was not opened before 1883 (Fig. 37.1). In this building, a small room in the northwestern corner (next to the 'small' lecture hall) was dedicated to house parts of the zoological collection. In 1884, Haeckel provided the total number of museum specimens (including those from the old city castle) as ca. 25,000. In the following years, the number of specimens rapidly increased, e.g., due to zoological expeditions by Haeckel, Richard Semon, Willy Kükenthal and others. In 1904, Haeckel estimated the number of specimens at 130,000, but the documentation of incoming (and outgoing) material was incomplete, and none of the objects were properly inventoried, i.e., individually and uniquely numbered and labelled. It must be assumed that many of these specimens have been lost or destroyed in the following decades or their scientific value has been lost due to insufficient labelling, irregular use for teaching, etc.

37.2 The Phyletisches Museum

In 1905, Haeckel wrote a letter to the curator (the head of the university administration) and suggested to build a 'biological museum'—with the shortage of space in his institute as first argument (Uschmann 1959). The Phyletisches Museum (an abbreviated term for 'Phylogenetic Museum') was founded already two years later. The building was opened and donated to the university in August 1908, but the first permanent exhibition was not shown earlier than 1912. Haeckel originally planned magazines in the basement of the museum that were, however, not realised due to other demands. In 1909, he noted in the entry book that 'it has been decided to reorganise the collections of the Zoological Museum and to transfer a large part of it (about half) to the new(...)Phyletisches Museum' (Fig. 37.2).

Haeckel's successor Ludwig Plate, director of the Phyletisches Museum and professor of zoology in Jena from 1909 to 1935, kept most of the material in the buildings of the Institute of Zoology (Chap. 38). A second zoology building, opened in 1912, provided new space for the collections in rooms neighbouring the new large lecture hall (figured in Chap. 38), allowing easy access for teaching purposes.

Fig. 37.1 The first zoology building in Wilhelminian style was opened in 1883. On the *left*, the second zoology building from 1912 which houses the zoological teaching collection until today (Photo: © with G. Brehm 2015)

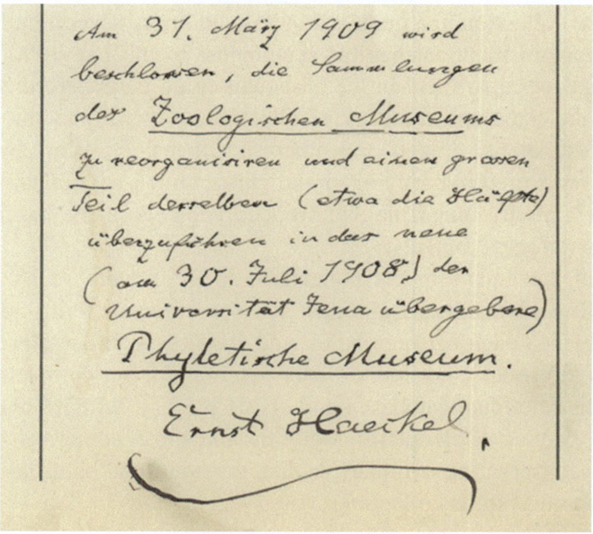

Am 31. März 1909 wird
beschlossen, die Sammlungen
des Zoologischen Museums
zu reorganisiren und einen grossen
Teil derselben (etwa die Hälfte)
überzuführen in das neue
(am 30. Juli 1908) der
Universität Jena übergebene)
Phyletische Museum.
Ernst Haeckel.

Fig. 37.2 A note by Haeckel made in 1909 in the museum catalogue

Material was also stored in a small room in the upper floor of the 1883 building. Plate seems to have missed to inventory the collections of the institute and the museum, despite substantial gains of material due to own expeditions and donations of material (Chap. 38).

Major changes started only in 1956 when strong efforts were made to reorganise all zoological collections and to clearly separate the teaching collection from a scientific museum collection in newly established storage rooms in the Phyletisches Museum. Amongst others, this allowed to save the most valuable specimens, including unique type material, from the potentially destructive demands of teaching. The first full inventory of the collections in the following decades is the work of Dietrich von Knorre (curator 1969–2003), which has become the fundament for the long-term custody and sustainable scientific exploration of the collections. Other collections such as the university's palaeontological collection were integrated as well as further donations, e.g., by preparator Eberhard Burkhardt (Fischer et al. 2008; von Knorre and Beutel 2006).

37.3 The Zoological Collections in 2015

In 2015, the collections are organised in different magazines located in the basement and in the second floor of the museum. The second floor houses a 120 m^2 magazine with insect cabinets and other dry materials in a former exhibition hall as well as the type collection in a metal cabinet. In addition, several built-in closets in the corridor contain mammal habitus specimens. In the late 1990's, the former 120 m^2 housekeeper apartment in the basement could be transformed in total to a magazine for the osteological and skin collections of mainly mammals and birds, but also mollusc shells, models and other specimens. A 39 m^2 'wet magazine' contains material stored in partly historical glasses filled with ethanol or formalin, and another 13 m^2 dry magazine and wooden cabinets in the basement corridor contain palaeontological objects.

The total of specimens in the collections in 2015 were 55.192 objects with molluscs, birds and mammals amongst the top three in terms of individually inventoried objects, each between 8.000 and 10.000 specimens. Insects and other small animals are usually not individually inventoried, but complete drawers and series represent individual database numbers (Fig. 37.3). Butterflies + moths (Lepidoptera) and beetles (Coleoptera) represent the most numerous groups in the collection. A conservative estimate is that the museum holds around 500.000 individual specimens in its collections.

Fig. 37.3 The entomological collection of the museum comprises many now regionally extirpated species such as the Hermit [*Chazara briseis* (L.)] and the Grayling (*Hipparchia semele* L.). Inventory number PMJ Hex 81 (complete drawer) (Photo: © with G. Brehm 2015)

37.4 Importance and Value of the Zoological Collections

Besides being the fundament of taxonomy, biological collections provide particularly useful information on morphology, distribution and chemical composition of species (Pyke and Ehrlich 2010). The Jena zoological collections contain many valuable specimens, both scientifically and historically, and we can only refer to some examples. The collections currently hold around 550 identified type specimens (Brehm and Fischer 2014); an overview is currently prepared by Dietrich von Knorre. New type specimens are regularly deposited, for example, new Neotropical geometrid moths from Ecuador and Costa Rica (Brehm 2015). Although Haeckel described more than 4.000 species of Radiolaria (Lazarus and Suzuki 2009) as well as species of jellyfish, sponges, etc., only very few type specimens had been deposited. Apart from type material, the wet magazine comprises important material (mostly marine) gathered through expeditions undertaken by Haeckel, Semon, Kükental, Plate and others. The palaeontological collections mainly encompass some important Triassic fossils and specimens from Quaternary sites in Thuringia (von Knorre and Beutel 2006). The dry collections comprise many regionally important collections of vertebrates, molluscs, insects and other organisms that are valuable, e.g., for long-term faunistic comparisons (Fig. 37.3). Since more than 20 years, the museum has a focus on the collection of German wildcats (*Felis silvestris silvestris* L.). More than 130 wildcats and house cats with pelage characters close to the wildcat have been collected since.

Specimens are long-term stored for scientific purposes, but they are also frequently used for exhibitions at the museum. Two prominent specimens shown in the permanent exhibition are 'Goethes Fossiler Stier' (*Bos primigenius* Bojanus) and a shoulder blade of a bowhead whale (*Balaena mysticetus* L.) which was painted with sceneries from whale hunting in 1646 (illustrated in Brehm and Fischer 2014). Most of the valuable material is, however, not displayed but stored in the museum magazines (see below). It is our responsibility to take care of all parts of the collections, to keep their scientific value and to conserve it successfully for future generations.

37.5 Biodiversity Crisis and Legal Impediments

It has often been falsely assumed in the past that zoological collection has a relatively little value because specimens can easily be replaced (von Knorre 2004). However, it has become clear that this is not the case, mostly because of the global biodiversity crisis (Hamer 2012) as well as increased legal impediments for sampling animals. This is not only valid for taxa that have always been rare such as the New Zealand kakapo (*Strigops habroptilus*) (Fig. 37.4), a critically threatened species. It is also true at a regional scale: For example, an insect collection sampled in Jena in the 1950's and 1960's contains a large number of species that have become rare or even regionally extirpated such as the Hermit [*Chazara briseis* (L.)] (Fig. 37.3). The zoological collections are thus archives of animal communities that have undergone fundamental changes, particularly in the last decades.

In parallel with the ongoing loss of biodiversity, bureaucracy and legislation have grown. For example, most butterfly species are legally protected in Germany today ('Bundesartenschutzverordnung'). Although this can be seen as a positive development at first sight, there are severe drawbacks: There is no scientific evidence that the formal legal protection has helped any butterfly species in its effective conservation, because collectors are not or very rarely responsible for butterfly species declines but instead habitat destruction and land use changes are. Legal protection also discourages young amateurs: If it is forbidden and unwanted by our society to touch a butterfly, who will collect and rear a caterpillar and set up a new collection that will be a valuable document of our present time in the future? Legislation in most countries differentiates too little between the unwanted commercial exploration of endangered species and non-commercial zoological research.

Fig. 37.4 The kakapo from New Zealand (*Strigops habroptilus* Gray, inventory number PMJ Aves 503) is one of the rarest bird species on Earth and critically threatened by invasive species such as the stoat (*Mustela erminea* L.) (Photo: © with G. Brehm 2013)

Fig. 37.5 Example of ethanol samples that were dried out before regular curation started. Scale bar: 1 cm (Photo: © with G. Brehm 2015)

37.6 Conservational Challenges

37.6.1 Regular Curation

Although it may seem trivial, every scientific collection requires a certain extent of curation and thus causes personal costs. This need increases in collections with specimens that are regularly used or studied (also by visitors), that are accessible and consumable to pest species and that frequently need to be handled, e.g., because ethanol needs to be refilled. In short, an uncurated collection is prone to chaos, damage (Fig. 37.5) and destruction.

37.6.2 Storage Environment

It is widely accepted that objects stored for long-term purposes should be kept under constant and relatively cool conditions (Hamer 2012). DIN ISO 11799 for archives recommends temperatures below 18 °C and relative humidity values

Fig. 37.6 Insect magazine with cabinets integrated in the outer wall (*left*) and tightly sealing insect drawers (*right*). The preparation of a blue peafowl (*Pavo cristatus* L.) demonstrates the shortage of space in the museum (Photo: © with G. Brehm 2015)

between ca. 30 and 50 % for most materials (paper, film, etc.). Similar regimes are standard in modern zoological magazines, e.g., in the Smithsonian Institution (Washington, DC) and the Natural History Museum (London). None of the rooms used as magazine in the Phyletisches Museum were originally designed for this purpose, and, as a consequence, none meets modern museum conservation standards.

Rooms in the basement tend to have a relatively high humidity in the summer months when humid air from outside cools down in the building (RH measured up to 56 %). On the contrary, humidity in winter months is low due to heated air, leading to a yearly cycle of increasing and decreasing humidity. While such a regime will not do any harm to fossils or mollusc shells, specimens such as bird and mammal skins might become fragile under too dry conditions and exposed to mould with too humid conditions. The western basement rooms also currently have problems of moderate water intrusion from outside. While temperatures remain relatively constant in the basement of the building (range ca. 18°–24 °C throughout the year), fluctuations are higher in the second floor magazine because the museum roof and the floor ceiling are still poorly insulated (range ca. 14°–29 °C throughout the year and daily fluctuations up to 6 K). Moreover, many insect drawers are still kept in cabinets that are integrated in the outer wall originally planned for exhibition purposes (Fig. 37.6). Cabinets containing valuable material should not be a functional part of the wall insulation. The university committed itself to replace these cabinets by freestanding ones in the next years.

Table 37.1 List of substances applied or possibly applied in the past in the Phyletisches Museum

Substance	Past applications	EU classification
Organic compounds		
Naphthalene	IM: possibly but not commonly applied	Harmful
Lindane	IM: possibly applied	Toxic
1,4-Dichlorobenzene	DM applied, IM possibly applied	Harmful
Nitrobenzene	IM: possibly applied	Toxic
Dichlorodiphenyltrichloroethane (DDT)	IM: possibly applied	Toxic
Organophosphates		
Dichlorvos/'Vapona'	Probably not applied	Very toxic
Chlorpyrifos	DM and IM: applied (Nexa Lotte ©)	Toxic
Pyrethroids		
Empenthrin	In use since 2005	None
Arsenic compounds		
Arsenic	Probably applied in many bird skins	Very toxic
Realgar	DM: possibly applied	Toxic
Mercury compounds		
Mercury (II) chloride	Possibly in old preparations	Very toxic

IM insect magazine; *DM* dry magazine

37.6.3 Pests and Poisons

Pest insects have been a problem for museum collections from the beginning, and most curators have experienced the damage that globally displaced pest insects can cause, e.g., the varied carpet beetle *Anthrenus verbasci* (L.) and other dermestid beetles (Linnie and Keatinge 2000). A wide range of toxic chemical products has been applied to treat zoological collections in order to prevent collected specimens from destruction (e.g., Händel 2001; Marte et al. 2006). Such applications have usually not been documented in the Phyletisches Museum, which means that a contamination with harmful substances is possible but largely unknown. Table 37.1 provides a list of substances that potentially have been applied in the zoological collections in Jena. Most of these substances are not in use in public museum collections any more, and integrated pest management (IPM) is commonly applied because the use of substances such as lindane, nitrobenzene or arsenic (Händel 2001; Marte et al. 2006; Töpfer et al. 2011) is regarded as a serious health risk (Table 37.1). Prevention measures at the Phyletisches Museum include the purchase of new tightly sealing insect drawers and cabinets and obligatory freezer quarantine for incoming material. A major problem is ventilation through open windows, allowing pest insects to enter the collection. An automated ventilation system for all magazines is highly desirable in order to prevent pest damage and to clean the rooms from contaminated air.

Acknowledgements We thank Gerta Puchert for the helpful comments as well as Egbert Friedrich and Rainer Plontke for further information about regionally extirpated butterfly species.

References

Brehm G (2015) Three new species of Hagnagora Druce, 1885 (Lepidoptera, Geometridae, Larentiinae) from Ecuador and Costa Rica and a concise revision of the genus. ZooKeys 537:131–156

Brehm G, Fischer MS (2014) Phyletisches Museum im Institut für Spezielle Zoologie und Evolutionsbiologie. In: Werneburg R, May E (eds) Thüringer Natur-Schätze. Schnell & Steiner, Regensburg, pp 209–220

Fischer MS, Brehm G, Hoßfeld U (2008) Das Phyletische Museum in Jena. Institut für Spezielle Zoologie und Evolutionsbiologie, Friedrich-Schiller-Universität Jena

Hamer M (2012) An assessment of zoological research collections in South Africa. S Afr J Sci 108:60–70

Händel J (2001) Lindan in der Insektensammlung—Risiken und Alternativen. Mitt Thür Entomologenverband 8:92–97

Lazarus D, Suzuki N (2009) Introduction to the reexamination of the Haeckel and Ehrenberg radiolarian collections. In: Tanimura Y, Aita Y (eds) Joint Haeckel and Ehrenberg project: reexamination of the Haeckel and Ehrenberg microfossil collections as a historical and scientific legacy. National Museum of Nature and Science Monographs 40, Tokyo, pp 23–34

Linnie MJ, Keatinge MJ (2000) Pest control in museums: toxicity of para-dichlorobenzene, "Vapona", and naphthalene against all stages in the life-cycle of museum pests, *Dermestes maculatus* Degeer, and *Anthrenus verbasci* (L.) (Coleoptera: Dermestidae). Int Biodeterior Biodegrad 45:1–13

Marte F, Péquignot A, Von Endt DW (2006) Arsenic in taxidermy collections: history, detection, and management. Collect Forum 21:143–150

Pyke GH, Ehrlich PR (2010) Biological collections and ecological/environmental research: a review, some observations and a look to the future. Biol Rev 85:247–266. doi:10.1111/j.1469-185X.2009.00098.x

Töpfer T, Gamauf A, Haring E (2011) Utility of arsenic-treated bird skins for DNA extraction. BMC Res Notes 4:197

Uschmann G (1959) Geschichte der Zoologie und der zoologischen Anstalten in Jena 1779–1919. Jena

von Knorre D (1983) Die zoologisch-paläontologischen Sammlungen des Phyletischen Museums. Friedrich-Schiller Universität Jena, Jena

von Knorre D (2004) Warum und zu welchem Zweck erhalten wir naturkundliche Sammlungen? Mainz Naturwiss Arch 42:301–313

von Knorre D, Beutel R (2006) Die geowissenschaftlichen Sammlungen im Phyletischen Museum in Jena. Beitr Geol Thüringen 13:61–72

Walch JEJ (1773) Die Naturgeschichte der Versteinerungen, zur Erläuterung der Knorrschen Sammlung von Merkwürdigkeiten der Natur. Nürnberg

Chapter 38
JENA: The Teaching Collection at the Zoological Institute of the University of Jena: Its Importance, Value and Conservational Problems

Gerta Puchert and Gunnar Brehm

Abstract A clear separation between a zoological teaching and a scientific museum collection at the University of Jena was not realised before the 1960's. We report the brief history of the teaching collection, its importance, value and problems. We illustrate selected objects and provide an overview of the material that is currently part of this collection.

Keywords Jena • Haeckel • Teaching collection • Zoology

38.1 Brief History and Overview

Zoological collections exist in Jena since the eighteenth century, and their early history is described in Chap. 37. Zoology professors Ernst Haeckel (1834–1919) and his successor Ludwig Plate (1862–1937) had obviously different plans with regard to the collection after the foundation of the Phyletisches Museum in 1908 (Fischer et al. 2008): While Haeckel wanted to transfer half of the existing material from the existing collections to the new museum, Plate stated in a speech, "Among the objects on display [in the Phyletisches Museum] most is new, since the best specimens were not allowed to be withdrawn from the zoological institute because of teaching" (Plate 1912). It remains uncertain which parts of the zoological collections were actually used for teaching or whether the collections were more or less generally used for this purpose. Their major depository next to the large zoology lecture hall at least suggests that objects were easily in reach for teaching. A major reorganisation started in 1956 under Manfred Gersch as a director and later with Dietrich von Knorre as curator (Chap. 37). In the following years, the collections were separated consequently into a teaching collection on the one

G. Puchert • G. Brehm (✉)
Institut für Spezielle Zoologie und Evolutionsbiologie mit Phyletischem Museum,
Erbertstraße 1, 07743 Jena, Germany
e-mail: gunnar.brehm@uni-jena.de

L.A. Beck (ed.), *Zoological Collections of Germany*, Natural History Collections,
DOI 10.1007/978-3-319-44321-8_38

Fig. 38.1 Storage room (next to the large lecture hall in the 1912 zoology building) for the teaching collection with the large skeleton of an African elephant (*left*) and the wet material collection (*right*). The most light-sensitive objects are stored in shade. Further, glass cabinets with skeletons and models are in the next room (Photo: © with G. Brehm 2015)

Fig. 38.2 Diverse zoological material stored in ethanol in the teaching collection (Photo: © with G. Brehm 2015)

hand (Figs. 38.1 and 38.2), still housed next to the large lecture hall in the zoology building, and a scientific museum collection on the other hand (in various magazines of the Phyletisches Museum) (Chap. 37). Today, the teaching collection largely contains material of lower scientific value, e.g., mostly unlabelled specimens, relatively common species and models of moderate value. Properly labelled specimens, rare species and models of high value were usually transferred to the scientific collection in the magazines of the museum. A first list of objects is due to K. Richter in the 1960's when parts of the teaching collection were moved. Since 1988 a new object list has documented the holdings of the teaching collection.

A collection of 330 zoological wall charts was transferred in 2004 to the Working Group Biology Didactics of the Friedrich-Schiller University, while in the other way around, some 200 specimens were donated to the teaching collection. Ten of the wall charts, probably originally drawn by Haeckel, are now deposited in the Ernst Haeckel House (Chap. 37).

38.2 Intensive Use and Arising Problems

Today, the teaching collection is still an essential part of lectures, seminars and practical training for biology students. PowerPoint presentations cannot replace three-dimensional original objects. However, the use of a collection for teaching is challenging in various aspects, for example, it requires a permanent curation, repair of damaged objects and purchase of new material, and it poses serious problems particularly to the more valuable objects. Other than properly stored museum material that is rarely moved, exposed to light, touched, etc., the use of specimens for teaching is often unsustainable and leads to irreparable damages of original specimens, or it requires regular restoration of models. Moreover, a historically grown teaching collection as in Jena can still contain hidden treasures such as unlabelled type specimens or otherwise scientifically or historically important material. For example, a fish specimen was found in the teaching collection in 2003, labelled *Pegasus chiropterus*, Nova spec. Haeckel. While such a label might suggest at first sight that this fish could be a type specimen designated by Haeckel, it is important to know that sometimes specimens have been labelled as "new species"—simply because they were personally unknown species (Dietrich von Knorre, personal communication). The *Pegasus* specimen is probably the one that was illustrated by Haeckel in the famous "Kunstformen der Natur" and has therefore a historical value. After its discovery, the specimen was immediately transferred to the scientific collection. However, since the taxon has not formally been described or marked as new species in his book, *P. chiropterus* has never been treated as a valid taxon by the scientific community.

Fig. 38.3 Damaged specimen of a gar species (*Lepisosteus* sp.). *Scale bar*: 10 cm (Photo: © with G. Brehm 2015)

The use of poisonous substances could also be problematic when humans are exposed, for example, to emitted gases or to bird skins treated with arsenic (Chap. 37). By means of selected examples, we illustrate some major problems for the objects arising from the intensive use in a teaching collection:

(a) Despite its hardness, the skin of a gar (*Lepisosteus* sp.) preparation is so severely damaged that the inner parts are now well visible (Fig. 38.3). The filling consists of cell glue and chalk, wrapped with hemp fibre. This indicates that the objects stem from the nineteenth century.
(b) Wax models were produced in the nineteenth and beginning of the twentieth century by Ziegler (Freiburg) and Sommer (Sonneberg) manufacturers and are still famous for their grace (Fig. 38.4). Despite their fragility, they "survived" their use in the teaching collection, but not without damage. Our illustrated wax model demonstrates the development of the skull of a short-beaked echidna (*Tachyglossus aculeatus*). It is still used for teaching, but was restaurated in 2013 by SOMSO Modelle GmbH, Coburg.
(c) Werner Spalteholz (1861–1940) developed a method to stain blood vessels with stained gelatine, combined with a complete dehydration of the tissue (Spalteholz 1914). The illustrated preparation of part of a human stomach (Fig. 38.5) is stored in a specific mixture of benzyl benzoate and methyl salicylate. Although the results are impressive visually, problems can arise through small leaks of the vessel that allow the potentially poisonous chemicals to effuse.
(d) A modern mounted skeleton of a goshawk (*Accipiter gentilis*) from the year 1996 (Fig. 38.6) is equipped with a flexible steel wire. This allows a "user-friendly" and comfortable use in teaching. Older and inflexible preparations can break more easily, particularly fragile bird bones. Modern legislation poses further challenges for such objects because birds of prey are generally legally protected in Germany.
(e) A folder of microscopic slides contain originals from the late nineteenth century from excursions made by Haeckel and students, e.g., to Messina/Italy (Fig. 38.7). Microscopic slides were frequently used for teaching and often

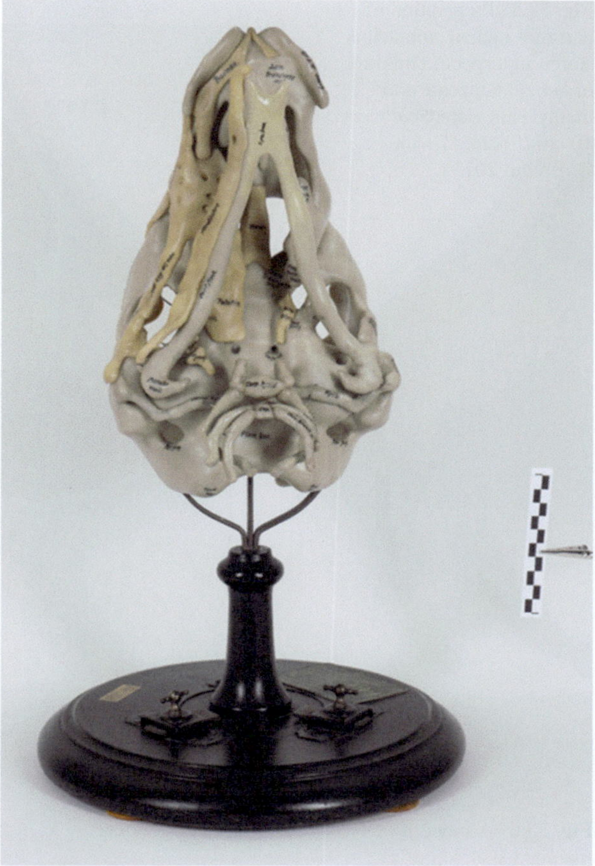

Fig. 38.4 The historical wax model demonstrates the development of the skull of a short-beaked echidna (*Tachyglossus aculeatus*). *Scale bar*: 10 cm (Photo: © with G. Brehm 2015)

included highly valuable or even type material. Due to inappropriate use, many slides broke and were discarded.

38.3 Housing, Maintenance and Inventory

The teaching collection is still mostly housed in three rooms neighbouring the large zoological lecture hall. Gerta Puchert was in charge of the collection until May 2016. She was supported by the technical staff of the Phyletisches Museum. Conservational problems are similar as described in Chap. 37.

The microscopic slides of the teaching collection are currently deposited in two drawers in the upper floor of the Phyletisches Museum (Room D207). Numbers of groups and series are indicated on the drawer doors and allow orientation. Loans from this collection to teaching courses are documented in a book where possible damages are noted. Apart from this collection, a new collection of histologic

Fig. 38.5 Preparation of part of a human stomach is stored in a specific mixture of benzyl benzoate and methyl salicylate. *Scale bar*: 10 cm (Photo: © with G. Brehm 2015)

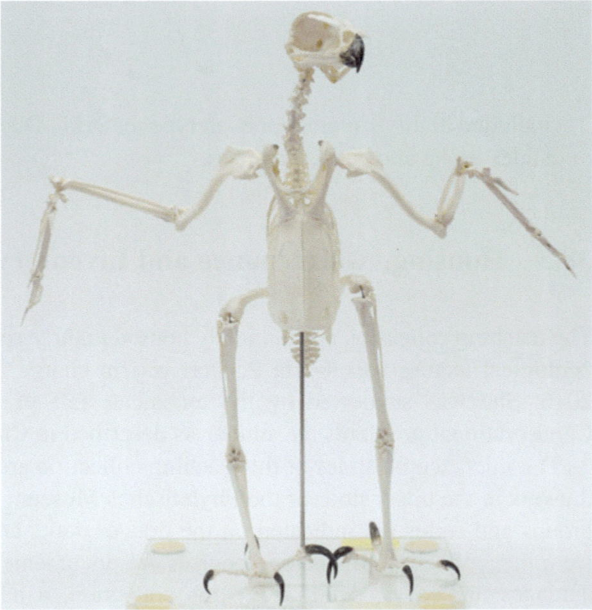

Fig. 38.6 A modern mounted skeleton of a goshawk (*Accipiter gentilis*) from the year 1996, equipped with a flexible steel wire (Photo: © with G. Brehm 2015)

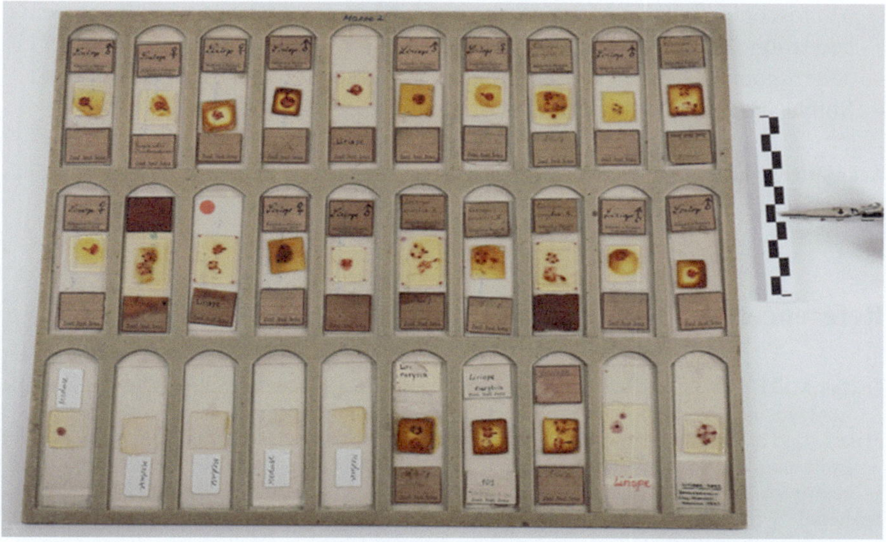

Fig. 38.7 A folder of microscopic slides containing originals from the late nineteenth century from excursions made by Haeckel and students, e.g., to Messina/Italy. *Scale bar*: 10 cm (Photo: © with G. Brehm 2015)

preparations for teaching was established which is deposited at the Institute for Zoology and which is not documented here.

A full inventory of the teaching collection has not been achieved, but lists of preparations of each cabinet exist. A yearly overview inventory has proven to be useful in order to re-establish the arrangement of preparations. It is documented when objects are segregated due to irreparable damage, etc.

An overview inventory of all parts of the teaching collection in 2014 resulted in a total stock of ca. 1450 macroscopic and 10,000 microscopic preparations. In detail, there are:

- 312 wet and dry preparations demonstrating organ systems and developmental biology, among them 14 "Spalteholz preparations".
- 573 wet and dry preparations and fossils from the area of systematic zoology.
- 36 mounted skeletons and 206 parts of skeletons.
- 203 dermoplastics in varying degrees of condition, among them 131 birds and 55 mammals.
- 32 displays of models and preparations (genetics, variability, development).
- 90 models and model series, among them 16 wax model series in varying degrees of condition from the manufacturers Ziegler (Freiburg) and Sommer (Sonneberg).
- On top of and behind cabinets, body models and locomotion models by Prof. Dr. Manfred Zoller, Berlin.
- Three model series: development of a water scavenger beetle (Hydrophilidae), of a lancelet (Branchiostomatidae) and of a frog (Ranidae).

– Microscopic slides: 120 series, mostly used in the first years of the students' education (3400 preparations).
– Microscopic slides: 24 series in folders as a reserve (2844 preparations).
– Single slides and small series for particular practical studies.

Acknowledgements We thank Dietrich von Knorre and Martin S. Fischer for proofreading this chapter.

References

Fischer MS, Brehm G, Hoßfeld U (2008) Das Phyletische Museum in Jena. Institut für Spezielle Zoologie und Evolutionsbiologie, Friedrich-Schiller-Universität Jena, Jena

Plate L (1912) Rede zur Einweihung des Erweiterungsbaues des Zoologischen Instituts und zur Eröffnung des Phyletischen Museums der Universität Jena, gehalten am 21. Mai 1912. Naturwiss Wochensch Neue Folge 11(30):465–470

Spalteholz W (1914) Über das Durchsichtigmachen von menschlichen und tierischen Präparaten und seine theoretischen Bedingungen nebst Anhang: Über Knochenfärbung. Hirzel, Leipzig

Chapter 39
KAISERSLAUTERN: Zoological Collection of the University of Kaiserslautern

Jürgen Kusch

Abstract The Zoological Collection of the University of Kaiserslautern was established in 1973 by Prof. Dr. Erwin Tretzel (28 May 1920–28 April 2001), who was a renowned arachnologist and ornithologist. Most of the specimens of the Kaiserslautern collection were prepared by the taxidermist Burghardt Hasenbein, beginning in March 1973 until his retirement in 2011. Today the Zoological Collection of the University of Kaiserslautern comprises about 35,000 specimens of approximately 4000 species. These species belong to invertebrate zoology (insects, arachnids, crustacean, molluscs and others) and vertebrate zoology (mainly birds and mammals, few reptiles, amphibians and fish). The majority of the zoological specimens originate from Germany. Geographically the emphasis is on species that come from the Palatinate region in southwestern Germany. This region belongs to the UNESCO biosphere reserve of the Palatinate Forest/North Vosges.

Specimens cover terrestrial, freshwater and marine habitats. Vertebrates include large numbers of most of the species of small mammals and birds occurring in Germany. While the majority of the vertebrate specimens are kept as cabinet skins, there are also many mounted specimens of different shapes and sizes shown in permanent exhibitions. An osteology collection of fish, amphibians, reptiles, birds and mammals contains skulls and skeletons, with many of them being mounted. The Zoological Collection of the University of Kaiserslautern offers resources mainly for teaching purposes but also for research objectives in biodiversity and ecology. Collections of butterflies and moths (Lepidoptera) from the Palatinate region, established by Rudolf Heuser between the 1920's and 1970's, are of particular scientific value. An extensive collection of caddisflies from the Palatinate Forest, mostly larvae, serves as a resource for current ecological research topics.

Keywords Zoological collection Kaiserslautern • Invertebrates • Vertebrates • Marine • Freshwater • Terrestrial • Lepidoptera

---

J. Kusch (✉)
Department of Biology, University of Kaiserslautern, Erwin-Schroedinger-Str. 14, 67663 Kaiserslautern, Germany
e-mail: stoeck@rhrk.uni-kl.de

L.A. Beck (ed.), *Zoological Collections of Germany*, Natural History Collections,
DOI 10.1007/978-3-319-44321-8_39

39.1 Objective

The Zoological Collection of the University of Kaiserslautern is committed to the collection and preservation of specimens for the purpose of teaching and research. It focuses on specimens and other material representatives for the animal kingdom and serves to ensure that resources for education are effectively utilized. The teaching collection is kept for use in courses at the university. Courses include biodiversity, identification of species, morphology, systematic zoology, general or aquatic ecology, functional ecology, applied ecology, environmental education as well as others. The teaching material comprises various animal phyla from diverse habitats, which are native or collected worldwide. Illustrative material (mounted specimens, skeletons, shells) of vertebrates and invertebrates is exhibited in several large glass showcases (Fig. 39.1).

Permanent research collections are also maintained for use by researchers within the university. These collections identify and support species records, biodiversity research and conservation in the UNESCO biosphere reserve of the Palatinate Forest/North Vosges, where the University of Kaiserslautern is located.

Fig. 39.1 Exhibition room with large glass cases showing mounted specimens of birds and mammals

39.2 Collection Areas

39.2.1 Vertebrates

A small fish collection is available, preserved in alcohol or as skeletons. In addition, there are several fish skeletons mounted in glass cases. Freeze-dried specimens and skeletons of amphibians and reptiles are available.

39.2.1.1 Mammals

Key aspects of the Zoological Collection of the University of Kaiserslautern are specimens of small mammals (Fig. 39.2). The collection represents most of the native species of Germany. A total of about 1500 mammal specimens belong to 103 species (including non-natives). For teaching purposes, most small mammals are available in dozens of skin and skull specimens (Fig. 39.3), e.g., insectivores (Erinaceidae, Soricidae, Talpidae), rodents (Arvicolidae, Cricetidae, Gliridae, Muridae, Sciuridae), bats (Chiroptera: Vespertilionidae) or carnivores (Mustelidae). Among these species, there are 35 specimens of the European

Fig. 39.2 Mounted specimens of garden dormouse (*left*) and of European hamster (*right*)

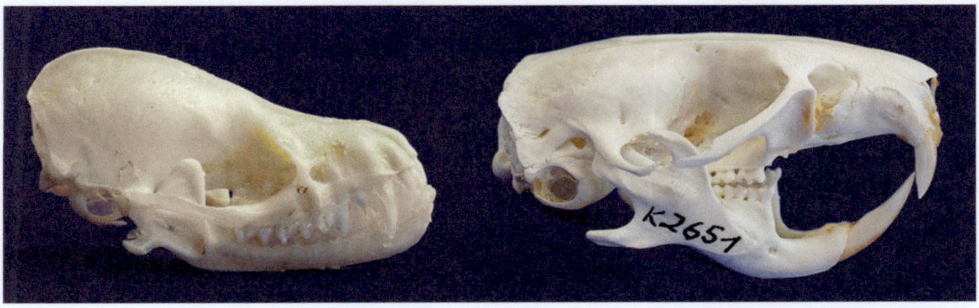

Fig. 39.3 Skulls of the greater mouse-eared bat (*Myotis myotis*; *left*) and European hamster (*right*) (not by scale)

hamster (*Cricetus cricetus*), for instance. These were caught by professional hunters in the 1970's to deplete their populations in cultivated areas. Today this species is strongly endangered in Germany and is protected by law.

Skulls are available from medium- to large-sized mammals, like carnivores (Canidae, Felidae, Mustelidae, Phocidae, Ursidae), lagomorphs, even-toed ungulates (Artiodactyla: Suidae, Bovidae, Cervidae) or odd-toed ungulates (Perissodactyla). The collection also holds skeletons of larger native species, e.g., brown bear (*Ursus arctos*) and red deer (*Cervus elaphus*), and skeletons of species of other parts of the world, e.g., red-necked pademelon (*Thylogale thetis*), southern tamandua (*Tamandua tetradactyla*), European bison (*Bison bonasus*), ring-tailed lemur (*Lemur catta*), bonnet macaque (*Macaca radiata*) or common chimpanzee (*Pan troglodytes*). Mounted specimens of small- and medium-sized species as well as skeletons are shown in glass showcases (Fig. 39.1).

39.2.1.2 Birds

The bird collection numbers about 213 species with 1137 specimens. It includes many examples of birds that are endangered or rare in Germany, such as peregrine falcon (*Falco peregrinus*), Tengmalm's owl (*Aegolius funereus*), hazel grouse (*Tetrastes bonasia*), lesser spotted woodpecker (*Picoides minor*), European bee-eater (*Merops apiaster*), ortolan (*Emberiza hortulana*) or bearded reedling (*Panurus biarmicus*). The collection houses mounted birds (Fig. 39.4) of most of the bird orders occurring in Germany. It is representative of the common German species. The majority of birds are skin specimens that are intended for teaching purposes. Most specimens stem from nearby locations of Kaiserslautern, yet many marine species are from northern Germany including Helgoland. The mounted collection also includes several glass showcases of high-quality taxidermy, exhibiting birds of different systematic groups and habitats. Three of these glass-fronted cases show various passerine birds.

Fig. 39.4 Mounted specimens of Tengmalm's owl (*left*) and Eurasian woodcock (*right*) (not by scale)

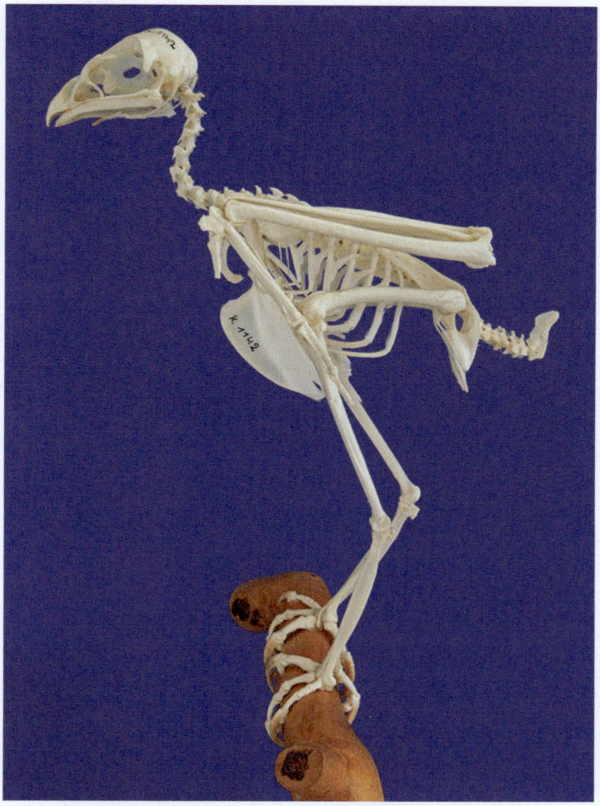

Fig. 39.5 Skeleton of the common buzzard (*Buteo buteo*)

The bird collection includes skulls and skeletons (Fig. 39.5), e.g., mounted skeletons of mute swan (*Cygnus olor*), sparrowhawk (*Accipiter nisus*) or hawfinch (*Coccothraustes coccothraustes*) or skeletons of foreign species like ostrich (*Struthio camelus*), greater rhea (*Rhea americana*) or African penguin (*Spheniscus demersus*).

39.2.2 Invertebrates

Alcohol-preserved specimens of crustacean are available (Fig. 39.6), and few are mounted, dry specimens. Crustaceans cover terrestrial, freshwater and marine species. Spiders of 320 native species are preserved in alcohol. Several spider species are available in a large number of specimens for teaching courses.

Shells of approximately 350 mollusc species were collected from terrestrial habitats, freshwater or marine coasts (Fig. 39.7). Marine species stem from the North Sea, Atlantic Ocean, Mediterranean Sea or tropical seas. Teaching material comprises a large number of species from Germany, with many shell examples of each species being available.

Fig. 39.6 Specimens of marine crustacean

Fig. 39.7 The common slipper shell (*Crepidula fornicata*), native to the western Atlantic Ocean, was introduced to Europe together with the eastern oyster

39.2.2.1 Insects

These collections comprise 28,000 specimens reflecting approximately 3000 species. They focus on the Palatinate region in southwestern Germany, while others belong to the Mediterranean area or the tropics. Specimens cover the major insect orders with a particular focus on Lepidoptera, Trichoptera and Coleoptera. The

Fig. 39.8 Specimens of insects (common scorpionfly, *Panorpa communis*, *left*; rose chafer, *Cetonia aurata*, *right*) are preserved in alcohol for basic teaching courses

Fig. 39.9 Lepidoptera were collected within the Palatinate region (Germany) between the 1920's and 1970's by Rudolf Heuser. The collection comprises small moths of diverse families, besides butterflies. Shown here on the *left* is the scotch grass-veneer (*Catoptria permutatella*, Pyralidae) and the bird-cherry ermine (*Yponomeuta evonymellus*, Yponomeutidae) on the *right*

acquisition of specimens over the last 40 years has allowed us to assemble a regionally important collection of insects from the UNESCO biosphere reserve of the Palatinate Forest/North Vosges. The collections support biodiversity research and conservation. Other insect collections are used for teaching (Fig. 39.8).

A comprehensive collection of butterflies and moths (Lepidoptera) from the Palatinate region was established by Rudolf Heuser between the 1920's and the 1970's (Heuser et al. 1957–1971) and is now of an outstanding scientific value. This collection includes, e.g., Crambidae, Gelechiidae, Pyralidae, Tortricidae, Yponomeutidae and many other families of small moths (Fig. 39.9).

A collection of caddisflies, predominantly in larval stages, was established in recent times (Fig. 39.10). It forms the basis for ecological research on climate change ecology or on adaptive mechanisms developed during species evolution

Fig. 39.10 Larval stages of caddisflies with their caddies from waters of the Palatinate Forest. From *left* to *right*: *Odontocerum albicorne*, *Sericostoma personatum*, *Drusus annulatus*, *Silo pallipes* (*above*), *Limnephilus rhombicus*, *Chaetopteryx villosa*, *Brachycentrus montanus*, *Glyphotaelius pellucidus* (*below*)

(Kusch 2015). The collection includes 84 species, mainly originating from streams of the Palatinate Forest with approximately 1000 specimens. The collection currently is a main focus.

39.3 Perspective

Forty years after the foundation of the Zoological Collection of the University of Kaiserslautern, it still fulfils extensive tasks in support of teaching and research at the University of Kaiserslautern. Today, due to lack of personnel, the vertebrate collections can no longer be renewed or expanded, which may, over time, worsen teaching tasks. Furthermore, the databased registration of collections constitutes a major problem. Most of the butterflies, but not moths, are digitally catalogued. Data of caddisflies (Trichoptera) and bees (Apidae) are completely digitized with geographical coordinates of localities. Other collections are catalogued on paper only. Unavailability of staff is also the main problem with this unpleasant situation.

Acknowledgements I thank Burghardt Hasenbein for the valuable help with summarizing details of the Zoological Collection of the University of Kaiserslautern. I am also grateful to Bärbel Becht-Hasenbein for a revision of the English manuscript.

References

Heuser R, Jöst H, Roesler R (1957–1971) Die Lepidopteren-Fauna der Pfalz, Teile I–V. Mitteilungen der Pollichia, Bad Dürkheim

Kusch J (2015) Interacting influences of climate factors and land cover types on the distribution of caddisflies (Trichoptera) in streams of a central European low mountain range. Insect Conserv Divers 8:92–101

Chapter 40
LEIPZIG: Naturkundemuseum Leipzig: Museum of Natural Sciences

Jana Domaratius and Ronald Schiller

Abstract The Naturkundemuseum Leipzig is one of the major institutions in Saxony, where the treasures of our local nature, the flora and fauna, habitats, their geological foundations and the beginnings of the settlement history of the region have been well documented and are still documented.

Keywords Regional collections • Collections of the Zoological Museum of the University of Leipzig • Herman H. ter Meer • Eduard Poeppig • Peter Simon Pallas • Extinct bird species

The Naturkundemuseum Leipzig was founded in 1906 as a "local natural history museum" by the former "Scientific Association of Leipzig's teachers". As a regional museum, the focus of the collecting activities was in the region of Northwest Saxony. Today there are geological, archaeological and biological collections and archives with photographs, maps and graphics for landscape history, which were founded at this time. Since 1930, the museum is an institution of the city of Leipzig. Particularly the zoological collections were significantly enhanced through the taking over of parts of the collections of the Zoological Museum of the University of Leipzig in the year 1968. So today preparations by Peter Simon Pallas and Eduard Poeppig as well as large animal dermoplastics of Herman H. ter Meer are part of the collection. Currently the museum is in a phase of restructuring. The museum is expected to be reopened at a new location in 2020. The profile and the main emphasis of the permanent exibition will be modernized and adjusted to the new environment. The scientific collections moves completely into the new building.

Currently there are nine employees working at the museum, including a botanist, a zoologist and a taxidermist for vertebrates. The house has a modern preparation workshop, 800 m^2 permanent exhibition space and 80 m^2 special exhibition area. It is financed by the city of Leipzig. In the permanent exhibition, large animal plastics,

J. Domaratius (✉) • R. Schiller
Lortzingstr. 3, 04105 Leipzig, Germany
e-mail: jana.domaratius@leipzig.de; http://www.naturkundemuseum.leipzig.de/; ronald.schiller@leipzig.de

L.A. Beck (ed.), *Zoological Collections of Germany*, Natural History Collections,
DOI 10.1007/978-3-319-44321-8_40

Fig. 40.1 View of the permanent exibition: Tertiary in North West Saxony

mainly manufactured by ter Meer, are arranged. Also the regional witnesses of geological history with a focus on Tertiary and Quaternary and the settlement history of the region by the people and the wildlife of Northwest Saxony are shown. Three to four special exhibitions per year are designed on various topics mainly by the employees themselves (Fig. 40.1).

In the vertebrate collections, there are approximately 1.600 species and subspecies which are evidence-based in about 15.000 pieces, including a number of historically important scientific objects. With 232 preparations, the Naturkundemuseum Leipzig has the largest inventory of dermoplastics by the taxidermist Herman H. ter Meer, who has worked from 1907 to 1934 in Leipzig (Becker 2004). Among there are preparations of great auk (*Alca impennis*), passenger pigeon (*Ectopistes migratorius*), huia (*Heteralocha acutirostris*) and owl parrot (*Strigops habroptilus*) that have been newly established by him. Some evidences of Peter Simon Pallas are the oldest objects in the collection (Meyer 2003). In addition, the museum has 38 bird and 6 mammal specimens of Eduard Poeppig (Joost and Meyer 2000).

Furthermore the stock includes an extensive historical bird egg collection (550 species and 5.073 scrim), mainly from the Palearctic, and even evidence material of protected species from the region, because dead material of endangered vertebrates has been collected, analysed and archived for decades. For mammalian fauna of Saxony (Hauer et al. 2009) and for many other regional faunistic publications and researches, contributed data were made available.

The main emphasis of the invertebrate collections is on insects and molluscs from Eastern and Central Germany. Overall, around 180.000 samples and individual objects are present, as well as 11.000 documents of native snails and shells and 75.000 specimens of Central European butterflies. More extensive regional collections are also available for beetles, bugs, dragonflies and grasshoppers. Data from the various collections were incorporated into a variety of regional faunistic publications and studies, for example, Gebert (2006), Reinhardt et al. (2007, 2011), Rößner (2012), Schönborn (2011) and Zeißler (1999). A current focus is to investigate the butterfly fauna of Northwest Saxony. For butterfly monitoring in Germany, the Naturkundemuseum Leipzig works together with the Centre for Environmental Research Leipzig—Halle.

The museum releases the "Publications Naturkundemuseum Leipzig". The library is in regular discussions and exchanges with other publishers of scientific relevant literature. The stock of publications, magazines and books is growing steadily.

In the museum, 11 working groups and clubs are established whose priorities include, for example, the regional processing of birdlife, botany or world of insects. Even a young group of geologically interested children are meeting once a month.

For the audience, several forms of events like public lectures and tours are organized. Regularly there are collaborations between different institutions like other museums, educational institutions or regional conservation organizations. Particularly guided tours in the exhibitions and excursions in nature belong to the schedule. For children and young people especially during public holidays, an extensive programme is offered. Two salaried museum educators are responsible for these programmes.

Interest groups, friends and supporters are regularly provided with all the information via email or newsletters. Various channels and social networks are frequently used. All information is contained on the website.

References

Becker C (2004) Wie ein zweites Leben—Der Tierbildner Hermann H. ter Meer. Passage-Verlag, Leipzig, p 119

Gebert J (2006) Die Sandlaufkäfer und Laufkäfer von Sachsen, Teil 1 (Carabidae: Cicindelini–Loricerini). In: Klausnitzer B, Reinhardt R (Hrsg.) Beiträge zur Insektenfauna Sachsens Band 4. Entomologische Nachrichten Berichte Beiheft 10. Dresden, p 180

Hauer S, Ansorge H, Zöphel U (2009) Naturschutz und Landschaftspflege—Atlas der Säugetiere Sachsens. Sächsisches Landesamt für Umwelt, Landwirtschaft und Geologie (Hrsg.). Dresden, p 416

Joost W, Meyer M (2000) Von Eduard Poeppig während seiner Amerikareise 1822 bis 1832 87 gesammelte Wirbeltiere im Naturkundemuseum Leipzig. Veröff Naturkundemus Leipzig 19:148–173

Meyer M (2003) Wirbeltierpräparate aus den zoologischen Sammlungen von Peter Simon Pallas (1741–1811) im Naturkundemuseum Leipzig. Veröff Naturkundemus Leipzig 22:46–59

Reinhardt R, Sbieschne H, Settele J, Fischer U, Fiedler G (2007) Tagfalter von Sachsen. In Klausnitzer B, Reinhardt R (Hrsg.) Beiträge zur Insektenfauna Sachsens Band 6. Entomologische Nachrichten Berichte Beiheft 11. Dresden, p 696

Reinhardt R, Fischer U, Graf F, Klemm R, Schiller R, Sobczyk T, Stöckel D (2011) Checkliste der Schmetterlinge Sachsens (Lepidoptera). 2. Auflage (Stand 31.12.2010). In: Klausnitzer B, Reinhardt R (Hrsg.) Beiträge zur Insektenfauna Sachsens Band 14. Mitteilungen Sächsischer Entomologen, Supplement 10:1–110

Rößner E (2012) Die Hirschkäfer und Blatthornkäfer Ostdeutschlands (Coleoptera: Scarabaeoidea). Verein der Freunde und Förderer des Naturkundemuseums Erfurt e.V. Erfurt, p 508

Schönborn C (2011) Schmetterlingsfauna Sachsen-Anhalts, Band 1—Spanner (Geometridae). Weissdorn-Verlag, Jena, p 352

Zeißler H (1999) Molluskenfauna von Nordwestsachsen. Veröff Naturkundemus Leipzig 17:1–95

Chapter 41
LEIPZIG: The Zoological Study and Research Collection of the University of Leipzig

Detlef Bernhard

Abstract The zoological collection of the University of Leipzig looks back on a long tradition and a chequered history. The museum was founded in 1837 with its exhibits quickly growing into an extensive and scientifically significant zoological collection. During the 1960's the collection was largely liquidated as most objects were transferred to other museums, while only a small fraction of the original exhibits remained at the university. That remaining part, however, was still quite impressive and could be restored by the beginning of the twentieth century. Today the objects of the zoological collection still serve teaching purposes as well as scientific research, especially the new additions such as DNA and tissue samples. Accordingly, the Institute for Biology in Leipzig is dedicated to its preservation and restoration.

Keywords Rudolf Leuckart • Carl Chun • Valdivia expedition • Herman H. ter Meer • Blaschka glass models

The Zoological Museum of the University of Leipzig was founded by Eduard Friedrich Poeppig (1798–1868) in 1837. Poeppig studied medicine and natural history at the University of Leipzig (Fig. 41.1). During the 10 years following his graduation, he travelled extensively throughout North and South America to explore the botany and zoology of the continent. His longest journey lasted 5 years (1827–1832) and took him across the Andes and downstream to the mouth of the Amazon. He later published an account of his journey in a book, which ranks among the most impressive descriptions of nature along with those by Humboldt (Müller 1998). After his return in 1832, he was appointed Professor of Natural Sciences and later became the first Professor of Zoology at the University of Leipzig. The numerous collections from his expeditions provided the foundation for the Zoological Museum in Leipzig. In the following years, the collections grew,

D. Bernhard (✉)
Talstr. 33, 04103 Leipzig, Germany
e-mail: zoosa@uni-leipzig.de; http://home.uni-leipzig.de/agspzoo/Zoologische%20Sammlung.html

L.A. Beck (ed.), *Zoological Collections of Germany*, Natural History Collections,
DOI 10.1007/978-3-319-44321-8_41

Fig. 41.1 Eduard Friedrich Poeppig (with permission from Archives of the University of Leipzig)

and only 10 years later (in 1847), they were moved into the Augusteum, the newly constructed main building of the University of Leipzig.

In 1869, after Poeppig's death, Rudolf Leuckart (1822–1898) was appointed Professor of Zoology and Director of the Zoological Museum in Leipzig (Fig. 41.2). He was an outstanding nineteenth-century zoologist and is widely recognized as the Founder of Parasitology. Meanwhile, the rooms in the centre of Leipzig had become insufficient for the large zoological collection. So Leuckart promoted plans for a new building, and in 1880, only 2 years after the approval, the new zoological institute and museum were inaugurated at the site where they still stand today (Fig. 41.3). At that time, the zoological collection comprised about 60,000 exhibits (Naumann 2009). Leuckart was an eminent teacher. He established a separate teaching collection, and he is also renowned for his beautifully painted wall charts, which were used for teaching around the world. Unfortunately, only a few of these charts are still part of the collection. Under Leuckart's guidance, both the Zoological Museum and the Zoological Institute gained worldwide recognition.

In 1898 Carl Chun (1852–1914) became Leuckart's successor. He was Leuckart's student and later became his assistant before being appointed Professor of Zoology in Königsberg and Breslau (Fig. 41.4). Carl Chun was interested in marine and deep-sea zoology. His name is particularly associated with the first German deep-sea expedition

Fig. 41.2 Rudolf Leuckart (with permission from Archives of the University of Leipzig)

Fig. 41.3 The entrance of the zoological institute (with permission from University of Leipzig. Photo: Marion Wenzel)

Fig. 41.4 Carl Chun. Archives of the University of Leipzig

on board the ship Valdivia in 1898–1899, which was initiated and led by him. Under Chun's guidance the zoological collection expanded, and a new studio for taxidermy was established. He succeeded in enlisting Herman H. ter Meer (1871–1934), one of the leading taxidermists of his time, who had been offered positions at museums worldwide. Ter Meer's first assignment in Leipzig was the preparation of objects collected during the Valdivia expedition, such as a scene of the Kerguelen Islands with two female elephant seals in its centre. He prepared hundreds of objects in Leipzig, most of them for the museum in Leipzig, but also for other museums. The most famous of these objects were big cats and great apes (see Becker 2004).

Under Poeppig, Leuckart and Chun, the Zoological Museum of the University of Leipzig gained major significance and worldwide recognition. Over the following years, the scientific focus of the professors at the Zoological Institute underwent a change, but the Zoological Museum retained its significance for decades to come. During World War II, the building that housed the institute and the museum remained undamaged in contrast to the original Botanical Institute, which was completely destroyed along with its notable herbarium. As a consequence the Botanical Institute was moved into the building of the Zoological Institute after the war. This necessitated the first reduction in space for the zoological collection. Further restrictions in later years led to an almost complete dissolution of the extensive and valuable collection during the 1960's. Most objects were transferred to the *Museum für Naturkunde* in Berlin, the *Museum für Tierkunde* in Dresden and the *Naturkundemuseum* in Leipzig. 232 exhibits by ter Meer, for example, including professionally restored mounts of the extinct great auk (*Pinguinus impennis*), passenger pigeon (*Ectopistes migratorius*) and huia (*Heteralocha acutirostris*) are now located at the *Naturkundemuseum* in Leipzig (Meyer 2007).

Some of the objects, however, were lost during that time. What was left of the collection was temporarily stored in cupboards, in cubicles or in the attic. Due to storage under these inadequate conditions, many objects became soiled or even damaged. Valuable pieces were left lying around neglected. It was with much surprise, for example, that type material of the glass sponge *Monorhaphis chuni*, stemming from the Valdivia expedition, was discovered in an old cabinet during the restoration of the collection in 2002.

With the reconstruction of the institute building at the beginning of the twentieth century, the entire remaining collection had to be packed up and transferred to interim storage. In April 2005, almost 40 years after its dissolution, the zoological collection was returned and put on display at its original location, owing to the help of many enthusiastic students. The collection is now divided into the exhibition area on the ground floor and the repository in the attic. The exhibition area comprises two large rooms, one showing examples of local fauna, while the other presents animal systematics (Fig. 41.5). In a smaller third room, models and selected objects are on display. Today the exhibition rooms serve educational purposes, but are open on request to the interested public or to school classes. Once a year the zoological collection is part of the "Public Night of Museums" in Leipzig and attracts up to 1200 visitors each year.

Restoration was followed by the complete archiving of all objects, which finally made the collection available for study purposes. Even in its present condition, the remnants of the formerly famous zoological collection are still impressive, as they number more than 2500 objects, the insect collection excluded. They comprise

Fig. 41.5 View into one of the exhibition rooms (with permission from University of Leipzig. Photo: Marion Wenzel)

taxidermy mounts, some by ter Meer, along with skeletons, skulls and numerous specimens preserved in fluid (Fig. 41.6). Even a handful of objects from the Valdivia expedition have survived along with several other historically significant exhibits, such as models made of gelatine, papier maché and wax, which came from the workshops of Paul Osterloh and Paul Loth in Leipzig at the end of the nineteenth century. Furthermore, a few Blaschka glass models can also be found in the collection. The collection covers virtually all groups of animals (Fig. 41.7), yet one half of the objects are vertebrates, among them about 300 taxidermy mounts of birds and mammals. Many birds come from purchases under Poeppig and Leuckart. There are objects from all regions of the earth. The oldest were acquired by Poeppig and date back to the end of the eighteenth century. The considerable insect collection comprising more than 100 boxes with primarily beetles, but also with bugs and Hymenoptera, should also be mentioned. The insects were collected by Hermann Dietze (1889–1980), a teacher from Leipzig, from the 1920's to the 1960's of the last century. They represent an important record of the regional fauna for present and future studies.

Due to their age and inadequate previous storage, many objects of the historical collection are in poor condition. Unfortunately, direct financial support for the collection has been withdrawn. Therefore, restorations as well as current expenses must be paid for by the scientific working group responsible for the collection. Owing to limited funds, however, some of the work must be put off to the future.

Fig. 41.6 Southern two-toed sloth (*Choloepus didactylus*). Taxidermy mount made by ter Meer (with permission from University of Leipzig. Photo: Marion Wenzel)

Fig. 41.7 *Hydractinia echinata*. Glass model made by Blaschka (with permission from University of Leipzig. Photo: Marion Wenzel)

Many objects of the historical zoological collection are still used in various courses for students of biology and medicine. Accordingly, the zoology collection is an indispensable component of both learning about the local fauna and of understanding morphology, evolution and the development of animals. In addition, this part of the collection provides data for local checklists, the Red Lists, and is also used for identifying difficult species by direct comparison. This, however, is restricted to a few insect groups due to the incompleteness of the remaining collection and the loss of inventory records.

A new collection, which was established in 2000, serves a variety of scientific purposes. This part of the current collection encompasses voucher specimens from several different research projects. Most of them originate from entomological and faunistic investigations in the Leipzig area (e.g., Unterseher et al. 2007). But they also include vertebrates, e.g., within the framework of a study of the mammals of Saxony (Hauer et al. 2009) or from investigations into the causes of death in various mammals (e.g., beavers, bats, otters). Also included in this part of the collection are tissue and DNA samples from numerous research projects. The DNA and tissue collection is the part of the zoological collection which is most frequently used for scientific purposes. It is also the one most commonly requested by the outside community.

Each year, between five and eight bachelor's, master's and doctoral theses which are in some way linked to the zoological collection are completed in the Zoological Institute. Currently, the new collection comprises about 5000 entries for vertebrates and about 2700 DNA and tissue samples.

The Zoological Museum of the University of Leipzig is an example of an exceptional rise followed by a deplorable decline, which led to the almost complete loss of the collection over several decades. Also, the collection is a good example of the revival of a small museum, which has found its place in a modern, future-oriented University Institute of Biology.

References

Becker C (2004) Wie ein zweites Leben: Der Tierbildner Herman H. ter Meer. Passage-Verlag, Leipzig

Hauer S, Ansorge H, Zöphel U (2009) Atlas der Säugetiere Sachsens. Sächsisches Landesamt für Umwelt, Landwirtschaft und Geologie, Dresden

Meyer M (2007) Herman H. Ter Meer—ein Präparator von Weltruf. In: Schlatter R (ed) Pro Natura—100 Jahre Naturkundemuseum Leipzig. Leipzig, p 90

Müller GK (1998) Eduard Pöppig—sein Leben. In: Morawetz W, Röser M (eds) Eduard Friedrich Poeppig, 1798–1868, Gelehrter und Naturforscher in Südamerika. Universitätsverlag, Leipzig, pp 1–9

Naumann WW (2009) Zoologie. In: von Hehl U, John U, Rudersdorf M (eds) Geschichte der Universität Leipzig 1409–2009, Band 4. Universitätsverlag, Leipzig, pp 1138–1161

Unterseher M, Morawetz W, Klotz S, Arndt E (eds) (2007) The canopy of a temperate floodplain forest—results from five years of research at the Leipzig canopy crane. Universität Leipzig, Merkur, Leipzig

Chapter 42
LÜBECK: Museum of Nature and Environment, Lübeck, Zoological Collections

Susanne Füting

Abstract The "Museum für Natur und Umwelt"—Museum of Nature and Environment—presents biological and geological themes with emphasis on natural history and biodiversity of the region. The former zoological collections of Lübeck were primarily based on the private collections of the physician and passionate naturalist Johann Julius Walbaum. In 1942, during World War II, the museum's building, together with its natural exhibitions and local and overseas collections, was destroyed.

In 1963, the museum of natural history was reopened as "Naturhistorisches Museum—Museum für Natur und Naturgeschichte in Schleswig-Holstein" in a new building, but at the same place.

Today the collections of the Museum of Nature and Environment, Lübeck, consist of a herbarium, zoological, geological and palaeontological collections. More than 200,000 objects represent local wildlife and natural history. The zoological collections comprise indigenous insects, including various historical beetle collections and a historical collection of butterflies, as well as molluscs, vertebrates and skeletons. Fossil whale skeletons originating from the primordial North Sea (Miocene) are unique and of international significance.

Keywords Museum für Natur und Umwelt Lübeck • Johann Julius Walbaum • Fossil whale skeletons • Miocene

42.1 Museum of Nature and Environment, Lübeck

The "Museum für Natur und Umwelt"—Museum of Nature and Environment—is a civic institution and belongs to an association of local museums called "die Lübecker Museen". The building is situated in the old town of Lübeck, next to the cathedral (Fig. 42.1).

S. Füting (✉)
Museum of Nature and Environment, Musterbahn 8, Mühlendamm 1-3, 23552 Lübeck, Germany
e-mail: susanne.fueting@luebeck.de; naturmuseum@luebeck.de; http://www.museum-fuer-natur-und-umwelt.de

L.A. Beck (ed.), *Zoological Collections of Germany*, Natural History Collections,
DOI 10.1007/978-3-319-44321-8_42

Fig. 42.1 View on museum and cathedral from above (with permission from Museum für Natur und Umwelt, Lübeck; photo: Susanne Füting)

Exhibitions present biological and geological themes on 1,300 square metres. The main emphasis is on natural history and biodiversity of the region, including the Baltic Sea area. Furthermore, temporary exhibitions highlight a wide range of current issues. Our key objectives are environmental and scientific education. We welcome about 32,000 visitors every year (Figs. 42.2 and 42.3).

More than 200,000 objects are included in our collections, documenting the regional flora, fauna and geology. Fossil whale skeletons originating from the primordial North Sea (Miocene) are unique and of international significance.

42.2 Historical Background

The zoological collections of Lübeck were primarily based on the private collections of the physician and passionate naturalist Johann Julius Walbaum (Fig. 42.4). His intention was to teach natural sciences and worship of nature. After his death, his children donated his legacy for educational purposes in 1799. Besides minerals, his collections included many zoological specimens such as fish, seashells, snails, corals, turtle shells, various mammals, birds and insects. Since then, the collections have constantly been expanded.

At first animals, plants and other objects collected worldwide were presented to the public in a "chamber of curiosities". It was not until 1893 that due to the legacy of the local businessman Georg-Ludwig Blohm, the collections could be shown in a new building. The "Museum am Dom" presented natural history, ethnology and art under one roof (Figs. 42.5 and 42.6).

In 1942, during World War II, bombs destroyed the museum's building together with the natural exhibitions and collections.

Afterwards, dedicated citizens of Lübeck started to build up new collections with focus on local nature. In 1963, the museum of natural history was reopened as "Naturhistorisches Museum—Museum für Natur und Naturgeschichte in Schleswig-Holstein" in a new building, but at the same place.

In 1993, geological eras including the ice ages and a large Tertiary collection as well as fossilized whales from the Miocene sea were displayed for the first time in a newly installed exhibition area. Educational work became more and more important so that a room for educational purposes was equipped, and many new educational programmes were introduced. In 1998, the museum's new concept was expressed by changing the name to "Museum für Natur und Umwelt" (see Eckloff and Müller 1999).

The "Merman's Realm" exhibition showing biosphere and inhabitants of the local rivers and the nearby Baltic Sea was opened to the public in 2008 (Fig. 42.7).

Many volunteers and citizen scientists contributed greatly to the collections. Besides helping to expand them continually, they also worked dedicatedly on the collections.

Fig. 42.2 "Museum für Natur und Umwelt" today Lübeck today (with permission from: Museum für Natur und Umwelt, Lübeck; photo: Wolfram Eckloff)

Fig. 42.3 Entrance of the "Museum für Natur und Umwelt" (with permission from: Museum für Natur und Umwelt, Lübeck; photo: Susanne Füting)

Fig. 42.4 Johann Julius Walbaum (with permission from: Museum für Natur und Umwelt, Lübeck; photo: Susanne Füting, Picture of a painting)

Fig. 42.5 Former "Museum am Dom" showing art, applied art, ethnology and natural history (with permission from: Museum für Natur und Umwelt, Lübeck; photo: Susanne Füting)

Fig. 42.6 Apes—exhibits of the former museum (with permission from: Museum für Natur und Umwelt, Lübeck; photo: Susanne Füting, original picture taken by Johannes Nöhring)

Fig. 42.7 View into exhibition "Merman's Realm"—here parts of collection of breeding and migratory birds (with permission from: Museum für Natur und Umwelt, Lübeck; photo: Michael Haydn)

42.3 Recent Collection

Today the collections of the Museum of Nature and Environment, Lübeck, consist of a herbarium, zoological, geological and palaeontological collections. More than 200,000 objects represent local wildlife and natural history. The zoological collections comprise indigenous insects, including various historical beetle collections and a historical collection of butterflies, as well as molluscs, vertebrates and skeletons.

42.3.1 Insects

With approximately 65,000 specimens, all orders of indigenous insects are well represented: beetles, butterflies, dragonflies, Orthoptera, Hymenoptera, bugs and flies from, e.g., Schleswig-Holstein, Hamburg, Lower Saxony and West Mecklenburg.

Beetles were collected by H. Saager, K. Nikoleizig and W. Ziegler. Historical collections were built up by P. Gussmann, L. Benick and G. Benick (Fig. 42.8). A historical collection of butterflies was the work of collector K. F. Marquardt. K. Lunau contributed historical collections of dragonflies and grasshoppers. J. van der Smissen studied and collected bees, wasps and ants especially in the Lübeck region. H. Saager examined Hymenoptera, flies, bugs and other orders of insects.

42.3.2 Molluscs

About 38,200 bivalves and snails are included in historical collections of E. Schermer and W. Hähnel.

Fig. 42.8 Collection of P. Gussmann (with permission from: Museum für Natur und Umwelt, Lübeck; photo: Uli Schmidts)

42.3.3 Vertebrates

Up to 1,700 birds and mammals are part of the collections. Several are shown in the exhibitions.

The different species of breeding and migratory birds of the Lübeck region are almost complete, even several rare vagrants are present (G.v. Studnitz).

42.3.4 Geological and Palaeontological Findings

The geological and palaeontological collections comprise about 15,900 objects including rocks, sands, minerals and fossils. More than 11-million-year-old findings, baleen and toothed whales as well as a shark from the Miocene, were excavated near the small village Groß Pampau and are shown at the Museum of Nature and Environment on the ground floor (see Höpfner 2014). They document the life in the primordial North Sea. Some of these exceptional fossils are holotypes (see f. e. Kazár and Hampe 2014).

42.4 Collections, Exhibitions and Education

Many specimens of the collections are displayed. The exhibitions offer insights into the natural history of Schleswig-Holstein and the rich flora and fauna of the Lübeck area. An exhibition of rocks and drifts invites the visitors to a time journey through the geological eras, and especially children can explore fossils at the staging of a palaeontological excavation site. The fossilized whales from the Miocene sea are significant exhibits. Furthermore, a 14-m-long skeleton of a recent sperm whale is shown (Fig. 42.9). In the nature experience exhibition "Merman's Realm", habitat dioramas and interactive media stations display the biosphere and inhabitants of the local rivers and the nearby Baltic Sea. Living pond turtles, grass snakes, fish and a colony of bees are among the exhibits. Visitors can observe the behaviour of bees and learn about their communication.

Fig. 42.9 Skeleton of recent sperm whale in the yard adjacent to the cathedral (with permission from: Museum für Natur und Umwelt, Lübeck; photo: Susanne Füting)

Fig. 42.10 Excavating whale bones (with permission from: Museum für Natur und Umwelt, Lübeck; photo: Susanne Füting)

Environmental education is a key aspect of all museum activities, and the scientific collections serve as a basis for educational actions. The museum is a certified educational institution of sustainability and offers numerous learning opportunities, i.e., lectures, seminars, vacation programmes, excursions and events.

42.5 Perspectives

At the moment there is no taxidermist to strengthen our museum team. We have to tackle this problem and find a viable solution. Digitalization of the collections has not been finished yet. So one of the challenges of the future will be to continue and complete digitalization and to cross-link Lübeck collections with other databases.

Currently, there is another palaeontological excavation in the gravel pit in Groß Pampau, 45 min from Lübeck City. The museum together with head of excavation Gerhard Höpfner and the excavation team are uncovering the bones of whales from the Miocene together with rich accompanying fauna. Interesting new results are expected (Fig. 42.10).

The museum is a new member of NORe, an association of natural history museums and natural science collections of the North and Baltic Sea area.

Acknowledgements The author would like to thank Klaus Schreiber and Oda Benthien (PhD student) for their support.

References

Eckloff W, Müller S (Hrsg.) (1999) Museum für Natur und Umwelt. Walbaum-Festschrift. Berichte des Vereins "Natur und Heimat" und des Naturhistorischen Museums zu Lübeck. Heft 25/26 Lübeck 1999

Höpfner G (2014) Aus der Urzeit aufgetaucht. Bilddokumentation über sensationelle Wal- und Haifunde aus Schleswig-Holstein. Berichte des Museums für Natur und Umwelt und des Naturwissenschaftlichen Vereins zu Lübeck. Heft 2 a/Lübeck 2014

Kazár E, Hampe O (2014) A new species of Kentriodon (Mammalia, Odontoceti, Delphinoidea from the middle/late Miocene of Groß Pampau (Schleswig-Holstein, North Germany). J Vertebr Paleontol 34(5):1216–1230

Chapter 43
MAINZ: The Zoological Collections of the Mainz Natural History Museum/State Collection of Natural History of Rhineland Palatinate

Carsten Renker, Bettina Henrich, and Uwe Hildebrand

Abstract The oldest parts of the collections at the 'Mainz Natural History Museum/State Collection of Natural History of Rhineland-Palatinate' date back to the year 1834 and had been assembled by the 'Rheinische Naturforschende Gesellschaft' (RNG). Unfortunately, major parts of these old collections were destroyed during a British air raid in 1945.

The museum itself dates back to the year 1910. A major step forward for the gathering of scientifically valuable collections was the foundation of the State Collection of Natural History of Rhineland-Palatinate in 1988. To date the zoological collections contain more than 700,000 specimens with a focus on the fauna of Rhineland-Palatinate.

Keywords Mainz Natural History Museum/State Collection of Natural History of Rhineland-Palatinate • Rheinische Naturforschende Gesellschaft • Javan rhinoceros (*Rhinoceros sondaicus*) • Pyrenean ibex (*Capra pyrenaica* pyrenaica) • Thylacine (*Thylacinus cynocephalus*) • Quagga (*Equus quagga* quagga)

43.1 Historical Background

The oldest parts of the collection date back to the year 1834, founding year of the 'Rheinische Naturforschende Gesellschaft' (RNG). After closure of the University of Mainz in 1798, when all parts on the western side of the river Rhine belonged to France, the citizens of Mainz tried to continue the Age of Enlightenment and struggled against the scientific vacuum by the foundation of this society, one of the oldest still existing natural history societies in Germany and Central Europe. The aim, defined in a preamble, is to collect and properly store and arrange natural

C. Renker (✉) • B. Henrich • U. Hildebrand
Naturhistorisches Museum Mainz/Landessammlung für Naturkunde Rheinland-Pfalz, Reichklarastr. 1 + 10, 55116 Mainz, Germany
e-mail: dr.carsten.renker@stadt.mainz.de; http://www.mainz.de/nhm/index.html

L.A. Beck (ed.), *Zoological Collections of Germany*, Natural History Collections,
DOI 10.1007/978-3-319-44321-8_43

history items, mainly these from the Rhineland, and to organize scientific presentations and talks (Würz 2009). Notary Dr. Carl Friedrich Bruch, founder member of the society, still had a large and international collection of birds at that time, which he sold to the society in 1837. His private collection contained at least 832 specimens of 412 bird species (Kunz 2009). Based on this collection, the society succeeded in collecting birds and mammals, mainly organized by Bruch. He always attempted to receive complete 'families' comprising one male, one female and one juvenile specimen of each species. In 1843 he published a list of mammals, listing 383 species with 520 mounted specimens, 105 skulls and 38 skeletons from all over the world, comprising, for example, 27 species of marsupials (Bruch 1843). The collections were exhibited right from the beginning at two different locations in Mainz before they were moved to the Electoral Palace (Kurfürstliches Schloss) in 1842, where they were shown for the next 60 years as the 'Naturalienkabinett of the RNG'.

Due to the rapidly increasing collections, responsibilities were gradually shifted from the society to the town administration, and first personnel were at least partially paid by the city of Mainz. At the beginning of the twentieth century, a Natural History Museum was arranged in the former monastery 'Reichklara', and based on a contract signed February 10, 1910, all collections of the society became property of the city of Mainz, including the responsibility to care for an orderly administration and accommodation of the collections forever and all time. The city of Mainz confirmed this contract in recent years.

More than 90 % of these first zoological collections were destroyed during a British air raid on February 27, 1945. Only small parts of the mammal collection (i.e., quagga, Javan rhinoceros, thylacine, Pyrenean ibex) survived the bombing. Many objects got lost or were later destroyed due to lacking curatorship between the end of World War II and the reopening of the museum in June 1962.

In the first years, from 1960 until 1979, the new museum vertebrate collections were mainly curated by the taxidermist Ernst Wadewitz, who restored, cleaned and saved many of the old specimens which had survived the air raid. A first inventory of the vertebrate collection after the destructions was conducted in 1980 with approx. 1100 objects.

The reorganization of the entomological collections started with the help of volunteers. Between 1963 and 1966, the pupil and student Hannes F. Paulus took the responsibility for this part of the collection and gave some short information concerning the remaining parts of material after World War II (Paulus 1966). Afterwards, another 14 years passed by before Ulrich Schmidt gained a first official position as scientist and curator of zoology at the museum from 1979 until 2006. Again, the main work in the entomological collections was done by a volunteer, Adalbert Frey, who worked in the collections between 1980 and 1992. Again substantial parts of the 'new' entomological collections, built up by Hannes Paulus, had to be thrown away, due to inadequate storing facilities and lacking curatorship. A major step forward was the foundation of the 'State Collection of Natural History of Rhineland-Palatinate' in 1988 (Fig. 43.1). This had been initiated already in 1979 by Franz Otto Neuffer, director of the museum from 1978 to 2003. Thanks to this

Fig. 43.1 The 'Tower of Time'—main entrance of the Mainz Natural History Museum until 2015

cooperation between the city of Mainz and the state government of Rhineland-Palatinate, subsidies from the state considerably support both the collecting work and scientific activities of the museum.

At the transition from the twentieth to the twenty-first century, the staff responsible for the zoological collections reached its maximum with two taxidermists, two permanent scientists (Ulrich Schmidt, Jürgen Jungbluth) and several trainees, supporting the work. During the last years, personnel have drastically been reduced ending up with two taxidermists, Uwe Hildebrand (since 1977) and Bettina Henrich (since 1989) and one scientist, Carsten Renker (since 2006), now on a 66 % position, being also responsible for the botanical collections.

43.2 Recent Collection

43.2.1 Structure

Due to steadily growing specimen and species numbers in the different parts of the collection, it is difficult or even impossible to give fully reliable numbers. Traditionally, the zoological collections are split up into three parts: (i) the vertebrate collection (inventoried with a W in front of the inventory number), (ii) the entomological collection (E) and (iii) the invertebrate collection (WL), which mainly contains molluscs.

Parts of the collections can be accessed via the 'Global Biodiversity Information Facility' (GBIF) under http://www.gbif.org/dataset/search?q=Naturhistorisches+Museum+Mainz where 9038 georeferenced zoological datasets are available. A more visually based web portal is 'Museum Digital Rheinland-Pfalz' (http://www.museum-digital.de/rlp/index.php?t=institution&instnr=13) showing 209 objects from the zoological collections.

43.2.1.1 The Vertebrate Collection

The most valuable and oldest parts of the collection still contain a set of extinct or almost extinct species (Plate 43.1):

- Three Pyrenean ibexes (*Capra pyrenaica pyrenaica*) (male, female, juvenile) purchased in 1835/1836 by Bruch, who recognized them as distinct species but left the first description to Heinrich Rudolf Schinz (Zurich), who described and figured the species (Schinz 1837, see also Brehm 1844).
- Three quaggas (*Equus quagga quagga*) (male, female, juvenile/foetus) purchased in 1842/1843 by Bruch. The juvenile was almost destroyed in World War II, but restored between 1980 and 1982 together with the two adults which were remounted on new manikins (Rau 1981, 1984, 2004). Samples of the skin tissue were used for the first sequencing of an extinct species (Higuchi et al. 1984) and inspired Michael Crichton to write his book 'Jurassic Park', published in 1990 (Frenz 2012: 164). Finally, discussions arose whether the quagga was just a subspecies of the still existing plains zebra (e.g., Groves and Bell 2004) or a distinct species (Leonard et al. 2005).
- One thylacine (*Thylacinus cynocephalus*) about which almost nothing is known and which was not mentioned by Bruch (1843). Accordingly, the specimen should have been integrated in the collection after 1843.
- One European bison (*Bison bonasus*) from Białowieża, a present of Nicholas I, Emperor of Russia, for Louis II von Hessen und bei Rhein, Grand Duke of Hessen-Darmstadt, who donated the specimen to the Rheinische Naturforschende Gesellschaft in 1841.
- One Javan rhinoceros (*Rhinoceros sondaicus*) from 1842. During restoration of this specimen at the beginning of the 1980's, it was found that the head belonged to a male, the back part to a female specimen. Both were tied together by horseshoe nails.

Plate 43.1 The 'Top Rarities' of the zoological collections at the Mainz Natural History Museum. (**a**) Javan rhinoceros (*Rhinoceros sondaicus*). (**b**) Male Pyrenean ibexes (*Capra pyrenaica pyrenaica*). (**c**) Thylacine (*Thylacinus cynocephalus*). (**d**) Female and (**e**) male quagga (*Equus quagga quagga*) (with permission from: Mainz Natural History Museum)

The vertebrate collection is the only part of the zoological collection which is almost fully inventoried and which contains at the moment (December 2014) *10,502 numbers*. The focus is on birds and mammals.

The birds are the largest part. Based on a 2006 species list, at least *530 bird species* are preserved within the collection, particularly species from Central Europe, but also species from other parts of the world; these are mainly based on specimens from zoos in the region. Most specimens are represented as mounted specimens; Central European species, available in larger numbers, are also stored as specimen skins. A large *oological collection* incorporated in 2011 contains eggs of *463 bird species*. Other items of the bird collection comprise bird's nests and skeletons.

The mammal collection contains *249 species* (status 2006), mainly mounted specimens, specimen skins and skulls or full skeletons. Reptiles, amphibians and fishes are represented by 55, 21 and 31 species, respectively.

Scientifically relevant collections within the bird section are the collection of *Carlo Freiherr von Erlanger* representing mounted bird species mainly from the area around Ingelheim but also from North Africa (mainly Tunisia), collected during the transition from the nineteenth to the twentieth century, some of which are nowadays extinct in the whole of Germany. This collection is a permanent loan by the 'Ingelheimer Museum bei der Kaiserpfalz', since 1993 curated in the Mainz Natural History Museum [Akronym: MNHM (Thiers 2015, preferred abbreviation), also NHMM (Evenhuis 2015)]. Altogether, approx. 700 specimens are represented in this collection (Hildebrand 2004).

The oological collection could be obtained in 2011 from *Jürgen Partenscky* (Karlsruhe), containing 931 clutches of 463 species, of which 708 clutches have been collected by *Wolfgang Makatsch* (Dresden). 890 of these clutches, representing 437 species, have been collected in all parts of Europe. Unfortunately, locality data are most often imprecise.

Scientifically relevant collections within the mammalian section are a collection of specimen skins and skulls of European small mammals (mainly shrews and rodents) collected by *Franz Malec* (Kassel) and donated in 2007. This collection contains 1541 specimens in 23 species.

In 1990 and 1993, a large collection of mainly skulls and skeletons, but also skins, was bought from *Siegfried Eckardt* (Frankfurt a. M.), which altogether contains 2396 objects.

In 2011 the museum received a collection of 38 freshwater fish skeletons from *Jutta Krey*, collected along the river Rhine. These were used to compare recent fishes with subfossil material from the archaeological excavation at Bedburg-Königshoven (North Rhine-Westphalia) in 1987/1988.

Due to the partnership between Rhineland-Palatinate and Rwanda (Africa), the museum, particularly *Ulrich Schmidt*, deputy director of the museum at this time, collected approx. 70 skulls, skeletons and mounted specimens of mammals and birds, but also bird's nests, insects and reptiles (in ethanol) between 1988 and 1992. Some of these specimens were used to equip the Natural History Museum Kigali, also known as 'Kandt House', e.g., male impala antelope, Grant's zebra foal and other zoological objects.

43.2.1.2 The Entomological Collection

A summary on the entomological collections was published in 2009 (Renker and Henrich 2009). A rough estimate ended up with 700,000 insect specimens at this time. Since this publication the following collections or parts of the following collections were integrated (collectors in alphabetical order): *Ernst Blum* (Lepidoptera, 2014), *Klaus Cölln* (Hymenoptera, continuously), *Adalbert Frey* (private collection, Lepidoptera, 2014), *Andrea Jakubzik* (Hymenoptera, continuously), *Friedhelm Nippel* (Lepidoptera, 2014), *Jochen Rodenkirchen* (Lepidoptera, 2014), *Alexander Walland* (Coleoptera, 2011), and *Ernst Zebe* (Coleoptera, Lepidoptera, Orthoptera, 2014), adding at least another 25,000 specimens to the collections. The only collection containing a significant number of type specimens within the whole zoology is the Heteroptera collection of *Hannes Günther* (Koschwitz 2007).

43.2.1.3 The Invertebrate Collection

The invertebrate collection mainly consists of molluscs. The focus is on Central Europe. Almost all species of land gastropods figured in Kerney et al. (1983) are available. Until 1989 the former curator Jürgen H. Jungbluth (co-author of this book) inventoried 5499 specimen series of molluscs. Up to this time, the collection mainly contained the following material: freshwater gastropods collected by *Ragnar Kinzelbach* and co-workers within the frame of different projects (e.g., Meinert and Kinzelbach 1985), unionids collected by *Jürgen H. Jungbluth* and collection *Ulrich Hecker*. Besides this valuable material, a representative set of marine gastropod and bivalve species representing at least all major marine families was bought from the natural history dealer *Jens Hemmen*, with the aim of allowing comparative studies between recent species and the fossil mollusc samples from the Cenozoic Mainz Basin.

During the last years, the extent of the mollusc collection was doubled concerning collection numbers by the integration of the collection by *Friedrich Ehrenfeld* from the first half of the twentieth century, which contains (self-collected) material from all parts of Germany (mainly land gastropods), exchange material from many colleagues (e.g., Kurt Büttner) but also many species from abroad (e.g., Sri Lanka, Madagascar, the Philippines and New Guinea) bought from the natural history dealer Hermann Rolle. Another large set of land gastropods was donated by *Dieter Weber*, collected in caves, caverns and chambers of Luxembourg, Saarland and Rhineland-Palatinate (Renker et al. 2013). Two donations in 2013 and 2014 by *Winfried Engl* enriched the collection with marine (micro-) molluscs from all European seas (mainly Rissoidae, Rissoinidae, Eulimidae, Aclididae, Epitoniidae, Cerithiopsidae, Mangeliidae, Drilliidae, Raphitomidae), which were completely missing in the collection up to this point.

Two other necessary topics within the invertebrate collection exist: (i) the alcohol collection of scorpions collected by *Ragnar Kinzelbach* (Kinzelbach 1982, 1984) which contains 276 series and (ii) the alcohol collection of invertebrates from the river Rhine (Kinzelbach 1985).

43.2.1.4 Infrastructure of the Zoological Collections

The available infrastructure of the whole museum is limited, consisting of a basically equipped workroom for taxidermy and a basically equipped room for microscopy. The magazine has been enlarged considerably in 2014. The library of the museum has an extended literature exchange, receiving almost 500 periodicals (covering all fields of natural sciences) in this frame. Another 500 periodicals are available based on donations. Availability of monographs is limited since the library has only a small budget. Due to the communal responsibility for the museum, it seems to be unlikely that the financial situation will improve within the next years because maintaining cultural institutions isn't a communal duty.

43.2.2 Research

Due to the limited personnel power, the research program is limited. Faunistic survey programs are supported wherever possible. Based on the activities of Jürgen Jungbluth and the 'Projektgruppe Molluskenkartierung', the land and freshwater molluscs of Rhineland-Palatinate were collected by Peter Subai and Paul Schnell based on a 10 × 10 km UTM grid between 1989 and 1996. Parts of this effort have been published by Vogt et al. (1994).

During the last years, the museum was involved in the publication of a dragonfly atlas for parts of Belgium, France, Luxembourg, Saarland and Rhineland-Palatinate (Trockur et al. 2010). Another project was the atlas of grasshoppers, crickets and mantids (Pfeifer et al. 2011).

The museum issues the 'Mainzer Naturwissenschaftliches Archiv' and its supplements on an annual basis. The volumes of the main series contain between 200 and 500 pages of articles covering all fields of natural sciences with a regional focus on Rhineland-Palatinate.

43.2.3 Educational Work

The main activity concerning the educational work of the museum for a larger audience can be seen in the (special) exhibitions, covering all fields of natural sciences. During the last years, the museum showed zoological exhibitions on rats ('Ratten', 2013/2014), poisonous animals ('Gifttiere', 2012/2013) and bats ('Flatterwochen! Fledermäuse—kleine Kobolde der Nacht', 2010/2011) as well as grasshoppers, crickets and mantids ('Heuschrecken—übersehene Schönheiten', 2010). A short survey of the exhibitions in previous years has been summarized by Frankenhäuser (2009).

The museum's educational service has two full-time employees, who offer a wide range of activities for all kinds of individuals and groups with a focus on kindergartens and primary schools. The museum is a well-known place for extra-curricular learning in the region.

Between 1987 and 2006, three Rwandese taxidermists could be trained at the museum: 1987, Jean-Baptiste Kayishema; 1990, Theogene Bosenibamwe; and 2006, Jean-Claude Igiraneza (Schmidt 2009).

References

Brehm CL (1844) Einige naturgeschichtliche Bemerkungen auf einer Reise an den Rhein im September und October 1842. Dritte Abtheilung. Isis 1844(7):484–500 [*Capra pyrenaica*: pp 484–485]

Bruch CF (1843) Verzeichniss der in dem ehemaligen kurfürstlichen Schlosse zu Mainz aufgestellten Säugethiere der rheinischen naturf. Gesellschaft. Erste Abtheilung. Säugethiere und deren Skelette. Mainz, 17 pp

Evenhuis NL (2015) Abbreviations for insect and spider collections of the world. http://hbs.bishopmuseum.org/codens/codens-inst.html. Last accessed 22 Jun 2015 [continuously updated]

Frankenhäuser H (2009) Sonderausstellungen im Naturhistorischen Museum. Mainzer Naturwiss Arch 47:455–461

Frenz L (2012) Lonesome George oder Das Verschwinden der Arten. Rowohlt, Berlin, 350 pp

Groves CP, Bell CH (2004) New investigations on the taxonomy of the zebras genus *Equus*, subgenus *Hippotigris*. Mamm Biol 69(3):182–196

Higuchi R, Bowman B, Freiberger M, Ryder OA, Wilson AC (1984) DNA sequences from the quagga, an extinct member of the horse family. Nature 312:282–284

Hildebrand U (2004) Carlo von Erlanger (Ingelheim) und die Präparatesammlung aus Ingelheim im Naturhistorischen Museum Mainz—Sammlungsbericht. Mainzer Naturwiss Arch 42:267–278

Kerney MP, Cameron RAD, Jungbluth JH (1983) Die Landschnecken Nord- und Mitteleuropas: Ein Bestimmungsbuch für Biologen und Naturfreunde. Verlag Paul Parey, Hamburg und Berlin, 384 pp

Kinzelbach R (1982) Die Skorpionssammlung des Naturhistorischen Museums der Stadt Mainz.—Teil I: Europa und Anatolien. Mainzer Naturwiss Arch 20:49–66

Kinzelbach R (1984) Die Skorpionssammlung des Naturhistorischen Museums der Stadt Mainz.—Teil II: Vorderasien. Mainzer Naturwiss Arch 22:97–106

Kinzelbach R (Hrsg.) (1985) Die Tierwelt des Rheins einst und jetzt. Symposium zum Jubiläum der Reinischen Naturforschenden Gesellschaft und des Naturhistorischen Museums Mainz am 9. November 1984. Mainzer Naturwiss Arch Beiheft 5:183 pp + 3 pls

Koschwitz U (2007) Dr. Johannes (Hannes) Günther. In: Renker C (Hrsg.) Festschrift zum 70. Geburtstag von Hannes Günther. Mainzer Naturwiss Arch Beiheft 31:7–19

Kunz A (2009) Das ornithologische Schaffen des Mainzer Notars Carl Friedrich Bruch (1789–1857) im Überblick. Mainzer Naturwiss Arch 47:89–106

Leonard JA, Rohland N, Glaberman S, Fleischer RC, Caccone A, Hofreiter M (2005) A rapid loss of stripes: the evolutionary history of the extinct quagga. Biol Lett 1(3):291–295

Meinert W, Kinzelbach R (1985) Die Limnischen Schnecken und Muscheln von Rheinland-Pfalz (Mollusca: Gastropoda et Bivalvia). Mainzer Naturwiss Arch Beiheft 4:58 pp + 65 maps

Paulus H (1966) Unser Insektenkabinett. Z Rheinischen Naturforsch Ges Mainz 4:85–87

Pfeiffer MA, Niehuis M, Renker C (Hrsg.) (2011) Die Fang- und Heuschrecken in Rheinland-Pfalz. Fauna Flora Rheinl Pfalz Beiheft 41:678 pp

Rau RE (1981) Zur Geschichte der Präparation des “Mainzer Quaggas”. Mainzer Naturwiss Arch 19:221–236

Rau RE (1984) Über die Restaurierung des Mainzer Quagga Fohlens. Mainzer Naturwiss Arch 22:231–238

Rau RE (2004) Der Mainzer Quagga-Fötus. Mainzer Naturwiss Arch 42:245–257

Renker C, Henrich B (2009) Die Entomologischen Sammlungen des Naturhistorischen Museums Mainz/Landessammlung für Naturkunde Rheinland-Pfalz. Mainzer Naturwiss Arch 47:395–447

Renker C, Weber D, Pohl A (2013) Schnecken aus Höhlen des Großherzogtums Luxemburg. Ferrantia 69:96–107

Schinz HR (1837) Bemerkungen ueber die Arten der Wilden Ziegen, besonders mit Beziehung auf den Sibirischen Steinbock, den Steinbock der Alpen und den Steinbock der Pyrenaeen. Neuchatel, 25 pp, 3 pls

Schmidt U (2009) Das Ruanda-Engagement des Naturhistorischen Museums (1982—heute). Mainzer Naturwiss Arch 47:463–477

Thiers B (2015) Index Herbariorum: a global directory of public herbaria and associated staff. New York Botanical Garden's Virtual Herbarium. http://sweetgum.nybg.org/ih/. Last accessed 27 Jul 2015 [continuously updated]

Trockur B, Boudot J-P, Fichefet V, Goffart P, Ott J, Proess R (2010) Atlas der Libellen/Atlas des libellules (Insecta Odonata). Fauna und Flora in der Großregion/Faune et Flore dans la Grande Région 1:201 pp

Vogt D, Hey-Reidt P, Groh K, Jungbluth JH (1994) Die Mollusken in Rheinland-Pfalz—Statusbericht 1994. Fauna und Flora in Rheinl Pfalz Beiheft 13:222 pp

Würz M (2009) 175 Jahre Rheinische Naturforschende Gesellschaft und 100 Jahre Naturhistorisches Museum Mainz. Mainzer Naturwiss Arch 47:35–88

Chapter 44
MARBURG: Zoological Collection of the Philipps University of Marburg

Elisabeth Hempel and Lothar A. Beck

Abstract The Zoological Collection of the Philipps University of Marburg was founded in 1817–1819 and has existed continuously to this day, surviving both world wars mostly undamaged. It is subdivided into a reference collection, a teaching collection and a public collection. The entire collection consists of approximately 40,000 specimens with over 20,000 insects and other arthropods, 3500 further invertebrates (marine species, terrestrial molluscs, parasites), over 650 birds and mammals, about 300 reptiles and amphibians and more than 630 skeletons, animal and human skulls. Furthermore, it owns 5800 microscope slides dating back to the nineteenth century as well as about 600 didactic wall maps. Among its special features are the complete mounted skeleton of an Asian elephant, several mummy skulls, many special molluscs and the coleopteran collection of C. F. Riehl. Today, the Zoological Collection is still used for teaching purposes and serves as a reference collection or resource for exhibitions. In addition, ongoing research is conducted within the collection, in connection with student theses or focusing on the history of the collection or on its specimens. In this context, for example, old department files are being processed and digitalised. At times, certain specimens of the collection participate in both regional and national exhibitions. Guided tours of the collection are available on request.

Keywords Asian elephant • Mummy skulls • Pearls • Leuckart wall maps • C.F. Riehl • B. Merrem

---

Acronym: MZC

E. Hempel • L.A. Beck (✉)
Department of Biology, Animal Evolution and Systematics and Didactics of Biology, Zoological Collection of the Philipps-University of Marburg, Karl-von-Frisch-Str. 8, 35032 Marburg, Germany
e-mail: beck@biologie.uni-marburg.de; http://www.uni-marburg.de/fb17/admin/infrastruktur/sammlungen/zoologische_sammlung/index_html

L.A. Beck (ed.), *Zoological Collections of Germany*, Natural History Collections,
DOI 10.1007/978-3-319-44321-8_44

44.1 Historical Background

The Marburg Zoological Collection was founded in the years 1817–1819 as part of the new Zoological Department at the Philipps University. **B. Merrem** (1761–1824), who, in 1817/1818, had contributed essentially to the collection's foundation by requesting and accumulating a set of objects, became the department's first director in 1819 (Beck 2014; Bohle 2015). The first specimens were doublets from the Electoral Museum in Kassel (Kurfürstliches Museum Kassel), which were made available to the Zoological Department by the Elector of Hesse. They comprised about 180 birds (mounted or as skins), a few mammals and approximately 70 reptiles and amphibians preserved in alcohol, some presumably dating back to the end of the eighteenth century. Fishes and invertebrates were poorly represented. In 1822, this original stock was supplemented with the collection of chief forestry commissioner (Oberforstmeister) **L. C. E. H. F. von Wildungen** (1754–1822), which mainly consisted of native birds and mammals.

Despite best efforts in the period following Merrem, it seemed impossible at first to enlarge the collection owing to the financial situation and the widespread view of that time that zoology was just an ancillary science. Thus, additions during that time mostly attributed to donations or small purchases, as well as to the studies on insects and other invertebrates by **M. Herold** (1790–1862). It was due to the great endeavour of another of Merrem's successors, **C. Claus** (1835–1899), that in 1867 the Zoological Collection was augmented by objects from the comparative anatomical collection of the university's Department of Anatomy (Aumüller 2012; Museum anatomicum 2015). For the most part, these were skeletons of mammals, birds and some reptiles. In the time that followed, the Zoological Collection purchased numerous specimens, e.g., the coleopteran collection of C. F. Riehl in 1873, doublets of the important collection of Marburg geologist W. Dunker in 1891 and a collection of bird skins of **C. Floericke**. The first catalogue, which was compiled in 1866, already listed about 20,000 insects and other arthropods, 3000 marine animals, 600 vertebrate skeletons and over 700 mounted birds and mammals. As a result of his research, **R. Greeff** (1829–1892) added a great number of marine invertebrates to the collection before under **E. Korschelt** (1858–1946), the Zoological Institute expanded and along with it the collection: the teaching collection was established, and a further floor was added to the building in Ketzerbach Street, thus allowing the collection to be opened to the public—like a museum of natural history—for many years to come. In this location, the Marburg Zoological Collection survived both world wars largely undamaged. There were rumours, however, that just after the war, in 1945, alcohol from jars containing preserved animals was drunk.

The good years that the institute had experienced while Korschelt was in charge did not continue during the 1950's and 1960's. The museum stayed closed, the skeleton of the Asian elephant was demounted and put away in a storeroom, and collection rooms were converted into research laboratories. Even after the institute had moved to the modern buildings at Marburg Lahnberge (hills east of Marburg) in 1974, there was a seemingly endless discussion as to whether most of the collection should be stored in the basement or whether it should not. Zoologists like **R. Remane**, **Chr. Kirchner**, **H. W. Bohle**, **H.-O. von Hagen** and many of their staff were able to prevent the worst.

Since 1998, Lothar A. Beck has been in charge of the Zoological Collection. He is supported by one technician. The budget is low (2000,-€), but the Department of Biology financed the renewal of all glass cases for birds and mammals as well as the safety cupboards for the wet collection some years ago. Beginning in 2000, the public appreciation of the collection strongly improved. That improvement was supported by the 475th anniversary of the Philipps University in 2002 and by a ministerial query concerning the collection's value (estimated at about one million euros). It was further promoted by the general discussion about collections and the appointment of a commissioner for all 34 of the university's collections. Over the years, the collection has suffered the loss of some of its specimens due to a variety of reasons, among them sale, removal by departing professors or even theft.

During the early nineteenth century, the collection and the Zoological Department were accommodated in a building in the town centre, which formerly housed the commander of the Teutonic Order. It was only in 1903 that the department moved to an adequate location in the building of the former Department of Anatomy in Ketzerbach Street. In 1974, the Zoological Department finally moved to its present-day location on the Lahnberge hills outside of town. The primary purpose of the Zoological Collection was to give a representative review of the animal kingdom, a goal, of course, that was far from being accomplished during the early years. From the very beginning, however, the collection with its specimens and its didactic wall maps has served as a valuable contribution to the teaching of natural sciences and medicine, as well as to diploma, bachelor and master courses of studies and to the training of secondary school teachers, which to this day is one of its main purposes.

44.2 Recent Collection

44.2.1 Structure

Today the Marburg Zoological Collection, though still comparatively small, offers a number of peculiarities (see below). It comprises about 40,000 specimens: over 20,000 insects and other arthropods, 3500 marine or other invertebrates, more than 650 birds and mammals, about 300 reptiles and amphibians and over 630 skeletons, animal and human skulls. While doing research for this text, about 5800 microscope slides were rediscovered. They are part of the collection and date back to the nineteenth century.

As it used to be in the beginning, one of the main purposes of the collection today is education. As the collection is relatively old, research focuses on its history, in addition to the ongoing study of particular specimens.

The collection is divided into a reference collection (350 m^2), a teaching collection (60 m^2) and a public collection (mainly at the large north foyer of the building) and some workshops.

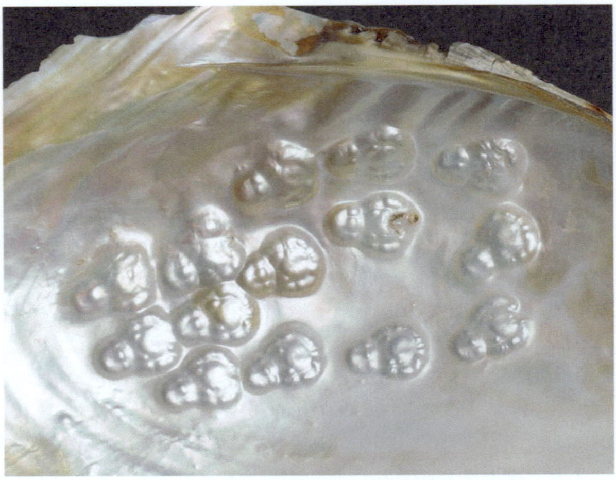

Fig. 44.1 Chinese freshwater mussel with artificial figurines of Buddha covered by the mother of pearl (Photo: © with MZC: Beitz, Mecke, Worth)

The reference collection comprises a dry collection, which, for the most part, houses invertebrates, and a wet collection with again mainly invertebrates. It also includes a significant amount of fishes, reptiles and amphibians. The **taxidermy cabinet** contains primarily native birds and mammals, with some specialties (like hummingbirds and a double-headed malformed calf). The **skeleton cabinet** features a similar inventory, mostly bird and mammal skeletons or skulls, along with some human skulls.

The teaching collection consists of mostly invertebrates, either dry or in alcohol, depending on the species preserved. Specimens from sections of the **wet collection**, the skeleton and the taxidermy cabinet are also used as demonstration material for teaching purposes. In addition, there are about 600 didactic **wall maps**, which are used mostly in courses and seminars.

The public collection is permanently on display in the department's foyer. Students and visitors can have a close look at a selection of skeletons, mummy skulls and showcases, which outline the research on pearl formation, which was conducted in the past (Fig. 44.1).

Since 2012, the Marburg Zoological Collection has been cross-linked to the SESAM database of the Senckenberg Museum in Frankfurt/Main. The presence of the collection on the website of the university lead to national and international requests for research material.

44.2.2 Special Features

The Marburg Zoological Collection includes a number of specimens that one would be astonished to find in a collection in Central Germany/Hesse. Among them are a kakapo, a kiwi, a short-beaked echidna and a duck-billed platypus (all stuffed, Fig. 44.2), a stuffed slow loris (Fig. 44.3), a phrenological skull (Fig. 44.9) and a roebuck with perruque head (Fig. 44.10). The biggest surprise, however, is the largest specimen (Fig. 44.4): a complete mounted skeleton (with tusks) of a male

Fig. 44.2 Stuffed kakapo, kiwi, short-beaked echidna and duck-billed platypus (Photo: © with MZC: Beitz, Mecke, Worth)

Fig. 44.3 This slow loris used to be a pet of a professor of the university, who donated it to the collection after the animal had passed away in 1995 (Photo: © with MZC: Beitz, Mecke, Worth)

Asian elephant (Bohle 2014). It has achieved almost iconic status among biology students in Marburg as it was mounted in one of the department's foyers in 2002 and has been standing there ever since. It was even furnished with its own web page outlining the history and construction. The elephant became part of the collection in 1863, when the animal had to be put down due to being in musth,[1] which means it

[1]Musth is a state of high aggressiveness due to very high testosterone levels, which elephant bulls develop to dominate over other male elephants.

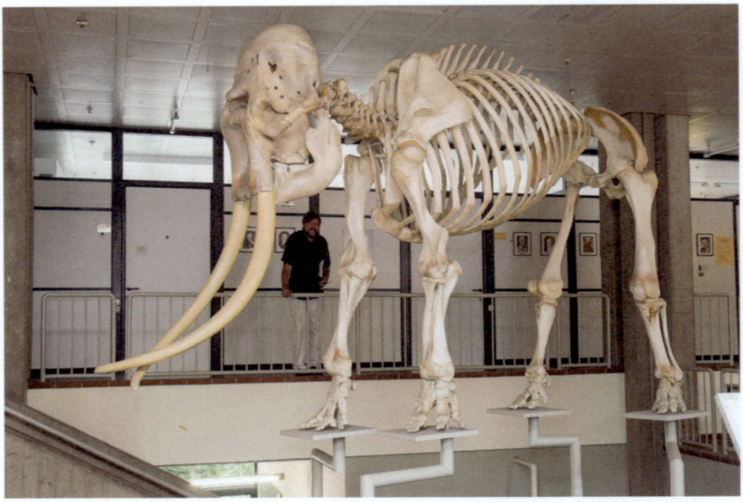

Fig. 44.4 The skeleton of the Asian elephant from 1863, called "Jack", mounted in 2002 in one of the foyers of the Department of Biology (Photo: © with MZC: Beitz)

had become unmanageable for the circus it belonged to (Eisenberg et al. 1971; Poole and Moss 1981). Students lovingly call it "Jack", after its circus name.

As the Marburg Zoological Collection was founded at the beginning of the nineteenth century, it holds specimens from that early period. In particular, it owns gastropods from well-known collectors, e.g., *Conchylia* species from **J. Pfeiffer** (1824), Cuban *Conchylia* from **J. Chr. Gundlach** (1859) and doublets from the *Conchylia* collection of **W. Dunker** (1891). In addition, the Zoological Collection possesses a number of specimens from the German deep-sea expedition "**Valdivia**" (1898/1899), which were provided by A. Brauer, a private lecturer in Marburg, who took part in the expedition.

Specimens that have recently attracted public attention are the 33 **mummy skulls**. The skulls are 2500–3000 years old and originate from Thebes, Abydos and Saqqara (Fig. 44.5) (Harbort 2010). They date back to the African traveller F. Mook and were purchased for the collection in 1881. Some of them are on public display in Marburg; others were recently part of exhibitions elsewhere. The skulls were featured in a television programme (ARTE in 2013). They were also the topic of three scientific theses, during the course of which they were studied using non-invasive techniques (such as computed tomography and radiography).

Impressive by its sheer elegance and beauty is the coleopteran collection of **C. F. Riehl** (Fig. 44.6), which presents mostly exotic beetles. A display depicting research on pearl formation constitutes a component of rather regional historical interest. Subjects were native limnic species such as the freshwater pearl mussel as well as Chinese and East Asian species (Fig. 44.1). It is but a relic, which shows that from the eighteenth century on, **pearl fishing industry** flourished in the Electorate of Hesse (Kurhessen). Today, however, it does not exist anymore.

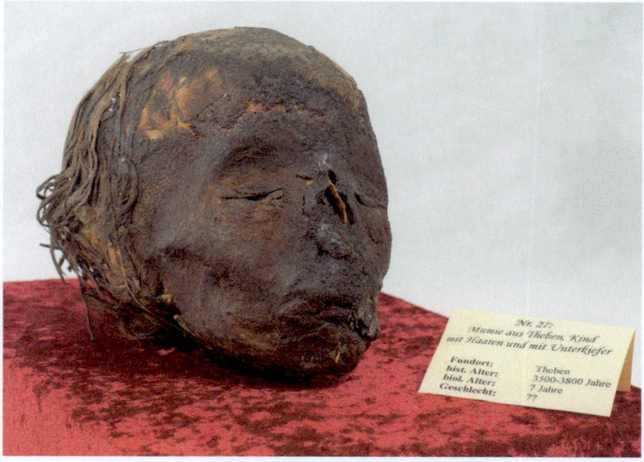

Fig. 44.5 A mummy skull of a child from Thebes with well-preserved hair. The skull is estimated to be about 2500–3000 years old (previously estimated to be 3500–3800 years old) (Photo: © with MZC: Beitz)

Fig. 44.6 One of the insect boxes with coleopterans from the collection of **C. F. Riehl** (moved to a new insect box) (Photo: © with MZC: Beitz, Mecke, Worth)

Fig. 44.7 Recently, the Marburg Zoological Collection was augmented by about 4500 gastropod specimens from the dissolved Marburg Geological-Paleontological Collection. Displayed is one of the specimens that were collected by Marburg geologist **W. Dunker** (Photo: © with MZC: Beitz)

44.2.3 Augmentation and Access

The Marburg Zoological Collection is still growing; more specimens or entire collections are added to its stock (Fig. 44.8, crucifix fish, Fig. 44.11 stuffed mammals from a secondary school collection). Recently, it was enlarged by a collection of marine and limnic gastropods (about 4500) from the Marburg Geological-Paleontological Collection (Fig. 44.7), which is being dissolved. It is furthermore complemented with scientific specimen copies. These are mostly freshwater fishes, freshwater fish hybrids and insects. As already mentioned, the collection's objects are widely used in courses and seminars that are held regularly in the rooms of the collection. Guided tours, e.g., for adult education or for high school field trips, can be arranged given prior notification—an option that is used on a regular basis. Some of the specimens are displayed in regional (Marburger Kunstverein) and occasionally even in national exhibitions ("Valdivia", Naturkundemuseum Leipzig; "Mummies of the World", Mannheim) (Fig. 44.5).

44.2.4 Research

Ongoing research is conducted in various divisions of the Marburg Zoological Collection, mainly in conjunction with student theses. In addition to zoological topics, historical matters are addressed in this context.

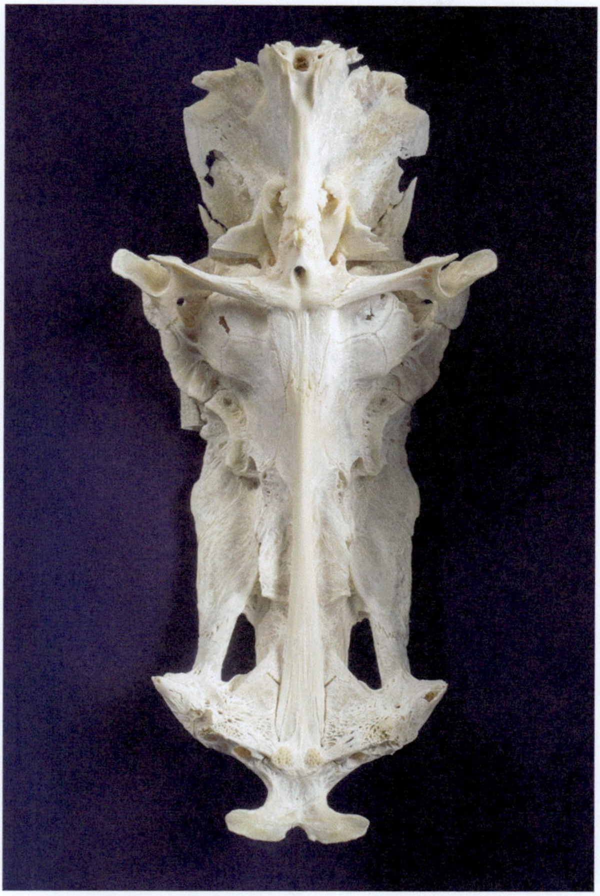

Fig. 44.8 The term "crucifix fish" originates from the perceived resemblance with an image of Jesus on the cross. It is, in fact, the inner part of the cranium of a member of the fish family Ariidae. The object is a rather new addition to the collection and was provided by the student Veronika Machnik in 2013 (Photo: © with MZC: Beitz)

In recent years, new theses have done research on objects, such as the primate skull collection of A. Remane or, as already mentioned, the mummy skulls. The collection owns several specimens from the **black smoker deep-sea fauna** of the Western Pacific (expedition "OLGA", 1989/90) which have been studied taxonomically and phylogenetically (Beck 2002; Schwarzpaul and Beck 2002; Sobjinski and Beck 2003). Phylogenetic studies have been conducted on **fiddler crabs** and crayfish by H.-O. von Hagen (Beinlich and von Hagen 2006). In general, a variety of faunistic and biodiversity studies have been performed over the years. Furthermore, research is being conducted on specimens from excursions to the Mediterranean as well as to the North Sea.

The collection owns about 100 wall maps by **R. Leuckart**, which have been studied in the light of didactics and were digitalised in the process.

At times, research is initiated by a request. In co-operation with the Department of Anatomy, we recently investigated whether our collection owns specimens from **S. T. Sömmering**, founder of the Anatomic Theatre Kassel. His collection was

Fig. 44.9 Phrenological skull (female), nineteenth century. Phrenology was a pseudoscientific theory based on the morphology of skull areas which were said to be symptoms for the strength of characters or moral strength (after F. J. Gall 1758–1828) (Photo: © with MZC: Beitz)

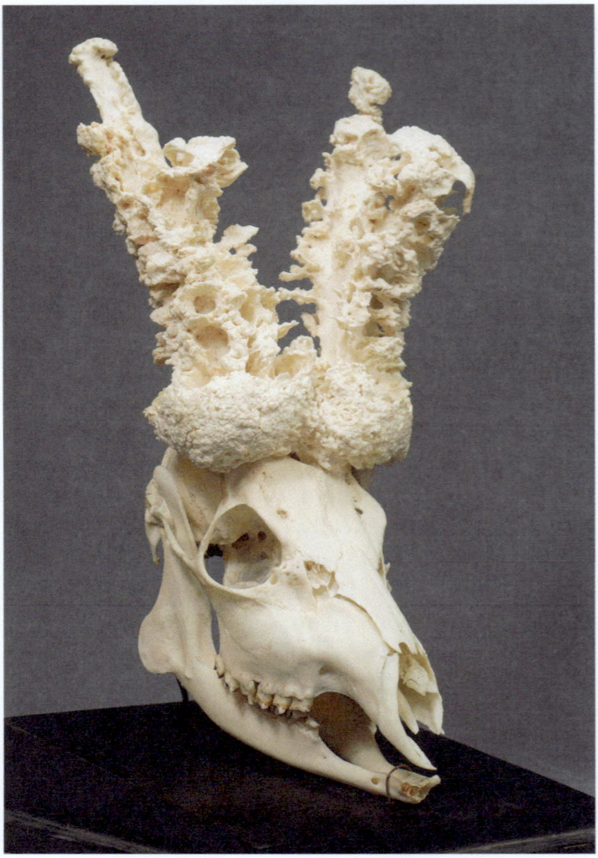

Fig. 44.10 Roebuck with perruque head. The abnormal antler growth is caused by lack of testosterone (Photo: © with MZC: Beitz, Mecke, Worth)

Fig. 44.11 The collection's taxidermy cabinet with some recently received specimens from a secondary school collection in front (Photo: © with MZC: Beitz)

passed on to the Institute of Anatomy in 1788. However, parts of the collection of the institute were transferred to the Marburg Zoological Collection (see above). Following intensive catalogue work, we were able to confirm that some of the material in question is indeed part of our collection.

Transliteration work has been going on for several years to further elucidate the history of our collection and, thereby, the fate of formerly owned specimens. In this process, old collection catalogues, collection labels and department files are being transliterated from former versions of German handwriting (e.g. Suetterlin script) and are then to be stored at the University Archives within the German public record office in Marburg.

Acknowledgements The authors would like to deeply thank H. Worth for providing recently acquired information and for her support in general. Thanks are due to Iris Krane who kindly read the English proof.

References

Aumüller G (2012) Überblick über die Geschichte der Anatomie in Marburg vom 16. bis ins 19. Jahrhundert. In: Grundmann G, Aumüller G (Hrsg.) Das Marburger Medizinhistorische Museum—Museum Anatomicum—Geschichte und Ausstellungsgegenstände. Marbg Stadtschriften Gesch Kultur 98:7–32

Beck LA (2002) *Hirtopelta tufari* sp. n., a new archaeogastropod hot vent limpet (Rhipidoglossa, Peltospiridae) harbouring endocytosymbiotic bacteria in its gill. Arch Molluskenkunde 132:23–34

Beck LA (2014) Zoologische Sammlung. In: Otterbeck C, Schachtner J (eds) Schätze der Wissenschaft—Die Sammlungen, Museen und Archive der Philipps-Universität Marburg. Jonas Verlag, Marburg, pp 76–83

Beinlich B, von Hagen HO (2006) Materials for a more stable subdivision of the genus *Uca* Leach. Zool Med Leiden 80:9–32

Bohle HW (2014) Das Marburger Elefantenskelett—Seine Geschichte als Teil einer Geschichte der vergleichenden Anatomie zwischen Medizin und Zoologie. In: Sahmland I, Grundmann K (eds) Tote Objekte lebendige Geschichten—Exponate aus Sammlungen der Philipps-Universität Marburg. Michael Imhof Verlag, Petersberg, pp 78–95

Bohle HW (2015) Von der Naturgeschichte zur Zoologie—Blasius Merrem und die Entwicklung der Zoologie an der Universität Marburg im 19.Jahrhundert (1807–1928). Waxmann Verlag, Münster, New York

Eisenberg J, McKay GM, Jainudeen MR (1971) Reproductive behavior of the Asiatic Elephant (*Elephas maximus maximus* L.). Behaviour 38:193–225

Harbort J (2010) Egyptian mummy skulls—the Marburg Zoological Collection. In: Wieczorek A, Rosendahl W (eds) Mummies of the world. Prestel, München, Berlin, London, New York, pp 322–325

Museum Anatomicum—Medizinhistorisches Museum der Philipps-Universität Marburg (2015) http://www.universitaetssammlungen.de/sammlung/107?setLocale=de. Last accessed 24 Jun 2015

Poole JH, Moss CJ (1981) Musth in the African elephant, *Loxodonta africana*. Nature 292:830–831

Schwarzpaul K, Beck LA (2002) Preliminary results: phylogeny of hydrothermal vent limpets ("Archaeogastropoda") based on morphological and 18S rDNA data. Cah Biol Mar (Paris) 43:381–385

Sobjinski K, Beck LA (2003) Phylogeny of the hydrothermal gastropod family Provannidae (Caenogastropoda). Zoology (Suppl. DZG/DGP Jahresversammlung Juni 2003)

Chapter 45
MÜNCHEN: The Zoologische Staatssammlung München (ZSM) of the Staatliche Naturwissenschaftliche Sammlungen Bayerns (SNSB)

Gerhard Haszprunar

Abstract The current review first provides a summary of the history of the zoological collections of the Freistaat Bayern, which are institutionally covered by the Zoologische Staatssammlung München (Zoological State Collections, ZSM) since 1811. Information is provided concerning the futuristic building of ZSM and its outstanding qualities. The huge collections of ZSM comprise about 25 million inventory entities and includes specimens of all animal phyla and of all continents and oceans. A short review concerns actual infrastructure, research focuses, and education activities of ZSM.

Keywords Zoological collections • Building facilities • Systematic research

45.1 Historical Summary

A detailed chronicle of the ZSM and its many collections has been published (ZSM-Staff 1992; Fittkau 1992). Here, only a summary is presented:

1 May 1807: The king of Bavaria, Maximilian I Joseph, established the Bayerische Akademie der Wissenschaften (founded 1759) as a national institution and ordered that all collection items of royal ownership are to be transferred to the Akademie der Wissenschaften and that these collections are to be deposited in the so-called Wilhelminum in the city of Munich (Neuhauser Straße) (Kraft and Huber 1992).

19 January 1811: Within the Wilhelminum an independent zoological-zootomical department was founded; the king appointed Dr. Johann Baptist Ritter von Spix

G. Haszprunar (✉)
Zoologische Staatssammlung München (SNSB-ZSM), Münchhausenstraße 21, 81247 Munich, Germany
e-mail: haszi@zsm.mwn.de

L.A. Beck (ed.), *Zoological Collections of Germany*, Natural History Collections,
DOI 10.1007/978-3-319-44321-8_45

as the first conservator (director). This is taken as the birth of the Zoologische Staatssammlung as an institution proper.

1817–1820: Together with the botanist Dr. Carl Friedrich Philipp von Martius, Ritter von Spix undertook a famous expedition to Brazil, the collected specimens of which comprise the scientific foundation of the collections of the ZSM. For a detailed biography of Ritter von Spix, see Schönitzer (2011a, b).

1827–1925: The ZSM was finally separated from the Akademie (Litten 1992) and assigned to the Ludwig Maximilians University (LMU) of Munich. During this time, the director of ZSM also held the chair of the Zoological Institute of the LMU, and there was a fruitful symbiosis between the Collection in the Wilhelminum (Doflein 1904) and the Institute of the LMU resulting in a significant increase of the collections and important publications (Balss 1926). World War I did not affect the collections.

1925–1927: In 1925, the new chair of the Zoological Institute of LMU, Nobel Prize winner Prof. Dr. Karl von Frisch, did not agree to chair also the ZSM. After 2 years of debate, Prof. Dr. Hans Krieg was elected as the new director of ZSM, which was thus again separated from the LMU.

The night from 24 to 25 April 1944 witnessed a catastrophic bombing of the Wilhelminum resulting in the complete destruction of the building and the loss of nearly all taxidermy specimens of the public exhibition (Kraft and Huber 1992). Fortunately, with the exception of fishes, the scientific collections were already deposited in mining tunnels outside of Munich, although many specimens were damaged or destroyed during hasty transports.

1946: The ZSM collections were housed in "provisional" rooms in the northern part of the Nymphenburg Castle, where the Nazis had formerly established a museum for hunting. These provisional rooms were used for nearly 40 years.

1984–1986: After many years of planning and construction (see Fechter 1992 for details), the collections of the ZSM were transferred into the new and generous building in Obermenzing, a district at the northwestern periphery of Munich (see below); the opening ceremony took place on 3 July 1986.

1990: Opening of the Museum Mensch und Natur in the previous rooms of the ZSM in the Nymphenburg Castle. Though being quite small (only about 2200 m^2 exhibition area), the museum shows attractive and important specimens from the ZSM like the extinct quagga or the brown bear "JJ1 alias Bruno" with more than five millions of visitors until now. Accordingly, the ZSM only offers occasionally special exhibitions.

1995: Re-establishment of the link between the ZSM and the LMU with the call of Prof. Dr. Gerhard Haszprunar as chair of Systematic Zoology and director of ZSM. In 2006, Prof. Haszprunar was elected also as director general of the SNSB, which includes four other natural history collections (botany, mineralogy, palaeoanatomy and anthropology and palaeontology and geology), the Botanic Garden Munich-Nymphenburg, the Museum Mensch und Natur, three further small museums in Munich (palaeontology, geology, mineralogy) and four local museums in Bamberg, Bayreuth, Eichstätt and Nördlingen.

January 2012: Official start of the campaign to widen the Museum Mensch und Natur towards a large (more than 7000 m^2 exhibition area) new Bavarian Museum of Natural History in the northern wing of the Nymphenburg Castle, which should also present the highlights of the ZSM collections. Currently, the project is established by the Bavarian state government, an international architectural competition has been carried out, and the planning of activities for construction and exhibitions has started.

45.2 The Current Building (Fig. 45.1)

Based on the outstanding plans of the architects Schmidt-Schicketanz and Partner, the current building of the ZSM is one of the most unusual natural history museums in the world (Ruthensteiner 1999; Fig. 45.1). Because of energetic reasons (all magazines are air conditioned; Schönitzer et al. 2009), the large building with its two levels expands mainly subterranean; the roof (covered by grass) is positioned only about 1.5 m above street level. There are two circular light areas in the centre being surrounded by the working rooms of the staff, which thus gain daylight and fresh air. The working rooms of the staff are always close to the respective magazines; thus, there are short distances. All in all there are more than 5.000 m^2 exclusively for storage systems including the library, and the whole building includes more than 11.000 m^2. Since the original equipment included simple storage racks, the actual expansion and growing of the various collections are managed by successively replacing the original systems with modern compactus systems, which increase the actual storage capacity of magazines between 50 and 80 %. Air condition and the overall dust-filtering system add to the high quality of storage facilities in the ZSM.

A natural history collection, which does not grow, will die. The premises of the ZSM include a large developable plot so that the building might be substantially enlarged in the future.

45.3 The Collections (Fig. 45.2)

With more than 20 millions of objects (label entities), the ZSM houses one of the three big zoological collections of Germany. The strength lies in entomology which covers about 90 % of all specimens and includes the world's largest lepidopteran collection with about 11 million specimens. Historical and actual focuses (Fig. 45.2) include beneath the Bavarian/mid-European material and the fauna of South America, Himalaya, Madagascar, Indonesia and Australia; the Mediterranean and Antarctica are focal points of the marine environment. Table 45.1

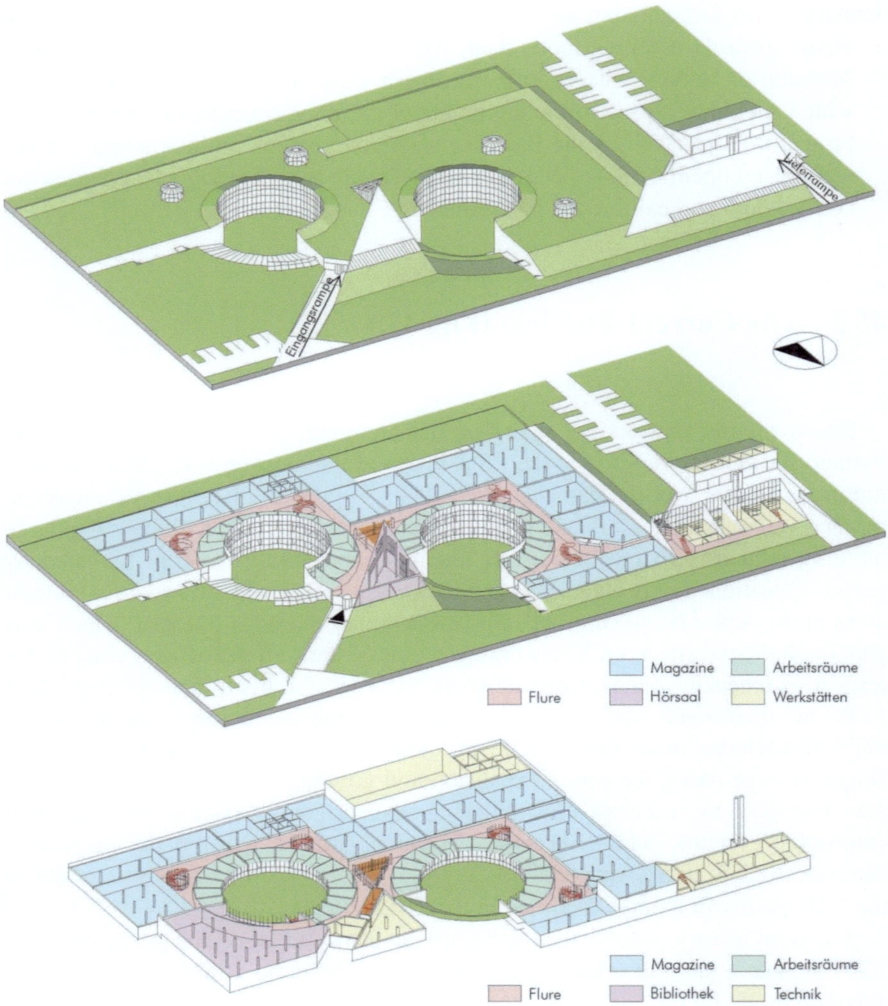

Fig. 45.1 The extraordinary building of the ZSM: external view/level 1/level 2 (from Michaela Ruthensteiner 1999 with permission of the author). The distance between the pillars is 5 m

provides an overview on the various collections (for details, see the respective web pages at www.zsm.mwn.de).

A good portion of the vertebrate collections and parts of the invertebrate taxa are already digitized. However, it will be impossible to digitize the many millions of insects in the next centuries at an individual scale. Accordingly, the entomologists of the ZSM digitize (1) on demand, if any specimen is processed (i.e., studied, loaned, etc.), and (2) en bloc by high-resolution photos of the whole boxes.

Fig. 45.2 Outstanding examples from the collections of the ZSM: (**a**) The "Royal Armadillo" brought alive to Munich from her Brazilian expedition by IKH Therese von Bayern (see Ruppel and Donoughue 2012) (Photo: © with Marianne Müller, ZSM). (**b**) A specimen of a male of the extinct passenger pigeon, *Extopistes migratorius* L., 1766 (Photo: © with Marianne Müller, ZSM). (**c**) One of the smallest reptiles, the endangered pygmy chameleon *Brookesia micra* from Madagascar (Photo: © with Dr. Glaw, ZSM). (**d**) A view of the lepidopteran magazine of ZSM—the largest butterfly/moth collection on earth (Photo: © with Dr. Axel Hausmann, ZSM)

In addition, the Staatssammlung für Anthropologie und Paläeoanatomie (SNSB-SAPM, www.sapm.mwn.de) houses about 52,000 human skeleton remains mostly from Bavaria and about 1.5 million archaeozoological items, in particular domestic mammals and birds mainly from Europe and Minor Asia. Moreover, the SNSB collections include about 1.5 million fossils, which are deposited in the magazines of the Bayerische Staatssammlung für Paläeontologie und Geologie (SNSB-BSPG, http://www.palmuc.de/bspg/) and of the three regional museums, the Naturkunde-Museum Bamberg (SNSB-NKMB, www.naturkundemuseum-bamberg.de), also including important historical collections of birds and insects), the Urwelt-Museum Oberfranken in Bayreuth (SNSB-UMO, www.urwelt-museum.de/) and the Jura-Museum Eichstätt (SNSB-JME, www.jura-museum.de/).

Table 45.1 Overview of the collections of SNSB-ZSM

Section	Label entities	Strengths and highlights
Evertebrata Varia	>20.000	Focus on Hydrozoa (types of Stechow) and Echinodermata (types of Döderlein—many from Japan)
Mollusca	>400.000	About 50,000 species; South America, Mediterranean, Antarctica, deep sea; great in Polyplacophora, opisthobranchs and marine soft bodies
Arthropoda Varia	>250.000	>1 Mio. items, large-type collection of Acari and terrestrial Isopoda; decapod Crustacea and Pantopoda of various large expeditions
Coleoptera	3–4.000.000	>100.000 species, about 10.000 primary types, actual focus on ground and water beetles
Diptera	500.000	>10 Mio. items, >14.500 species; ca. 1300 types; 2200 insect boxes, ca. 140,000 alcohol items; >60,000 microslides; focus on Chironomidae
Hemiptera	>400.000	>13,000 species mainly Heteroptera, 2000 species Orthoptera; types of Seidenstücker
Hymenoptera	>3.000.000	About 6500 primary types, mostly "Parasitica"
Insecta Varia	>300.000	>3 Mio. items (80 % in alcohol), about 8.000 species; >350 types; 25,000 micropreparates; focus on aquatic insects, Blattodea, Siphonaptera
Lepidoptera	>11.000.000	Largest collection of the world with substantial parts of the Museum Witt; current focus on Geometridae, Rhopalocera and Bombyces
Ichthyology	>40.000	ca. 7000 species, completely digitized, many correlated tissue and DNA samples
Herpetology	>150.000	35,000 Amphibia, >115,000 Reptilia; important collection of tadpoles; current focus on frogs and reptiles of Madagascar
Ornithology	>60.000	About 6000 species, particularly from South America and Europe; 600 type specimens; also eggs, feather and skull collections
Mammalogy	>40.000	Largely digitized; focus on Bavaria, Kaukasus, Southeast Asia, Africa, Argentina; important collection of primates, particularly orangs
DNA bank	ca. 70.000	Including samples of all sections, mainly fish and insect species
Library	>280.000	Including libraries of local entomological and ornithological societies, open for public

45.4 Infrastructure

The ZSM has a permanent staff of 16 scientists, being assisted by 14 technicians. Together with the administration and general functions (lab technicians, IT, building services, etc.), the staff currently are 50 heads.

The ZSM is equipped with a modern histological/ultrastructural and molecular lab and also houses high-tech facilities for fluorescence microscopy, scanning electron microscopy (SEM) and microcomputer tomography (μCT). Confocal laser scanning microscopy (CLSM) and transmission electron microscopy (TEM) as well as a large sequencing facility are available via the LMU. The extensive library (Diller 2006) is one of the backbones of the scientific work at the ZSM and of many external users.

Thanks to the initiative of Dr. Juliane Diller (formerly Koepcke), the current vice-director of ZSM, the ZSM scientist can use an excellent site for actual and long-time research in the tropical rain forest. The research station "Panguana" near the Andes Mountains in the lowland of Peru (see www.zsm.mwn.de/panguana/ for details) is the scientific and educational centre of a nationally protected area of about 7 km^2 mostly covered by virgin jungle.

45.5 Research (Fig. 45.3)

The ZSM is highly productive. In 2014 16 permanent scientists and about 15 third-party-financed research associates of the ZSM scientists have published no less than five monographs, 100 peer-reviewed and indexed articles, 28 further scientific contributions as full papers, and 35 short communications.

Aside from numerous special research projects on the various taxa, the ZSM is among the world-leading institutions concerning the International Barcode of Life Initiative (iBOL: see www.ibol.org) (Fig. 45.3). The main focus is the Bavarian and German fauna, where more than 16.000 species have been barcoded through the last

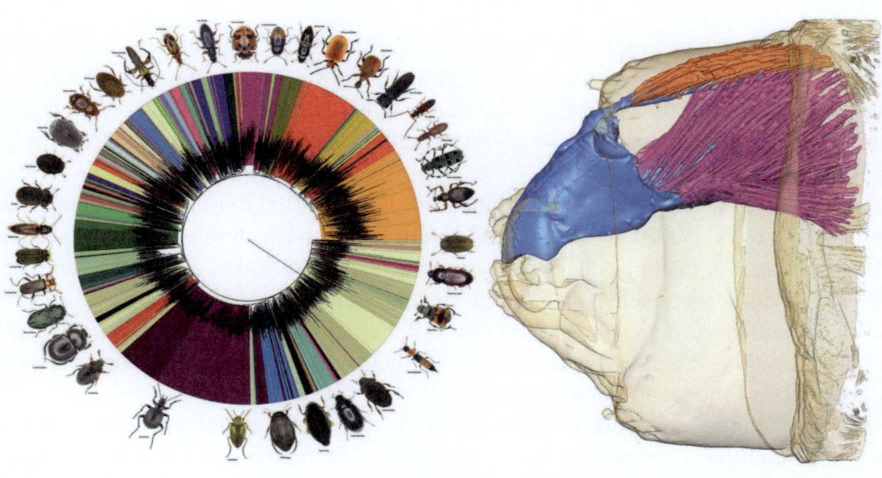

Fig. 45.3 Two focuses of research in the ZSM. (***left***) Barcoding tree of 3.500 Bavarian beetles, the largest barcoding data set of determined species ever done (from Hendrich et al. 2015 with permission). (***right***) 3D visualization of the mandibular apparatus of a longhorn beetle larva with cuticular element (*blue*) and muscles (*pink*/*red*) based on µCT technology (Courtesy of Dr. Bernhard Ruthensteiner, ZSM)

years. Moreover, ZSM staff contributed more than 120.000 specimens to the global project. Indeed, the barcoding activities have already resulted in several substantial papers on the subject, have increased significantly the collections and also have generated a very positive impact on cooperation within the ZSM and other institutions at national and international level.

In principle, ZSM scientists follow the concept of "Integrative Taxonomy" (e.g., Padial et al. 2010; Schlick-Steiner et al. 2010; Riedel et al. 2013; Pante et al. 2015) in order to combine phenotypic and genotypic data towards a synthetic taxonomic decision. Based on a robust taxonomy, the ZSM scientists also infer speciation processes and phylogeny and consider ecological and biogeographical aspects, thus covering the whole range of current biodiversity research. ZSM scientists also curate the node "Wirbellose II: Chelicerata and Mollusca" of the German GBIF network (www.gbif.de).

Concerning methodology, the ZSM has a focus on 3D visualization at all levels (ultrastructure to macroscopy, Fig. 45.3) and high-resolution scanning of insect boxes for digitization of collections (Fig. 45.3). We profit from the outstanding quality of the SNSB IT centre (head, Dr. Dagmar Triebel), which provides an excellent software framework for all digitization projects of the SNSB collections.

The connection to the university via the director's chair and by teaching activities done by several staff members motivate many students to write their final thesis at all levels or dissertation in the ZSM.

45.6 Educational Work

The ZSM houses two biological societies, the Münchener Entomologische Gesellschaft e.V. (MEG, http://meg-bayern.de/) and the Ornithologische Gesellschaft in Bayern e.V. (OG, http://og-bayern.de/), each with several hundreds of professional and amateur scientists. These societies significantly add activities and multiply education events of the ZSM.

As outlined above, many public education activities, exhibitions and lectures are organized by the Museum Mensch und Natur, the central exhibition platform of all Bavarian natural history collections with more than 200,000 visitors per year. The ZSM provides specimens for exhibitions and adds the expert knowledge of the curators and staff members. In addition to the exhibitions at the Museum Mensch und Natur, the ZSM itself welcomes about 8000 visitors per year at a day of open doors, guided tours and public lectures.

The scientific collections of the SNSB usually do not use social media for sharing scientific information because of copyright reasons, but Facebook pages are used by the public museums of the SNSB as a platform for distributing information on exhibitions and events. Exhibitions and news are also published via the main SNSB homepage, www.snsb.de.

SNSB press releases are distributed via several press distribution lists, aimed at our target groups. For scientific information, we also use the newsletters of the

Gesellschaft für Biologische Systematik (GfBS, www.gfbs-home.de) and of the Verband der Biologie (VBIO, https://www.vbio.de/) as well as the Informationsdienst Wissenschaft, www.idw-online.de, a German scientific information service with currently 940 member institutions and 33.000 subscribers including 7.300 journalists.

References

Balss H (1926) Die Zoologische Staatssammlung und das Zoologische Institut. In: Müller KA (ed) Die wissenschaftlichen Anstalten der Ludwig-Maximilians-Universität zu München. Chronik zur Jahrhundertfeier. Oldenbourg und Wolf, München, pp 300–315

Diller J (2006) Die Bibliothek der Zoologischen Staatssammlung München. Bibliotheksforum Bayern 34(1/2):26–40

Doflein F (1904) Sechs Wanderungen durch die Münchener Zoologische Staatssammlung. Verlag Valentin Höfling, München, 1–47

Fechter H (1992) Der Neubau der Zoologischen Staatssammlung München. In: Chronik der Zoologischen Staatssammlung München. Spixiana Suppl 17:176–188

Fittkau J (1992) Vom Naturalienkabinett zum modernen Forschungsinstitut: Geschichte und Bedeutung der Zoologischen Staatssammlung. In: Chronik der Zoologischen Staatssammlung München. Spixiana Suppl 17:24–34

Hendrich L, Morinière J, Haszprunar G, Hebert PDN, Hausmann A, Köhler F (2015) A comprehensive DNA barcode database for Central European beetles: Adding more than 3,500 identified species to BOLD. Mol Ecol Resour 15(4):795–818 (several extensive supplements)

Kraft R, Huber W (1992) Die Zoologische Schausammlung in der Alten Akademie in München 1809–1944. In: Chronik der Zoologischen Staatssammlung München. Spixiana Suppl 17:189–200

Litten F (1992) Die Trennung der Verwaltung der wissenschaftlichen Sammlungen des Staates von der Bayerischen Akademie der Wissenschaften. Z Bayer Landesgeschichte 55(2):411–420

Padial JM, Miralles A, de la Riva I, Vences M (2010) The integrative future of taxonomy. Front Zool 7(16):1–14

Pante E, Schoelinck C, Puillandre N (2015) From integrative taxonomy to species description: one step beyond. Syst Biol 64(1):152–160

Riedel A, Sagata K, Suhardjono YR, Tänzler R, Balke M (2013) Integrative taxonomy on the fast track—towards more sustainability in biodiversity research. Front Zool 10(15):1–9

Ruppel B, Donoughue E (2012) Das Gürteltier der Prinzessin. Ein Reiseandenken der Therese von Bayern in der Zoologischen Staatssammlung München. Aviso 4:8–9

Ruthensteiner M (1999) Die Zoologische Staatssammlung München—Gebäudeanalyse. Report for the Institut für Gebäudelehre der TU Wien. 1–42

Schlick-Steiner BC, Steiner FM, Seifert B, Stauffer C, Christian E, Crozier RH (2010) Integrative taxonomy: a multisource approach to exploring biodiversity. Annu Rev Entomol 55:421–438

Schönitzer K (2011a) Johann Baptist von Spix (1781–1826). Brasilienforschung und Gründer der Zoologischen Staatssammlung München. Naturwiss Rundschau 64(9):466–471

Schönitzer K (2011b) Ein Leben für die Zoologie. Die Reisen und Forschungen des Johann Baptist Ritter von Spix. Allitera Verlag, edition monacensia, 1–224. ISBN:978-3-86906-179-5

Schönitzer K, Wolf M, Liu L-Y (2009) Energy efficiency by means of architecture and engineering: The Bavarian State Collection of Zoology. Museol Q (Taichung) 23(4):23–42

ZSM-Staff (1992) Chronik der Zoologischen Staatssammlung München. Festschrift zur Verabschiedung des Direktors der Zoologischen Staatssammlung München Prof. Dr. Ernst Josef Fittkau 1976–1992. Spixiana Suppl 17:1–248

Chapter 46
MÜNSTER: The Zoological Collections of the LWL-Museum of Natural History in Münster

Jan Ole Kriegs and Heinrich Terlutter

Abstract The zoological collections of the LWL-Museum of Natural History in Münster date back to the mid-nineteenth century. Since then, the focus on collecting objects of importance to the natural history of the region "Westphalia" remains. The collection contains more than one million specimens of insects; the majority belongs to the orders Coleoptera, Lepidoptera and Hymenoptera. Among the different kinds of invertebrate collections, there is a collection of 100,000 spiders. The vertebrate collections, for example, contain more than 26,000 objects within the bird collection and 13,000 objects within the mammalian collection.

Keywords Museum • Natural history • Zoological collections • Münster • Westphalia

46.1 Introduction

The first idea to establish a museum of natural history in Münster with focus on the region of Westphalia and the adjacent parts came up in the early 1870's. In the year 1872, a scientific society called "Westfälischer Provinzialverein für Wissenschaft und Kunst" was founded. Its aim was to build up a natural history museum based on the private collections of its members.

These plans were promoted by the founding director, Prof. Hermann Landois, who was also the director of the local zoological garden in Münster during that time. The administration of the Prussian province of Westphalia was convinced and agreed to fund the first building as well as the positions of scientists and taxidermists later on. The museum was finally opened in 1892. Since then, the focus on collecting objects of importance to the natural history of Westphalia remains. However on a smaller scale, objects from other countries and continents are collected as reference material and as exhibits for exhibitions.

J.O. Kriegs • H. Terlutter (✉)
LWL-State Museum of Natural History, Sentruper Str. 285, 48165 Münster, Germany
e-mail: jan_ole.kriegs@lwl.org; heinrich.terlutter@lwl.org

L.A. Beck (ed.), *Zoological Collections of Germany*, Natural History Collections,
DOI 10.1007/978-3-319-44321-8_46

Landois founded the collections of (I) animal specimen of all classes found in the province of Westphalia, (II) plant specimen of all classes found in the province of Westphalia, (III) minerals and fossils representing the geology and paleontology of the province of Westphalia, (IV) reference collections from other provinces or countries and (V) a scientific library (Ditt 1992). As written above, all collections were founded based upon private collections that were donated as gifts or as heritage by local private naturalists. Since then, numerous private collections of importance for the natural history of the region have been donated to or bought by the museum (Franzisket 1967). Many collections, handed out by universities and other institutions, contain reference material collected in ecological studies. The collections are systematically organized using a standardized classification scheme based on the systematic knowledge in the year 2000. Up to now, the vertebrate and the spider collections are documented in a digital database. Although the museum moved to a modern building with more storage space in 1982 (Hendricks 1992), today large portions of the collections have to be stored in hired depositories. A new depository building with plenty of space is scheduled to be finished in 2018. The main focus of collecting specimens from the region of Westphalia as well as collecting reference objects remains and is manifested in the current museums mission statement (www.lwl.org/LWL/Kultur/lwl-naturkunde/portal/).

The collections are frequently used by ecologists to investigate the Westphalian biodiversity and have been used as the fundament for numerous publications. Some of these are "Westfalens Tierleben in Wort und Bild" (Landois 1883), the "Avifauna von Westfalen" (Peitzmeier 1979) about the birds occurring in the region, "Die Säugetiere Westfalens" (1984) and "Online-Atlas der Säugetiere Nordrhein-Westfalens" (AG Säugetierkunde in NRW 2015) about the distribution of mammals in the region, "Die Käfer Westfalens" (Westhoff 1881/1882) and "Coleoptera Westfalica" (Ant 1971 and following volumes) about beetles as well as "Lepidoptera Westfalica" (Weigt 1982) about butterflies. Specimens of many vertebrate and insect groups in the collection are important basic references for red data lists (Figs. 46.1 and 46.2).

46.2 The Collections

Insect Collections

The size of these large collections can only be estimated. They reach by far more than one million specimens. In 2001, Martin Berger presented a detailed paper about the historical and current status of the insect collections in the museum. Since then, many further private collections were donated to the museum.

Coleoptera

The estimated number amounts to more than 550.000 specimens. The first collections came from Eduard Suffrian (1805–1876), Fritz Westhoff (1857–1896), Friedrich Peetz (1870–1942), Karl Barner (1881–1959) and Helmut Beyer

Fig. 46.1 View into the depository of the zoological collections (Photo: © with LWL/Steinweg)

Fig. 46.2 View into the depository of the zoological collections (Photo: © with LWL/Steinweg)

(1905–1989). Since 2001, the following collections have been donated to the museum: Ernst Weise (1903–1973), Rainer Feldmann (1933–2014), Karl Fügner (1842–1916), H. Gotovik, Hermann Heddergott (1913–2002), Klaus Renner (1936), Herbert Ant (1933–2010), Theodor Kock, Ulrich Holste (1946–2011) and some other smaller collections.

Lepidoptera
The estimated number amounts to more than 500,000 specimens. The butterfly collection started with the donation of the collection of von Droste-Kerkerinck-Stapel (1808–1872) in 1873. Some substantial collections came from Franz Vornefeld (1871–1958), Viktor Schultz (1891–1963), Alfred Meise (1904–1999), Gerhard Tannert (1914–1997), Hanns Zielaskowski (1889–1963), Bernhard Zukowsky (1886–1949) and August Hillmann (1888–1972). Since 2001, the following collections were donated to or bought by the museum: Hans-Joachim Weigt, Dietrich Gantzhorn (1939), Joseph Haves (1899–1972), Hermann Heddergott (1913–2002), Heinrich Linke (1917–2003), Heinrich Biermann and some smaller collections.

Hymenoptera
In addition to the more than 25.000 specimens of the existing collection, the museum bought the hymenopteran collection of Michael Kuhlmann (1968) which is in 57 store boxes (about 8.200 specimens).

Diptera
From this insect order, only the families Syrphidae and Asilidae are organized systematically. In addition to the existing material, the museum received a part of the dipteran collection of Karl-Ernst Lauterbach (41 store boxes).

Detailed information on collected specimens of other insect groups are given by Berger (2001).

Spiders
The spider collection consists of more than 100,000 specimens of nearly 600 species. The collection is completely documented in a digital database.

Molluscs
The mollusc collection contains specimens of mostly terrestrial and freshwater species. They were collected by Hermann Löns (1866–1914) and Bernhard Rensch (1900–1990), who collected nearly 600 species, Herbert Ant (1933–2010) and Helmut Beyer (1905–1989) (Rehage 2011). There are also specimens from a large number of marine species from different collectors.

46.3 Vertebrate Collections

Mammals

The mammal collection contains more than 13,000 specimens consisting of skins, mounted animals, skulls and full skeletons. Contributors to the mammalian collection include Martin Berger (1936–2010), Bernd von Bülow (1933), Ludwig Franzisket (1917–1988), Friedrich Goethe (1911–2003), Heinz-Otto Rehage (1934), Bernhard Rensch (1900–1990) and Joachim Zabel (1907–1975). The mammalian collections of the Ruhrlandmuseum Essen and of the Zoological Institute of the University of Münster were handed out to the LWL-Museum of Natural History in 1984 and 2003, respectively.

Birds

The bird collection is composed of more than 26.000 specimens including 16.000 skins and mounted skins, several skulls and skeletons and more than 10.000 eggs and clutches (Figs. 46.3 and 46.4). The collectors include Wilhelm Heuvel (1913–1987), Ludwig Franzisket (1917–1988), Hermann Landois (1835–1905), Hermann Müller (1829–1883), Heinz-Otto Rehage (1934), Hermann Reichling (1890–1948), Bernhard Rensch (1900–1990) and Fürst Leopold zu Salm-Salm (1838–1908). The bird collections of the Ruhrlandmuseum Essen and of the Zoological Institute of the University of Münster were handed out to the LWL-Museum of Natural History in 1984 and 2003, respectively.

Fig. 46.3 Eurasian greenfinch, skins. For reasons of pest control, skins and mounted specimens are frozen regularly and otherwise stored under airtight conditions with transfluthrin (Photo: © with LWL/Steinweg)

Fig. 46.4 Two out of seven specimens of the grey partridges of the extinct subspecies *Perdix perdix sphagnetorum* (Altum, 1894) within the collection, female (*left*) and male (*right*). These dark phenotypes have been endemic in areas of heath and moorland in the northeastern Netherlands and northwestern Germany. They have become extinct in the late 1920's (Photo: © with LWL/Steinweg)

Other Collections

Several hundred mammal, bird, mollusc, amphibian and reptile specimens are preserved in fluid (mainly in ethanol). Since 2009, about 2000 tissue samples of vertebrates have been taken; these are stored at −80 °C for future genetic analyses. The museum keeps huge data collections as well. The Animal Sound Archive contains more than 10,000 sound recordings of birds and mammals from Westphalia. The Mammalian Atlas Project stores about 40,000 data sets about the spatio-temporal distribution of mammals in North Rhine-Westphalia.

References

Ant H (1971) Coleoptera Westfalica. Abh Landesmus Naturkd Münster 33(2):1–64

Berger M (2001) Die Insektensammlungen im Westfälischen Museum für Naturkunde Münster und ihre Sammler. Abh Westf Mus Naturk 63(3):1–168

Ditt K (1992) Natur wird Kulturgut. Das Provinzialmuseum für Naturkunde in Münster 1892 bis 1945. Abh Westf Mus Naturk 54(Beiheft):5_50

Franzisket L (1967) Die Geschichte des Westfälischen Landesmuseums für Naturkunde. Abh Landesmus Naturk Münster 29(1):3–26

Hendricks A (1992) Geschichte des Westfälischen Museums für Naturkunde 1945–1992. Abh Westf Mus Naturk 54(Beiheft):5_50

Landois H (Hrsg.) (1883) Westfalens Tierleben in Wort und Bild. Bd. I. Säugetiere. Paderborn

Peitzmeier J (Hrsg.) (1979) Avifauna von Westfalen.—2. Aufl. Abh Landesmus Naturk Münster 41(3/4):1–576

Rehage HO (2011) Die Sammlungen von Prof. Dr. Herbert Ant im LWL-Museum für Naturkunde in Münster. Natur Heimat 71:109–123

Schröpfer R, Feldmann R, Vierhaus H (1984) Die Säugetiere Westfalens. Abh Westf Mus Naturk 46(4):1–393

Weigt H-J (1982) Lepidoptera Westfalica. Abh Westf Mus Naturk 44(1):1–111

Westhoff F (1881/1882) Die Käfer Westfalens. Verh Naturhist Ver (Rheinlande und Westfalen) 38:23 S

Chapter 47
OLDENBURG: 180 Years of Local Collection and Research in Oldenburg

Christina Barilaro

Abstract The Landesmuseum Natur und Mensch Oldenburg in the federal state of Niedersachsen was founded in 1836 and is a state museum since 1947. It contains three departments (natural history, ethnology, archaeology) whose collections enlarge through acquisitions, donations, inheritance and active collecting. The collection of natural history, whose focal point is the local flora and fauna, contains about 300,000 objects of which about 15 % are recorded digitally yet. The collection is an important source for exhibitions, education and research regarding internal projects as well as external enquiries at home and abroad.

Keywords Landesmuseum/State museum • Ivan Antonovitsch Kuprianov • Jean Louis Cabanis • Wilhelm Blasius • Carl Constantin Platen • Great Auk + egg

47.1 A Brief Journey Through Time

The Landesmuseum Natur und Mensch Oldenburg (State Museum of Nature and Man Oldenburg) was founded in 1836 under the name "Naturhistorisches Museum" (Museum of Natural History) and is one of the oldest museums in Germany (Bengen 2001). In 1947 the museum was transformed into a state museum. Since then, it is government financed and funded by the federal state of Niedersachsen (Lower Saxony). After several renaming, in 2000 the museum got its current name "Landesmuseum Natur und Mensch Oldenburg".

The history of the Museum dates back to the time when Oldenburg was a grand duchy (since 1815). In 1835 Grand Duke Paul Friedrich August von Oldenburg (1783–1853) acquired a collection of 9800 insects and 490 mounted local birds from district medical officer Dr. Ernst Otto Oppermann. The acquisition of native species was unusual in these days, because actually the exhibits should represent the good connections to manor houses in foreign countries.

C. Barilaro (✉)
Niedersächsische Landesmuseen Oldenburg, Landesmuseum Natur und Mensch Oldenburg, Damm 38-44, 26135 Oldenburg, Germany
e-mail: c.barilaro@landesmuseen-ol.de; http://www.naturundmensch.de

L.A. Beck (ed.), *Zoological Collections of Germany*, Natural History Collections,
DOI 10.1007/978-3-319-44321-8_47

Fig. 47.1 Donation of the Russian Tsar Nicholas I.: Tufted Puffin (*Fratercula cirrhata*, inventory no. AVE4575) and presumptive Brandt's Cormorant (*Phalacrocorax cf. penicillatus*, inventory no. AVE974) (with permission from: Landesmuseum Natur und Mensch Oldenburg; photo: W. Kehmeier)

The collection of Oppermann was the basis for the foundation of the museum in 1836. It was located at Huntestraße 8 in the so-called Bollmann'sche Stuben. According to the spirit of the age, mainly exotic and uncommon objects were presented, like objects of natural history and ethnology of Russian America, which were collected by the Russian naval officer Ivan Antonovitsch Kuprianov (1799–1857) and got into the collection due to family ties to the Russian Tsar Nicholas I. (1796–1855) (Fig. 47.1).

After another relocation and due to further space requirements, the museum got a new building at today's place in 1880. Besides exhibition space, the building accommodates first depots and a room for educational purpose.

During World War II, on September 22, 1943, the museum and mainly the neighbouring library were struck by bombs. Subsequently, undamaged books and objects were removed. After provisional renovation of the museum, the collections were turned back in 1946. In the 1960's, construction works began with the objective of expanding in the adjoining vacant building of the library. And finally on May 20, 1974, the museum reopened. Today the building of the former library provides space, e.g., for special exhibitions, the lecture hall, workshops and offices, whereas the original building shelters the permanent exhibitions (Fig. 47.2).

Today the Landesmuseum Natur und Mensch, the neighbouring Landesmuseum für Kunst und Kulturgeschichte (State Museum of Art and Cultural History) and the Schlossgarten (Palace Garden) belong together to the corporation "Niedersächsische Landesmuseen Oldenburg" (Lower Saxony State Museum

Fig. 47.2 The Landesmuseum Natur und Mensch Oldenburg at today's place (*white* building in the front, original museum; *pink* building in the background, former library) (with permission from: Landesmuseum Natur und Mensch Oldenburg; photo: W. Kehmeier)

Oldenburg). The corporation is funded by the federal state of Niedersachsen. In addition to the state funding, the Landesmuseum Natur und Mensch acquires third-party funds and sponsors for research projects, exhibitions and educational projects.

Altogether there are approximately 35 employees at the Landesmuseum Natur und Mensch. Besides two permanent scientific staff members, namely, the director and the head of natural history, there are two temporary research fellows who are replaced every 2 years. One is employed in the Department of Natural History and the other in the Department of Ethnology. During the employment, the junior staff gets deeper insight in the work related to the museum. For example, they are responsible for parts of the collection and its inventory or for the planning and elaborating of special exhibitions. In addition, each year the museum employs one graduate for a voluntary gap year (cultural). This employee mainly assists the section of marketing and museum education. According to third-party-funded projects, the museum employs further scientific staff for research projects.

47.2 The Collection of Natural History

Almost from the beginning, the museum contains three departments: natural history, ethnology and archaeology. The focal point of the collection of natural history dates back to the first curator (1837–1878) and later director (1879–1895) Carl

Friedrich Wiepken. He focused on local fauna, which was unusual in these days. Since then, the emphasis of the collections is placed on local flora and fauna from the Weser-Ems-Gebiet (area of Weser-Ems) in the northwest of Niedersachsen.

Most objects in the collection of natural history are zoological, and among them insects and birds are the lion's shares. There are about 15,000 ornithological items, including round about 260 nests and 7900 eggs. The collection of birds is entirely recorded digitally.

Due to friendly relations with ornithologists like Jean Louis Cabanis (1816–1906) and Wilhelm Blasius (1845–1912), also exotic birds belong to the collection of the museum. That way a syntype of *Strix seloputo wiepkeni* (W. Blasius, 1888), which is named after Carl Friedrich Wiepken, got into the museum's collection (inventory no. AVE14002). In addition, the museum owes a syntype of *Prioniturus platenae* (W. Blasius, 1888) (inventory no. AVE993) to the friendship between Blasius and Carl Constantin Platen (1843–1899), of which the latter collected in Indonesia and the Philippines. Further special objects in the collection of birds are the mounted Great Auk (*Pinguinus impennis* (Linnaeus, 1758), inventory no. AVE8086) and an egg of the same species (inventory no. AVE8365).

The collection of insects includes more than 200,000 specimens, of which round about 10 % are registered digitally yet. Butterflies and beetles are especially represented. Besides the collection of Oppermann and further collecting of Wiepken, the museum got more than 30,000 butterfly items (including approx. 2200 Macrolepidoptera; 24,250 Microlepidoptera; 4000 genital preparations) from Oswald Tiedemann (1913–1999) in 1995/1996 (Meyer 1998). Georg Kerstens (1903–1982) bequeathed the museum his collection of approximately 70,000 beetles, mainly Staphylinidae (Rose 2004). Further coleopterists with importance for the museum's collection were, e.g., Dr. med. Ernst Friedrich August Röben (1843–1912), Dr. med Wilhelm Paasch (1874–1959) and Johannes Kühn (1904–1989) (Erbeling 1991). Recently, the museum focuses its active collecting on dragonflies, locusts and beetles.

Further invertebrates are represented by molluscs (approx. 10,000), mainly snails and bivalves.

The collection of natural history includes comparatively few mammals (approx. 3000), reptiles (approx. 1000), amphibians (approx. 400) and fishes (approx. 500). Reptiles, amphibians and fishes are completely registered digitally, whereas only 1050 mammals are recorded yet. The collection of fishes includes furthermore a type specimen, namely, an 8.5 cm-long tooth of the shark *Carcharodon megalodon*, which is extinct since three million years. The collection of reptiles includes two syntypes of *Lycodon ruhstrati ruhstrati* (Fischer, 1886) from Taiwan (inventory nos. REP918, REP919) (Fuhrmann and Kucharzewski 2008; Kucharzewski et al. 2009) (Fig. 47.3). They got in the collection by Ernst Ruhstrat in the latter half of the nineteenth century.

Besides the zoological objects itself, the museum takes tissue samples of every bird and mammal since 1998. Today, the collection of tissue samples contains

Fig. 47.3 Syntypes of *Lycodon ruhstrati* (with permission from: Landesmuseum Natur und Mensch Oldenburg; photo: C. Barilaro)

446 samples of birds and 413 of mammals. Reptiles, amphibians and fishes are stored in whole in alcohol. Thus, tissue samples can be taken when necessary.

The herbarium of the museum (official acronym, LMO) includes approximately 40,000 specimen, of which 6000 are registered digitally yet. The collection is based on the 2300 specimens collected by Johann Friedrich Trentepohl (1748–1806) in the Grand Duchy of Oldenburg (Becker 2014). Further important collections are, e.g., 405 specimens of grass and meadow plants from Oldenburg and Osnabrück collected by Jacob Ludwig Meyer (1802–1869) and Karl Hagena (1806–1882), 100 specimens of conifers from Wilhelm Hochstetter (1825–1881) and 934 specimens of lichen from the Grand Duchy of Oldenburg collected by Heinrich Sandstede (1859–1951). Furthermore, the museum shelters the herbarium of Prof. Dr. Gerhard Wiegleb (born 1948), which contains 7000 specimens, mainly aquatic plants, from, e.g., Northern Germany, further European countries and Japan (Will 2014).

The geological collection includes about 15,000 geological objects. It features, e.g., 326 lacquer soil profiles of Prof. Dr. Drs. h.c. Reinhold Tüxen (1899–1980), which is the biggest collection of these profiles in Germany (Obermöller 2007). Tüxen developed the profiles in the scope of a vegetation mapping in Northern

Germany between 1952 and 1961. Thus, he focused on phytosociology, but his work provides furthermore information about geology, agrology and archaeology.

A lent collection of 500 gemstones from Hans Lüschen is the basis for the permanent exhibition of minerals and gemstones in the cellar vault of the museum (Becker 2014).

Most plant fossils are momentary at the Senckenberg Museum in Frankfurt for more detailed investigations.

In 2011, the Hofmann couple bequeathed the museum a substantial collection of fossils, mainly zoological objects. This collection was for the time being the last important incoming in the geological section. At the moment the objects are registered digitally.

47.3 Focal Points of Exhibitions

The main part of the collection of natural history is stored in internal and external depots. Only parts of the collections are shown in the permanent exhibition, which ranges over 1200 m^2 and is subdivided into five parts. The exhibition's moor, coast and marshland and geest show the characteristics of the regional landscape of the northwest of Niedersachsen. They focus on the interrelation of nature and man and therefore display archaeological objects as well.

The exhibition concerning the local river Hunte includes freshwater and seawater aquaria with living animals from this region (e.g, insects, crustaceans, amphibians, reptiles and fishes). The human influence is also considered, for example, by showing living goldfishes. From the latter half of the nineteenth century to the beginning of the twentieth century, Oldenburg was an important centre for their breeding. Through escapes of the ponds, goldfishes were already found living in local freshwater in 1870. Today there are still goldfishes in the wild, maybe as offspring or recent maroons.

Finally, the last permanent exhibition deals with minerals and gemstone, which are displayed in the cellar vault of the museum.

The space for special exhibitions ranges over 600 m^2 and is subdivided in a room (approx. 400 m^2) and a gallery above (approx. 200 m^2). Thus, it is possible to show two different special exhibitions at a time or one larger one. In general, the special exhibitions change twice a year. The displayed topics are generally related to the collections or to the region, involving the whole world.

47.4 The Collection and Its Exploration

The collection of natural history is the basis for research in different ways. Some projects are internally realised, like the joint project Vernetzung und Erschließung zoologischer Museumssammlungen—am Beispiel der paläarktischen Käfer. It is

Fig. 47.4 Insight into the collection of ground beetles (Carabidae) at the Landesmuseum Natur und Mensch Oldenburg (with permission from: Landesmuseum Natur und Mensch Oldenburg; photo: C. Barilaro)

funded by the Deutsche Forschungsgemeinschaft (DFG, JO-134/14-1), and five museums of the NORe Consortium (Museumsverbund der Nord- und Ostsee Region) are involved (Staatliches Naturhistorisches Museum Braunschweig, Naturkundemuseum Bielefeld, Zoologisches Institut und Museum der Universität Greifswald, Zoologisches Museum der Universität Hamburg, Landesmuseum Natur und Mensch Oldenburg). The project focuses on the ground beetles (Carabidae) from the Western Palaearctic. Each museum records and thereby revises their respective specimens. The aim is to digitalize and link the collections and to make the final database accessible to the world via the Internet in order to foster further research (Fig. 47.4).

The project started in March 2013 for a period of 3 years. At the Landesmuseum Natur und Mensch, a scientific staff is employed to record and revise this part of the collection. It contains approximately 13,500 specimens. Until July 2015 round about 11,000 specimens are already digitalized.

The bulk of this collection dates back to the nineteenth and twentieth centuries and is based mainly on the collecting of Carl Friedrich Wiepken (1815–1897), Dr. med. Wilhelm Paasch and Johannes Kühn. In the recent past, only few specimens were collected, for example, in the course of the special exhibition "natur. wert.schätzen", which was shown in summer 2014 and compared the fauna of the man-made palace garden in Oldenburg and an unspoiled alluvial forest, its former habitat.

In addition to research concerning the objects in the collection, the scientific staff could be involved in field studies, like what happened with the Tooth-billed

Pigeon *Didunculus strigirostris* (Jardine, 1845) in Samoa (Beichle and Baumann 2001).

Besides this internal research, the collection provides specimens, tissue samples and data for requests of external scientists and institutions. For example, the museum cooperates with the Forschungs- und Technologiezentrum Westküste (FTZ, Research and Technology Centre) in Büsum in the scope of the monitoring of plastic ingestion by the Northern Fulmar (*Fulmarus glacialis*) in the North Sea (see van Franeker et al. 2011). Before the integration into the museum's collection, each Northern Fulmar is sent to the FTZ for the analysis of the stomach contents.

Moreover, in 2015 the museum provides tissue samples for a project at the Senckenberg Naturhistorische Sammlungen Dresden dealing with the distribution and phylogenetic relationship of Grass Snakes (*Natrix natrix*; inventory nos. REP107, REP128, REP496, REP553, REP957, REP960, REP961, REP962, REP963) (see Kindler et al. 2013, 2014). Furthermore, a skin sample of the Great Auk (*Pinguinus impennis*, inventory no. AVE 8086) was taken for a project at Bangor University, Wales, and the University of Copenhagen, Denmark, which focuses to identify the skins of the last living great auks ever seen by man by the use of genetic analysis.

Occasionally research arises from the daily work. For instance, during the inventory of the collection of birds into the digital database, it turned out that the collection contains a mounted Snowy Egret (*Egretta thula*, inventory no. AVE1213) (Fig. 47.5). Beforehand, it was determined and handled as a Little Egret (*Egretta garzetta*). After the revision, it became clear that this individual is the first record of this species in Germany and the Western Palaearctic (Gottschling et al. 2005).

With the neighbouring school, Graf-Anton-Günther-Schule Oldenburg, furthermore, an occasional cooperation regarding the project "Jugend forscht" exists. In 2014, for example, two students analysed the flocking behaviour of fishes in the museum's aquarium and reached the first place in the regional final.

47.5 Spreading Knowledge

Results of own research and contemporary issues regarding the museum are published in written form through magazines and journals, e.g., the "Museumsjournal Natur und Mensch", published by the museum; the "Oldenburger Jahrbuch", published by the Oldenburger Landesverein für Geschichte, Natur- und Heimatkunde e. V.; or "DROSERA Naturkundliche Mitteilungen aus Nordwestdeutschland", which was published by the museum in cooperation with the Carl von Ossietzky Universität Oldenburg from 1976 to 2002. Today's editors are the Verein zur Förderung Naturkundlicher Untersuchungen in Nordwestdeutschland e. V. and the Universitätsbibliothek Oldenburg.

Fig. 47.5 Snowy Egret (*Egretta thula*). First recorded in the Western Palaearctic (with permission from: Landesmuseum Natur und Mensch Oldenburg; photo: W. Kehmeier)

Further information regarding special exhibitions is given in appendant catalogues. Via lectures at conferences and workshops, the museum spread information about new exhibitions, the collections, the educational opportunities and other contemporary issues concerning the museum.

Besides these ways of publication, the permanent and special exhibitions are used to inform people. The museum reaches especially families and schools but also other parts of the population. The permanent exhibitions mainly concern the regional landscape. By the use of archaeological objects and objects of natural history, the exhibitions show how people live and handle the characteristics of this environment. How do people and animals live at the coast with the risk of storm floods? What do people do to make the meagre soil of the geest fertile, and what are the consequences for nature? What is so special about the moor biotope, and what do the archaeological findings tell us about our ancestors?

Text panels, audio guides or special guided tours and educational workshops serve to engross in the topics of the permanent exhibitions. Furthermore, an interactive digital book gives the opportunity to exchange experiences about the permanent exhibitions and the regional landscapes. This digital book is developed in cooperation with senior citizen and young migrants within a third-party project funded by the Deutsche Bundesstiftung Umwelt (DBU, 32433/01-43/2).

The presented special exhibitions originate either from external or from internal. In certain cases, during the planning of exhibitions, the museum collaborates with external partners, like universities, nature conservation organisations and the neighbouring school.

The special exhibitions mainly refer to the museum's collection or to the region and include matter which is up to date. Besides factual information, the concepts always provide interactive elements. The visitors shall have the opportunity to experience the exhibition in different sensuous ways. This should facilitate the acquisition of knowledge.

47.6 The Museum's Heart

Collections constitute the core of museums and provide the basis for the museum's key tasks: collecting, conserving, researching, exhibiting and communicating. Every item serves as an analogue archive and holds information for various desiderata. Especially zoological collections are of particular importance regarding the distribution and development of species over time. They give insights in environmental variation, e.g., due to climate change or human action. Thus, the collections provide an opportunity for research and education regarding matters of natural history. The Landesmuseum Natur und Mensch Oldenburg performs this task for the flora and fauna of the northwest of Niedersachsen. Its collection represents this region and its changes since almost two centuries and will be continued in the future.

References

Becker P (2014) Zustand und Ausbau der Naturkundlichen Sammlungen im Landesmuseum Natur und Mensch. In: Becker C (ed) Begeisterung für die Vielfalt der Natur: Naturkunde als Fundament des Naturschutzes—historische Facetten im Wirkungsfeld des Landesmuseums Natur und Mensch (Oldenburg). Isensee Verlag, Oldenburg, pp 215–223

Beichle UR, Baumann S (2001) Ethological studies as a basic need in conservation of Samoan endemic landbirds. Vogelkundliche Berichte aus Niedersachsen 33(2):191–195

Bengen E (2001) Vom Großherzoglichen Naturhistorischen Museum zum Landesmuseum für Natur und Mensch. Oldenburger Jahrbuch 101:207–234

Erbeling L (1991) Die Käfersammlungen im Staatlichen Museum für Naturkunde und Vorgeschichte in Oldenburg. DROSERA '91(1/2):21–24

Fuhrmann K, Kucharzewski C (2008) Die herpetologische Sammlung im Landesmuseum für Natur und Mensch, Oldenburg (Amphibia et Reptilia). Museumsjournal Natur und Mensch 2007(03):7–35

Gottschling M, Fuhrmann K, Beichle U (2005) Ein alter Beleg des Schmuckreihers *Egretta thula* als Erstnachweis für Deutschland und die Westpaläarktis: [First record of Snowy Egret *Egretta thula* for Germany and the Western Palaearctic]. Limicola 19:225–232

Kindler C, Böhme W, Corti C, Gvoždík V, Jablonski D, Jandzik D, Metallinou M, Široký P, Fritz U (2013) Mitochondrial phylogeography, contact zones and taxonomy of grass snakes (*Natrix natrix*, *N. megalocephala*). Zoologica Scripta 42(5):458–472

Kindler C, Bringsøe H, Fritz U (2014) Phylogeography of grass snakes (*Natrix natrix*) all around the Baltic Sea: implications for the Holocene colonization of Fennoscandia. Amphibia-Reptilia 35:413–424

Kucharzewski C, Tillack F, Fuhrmann K (2009) Rediscovery and redescription of the type specimens of *Ophites ruhstrati* Fischer, 1886. HERPETOZOA 22(3/4):181–187

Meyer KO (1998) Die Lepidopteren-Sammlung Tiedemann: Neuzugang in der Insektensammlung des Oldenburger Museums. DROSERA '98(1):73–78

Obermöller M (2007) Boden schreibt Geschichte: Lackprofile—Erdgeschichtliche Abziehbilder. Schriftenreihe des Landesmuseums für Natur und Mensch, vol 52. Primus Verlag, Darmstadt

Rose A (2004) Die Kurzflügler (Coleoptera, Staphylinidae excl. Scaphidiinae, Dasycerinae et Pselaphinae) der Sammlung 'Georg Kerstens', nebst aktualisiertem Verbreitungsstatus aller Arten des Weser-Ems-Gebietes. DROSERA 2004:137–158

van Franeker JA, Blaize C, Danielsen J, Fairclough K, Gollan J, Guse N, Hansen P, Heubeck M, Jensen J, Le Guillou G, Olsen B, Olsen K, Pedersen J, Stienen EW, Turner DM (2011) Monitoring plastic ingestion by the northern fulmar *Fulmarus glacialis* in the North Sea. Environmental Pollution 159:2609–2615

Will B (2014) Biografische Skizzen—Teil 3. In: Becker C (ed) Begeisterung für die Vielfalt der Natur: Naturkunde als Fundament des Naturschutzes—historische Facetten im Wirkungsfeld des Landesmuseums Natur und Mensch (Oldenburg). Isensee Verlag, Oldenburg, pp 247–287

Chapter 48
POTSDAM: Naturkundemuseum Potsdam: Natural History Museum of Potsdam Documentation of Regional Biodiversity and 50 Years of Experience with a Freshwater Aquarium

Detlef Knuth, Ina Pokorny, and Dirk Berger

Abstract In 1909, citizens of Potsdam and Berlin founded a municipal museum in Potsdam. This also included collections on natural history, which were lost during the bombardment of Potsdam at the end of the Second World War. In 1954, the museum was organized again under the name "District Museum of Potsdam" in the historic building, known as "Ständehaus der Zauche." In 1965, the first freshwater aquarium for native fish was opened. After a long period of reconstruction, the Natural History Museum Potsdam reopened in the former building in 2001. From the beginning, the collection focused on the wildlife of Brandenburg. Today, the museum holds a collection of about 351,000 objects, representing all classes of animals of the Central European fauna. Since 50 years, research focuses on freshwater fauna, mainly on zoogeographic aspects of freshwater fishes. In recent years, the impact of human activity on biodiversity became a crucial research objective at the museums.

Keywords Freshwater aquarium • Regional collection • Potsdam • Brandenburg • West Palearctic • Exhibition

48.1 General Information

48.1.1 Brief Historical Overview

The history of the Natural History Museum Potsdam began in 1909 with the foundation of a Museum Association with the aim to maintain, preserve, and

D. Knuth (✉) • I. Pokorny • D. Berger
Naturkundemuseum Potsdam (NKMP), Breite Straße 11/13, 14467 Potsdam, Germany
e-mail: naturkundemuseum@rathaus.potsdam.de; http://www.naturkundemuseum-potsdam.de; ina.pokorny@rathaus.potsdam.de; dirk.berger@rathaus.potsdam.de

L.A. Beck (ed.), *Zoological Collections of Germany*, Natural History Collections,
DOI 10.1007/978-3-319-44321-8_48

expand all collections that are of historical, cultural, scientific, artistic, and other general significance for Potsdam and surrounding areas. Already in 1909, an exhibition at the Old Town Hall Potsdam was presented. In 1931, the Department of Natural History was opened in the City Palace. During the bombardment of Potsdam in April 1945, all natural history collections were lost. Nevertheless, the museum was reopened in 1946 in the Royal Stables. In 1953, the museum reorganized under the name “District Museum of Potsdam.” Simultaneously, the construction of the natural history collections began, and the Department of Natural History was reestablished in 1956. In 1960 the Department of Natural History and the Department of History and Arts moved into the partly reconstructed residential building, known as “Ständehaus der Zauche” that was built in 1770. In 1964, the publication series “Contributions to the Fauna of the Mark” (Beiträge zur Tierwelt der Mark) began. The year 1965 was of crucial importance. A first freshwater aquarium with eight aquariums was opened that introduced visitors to the native fish fauna.

In 1982, the museum got the official name “Potsdam Museum” that incorporated the Department of Natural History and the Department of History and Arts. After a general construction of the building, the museum reopened in 1983 with the “Aquarium Fishes of Havel Waters.” It was also completely reconstructed and became a visitor attraction. From the beginning a special feature of the aquarium was a large European catfish, with more than 1.6 m in length. It quickly became the mascot. In 1984, the permanent exhibition on the “Nature and the Environment of the Havel Region” was opened. After the fall of the Wall in 1989, a second general reconstruction of the building was carried out and lasted from 1998 to 2001. Within these years, the Department of Natural History underwent dramatic changes, as the city of Potsdam decided to close the department, due to the tense financial situation in the post-GDR period. A broad publicity campaign and only one vote more within the City Council rescued the natural history in Potsdam. For the new start, the Department of History and Arts moved to another place, and the former building was reopened in 2001 as the Natural History Museum of Potsdam. The “Aquarium Fishes of Brandenburg” was updated to the latest technical standard, and the permanent exhibition “Endangered Fauna—Exterminated and Endangered Wild Animals of Brandenburg” was presented. Since then, the museum presents permanent and special exhibitions with local and global topics. In 2009, the Natural History Museum celebrated its 100th, and in 2015 the aquarium celebrated its 50th anniversary (Paepke 2015; Knuth 2015; Rothe and Knuth 2015).

48.1.2 Future Perspectives

In the 1990’s when the closure of the museum was discussed, the museum lost two scientific employees, who worked mainly on freshwater ecosystems of Brandenburg. Today, the scientific work is still underrepresented and recovers only slowly from the loss of manpower. The main aspect of the next years is to reestablish the scientific positions and the scientific output. A first afford in 2015 was the reactivation of the museum publication series under the name of “Veröffentlichungen des Naturkundemuseums Potsdam, Heft 1, 2015.” The first issue deals with the 50th anniversary of the aquarium. The museum is today continuing to be financed by the city of Potsdam. During the last decades, some other museums in the estate of Brandenburg were closed, mainly due to lack financial support. That is why the Natural History Museum is the only museum left in Brandenburg with permanent exhibitions on natural history. In conclusion, the estate of Brandenburg carries a special responsibility to maintain the last museum. Currently, the city of Potsdam is in negotiation with the estate to get a cofinancing to promote scientific and educational work. Another future aspect is the expansion of the exhibition halls. Due to the lack of space (currently about 600 m^2), scientific topics can only be presented in a limited dimension.

48.1.3 Conditions and Infrastructure

The Natural History Museum of Potsdam has two historical buildings in the city center of Potsdam (Fig. 48.1). The exhibition building contains the aquarium (190 m^2), exhibition rooms (600 m^2), two conference rooms, offices, and depots for sensitive collections (162 m^2). The administration building holds offices, three collection depots (75 m^2), the taxidermy laboratory, and two depots for the ethanol collection (42 m^2). A small building on the campus serves as quarantine for the aquarium. The museum is financed by the city of Potsdam with about 1.5 mio euro yearly and, depending on projects, by the estate of Brandenburg. The museum employs 17 permanent employees, including service and technical staff, the director, two scientists, a collection manager, a taxidermist, an aquarium manager, an administrator, an employee for public relations, and a scientific trainee. The museum keeps its own laboratories for zoological taxidermy and collection preparation. Most of the taxidermist work is done here, except the skin tanning. A large depot is situated outside of Potsdam with an area of about 600 m^2, keeping collections and technical equipment.

48.2 Collection and Research

Currently, the Natural History Museum of Potsdam holds a collection of over 351,000 objects (Table 48.1), containing mainly recent zoological material. The first collecting activities date back to the 1950’s and 1960’s. Another intensive collecting period was in the 1980’s and 1990’s, as well as current collection expansion. Furthermore, historical collections of Brandenburg (given by private collectors or other museums, which close their natural history departments) have been integrated in the collections of the Natural History Museum of Potsdam.

Fig. 48.1 The National History Museum of Potsdam with the historical exhibition building (Ständehaus der Zauche) to the *left* and the administration building, the former barrack building of the Ständehaus, to the *right*

Table 48.1 Overview of the collection of the Natural History Museum of Potsdam (status 2016)

Collection	Number
Insects	280,000
Spiders	2200
Crustaceans	1200
Mollusks	11,000
Corals	50
Fishes	24,000
Amphibians and reptiles	900
Birds	20,500
Mammals	2500
Geology/mineralogy/paleontology	6500
Botanical collection	950
Others (models, technical equipment)	800
Books, photos, archives	19,000
Sum	369,600

48.2.1 Collection on Entomological Objects

The entomological collections are the most extensive part of the zoological collections. More than 280,000 insect specimens (about 170,000 Lepidoptera, 80,000 Coleoptera, and 30,000 further orders) are preserved here. The majorities of the entomological collections are acquisitions or donations of members of the past Entomological Association of Potsdam, recent entomologists, or were transferred from universities and research institutions from Brandenburg and Berlin. In addition, materials from own collecting activities and scientific projects complete the collections. Mentionable are the collection of the honorary citizen of Potsdam Max Volmer that comprises 9450 lepidopterans—from the Palaearctic region, Germany, the Alpine countries, Italy, Bulgaria, Russia, Spain, and Tunisia—and single specimens from Lapland, Japan, and Portugal; the regional collection of Rüdiger Ohnesorge with 20,970 lepidopterans; the collection of Erdmann Griep, the last secretary of the Entomological Association of Potsdam, that documents the coleopteran fauna of the region of Potsdam for more than three decades, with 34,200 specimens; and the collection of Karl-Heinrich von Eckartsberg (28,300 coleopterans).

A recent acquisition is the collection of Joachim Oehlke (11,800 lepidopterans) and of Lutz Lehmann which contains 59,285 diurnal butterflies and moths mainly from the Palaearctic, but also from North Africa, Central Asia, Middle East, and North America. Other collections which are deposited at the Natural History Museum of Potsdam are the collections of Heinrich Auel (3940 insects), Willy Dornbusch (5430 lepidopterans), Frank Ellner (2480 lepidopterans), Ewald Foerster (13,070 lepidopterans, coleopterans, growth stages, and parasitoids), Kurt Gericke (17,740 lepidopterans, 2130 coleopterans, 2430 others), Bernhard Hassenstein (1920 coleopterans), Ludwig Kempf (5440 coleopterans, 200 others), Paul Knaack (3900 lepidopterans), Claus-Peter Kockel (10,390 lepidopterans, 2310 coleopterans, 850 others), and Otto Krumme (2670 lepidopterans) and smaller collections (mainly lepidopterans). The systematic rearrangement of a principal systematic collection is in progress, whereby some collections are maintained as individual collection.

48.2.2 Collection on Crustaceans

The collection on crustaceans focuses on European freshwater decapods. A monitoring on European crayfish *Astacus astacus* in Brandenburg from 1990 to 2010 (Knuth 1999) led to a collection of 120 individuals from Eastern Germany (Thuringia, Saxony-Anhalt, Saxony, Mecklenburg-Western Pomerania). Individuals from Poland, Sweden, and Russia allow genetic comparison. The collection is accomplished by 300 individuals of other freshwater crustaceans (Astacidae, Cambaridae, Procambaridae, and Chinese mitten crab *Eriocheir sinensis*). The collection on Branchiopoda contains 700 specimens from Northern and Eastern Germany.

48.2.3 Collection on Freshwater Fishes

The fish collection is the largest vertebrate collection. Since 1956, freshwater fish species from lakes and rivers in Brandenburg are collected and preserved in ethanol (Fig. 48.2). Currently, the collection comprises about 24,500 individuals and is further expanded. Due to scientific research, the three-spined stickleback *Gasterosteus aculeatus* (8000 individuals collected by Hans-Joachim Paepke) and the European smelt *Osmerus eperlanus* f. *spirinchus* (11,000 individuals collected by Detlef Knuth) are represented in large numbers. The fish collection does not only give an excellent overview of the fish fauna of Brandenburg; it also documents the change in fish fauna induced by human activities. Noteworthy are several new records of fish species in Brandenburg, e.g., European bullhead (*Cottus gobio*), Common minnow *Phoxinus phoxinus* (Feiler 1966), White-finned gudgeon *Romanogobio belingi* (Rothe 1998), and Schneider *Alburnoides bipunctatus* (Rothe 2002).

48.2.4 Collection on Birds

The collection on birds includes 1500 dermoplastics, 2000 skins, 3000 skeletons, and 14,000 eggs. Since the 1980's, the museum was committed to collect endangered bird species, such as Peregrine falcon *Falco peregrinus*, European roller *Coracias garrulus*, Black stork *Ciconia nigra*, Great bustard *Otis tarda*, Hen harrier

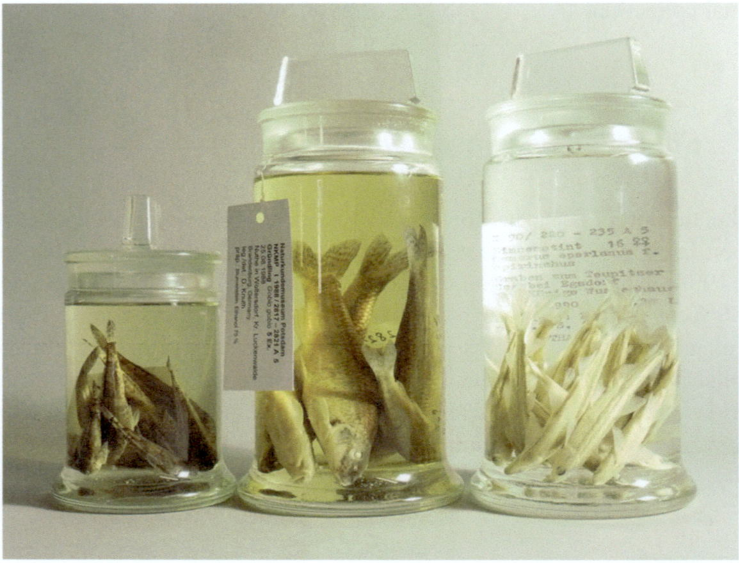

Fig. 48.2 Fish specimens mainly caught during biodiversity studies throughout Brandenburg, stored in the museum's wet collections

Circus cyaneus, and Montagu's harrier *Circus pygargus* by GDR law. Today, the bird collection is accomplished by birds that are found dead by employees of the Ministry of Agriculture and Environment of Brandenburg and their offices and by volunteers. Highlights of the collection are about 100 individuals of Great bustard *Otis tarda*, 40 individuals of White-tailed eagle *Haliaeetus albicilla*, and 55 individuals of Common buzzard *Buteo buteo*. With the acquisition of the egg collection of J. Kleinfeld in 2003, the museum received more than 7000 eggs from 511 bird species from Europe, which were collected between 1970 and 2000. For the exhibition "Storks of the Earth," 17 of 19 stork species were included in the collection. The ornithological collection of the museum contains comprehensive material of the biological diversity of birds in Brandenburg.

48.2.5 Collection on Mammals

Currently, the collection on mammals contains about 2500 individuals. Besides a systematic collection on mammals of the Western Palaearctic, the documentation focuses on endangered species in Brandenburg. In close cooperation with the Ministry of Agriculture and Environment of Brandenburg, the museum is the central institution for the collection of wolf, beaver, otter, and bats. This includes skins, skeletons, skulls, tissue samples, PEG objects, and animal traces. In particular, the museum preserves 50 autochthonous specimens of Eurasian wolf (*Canis lupus lupus*, status September 2016), which recolonized Germany from Poland during the last two decades. Therefore, the museum keeps the largest collection of autochthonous wolves in Germany. Furthermore, about 180 specimens of Eurasian beaver *Castor fiber albicus* (mainly from the endangered subspecies) and about 200 specimens of European otter *Lutra lutra* are part of the collection. The collection does not only include skulls, complete skeletons, and skins but also a depot of tissue samples for genetic or other molecular biological analysis. Some rare species, which are nowadays extinct in wide parts of Germany, are part of the collection, e.g., the last European hamster *Cricetus cricetus* from Brandenburg.

48.2.6 Research

For more than 50 years, the biodiversity of freshwater habitats is the main research area of the museum. As a result, new knowledge of the distribution and status of species such as the European pond turtle (Paepke 1977), Three-spined stickleback (Paepke 1970, 1971), European smelt (Knuth 1992b), and European crayfish (Knuth and Mietz 1993; Knuth 1999) were documented, and extensive collections were built up. The research on freshwater fish species was an important basis for the development of the first Red List of Fishes of Brandenburg (Knuth 1992a; Knuth and Rothe 1998). The research activities of the museum not only initiated fish protection in the GDR and in Germany but also promoted the protection of river systems, known as Fließgewässer-Verbundsystem Brandenburg (Braasch et al. 1993, 1994). As part of the project "Seenkataster Brandenburg," the museum investigated the fauna and the anthropogenic influence on freshwater communities of small lakes and ponds (Knuth 1994).

Currently, the museum works on the biodiversity of the agricultural drainage system in Brandenburg. Drainage ditches are spread over large parts of Brandenburg in order to draw back water from agricultural land. Thereby, many natural freshwater habitats have been destroyed within the last two centuries. The project focuses on the documentation of biological diversity of the ditches. The first results indicate that the ditches are important refuges for some threatened species, such as the European weather loach *Misgurnus fossilis*, the Common spadefoot *Pelobates fuscus*, the European fire-bellied toad *Bombina bombina*, and various water beetles.

The collection also serves in research with cooperation partners such as the University of Potsdam, the Humboldt University of Berlin, the University of Konstanz, other museums, and the Ministry for Rural Development, Environment and Agriculture of the Federal State of Brandenburg. The potential of the recent collections is obvious. It can not only reveal loss or changes in biodiversity due to land use transition since the 1950's, but it can also be useful to answer questions on climate change and the influence of the German "Energiewende" and its impact on land use and biodiversity. The collection of endangered species documents the loss of species and in recent times the recolonization of formerly endangered or extinct species.

48.3 Cooperation

The Natural History Museum is a member of the Museum Association of Brandenburg and the German Association of Museums, where the natural historic museums work in close scientific exchange. In education, the museum works closely together with different schools in Potsdam, the Department of Didactics at the University of Potsdam, the University of Applied Sciences Potsdam, the Eberswalde University for Sustainable Development, and numerous NGOs that deal with nature conservation and environmental education. In research and collection expansion, the museum cooperates with other museums and the different offices of the Ministry for Rural Development, Environment and Agriculture of the Federal State of Brandenburg, especially with Bird Conservation Centre (Staatliche Vogelschutzwarte). In freshwater research, the Soil and Water Association and nature lovers, citizens, and fishers are irreplaceable partners.

48.4 Educational Work

The Natural History Museum of Potsdam is the last existing museum in the estate of Brandenburg with permanent exhibitions and an education program of natural history. Therefore, it has a special mission, not only to entertain and teach local citizens of Potsdam but also to spread knowledge into the whole of Brandenburg.

48.4.1 Permanent and Special Exhibitions

The permanent exhibition on “Wildlife of Brandenburg” shows local aspects of changes in biodiversity due to land use changes over the last centuries. This includes the loss of natural habitats and their specific species. Furthermore, aspects of ongoing recolonization of formerly extinct or highly endangered species are presented. This includes, for instance, European wolf *Canis lupus lupus*, Eurasian beaver *Castor fiber*, European otter *Lutra lutra*, Great bustard *Otis tarda*, White-tailed eagle *Haliaeetus albicilla*, Eurasian eagle-owl *Bubo bubo*, European pond turtle *Emys orbicularis*, and European sturgeon *Acipenser sturio*.

Local aspects of biodiversity are part of the exhibitions “Wildlife in Gardens” and “Wildlife in the UNESCO World Heritage,” which show the extraordinary fauna of Sanssouci Park (Fig. 48.3). This exhibition represents parts of the city

Fig. 48.3 The exhibitions of the Natural History Museum are distributed over four floors in the exhibition building. Impression of the exhibition “Wildlife in the UNESCO World Heritage,” a current special exhibition that deals with the biodiversity of the historical and cultural places of Potsdam

ecology of Potsdam and is based on a modern educational concept for the presentation of natural history topics (Knuth 2014). Global aspects of biological diversity are present in the exhibitions “In Human Traces—Biological Invasions” and “Storks of the Earth.” The latter is currently touring through Germany.

Since 1965 the museum harbors a freshwater aquarium. At the beginning, in eight aquariums with a water volume of 3000 l, native fish species were presented. A modern aquarium complex with 16 aquariums and with a water volume of 36,000 l was opened after an extensive reconstruction of the museum building in 1983. In 2001, the last modernization of the aquarium, the “Aquarium Fishes of Brandenburg,” leads to 22 aquariums with a water volume of more than 30,000 l. Constantly, more than 40 native fish species and various invertebrates such as crabs, aquatic insects, and molluscs are shown (Fig. 48.4).

48.4.2 Didactic Concept

The main educational aspect is the presentation of the original object to reach visitors of all ages and of all educational backgrounds. Therefore, all material is prepared in the most natural way (Fig. 48.5) with the help of modern taxidermy methods, like the PEG technology (plastination). The museum avoids the exhibition

Fig. 48.4 View of the “Aquarium Fishes of Brandenburg.” The freshwater aquarium is situated in the basement of the exhibition building. Water quality, maintaining low temperatures in a cold water aquarium, and aesthetically furnished aquariums are important requirements to present the native life of the waters of Brandenburg

Fig. 48.5 The yellow-necked mouse *Apodemus flavicollis*—a naturalistic mounted specimen using PEG technology

of old and unnatural objects and replaces objects regularly to keep the aesthetical viewpoint high. The didactic concept includes lectures, events, and guided tours for adults, pensioners, families, groups of children (schools, kindergarten), and also disabled people. All exhibitions follow the concept of the presentation of the natural object. Therefore, text blocks are reduced and natural relationships are presented in biological groups. The integration of exhibition entertainment (interactive screens, listening stations, guided tours via headphones) is minimized to prevent the visitor from losing the attention from the original object.

About 55 % of the visitors come from Potsdam and Berlin, about 40 % from Brandenburg and Germany, and about 5 % from other nations. As the museum expects an increase in nonnative speakers, all new texts are displayed in German and English.

References

Braasch D, Scharf R, Knuth D (1993) Zur Erfassung und Bewertung sensibler Fließgewässer im Land Brandenburg. Naturschutz Landschaftspflege Brandenburg 11:104–105

Braasch D, Scharf R, Knuth D (1994) Konzeption eines naturschutzbezogenen Fließgewässer-Biotopverbundsystems im Land Brandenburg. Naturschutz Landschaftspflege Brandenburg 2 (2):31–36

Feiler M (1966) Groppe (*Cottus gobio*) und Elritze (*Phoxinus laevis* Ag.)—zwei neue Fischarten für das westliche Brandenburg. Beiträge zur Tierwelt der Mark III:51–55

Knuth D (1992a) Rote Liste Rundmäuler (Cyclostomata) und Fische (Pisces). In: MUNR—Ministerium für Umwelt, Naturschutz und Raumordnung des Landes Brandenburg, Gefährdete Tiere im Land Brandenburg—Rote Listen. Unze-Verlag Potsdam, pp 36–38

Knuth D (1992b) Die Verbreitung des Binnenstintes (*Osmerus eperlanus* f. *spirinchus*) in Brandenburg unter besonderer Betrachtung von populationsdynamischen Aspekten im Teupitzer Seengebiet sowie die Bedeutung eines Fließgewässerschutzsystems für den Erhalt charakteristischer Lebensräume und ihrer Naturausstattung. Dissertation A Math Nat Fakultät der Humboldt-Universität zu Berlin, p 118, 13 November 1992

Knuth D (1994) Ergebnisse der ichthyologischen Untersuchung an Kleinseen im Rahmen des Seenkatasters Brandenburg 1992—Fischartenkataster Brandenburg. Beitr Angew Gewässerökologie Norddeutschlands 1:130–144

Knuth D (1999) Neue Kenntnisse zur Verbreitung des Edelkrebses *Astacus astacus* in Brandenburg. Beiträge zur Tierwelt der Mark XIV:99–108

Knuth D (2014) Climate change and loss biodiversity require knowledge transfer in civil society, In: Historic gardens and climate change, recommendations for preservation. Edition Leipzig, pp 24–25

Knuth D (2015) Aquarium "Fische der Havelgewässer" 1978 bis 1999, Ein Kaltwasseraquarium initiiert den Fischarten- und Fließgewässerschutz in Brandenburg. Veröff Naturkundemus Potsdam 1:13–32 (Festschrift 50 Jahre Aquarium, Landeshauptstadt Potsdam)

Knuth D, Mietz O (1993) Verbreitung, Gefährdung, Gewässeransprüche und Erhaltung des Edelkrebses *Astacus astacus* in Brandenburg. Naturschutz Landschaftspflege Brandenburg 3 (2):16–21

Knuth D, Rothe U (1998) Rote Liste der Rundmäuler und Fische des Landes Brandenburg. Naturschutz Landschaftspflege Brandenburg 4 Beilage:1–19

Paepke H-J (1970) Studien zur Ökologie, Variabilität und Populationsstruktur des Dreistachligen und Neunstachligen Stichlings II. Die Variabilität der Lateralbeschilderung von *Gasterosteus aculeatus* L. in einer brandenburgischen Intergradationszone und ihre zoogeographisch-historischen Hintergründe. Beiträge zur Tierwelt der Mark VII:5–48

Paepke H-J (1971) Studien zur Ökologie, Variabilität und Populationsstruktur des Dreistachligen und Neunstachligen Stichlings III. Größen-, Alters-, und Geschlechtsunterschiede in brandenburgischen Populationen von *Gasterosteus aculeatus* L. Beiträge zur Tierwelt der Mark VIII:5–22

Paepke H-J (1977) Zur gegenwärtigen Verbreitung der Europäischen Sumpfschildkröte (*Emys orbicularis*) in den brandenburgischen Bezirken Potsdam, Frankfurt/Oder, Cottbus und in Berlin (Reptilia, Emydidae). Mitt Zool Mus Berlin 53(1):173–185

Paepke H-J (2015) Aquarium "Fische der Havelseen" 1964 bis 1977, Was macht der Wels?—Erinnerungen an die Anfangsjahre des Potsdamer Museumsaquariums. Veröff Naturkundemus Potsdam 1:5–12 (Festschrift 50 Jahre Aquarium, Landeshauptstadt Potsdam)

Rothe U (1998) Erstnachweise des Weißflossengründlings (*Gobio albipinnatus* Lukasch 1933) in Ostdeutschland. Beiträge zur Tierwelt der Mark XIII:47–52

Rothe U (2002) Der Schneider *Alburnoides bipunctatus* (Bloch, 1782) [Pisces, Cyprinidae]—erstmals in Brandenburg nachgewiesen. Zoosyst Evol 78(1):183–185

Rothe U, Knuth D (2015) Aquarium "Fische Brandenburgs" 2000 bis 2015, Moderne Ausstattung und qualifiziertes Aquarienpersonal garantieren einen vitalen Fischbestand. Veröff Naturkundemus Potsdam 1:33–52 (Festschrift 50 Jahre Aquarium, Landeshauptstadt Potsdam)

Chapter 49
ROSTOCK: The Zoological Collection of the University of Rostock

Stefan Richter and Andreas Bick

Abstract The Zoological Collection of the University of Rostock can be traced back to 1775. Since 1882, it is based at the Zoological Institute building in the center of Rostock. Today, the collection still contributes to the university's zoology teaching program. It serves as an inventory of the fauna of Mecklenburg-Vorpommern and the Baltic Sea, particularly for fish, marine bristle worms, and other invertebrates. The collection also includes important documents of the world's biodiversity. The Zoological Collection additionally functions as a small but well-visited zoological "museum" in the center of Rostock. The exact number of specimens or series of specimens in the Zoological Collection is unknown. We estimate there to be 150,000 series. Historical material is being reedited and put at the service of the biodiversity research that the department's scientists are involved in. The collection is thus an integral part of the Department of General and Systematic Zoology and an important asset.

Keywords Pfeilstorch • Arrow stork • O. Tychsen • F.E. Schulze • F. von Müller

49.1 Historical Background

The origin of the Zoological Collection of the University of Rostock can be traced back to 1775 when Oluf Gerhard Tychsen (1734–1815) put together a natural history collection for the University of Bützow which, in 1789, was (re)unified with the University of Rostock. Other parts of the collection came from the "Naturalienkabinett" or collection of curiosities, owned by the duke of Mecklenburg, Friedrich Franz I. The earliest objects still preserved today date back to the seventeenth century. In the nineteenth century, the collection was primarily used for teaching students of zoology. In 1865, Franz Eilhard Schulze (1840–1921) became the curator of both the Zoological Collection and the university's anatomical

S. Richter (✉) • A. Bick
Allgemeine and Spezielle Zoologie, Institut für Biowissenschaften, Universitätsplatz 2, 18055 Rostock, Germany
e-mail: stefan.richter@uni-rostock.de; http://www.zoologie.uni-rostock.de/sammlung/; andreas.bick@uni-rostock.de

L.A. Beck (ed.), *Zoological Collections of Germany*, Natural History Collections,
DOI 10.1007/978-3-319-44321-8_49

collection. The two collections were fused in 1871 when F.E. Schulze founded the Zoological Institute within the Faculty of Philosophy and became its first director. The professors who succeeded F.E. Schulze realized that the Zoological Collection in Rostock could never compete with the much larger and partly public collections in Berlin or Hamburg. Nevertheless, Alexander Goette (1840–1922), who was the director of the Zoological Institute [by now based in its own building (Fig. 49.1)] between 1882 and 1886, started to create an exhibition for educational purposes which was directed at both the students and the public. The exhibition was intended to document the **life** of the fauna in Mecklenburg. By the turn of the twentieth century, the entire second floor of the building was occupied by the Zoological Collection, a situation we have recently reestablished. Between then and the mid-twentieth century, various professors came and went as directors of the institute and heads of the collection, most notably the future Nobel Laureates Hans Spemann (1869–1941) and Karl von Frisch (1886–1982). During the second half of the twentieth century, the collection largely faded from notice, until in 1995 R. Kinzelbach (*1941) brought it back into public focus. For further information on the history of the Zoological Collection we refer to Braun (1891), Will (1907), and Kinzelbach (2000, 2013).

Today, the collection still contributes to the university's zoology teaching program. In addition, it serves as an important inventory of the fauna of Mecklenburg-Vorpommern and the Baltic Sea, particularly for fish, marine bristle worms, and other invertebrates. The collection also includes important documents of the world's biodiversity. The Zoological Collection additionally functions as a small but well-visited zoological "museum" in the center of Rostock.

Fig. 49.1 The Zoological Institute where the collection is located in the center of the Hanseatic city of Rostock (Photo: © with Zoological Collection, University of Rostock)

One of the highlights of the collection is the famous Mecklenburg white stork ("arrow stork") of 1822, which provided some of the first evidence of the long-distance migration of birds (Fig. 49.2) (Kinzelbach 2013). This stork was also the inspiration for the logo of our collection. Other highlights are Australian mammals and birds contributed by the botanist Sir Ferdinand von Müller (1825–1896), reptiles and amphibians contributed by the businessman Theodor Cordua (1796–1857), two ichthyosaurs from the Jurassic period found in the 180-million-year-old Posidonia Shale, and the reference specimen from the earliest scientific anatomical work on the Amazonian manatee by Hermann Friedrich Stannius (1808–1883).

One of the oldest invertebrate collections is that of mollusks. Some of the marine gastropods and bivalves in this collection are thought to be type material used by Heinrich Friedrich Link (1767–1851), but they have not yet been identified for certain. Some invertebrates from the first German deep-sea expedition with the research vessel "Valdivia" are also deposited in the invertebrate collection. We also have several collections of tropical and European butterflies and other insects.

Our collection includes important models of animals and of cleavage stages, including glass models by the world-renowned Bohemian manufacturer Blaschka and wax models by the German manufacturer Ziegler. The collection also boasts an abundance of beautifully illustrated animal wall charts created by Rudolf Leuckart (1822–1898) with the help of Rostock professors. Recent additions to the collection include a mounted polar bear as a gift from Rostock Zoo (Fig. 49.3).

Fig. 49.2 The Mecklenburg *arrow stork* (Pfeilstorch) provided some of the first evidence of the long-distance migration of birds and was also the inspiration for the logo of our collection (Photo: © with ITMZ, University of Rostock)

Fig. 49.3 The mounted polar bear "Churchill" at the opening of the special exhibition on the fauna of the polar regions in April 2014 (Photo: © with ITMZ, University of Rostock)

49.2 Focus of Collection

The exact number of specimens or series of specimens in the Zoological Collection is unknown. We estimate there to be 150,000 series. Some of the specimens are now included in the SESAM database maintained by the Senckenberg Institute, Frankfurt. The emphasis of our collection is on aquatic animals, mollusks, insects, fishes, and birds. The number of vertebrates in the collection stands at approximately 5000, and many of these specimens are mounted. The invertebrate collection is made up of around 50,000 mollusks (Fig. 49.4) and 100,000 insects. The most important reference collections from recent times are of marine bristle worms, crustaceans, scorpions, and fishes. The bird collection has recently benefited from material

Fig. 49.4 The collection of mollusks is one of the oldest parts of our collection and holds around 50,000 specimens (Photo: © with ITMZ, University of Rostock)

received in the course of the disbanding of other historical collections and from donations of a number of skeletons (Kinzelbach 2016).

49.3 Infrastructure

The collection is part of the Department of General and Systematic Zoology at the Institute of Biosciences. The staff responsible for the collection is attached to this department. The professor of General and Systematic Zoology is simultaneously the head of the Zoological Collection. The collection's staff includes a curator and one technician, though other members of the department also work in the collection. The collection benefits too from the many voluntary hours put in by the former head of the department. Financially, the collection is modestly supported by the university. The second floor of the Zoological Institute is now almost entirely occupied by the exhibition. There is only a partial separation between the exhibition and the collection.

49.4 Research

The collection is currently being expanded to take account of the research now being conducted in our department, particularly that into crustaceans, polychaetes, and fishes. Historical material is being reedited and put at the service of the biodiversity research that the department's scientists are involved in. The collection is thus an integral part of the Department of General and Systematic Zoology and an important asset. The scientific focus of the department is on evolutionary morphology, the systematics and taxonomy of arthropods and polychaetes, and fisheries biology. The collection contains material from all over the world which is used for taxonomic study—including the description of new species. The material has formed the basis of numerous graduate and undergraduate dissertations, Ph.D. theses, and professorial dissertations produced in the department in recent years.

The society *Verein der Freunde und Förderer der Zoologischen Sammlung Rostock e.V.* was set up in 2008 with the charitable aim of promoting the collection and providing support in areas including the documentation and upkeep of its material, the staging of exhibitions, the modernization of the repository, the publication of information, and the assimilation and upkeep of gifts and bequests.

49.5 Educational Work

Large parts of the collection are organized in accordance with traditional systematics (Fig. 49.5). One of our main intentions is to show the collection as it was originally organized for teaching purposes. The collection is housed in a nineteenth-

Fig. 49.5 Collection of invertebrates arranged according to traditional systematics—from sponges to insects (Photo: © with Zoological Collection, University of Rostock)

century building, lending it a unique atmosphere and making it one of the most visited attractions when Museum Nights are held in Rostock. We are now in the process of carefully reorganizing parts of the collection according to the biogeographic origin of our specimens. Since 2003 a small exhibition room has existed on the ground floor in which special exhibitions are shown which change annually. For the last 3 years, the special exhibition has been planned and arranged by Masters students. School classes regularly visit the collection and are shown around by its staff. At the moment, the collection is open to the public on weekdays only. The Zoological Collection is on Facebook and has its own website.

References

Braun M (1891) Geschichte der Zoologie und der zoologischen Sammlung an hiesiger Universität. Arch d Freunde d Naturgesch in Mecklenburg 44:III–VII. Güstrow

Kinzelbach R (2000) Die Zoologie an der Universität Rostock: Von der Wunderkammer zu Biodiversitätsforschung und Biotechnologie. Eine historische Übersicht. Verh Gesch Theorie Biol 5:159–181. Berlin (VWB 2000)

Kinzelbach R (2013) Das neue Buch vom Pfeilstorch. Basiliskenpresse, Rangsdorf, p 104

Kinzelbach R (2016) Die Vögel—Aves. Verzeichnis des Bestandes 2016. Katalog Zool Sammlung Univ Rostock 3. Auflage, p 199

Will L (1907) Begrüßungsansprache. Verhandlungen der Deutschen Zoologischen Gesellschaft auf der siebzehnten Jahresversammlung zu Rostock und Lübeck, den 21. bis 23. Mai 1907. Engelmann, Leipzig, pp 13–16

Chapter 50
SCHIFFWEILER: Zentrum für Biodokumentation des Saarlandes (ZfBS)

Andreas Werno

Abstract The Centre of Biodocumentation (Fig. 50.1) is a separate unit within the Ministry of Environment and Consumer Protection of the Saarland (MUV). Protection, care and improvement of the natural history collections of the Saarland are one of the main objectives. The collection of vertebrates mainly consists of preparations of the former collection of the University of the Saarland with some zoological rarities like the skeleton of the dodo (Fig. 50.2).

The invertebrate collection (insects) represents the largest part of the zoological collection with more than 10,000 species, many types and more than one million specimens from all over the world. The collection contains among others the large collections of Prof. Dr. G. de Lattin, Prof. Dr. G. Mosbacher, Dr. W. Marten, Dr. E. Diehl, etc.

Keywords Center for Biodocumentation • Delattinia • Dodo • Entomological Collection of de Lattin • Saarland • Germany

The Centre of Biodocumentation (ZfB) began its pioneering work at the future location of Reden by the end of 2002. The collections of natural science of the University of the Saarland were transferred to the ZfB in a form of a long-term loan in the beginning of 2003. The ZfB received later on further larger zoological private collections. The Centre of Biodocumentation is a separate unit of the Department (D/6) of Nature Conservation (Department of Nature conservation) within the Ministry of Environment and Consumer Protection (MUV) since September 1, 2014.

One of the main targets of the Centre of Biodocumentation is the protection, care and improvement of the collections of natural science of the Saarland.

A. Werno (✉)
Zentrum für Biodokumentation des Saarlandes (ZfB), Am Bergwerk Reden 11, 66578 Schiffweiler, Germany
e-mail: a.werno@umwelt.saarland.de

L.A. Beck (ed.), *Zoological Collections of Germany*, Natural History Collections,
DOI 10.1007/978-3-319-44321-8_50

Fig. 50.1 Center for Biodocumentation (with permission from: ZfBS)

Fig. 50.2 Dodo (*Raphus cucullatus*) (with permission from: ZfBS)

50.1 Zoological Collection Vertebrata

The collection of vertebrates gives an overview about the European and extra-European amphibians, reptiles, birds and mammals as well as the marine fauna. There are preparations in wax on the ontogenetic history of fishes, frogs and birds; half-side preparations of some vertebrates, skulls and front extremities with colour markings with respect to comparative anatomy of vertebrates; and torso models to the anatomy of men. Especially extensive are the European preparations of birds. The magazines contain in addition birds, rodents, ungulates, primates, carnivores, whales and different other mammals. There are cabinets with completely mounted skeletons, skeleton parts and skulls of vertebrates. Further parts of the collections are, for example, some zoological rarities like the skeleton of the dodo (*Raphus cucullatus*).

Presently recorded are the following exhibits:

Nine hundred eighty preparations of birds, 236 bird skins, 149 bird eggs or nests of bird eggs, 75 skulls of birds and 22 skeletons of birds or rather parts of skeletons (1462 exhibits altogether) and 375 skulls of mammals, 139 preparations of mammals and 72 skeletons of mammals or rather parts of skeletons (586 exhibits altogether).

Forty-five preparations of reptiles, 26 reptile skulls and 18 reptile skeletons or rather parts of skeletons, 89 reptile exhibits altogether; 24 amphibian exhibits; 55 fish exhibits; and 93 exhibits of corals and not nearer documented crustaceans, crabs, starfishes and mussels (Figs. 50.3, 50.4 and 50.5).

Fig. 50.3 Zwergkiwi (*Apteryx owenii*)

Fig. 50.4 Schnabeligel (*Tachyglossus aculeatus*) (with permission from: ZfBS)

Fig. 50.5 Schnabeltier (*Ornithorhynchus anatinus*) (with permission from: ZfBS)

50.2 Zoological Collection Invertebrata

The invertebrate collection (insects) represents with about 10,000 species and more than 1,000,000 specimens of Lepidoptera and 5000 species and more than 100,000 specimens of beetles from all over the world, the largest part of the zoological collection. Completely digitally recorded of it are presently about 210,000 Lepidoptera (butterflies and moths) and 67,000 Coleoptera (beetles).

The entomological collection consists of the former zoological collections of the University of the Saarland (Prof. Dr. Gustaf de Lattin) and the extensive collections of the natural history society “Delattinia (Naturforschende Gesellschaft des Saarlandes e.V.)”. The entomological collection is arranged in three parts for all insect orders: into a regional collection (Saarland with adjacent areas), into a collection of “Macaronesia” (Canaries, Madeira, Azores, Cape Verde Islands and coastal zone of Morocco) and into a worldwide collection. These collections contain numerous types of butterflies and moths (Lepidoptera) (Fig. 50.6) and beetles (Coleoptera). Many types exist especially from the South American collections of Marten, Spitz, Fassl and Richter from 1875 to 1929 (Figs. 50.7, 50.8 and 50.9).

Fig. 50.6 Typus von *Micropterix stuebneri* (Zeller, Werno & Kurz, 2013) (with permission from: ZfBS)

Fig. 50.7 Typus von *Pheles strigosa f. trichroma* (Staudinger, 1876) (with permission from: ZfBS)

Fig. 50.8 Typus von *Heliconius phyllis f. krügeri* (Neustetter, 1925) (with permission from: ZfBS)

Fig. 50.9 Typus von *Heliconius antiochus alba ab. trimaculata* (Krüger, 1933) (with permission from: ZfBS)

50.3 Collections of Delattinia

The collections belonging to Delattinia consist mainly of butterflies, moths and beetles but also of some other insect groups:

1. Collection G. Mosbacher (St. Ingbert): about 100,000 beetles in 120 drawers and about 40,000 butterflies and moths in 120 drawers and other insects in about 10 drawers.
2. Collection E. Dewes (Dudweiler): about 5000 butterflies and moths and 5000 beetles from Europe.
3. Collection H.E. Back (Bonn): about 10,000 micro- and macro-Lepidoptera in Europe/Africa/Asia/South America in about 70 drawers.

4. Collection R. Koshofer (Bonn): about 5000 beetles in Rhineland-Palatinate in 24 drawers.
5. Collection M. Meyer (Keßlingen): about 60,000 butterflies and moths in Europe in 200 drawers, special collection in Macaronesia/Morocco in about 90 drawers and 4 drawers in Ecuador and Australia.
6. Collection E. Diehl, E.L. Braun and H. Schreiber: about 150 drawers in Madagascar and Sumatra and special collection of sphingids of the world.
7. Collection Sand (Hassel): Lepidoptera in Europe in 30 drawers and butterflies and moths worldwide in 40 drawers.
8. Collection Dr. Kräßig (Mannheim): Lepidoptera in Central Europe in 70 drawers.
9. Collection H. Martin (Saarbrücken): about 10,000 butterflies in the Western Palaearctic in 110 drawers.
10. Collection W. Schmidt-Koehl (Saarbrücken): about 15,000 Lepidoptera in Saarland.
11. Further smaller collections of H. Hertz (Homburg), A. Röder (Losheim), H. Junk (Siersburg), H. György (Homburg), W. Beermann (Waldshut), etc.

50.4 Collections of the University of the Saarland (Entomological collection of de Lattin)

1. Collection de Lattin: about 300,000 insects worldwide, mainly Lepidoptera in the Western Palaearctic especially. Many smaller collections were integrated into this collection, e.g., Antoni, Werner, Werny, Riedel (Saarland), Wimmer (Bavaria), K. Ulrich (Hesse), Dr. Kosswig (Turkey), Leinfest (Persia), etc.
2. Collection Dr. Marten (with Spitz, Fassl, Richter): about 40,000 butterflies and moths of South America with many types.
3. Collection Dr. E. Diehl: about 10,000 butterflies and moths in Sumatra, Madagascar, etc.

The collection contains moreover numerous insect material from case studies and projects from Europe, Africa and South America of the former Institute of Biogeography (Prof. Paul Müller), which partly still "rests" unprepared in drawers (about 200,000 Lepidoptera; 10,000 Coleoptera; etc.) (Figs. 50.10 and 50.11).

Fig. 50.10 *Archiv Lepidoptera* (with permission from: ZfBS)

Fig. 50.11 *Ornithoptera alexandrae* (Rothschild, 1907) (with permission from: ZfBS)

Chapter 51
SIEGEN: University of Siegen Zoological Collections Teach Biodiversity Better than Books and Bytes

Klaudia Witte, Sven Dienstbach, Urs Christian Gießelmann, and Arndt Horst Johann Wellbrock

"You protect what you love. You love what you know. You know what you're taught."

Abstract One way to help to preserve biodiversity is to introduce students in teachers' training programs to the magnificent diversity of animals they are surrounded by every day. By intensively using and integrating our zoological collection in many different courses and lectures, we make local biodiversity one of our main topics within the teachers' training program. By this we hope that we can contribute an important part to motivate young people to stand up for nature conservation and to sustain biodiversity.

Keywords Education • Animal conservation • Artur Franz collection • Native species • True-to-life platics

51.1 Introduction

One of the major issues of our time is the worldwide rapidly declining biodiversity. Out of the 82,954 taxa reviewed by the International Union for Conservation of Nature and Natural Resources (IUCN 2016), 28.8 % of the species are referred to as "threatened" species. The most disastrous reasons for these dramatic worldwide extinction processes form the so-called evil quartet: overkill, habitat destruction, impact of introduced species, and chains of extinction due to climate change and synergetic effects (Brook et al. 2008; Caughley 1994; Diamond 1984, 1989; Seddon 2010; Sodhi et al. 2008).

K. Witte (✉) • S. Dienstbach • U.C. Gießelmann • A.H.J. Wellbrock
Department of Chemistry-Biology, Faculty of Science and Technology, Institute of Biology, University of Siegen, Adolf-Reichwein-Str. 2, 57068 Siegen, Germany
e-mail: witte@biologie.uni-siegen.de

L.A. Beck (ed.), *Zoological Collections of Germany*, Natural History Collections,
DOI 10.1007/978-3-319-44321-8_51

This trend of declining biodiversity does not only hold for rain forests or coral reefs in the tropics but also for many different habitats in Central Europe, including those in Germany. The German Federal Nature Conservation Agency (BfN) lists 207 (43 %) out of 478 studied vertebrate species (Haupt et al. 2009) and 2704 (46 %) out of 5910 studied invertebrate species (Binot-Hafke et al. 2011) as threatened in Germany. Fortunately, some former highly endangered species show increasing population size due to successful conservation programs, e.g., the European beaver (*Castor fiber*, *Linnaeus 1758,* Nolet and Rosell 1998). However, such programs are expensive and often generate conflicts with residents, especially in densely populated countries like Germany (Henle et al. 2008; Redpath et al. 2013). Thus, in order to establish sustainable conservation programs, it is necessary to convince people that biodiversity has to be protected at their own doorstep as well. Following the African proverb *You protoect what you love. You love what your know. You know what you're taught*, it is vital for conservation of national biodiversity to introduce people to the magnificent diversity of plants and animals they are surrounded by every day. They need to be told that it is not only necessary to protect the Bengal tiger (*Panthera tigris tigris*, *Linnaeus 1758*) but also, for example, the Bechstein's bat (*Myotis bechsteinii*, *Kuhl 1817*), the wildcat (*Felis silvestris*, *Schreber 1775*), the great capricorn beetle (*Cerambyx cerdo*, *Linnaeus 1758*), the red kite (*Milvus milvus*, *Linnaeus 1758*), or other species for which Germany bears a special responsibility (Petersen et al. 2003, 2004).

Although it is necessary to inform the current generation about the importance of the protection of native species, it is also essential for the sustainability of conservation programs to improve the awareness for national biodiversity in future generations. The best way to do this is to teach children early in school (primary school) to recognize common animals and plants by themselves (Schlegel et al. 2015). Children that get to know animals and plants by their name often develop a great enthusiasm for nature and biodiversity. Later in school (secondary school), this enthusiasm could be promoted by more detailed information about the behavior, physiological abilities, etc. of species the children already know and by introducing them to less common and endangered native species (Randler 2008). To perform this task, teachers well trained in identifying animals and plants are needed. They hold a key position in our society because they are "multiplicators" of our knowledge in general. Thus, the task of improving awareness for national biodiversity in future generations starts at the universities that train future teachers in training programs. Here, zoological collections play an important role. Whereas plants are relatively easy to find in their natural habitat, it is often hard and laborious to find and catch animals for identification. Therefore, it is much more effective to maintain a zoological collection providing a broad selection of prepared common species which can be used in identification courses and other courses in evolution, ecology, and behavior.

51.2 The University of Siegen

The University of Siegen is a young, modern, and innovative university located within the region "Südwestfalen" in North Rhine-Westphalia. It is deep-seated in "Südwestfalen," and it is nationally and internationally cross-linked with other partners. The university is guided by the central principle "creating a humane future" and has four faculties offering a diverse range of disciplines: Faculty I, Arts; Faculty II, Education, Architecture, and Arts; Faculty III, School of Economic disciplines; and Faculty IV, Science and Technology. Additionally, there are numerous associated scientific centers and institutes. The University of Siegen offers a great chance for interdisciplinary research including both basic and applied research. Today, the University of Siegen hosts more than 19,200 students.

51.3 The Institute of Biology at the University of Siegen

The Institute of Biology (Department of Chemistry-Biology within Faculty of Science and Technology) covers and offers a broad range of topics in organismic and molecular biology in research and teaching. At present, 25 persons are employed at the institute: two full professors, ten (senior) lecturer and teaching assistants, seven postdoctoral research fellows/Ph.D. students, four technicians, and two secretaries. The members of the institute belong to three different sections: "Organismic Biology" (head, Prof. Dr. Klaudia Witte), "Molecular Biology" (head, Prof. Dr. Hans-Michael Merzendorfer), and "Biology Didactics" (head, Dr. Hagen Kunz). At present, about 600 students study biology in different teachers' training programs (Bachelor degree of Science or Arts, Master of Education). The institute offers degree programs exclusively for teaching posts at different types of schools like primary school and secondary school (Grund-, Haupt-, Real-, Gesamtschule, and Gymnasium). Thus, we primarily focus on teaching classes for future teachers. Teachers fulfill a variety of important functions in our society. They do not only transfer our knowledge into next generations; they act as mediators between science and society, and have to understand and assess new scientific knowledge and explain it to pupils. Thus, biology teachers play a key role in the transmission of know-how and knowledge regarding biological topics. One of these biological topics is the rapid decline in biodiversity nowadays.

51.4 Zoological Collections as Teacher's Teaching Material

Since teachers act as "multiplicators" of knowledge, we want to impart profound knowledge about local native animals and plants to our students. Therefore, we intensively use and integrate our zoological collection in many different courses and lectures to make local biodiversity one of our main topics within the teachers' training program. Additionally, we offer students the opportunity to work on special topics of our collection within a bachelor or a master thesis.

In Siegen, the zoological collection is predominantly used in courses within the teachers' training programs. The contents of these training programs are guided by the standards derived from the conference of the educational ministers of the German provinces (KMK 2005; Ministerium für Schule und Weiterbildung des Landes Nordrhein-Westfalen 2005). One guiding principle is to give options for actions to the pupils regarding an environmentally and nature-compatible participation in terms of sustainability. Among other things, pupils should be able to identify specimen by using identification keys (core curriculum, Ministerium für Schule und Weiterbildung des Landes Nordrhein-Westfalen 2005). The Institute of Biology implements these guidelines considering current educational and didactic insights by using the zoological collection in different courses in zoology, ecology, behavior, and evolution. "Learning by handling specimens or seeing 'the real thing' is considerably better than seeing just an illustration of it" (De Clercq et al. 2003). Especially in modern times, most future teachers lack actual experiences with indigenous animals. Our opinion is, therefore, that by using the specimen of our zoological collection in several courses, the intrinsic motivation of students is supported, which positively influences the performance of these students (Lukesch 1997). Furthermore, contents which are difficult to explain verbally or digitally are much more perspicuous if real objects are used that can be touched and handled because such experiences are stored in the implicit (not declarative) memory (Wisniewski 2013). Additionally, according to the dual-coding theory, the delivery of content using objects is superior for learning compared to the delivery by only using words (Paivio 2007). Objects are "[. . .] considered to offer more insight than looking at drawings in a book [. . .]" (Lourenco 2002), and additionally our objects can be touched, which is not the case for most museum specimen, giving our students the unique opportunity to deepen their knowledge in the field of zoology with original specimen.

51.5 The Siegen Zoological Collection

The Siegen Zoological Collection includes samples of about 2900 invertebrate and 390 vertebrate species (Table 51.1). The collection of vertebrate specimen covers two agnathan species, 52 fish species, 16 lissamphibians, 18 reptiles, 124 bird

Table 51.1 Overview of the Siegen Zoological Collection and number of specimens per taxon

Taxon	Number of species	Taxon	Number of species
Porifera	12	Echinodermata	26
Cnidaria	31	Bryozoa	5
Platyhelminthes	8	Brachiopoda	1
Mollusca		Tunicata	5
Polyplacophora	1	Acrania	1
Scaphopoda	2	Petromyzontida	2
Gastropoda	161	Teleostei	52
Bivalvia	81	Lissamphibia	16
Cephalopoda	9	Reptiles	18
Annelida	30	Birds	
Chelicerata	25	Lifelike mounts, study skins	124
Myriapoda	10	eggs (1022)	122
Crustacea	42	Mammals (lifelike mounts, skulls, skeletons)	66
Insecta			
Ephemeroptera	30		
Odonata	16		
Plecoptera	12		
Saltatoria	7		
Hemiptera	38		
Megaloptera	3		
Lepidoptera	1964		
Trichoptera	46		
Diptera	28		
Hymenoptera	133		
Coleoptera	207		

species and 1022 bird eggs from 122 different species, and 66 mammals. Invertebrate samples consist primarily of more than 2480 insect species with a main focus on butterflies, beetles, and hymenopterans. Additionally, specimens of other most relevant taxa are present. Our zoological collection has been registered in the database for the "University Collections in Germany" (Humboldt University).

51.5.1 The Artur Franz Collection

In 2012, the Institute of Biology received the natural collection of Artur Franz, recipient of the Federal Cross of Merit, as a permanent loan, containing more than 1660 lepidopteran and 200 coleopteran species as well as more than 1000 bird eggs from 112 bird species. Artur Franz (1924–2003) was a well-known and valued naturalist and self-educated scientist and one of the founders of natural conservation in the "Siegerland" and in neighboring areas. Ever since he was a child, he collected insects and observed birds and mammals. He dedicated his whole life to nature conservation and was the founder of the first society for nature conservation (Bund für Naturschutz und Vogelkunde, BNV) in the "Siegerland." He was a charismatic person and taught and inspired a lot of children to appreciate, secure, and act for biodiversity by themselves. In fact, some of "his pupils" became biologists in prominent positions later in life. His collection is very special, because it includes very old specimen from this region, like a western capercaillie (*Tetrao urogallus, Linnaeus 1758*), which is more than 150 years old. It also contains two specimens of the very rare hazel grouse (*Tetrastes bonasia, Linnaeus 1758*), a typical grouse specimen from the "Hauberg," a special but common form of coppicing in the "Siegerland." His collection of Lepidoptera (Fig. 51.1) shows an impressive high variation within native species from about 60 years ago. Some of these species even do not exist anymore. However, those that still exist can be used to compare the previous variation with the variation pattern today. Some of the bird eggs are even more than 100 years old (Fig. 51.1).

51.5.2 True-to-Life Plastics

In addition to the collection of natural specimen, we also have a collection of true-to-life plastics of 14 native amphibian species, seven native reptile species, two bat species, and 17 true-to-life plastic skulls or bones of hominids, manufactured by the company "Marcus Sommer SOMSO Modelle GmbH," Coburg, Germany. These models can be used for species identification and are useful objects to understand the evolution of hominids. These true-to-life plastics are robust, do not change color due to solvents, can be hold in hands, and are three-dimensional.

51.6 Further Applications of the Siegen Zoological Collection

Besides the use of the zoological collection in multiple courses within the teachers' training program, we use this collection for special lectures (e.g., Children's University, "Kinderuni"), events, and exhibitions as well. In cooperation with local schools, museums, or local societies for nature conservation, we provide parts of our collection for exhibitions and further education.

We also started with the production of sustainable collection material like insects embedded in acryl, skeleton preparation, and taxidermy to enlarge our collection and to provide robust material for different courses and exhibitions. In

Fig. 51.1 At the *top*, Lepidopteran specimens (Zygaenidae, Arctiidae) of the Artur Franz Collection; *below*, one tray of the bird egg collection of Artur Franz showing eggs from different species of trushes and sylviid warblers (photos: © with S. Gierszewski)

special courses, we teach our students how to prepare specimen for building up an own collection or a collection for the school later on.

We intend to stage an annual public exhibition at the International Day for Biological Diversity (22th May) with parts of our zoological collection. In combination with that, we will train our students to plan and perform a biological exhibition. Additionally, we intend to provide a public data bank containing all specimen of our collection on our website, and we will start building a network between local museums, nature conservation societies, and other societies to provide interesting zoological material for further education and exhibitions.

We hope that we can contribute an important part to motivate young people to stand up for nature conservation and to sustain biodiversity.

Acknowledgment We thank Marliese Müller for assembling this zoological collection. We also thank Sascha Stettner who permanently lent us the Artur Franz Collection. We are grateful to Ulrich Banken (macrozoobenthos), the Biological Station Siegen-Wittgenstein (bird specimen), the Forschungsmuseum Koenig (fish specimen), Markus Fuhrmann (hymenopteran specimen), and Harald Sioli (cnidarian, mollusk, and fossil specimen). We thank Nina Kniel for proofreading the manuscript and Stefanie Gierszewski for providing photos.

References

Binot-Hafke M, Balzer S, Becker N, Gruttke H, Haupt H, Hofbauer N, Ludwig G, Matzke-Hajek G, Strauch M (eds) (2011) Rote Liste gefährdeter Tiere, Pflanzen und Pilze Deutschlands; Band 3: Wirbellose Tiere (Teil 1); Bundesamt für Naturschutz, Bonn-Bad Godesberg. Natursch Biol Vielfalt 70(3)

Brook BW, Sodhi NS, Bradshaw CJ (2008) Synergies among extinction drivers under global change. Trends Ecol Evol 23:453–460

Caughley G (1994) Directions in conservation biology. J Anim Ecol 63:215–244

De Clercq S, Steven WG, Lourenco Marta C (2003) A globe is just another tool. Understanding the role of objects in university collections. ICOM Study Ser 11:4–6

Diamond JM (1984) 'Normal' extinction of isolated populations. In: Nitecki MH (ed) Extinctions. Chicago University Press, Chicago, IL, pp 191–246

Diamond JM (1989) Overview of recent extinctions. In: Western D, Pearl M (eds) Conservation for the twenty-first century. Oxford University Press, Oxford/New York, pp 37–41

Haupt H, Ludwig G, Gruttke H, Binot-Hafke M, Otto C, Pauly A (eds) (2009) Rote Liste gefährdeter Tiere, Pflanzen und Pilze Deutschlands; Band 1: Wirbeltiere; Bundesamt für Naturschutz, Bonn-Bad Godesberg. Natursch Biol Vielfalt 70(1)

Henle K, Alard D, Clitherow J, Cobb P, Firbank L, Kull T, McCracken D, Moritz RFA, Niemelä J, Rebane M, Wascher D, Watt A, Young J (2008) Identifying and managing the conflicts between agriculture and biodiversity conservation in Europe—a review. Agric Ecosyst Environ 124:60–71

IUCN (2016) The IUCN Red List of threatended species. Version 2016-2

KMK [Kultusministerkonferenz] (2005) Bildungsstandards im Fach Biologie für den mittleren Schulabschluss (Jahrgangsstufe 10). Beschluss vom 16.12.2004. Luchterhand, München, Neuwied

Lourenco MC (2002) Are university collections and museum still meaningful? Outline of a research project. Museologia 2:51–60

Lukesch H (1997) Einführung in die pädagogische Psychologie, 3rd edn. Roderer, Regensburg

Ministerium für Schule und Weiterbildung des Landes Nordrhein-Westfalen (2005) Kernlehrplan für das Gymnasium—Sekundarstufe I in Nordrhein Westfalen—Biologie (2008). Ritterbach, Frechen

Nolet BA, Rosell F (1998) Come back of the beaver *Castor fiber*: an overview of old and new conservation problems. Biol Conserv 83:165–173

Paivio A (2007) Mind and its evolution. A dual coding theoretical approach. L Erlbaum Associates, Mahwah, NJ

Petersen B, Ellwanger G, Biewald G, Hauke U, Ludwig G, Pretscher P, Eckhard S, Ssymank A (eds) (2003) Das europäische Schutzgebietssystem Natura 2000. Ökologie und Verbreitung von Arten der FFH-Richtlinie in Deutschland. Band 1: Pflanzen und Wirbellose. Schrreihe Landschpfl Natsch 69:1–743

Petersen B, Ellwanger G, Bless R, Boye P, Schröder E, Ssymank A (eds) (2004) Das europäische Schutzgebietssystem Natura 2000. Ökologie und Verbreitung von Arten der FFH-Richtlinie in Deutschland. Band 2: Wirbeltiere. Schrreihe Landschpfl Natsch 69:1–693

Randler C (2008) Teaching species identification—a prerequisite for learning biodiversity and understanding ecology. Eurasia J Math Sci Tech Educ 4:223–231

Redpath SM, Young J, Evely A, Adams WM, Sutherland WJ, Whitehouse A, Amar A, Lambert RA, Linnell JDC, Watt A, Gutiérrez RJ (2013) Understanding and managing conservation conflicts. Trends Ecol Evol 28:100–109

Schlegel J, Breuer G, Rupf R (2015) Local insects as flagship species to promote nature conservation? A survey among primary school children on their attitudes toward invertebrates. Anthrozoös 28(2):229–245

Seddon PJ (2010) From reintroduction to assisted colonization: moving along the conservation translocation spectrum. Restor Ecol 18:796–802

Sodhi NS, Bickford D, Diesmos AC, Lee TM, Koh LP, Brook BW, Sekercioglu CH, Bradshaw CJ (2008) Measuring the meltdown: drivers of global amphibian extinction and decline. PLoS One 3:e1636

Wisniewski B (2013) Psychologie für die Lehrerbildung. Julius Klinkhardt, Bad Heilbrunn

Chapter 52
STRALSUND: The German Oceanographic Museum

G.-B. Reinicke

Abstract The foundation German Oceanographic Museum in Stralsund (Stiftung Deutsches Meeresmuseum, DMM) today is established as a scientific, educational and collective national museum institution. It maintains four exhibition facilities attracting about 800,000–1 million visitors annually. The exhibitions and marine aquaria impart knowledge about marine environments, organisms, processes and ecological contexts of life in the seas, as well as interactions of man with the marine ambience and resources. As a research museum, it houses zoological (and botanical) collections of marine and coastal organisms, documents of coastal geology, marine science technologies and artisanal coastal fisheries.

The rather young museum—founded in 1951 during the era of the German Democratic Republic (GDR)—works with an explicit emphasis on attractive educational exhibitions and aquaria. Research activities were focused on distinct subject sectors like marine mammals, fish systematics or coral reef ecology—the collections amount to approx. 42,000 items (in 2017). Large subunits are represented by the collections of fish, coelenterates, crustaceans and shelled molluscs, with a further significant number of Baltic marine mammals. Smaller convolutes include shorebirds, marine turtles, cephalopods and various invertebrate taxa.

Keywords Baltic • Marine mammals • Fishes • Crustaceans • Molluscs • Coral

52.1 The Museum's Collection History in Brief

The museum was founded only in 1951 by Prof. Dr. Otto Dibbelt as a municipal nature museum. Its natural history collections and exhibits originated from aged gymnasium funds, from various acquisitions and from the private collection Dibbelt had accumulated as a teacher from 1920 to 1940 in Kolberg (Kołobrzeg, PL) to establish a local museum. The overall convolute was a diverse accumulation of

G.-B. Reinicke (✉)
German Oceanographic Museum/Stiftung Deutsches Meeresmuseum, Museum für Meereskunde und Fischerei Aquarium, Katharinenberg 14-20, 18439 Stralsund, Germany
e-mail: info@meeresmuseum.de; http://www.deutsches-meeresmuseum.de

L.A. Beck (ed.), *Zoological Collections of Germany*, Natural History Collections,
DOI 10.1007/978-3-319-44321-8_52

mixed quality and value. It contained a couple of noteworthy objects like the skeleton of a fin whale that stranded in the mouth of the Dievenow (Dziwna) in 1899. In 1940, Dibbelt was displaced to Franzburg, and in 1944, his collection was relocated to the town hall of Stralsund (Schulze 2006). Dibbelt died in 1956, aged 75. His collections first exhibited in the "Natur-Museum" became the nucleus of the "Meereskundemuseum" in Stralsund, which significantly evolved during the GDR era.

Taking over as director in 1956, Dr. Sonnfried Streicher began to purposefully develop the museum's marine profile, focusing on research activities, collections and exhibitions towards life in and close to the marine environment. With the museum situated on the Baltic coast, the regional subjects of artisanal and coastal fisheries were included and became integral elements of the museum's profile as the "GDR Museum of Marine Research and Fisheries" in 1974. Housed in the medieval St. Catherine's Monastery founded in 1251, the exhibitions were extended into the rebuilt hall church since 1974 (Fig. 52.1). Already in the 1980's, the museum had attracted up to 900,000 visitors per year. After the end of the GDR and the subsequent reunification of Germany, the municipal museum was turned into a private law foundation and was acknowledged as a collective national museum.

In 1991 and 1999, the museum regionally expanded with the opening of two smaller exhibition locations before opening a second large facility at the Stralsund portside, the OZEANEUM, in 2008 (Fig. 52.2). Visitors' numbers then surpassed 1.2 Mio. guests in 2009.

With the marine focus pursued since 1956, the collections and exhibitions were consequently profiled. Exhibits of terrestrial origin got handed to other museums or were occasionally exchanged for marine objects. Purposeful collecting extended the inventory during 1974–1983 alone by 10,076 well-documented zoological collection units. The museum's field campaigns, e.g., two ACROPORA expeditions to the Red Sea in 1976 and 1979, as well as the work of the GDR national research, fishing and trading fleets yielded significant collections that also served as an explicit fundus for the layout of exhibitions. A key exhibit first presented in 1982 was the 4.5 m-high model of a coral pinnacle copied from original habitats in the Red Sea (Fig. 52.3).

The DMM collections today represent about 2 % of marine life collections in German museums (Türkay 2015, pers. comm.). They are stored mainly in the wet collections and in historic storehouse magazines. An extensive convolute of over 50 artisanal fishing boats from the Baltic shores is kept in open sheds and a modern exhibition storehouse.

52.2 Vertebrate Collections

Fish, shorebirds and marine mammals constitute the core of the vertebrate collections. A small convolute of marine turtles amounts about 180 collection units.

Fig. 52.1 The DMM's parent house is located in St. Catherine's Monastery with its spacious hall church (with permission from: Stiftung Deutsches Meeresmuseum)

Fig. 52.2 Exhibition of marine biodiversity in the OZEANEUM in Stralsund; Source: Johannes-Maria Schlorke Photography (with permission from: Stiftung Deutsches Meeresmuseum)

Fig. 52.3 The DMM's model of a coral pinnacle after the extensive revision during 2012–2014 (with permission from: Stiftung Deutsches Meeresmuseum)

52.2.1 Fishes (Pisces)

In the collection, currently about 4400 units are registered, mainly comprising of over 1000 species. Most of the material is kept as wet collection, mainly fixed in formalin and stored in 70 % ethanol. Over 500 units in the dry collection include specimens prepared for various exhibition purposes in the museum's taxidermy workshop.

Older fish collections include material collected in cooperation with the GDR's coastal and high-seas fishing fleets as well as the museum's director S. Streicher from various locations: off Mozambique, the North Atlantic, off West Africa and the Baltic Sea. An important collection is the fishes from the museum's Red Sea expeditions in 1976 and 1979, originating to a large part from locations around Port Sudan. After 1990, the collections focused on the Baltic and North Sea waters as well as the Mediterranean northern coasts. Significant enlargement of the collection came from T. Moritz's personal collection of African fishes, as well as from a recent collection trip to Taiwan. A larger collection of cleared and double-stained fish specimens for systematic and anatomical studies is presently built up, as well as a collection of tissues for molecular studies, presently containing about 1300 samples.

52.2.2 Birds (Aves)

The DMM bird collection comprises 1160 mounted specimens, 130 bantlings, 200 skeletons and skulls as well as close to 6000 eggs. About 260 of the mounted specimens are displayed in the exhibitions and were prepared in the museum's taxidermy workshop. Most of the material, mostly shore- and seabirds, originates from the southern Baltic.

In the context of ongoing climate changes, especially old records of historic occurrence are of noteworthy significance. Today's regionally rare specimens of ruff (*Philomachus pugnax*) and dunlin (*Calidris alpina schinzii*) tell stories of bird diversity in past decades, when these were regular and abundant breeding species.

For gulls (Laridae), a collection of over 4000 blood and tissue samples was established in 2006. DNA sequences were used for molecular analyses of systematics and speciation in this taxon (Liebers-Helbig et al. 2010). However, type material is not represented in the DMM bird collection.

52.2.3 Marine Mammals (Mammalia)

The whale collection comprises about 850 units. Of these 150 represent large cetaceans and dolphins stranded on German shores. This includes significant historical items transferred from the Zoological Museum of the University of Greifswald in 1968. Thanks to those former zoologists, the skeleton of a young fin whale (*Balaenoptera physalus*) which stranded on the western shores of Rügen Island in 1825 (Fig. 52.4), an orca whale (*Orcinus orca*) that beached in Neu-Mukran/Rügen in 1851 and a bottlenose whale (*Hyperoodon ampullatus*) that beached north of Stralsund in 1877 were recovered. Further to these, the collection keeps a number of historical documents of the early German whale research in Greifswald.

The collection of harbour porpoise finds (*Phocoena phocoena*) on German shores amounts about 700 units (in 2017) and documents the carcass monitoring along the coastline of Mecklenburg-Vorpommern carried out since 1955. Since 1995, finds from Schleswig-Holstein Baltic and North Sea coasts were included. The collection keeps skulls and entire skeletons of both sexes and diverse age groups representing the find locations of the North Sea and Baltic realm, including the available scientific documentation about the specimens (Fig. 52.5). Type material is not represented in the DMM marine mammal collection.

The section of Pinnipedia and Sirenia (seals and sea cows) houses about 400 units (in 2017) of mainly earless seals (Phocidae) as well as eared seals (Otariidae) and walruses (Odobenidae) and includes three units of sea cows from all oceans. The main focus on the regional fauna documents the regular carcasses monitoring on Mecklenburg-Vorpommern shores. Skulls, complete skeletons, taxidermy models and organ tissue wet samples are kept including the available

Fig. 52.4 Hands on—the fin whale skeleton of 1825 is among the most spectacular exhibits of the museum (with permission from: Stiftung Deutsches Meeresmuseum)

scientific documentation about most specimens. Some further units represent other locations in the Baltic region.

Fig. 52.5 The DMM's dry collection of small marine mammals and marine turtles (with permission from: Stiftung Deutsches Meeresmuseum)

52.3 Invertebrate Collections

Shelled molluscs, coelenterates (especially Scleractinia and Octocorallia), crustaceans and echinoderms constitute the core of the invertebrate collections.

52.3.1 Coelenterates (Coelenterata)

The core of about 85 % in the DMM coelenterate collection comprises c. 3120 dry samples of stony coral skeletons (Scleractinia). Geographic focus areas cover the Indo-Pacific region with emphasis on the Red Sea with collections of the DMM's expeditions in 1976 and 1979 and of Schuhmacher, from Yemen (collection Eisinger) and the Philippines (collection Kühlmann). Also the Caribbean region is represented by collections of two German coral researchers in the second half of the twentieth century, H. Schuhmacher and D. H. H. Kühlmann.

Further, 15 % of the collection includes wet collections of diverse taxa, especially octocorals (Alcyonacea) from the Red Sea and subantarctic Atlantic waters. Some minor convolutes include anthozoans, hydrozoans and scyphozoans from the North Sea and North Atlantic locations. Type material is not represented in the DMM coelenterate collection.

Fig. 52.6 Detail from the wet collection of cephalopods (with permission from: Stiftung Deutsches Meeresmuseum)

52.3.2 Molluscs (Mollusca)

The shell collection includes especially gastropods and bivalves, in total over 10,500 units comprising over 2000 species. About 3 % of samples are preserved as wet collections, including c. 200 units of Cephalopoda (Fig. 52.6). Geographic focus areas cover the Red Sea (collection of the museum expeditions 1976 and 1979) including the Gulf of Aden, extensive stretches of the Mediterranean and the French Atlantic coastlines (collection of L. + R. Enzenross during 1991–2005), as well as northern European coastal regions and diverse worldwide material.

Old collections date back to the acquisitions by O. Dibbelt or entered the museum during the profiling in the 1960's. They include material of J. W. Tetschke (1796–1865, a teacher in Stralsund), material referred to H. v. Maltzahn (Waren/Berlin) and F. Borcherding (1849–1924, Vegesack). A collection of Red Sea Mollusca collected by C. F. Jickeli (1850–1925) was transferred from the museum in Görlitz in 1987 and material collected by E. Sturmhövel (1904–1985) mainly from the Baltic shores—these, however, go without extensive documentation. Type material is not present in the DMM mollusc collection.

52.3.3 Crustaceans (Crustacea)

The collection of crustaceans comprises about 4000 units (550 species) mostly preserved as wet collection in formalin or ethanol (Fig. 52.7). Over 770 specimens

Fig. 52.7 Part of the DMM's wet collection of fish, coelenterates, molluscs and crustaceans (with permission from: Stiftung Deutsches Meeresmuseum)

are kept as dry samples. Geographic focus areas of the collection cover the Red Sea (collection of the ACROPORA expeditions in 1976 and 1979) and extensive stretches of the Mediterranean and the French Atlantic coastlines (collection of L. and R. Enzenross during 1991–2005), which include reference samples of Lessepsian migrants through the Gulf of Suez into the Mediterranean Sea.

Further the collection holds documents of Pacific and Atlantic neozoan invaders to European waters. Immigrant species to the coasts of Mecklenburg-Vorpommern receive particular attention.

52.4 Other Small Convolutes of Natural History Collections

52.4.1 Echinoderms (Echinodermata)

Dry and wet collections of echinoderms comprise c. 420 units from various locations on European shores and the Red Sea. One separate convolute of 60 units originates from the shallow waters around the island of Soqotra (Yemen, Indian Ocean).

Fig. 52.8 Legend has it that the catch of this leatherback turtle near Stralsund in 1965 initiated the focus as an oceanographic museum (with permission from: Stiftung Deutsches Meeresmuseum)

52.4.2 Marine Turtles (Cheloniidae)

A small collection of marine turtles highlights one special focus of the museum's work. Three species kept alive on display in the aquarium are supplemented by about 180 units representing this taxon. Among them are the remains of the only leatherback turtle (*Dermochelys coriacea*) ever recorded from the Baltic Sea that got caught in a fishing net close to Stralsund in the Prohner Wieck in 1965. A model copy of the specimen is presented in the exhibition (Fig. 52.8).

52.4.3 Plant Collection (Herbarium)

Further to the zoological collections, the DMM houses herbaria of about 780 aquatic algae (Rhodophyceae, Phaeophyceae, Chlorophyceae, marine fungi) and phanerogams from the Baltic and North Sea, as well as some units from Mediterranean, Atlantic and tropical waters. The Stralsund homeland herbarium of O. Bürgener (1876–1966) comprises another 3285 files of fern and seed plants (1313), liverworts (218), mosses (914) and fungi from Western Pomerania including the islands of Rügen and Usedom (Hoppe 2006).

52.4.4 Geology-Palaeontology

The museum's collections include about 3500 units of fossil and geological specimens and originate from the museum's founder O. Dibblet's collections in the Stralsund gymnasium before 1951. After 1978, the collection was focused towards the following aspects: sedimentary rocks of marine origin including usable minerals and rocks from seafloors and coasts, fossils and sedimentary rocks from the Baltic shores and cliffs, glacial drift deposits containing Pleistocene fossils from the Baltic region and fossils of marine origin including characteristic representatives of major geological eras and marine taxa.

52.5 Current Situation and Perspectives

The DMM natural history collections are maintained by five museum curators and one technician in cooperation with the taxidermy workshop. New collections are acquired in the course of ongoing projects and fieldwork, occasionally transferred from other collectors. However, capacities of staff are in conflict with various other duty obligations, and storage space for collections is rather exhausted. In the long term, a new magazine facility is inevitable for responsible care and preservation.

52.6 Digital Exploitation, Access and Contact

Most of the DMM collections are digitally listed and can be researched via personal consultation. Ongoing inventory revisions prepare for the presentation and accessibility through internet portals. Some historical convolutes await further revision and special systematic research. Further information can be provided by the scientific staff of the DMM, who also supplied many details for this manuscript.

References

Hoppe E (2006) Zur Geschichte der botanischen Sammlungen des Deutschen Meeresmuseums. Meer Mus 19:98–108

Liebers-Helbig D, Sternkopf V, Helbig A-J, de Knijff P (2010) The herring gull complex (*Larus argentatus–fuscus–cachinnans*) as a model group for recent Holarctic vertebrate radiations. In: Glaubrecht M (ed) Evolution in action. Springer, Berlin, Heidelberg, pp 351–371

Schulze G (2006) Professor Otto Dibbelt—Leben und Lebenswerk. Meer Mus 19:7–55

Türkay M (2015) Biologische marine Objektsammlungen in Deutschland (Stand März 2015), pp 1–3 (unpubl.)

Chapter 53
STUTTGART: The Zoological Collections of the Stuttgart State Museum of Natural History

Stefan Merker and Arnold H. Staniczek

Abstract Emerging from the sixteenth century "Cabinet of Arts and Natural Curiosities" of the Dukes of Württemberg, the Stuttgart State Museum of Natural History (SMNS) ranks among the oldest and most prominent natural history museums in Europe. With its collections, libraries, laboratories, and other research facilities, the museum comprises two building complexes both with extensive exhibition and storage areas. The permanent exhibitions at Museum Schloss Rosenstein and Museum am Löwentor are tightly interwoven in form and content. Today, the SMNS collections hold more than 12 million specimens and other natural objects as well as numerous associated data on taxonomy, genetics, ecology, and geography. Together these comprehensive records and archives of life on Earth and its history constitute a large-scale research infrastructure used by research scientists and the international community. Characteristic of the SMNS is the tight collaboration between its paleontologists and researchers working on extant fauna.

Keywords SMNS • Museum am Löwentor • Schloss Rosenstein • Naturkundemuseum • Willi Hennig • Württemberg

53.1 The Museum's Infrastructure

The museum is carried by the Federal State of Baden-Württemberg. In 2014, its annual balance amounted to 8.3 million euro, of which government funding made up 7.0 million euro. The remaining budget was acquired by external funding and

S. Merker
Department of Zoology, State Museum of Natural History, Rosenstein 1, Stuttgart 70191, Germany
e-mail: stefan.merker@smns-bw.de

A.H. Staniczek (✉)
Department of Entomology, State Museum of Natural History, Rosenstein 1, Stuttgart 70191, Germany
e-mail: arnold.staniczek@smns-bw.de

L.A. Beck (ed.), *Zoological Collections of Germany*, Natural History Collections,
DOI 10.1007/978-3-319-44321-8_53

entrance fees to the exhibitions (Eder 2015). An independent scientific advisory council supports the SMNS in its strategies and policies. The museum is also backed up by the "Gesellschaft zur Förderung des Naturkundemuseums Stuttgart e.V.", a society established in 1912 by friends of the museum to financially and ideationally support the mission of the SMNS, in particular the development of its collections.

As of 2015, the SMNS employs a permanent staff of 96 persons. These comprise scientific, technical, and administrative positions. Additionally, 49 persons including Ph.D. students and scientific and technical trainees are contracted on a temporary basis. More than 200 honorary assistants—some of them former staff but many of them dedicated laymen—commit parts of their spare time to work in the museum.

The staff is organized into four research departments (botany, entomology, paleontology, and zoology) and departments of administration, exhibitions, as well as education and public relations.

The museum's staff, collections, and exhibitions are accommodated in two main complexes of buildings: the "Museum am Löwentor" (Fig. 53.1) and the "Museum Schloss Rosenstein" (Fig. 53.2). Two additional locations provide space for provisional outsourcing of various voluminous collections and only temporarily used equipment, e.g., for special exhibitions. These locations—all within walking distance—provide space of 24,000 m^2 in total.

The *Museum am Löwentor* (opened 1985, 15,800 m^2) houses the permanent paleontological exhibition and provides room for temporary special exhibitions.

Fig. 53.1 Museum am Löwentor, exhibition hall (with permission from: SMNS; photo: R. Baumann)

Fig. 53.2 Museum Schloss Rosenstein (with permission from: SMNS; photo: U. Schmid)

Furthermore, most laboratories and equipment, seminar rooms, the main library, most offices, the museum administration, and the major part of the collections are located here (herbarium, insect collection, large parts of the paleontological collection, as well as the malacological, mammalogical, and osteological collections).

The *Museum Schloss Rosenstein* (built in the 1820's, part of the SMNS since 1954, 5000 m^2) features the permanent exhibition of extant fauna and flora and also provides room for temporary exhibitions. Additionally, it houses the ornithological collection, laboratories for taxidermy, seminar rooms, and several offices.

The SMNS maintains numerous, well-equipped scientific laboratories to support research and exhibitions. Apart from labs specifically serving the needs of paleontological research (sieving, acid etching, hydrofluoric acid maceration, grinding, rock cutting, and sandblasting), the following labs and devices are available at the SMNS:

Molecular Laboratory (Fig. 53.3)

DNA/RNA extraction, amplification, and analysis are facilitated by state-of-the-art PCR equipment as well as modern robotics and a fragment analyzer. Biological tissues and extracted DNA are stored in ultralow temperature freezers (−80 °C).

Nitrogen Chamber For effective pest control, museum objects—especially new acquisitions—are treated in a large nitrogen chamber (12 m^3) or are temporarily stored in ultralow temperature freezers.

Degreasing Facility A computer-controlled degreasing device (volume 150 l, closed circuit) allows for degreasing of bones.

Fig. 53.3 Molecular laboratory (with permission from: SMNS; photo: A. Staniczek)

Taxidermy Laboratories (Fig. 53.4)
Spacious and well-equipped taxidermy labs allow for high-quality preparation, restoration, and modeling (Fig. 53.5) of zoological objects. Besides traditional techniques, the lab also permits polyethylene glycol (PEG) treatment to obtain a lifelike appearance and preservation of animals. Over the past years, SMNS taxidermists have been awarded very prestigious prizes in European and world championships in their field (Fig. 53.6).

Histology Laboratory Serial thin sections are obtained with a Leica 1516 rotary microtome and subsequent staining.

Micro-CT Scanner (Fig. 53.7)
A Bruker Skyscan 1272 tomographic device generates 3D renderings of biological objects and is mainly used to investigate internal structures in a wide range of taxa.

Optical Laboratories Optical instruments at the SMNS include a digital microscope (Keyence VHX-500D) and a fully apochromatic zoom system (Leica Z16 APO A macroscope) to generate focus-stacked images. A scanning electron microscope (Zeiss Evo LS 15) allows the study of (type) material at different temperatures, humidity, and pressure without prior sputter coating (Fig. 53.7).

High-Resolution Scanners Specialized scanners deliver high-resolution images of insect boxes and herbarium sheets for efficient digitization of collections.

Fig. 53.4 Taxidermy laboratory (with permission from: SMNS; photo: A. Staniczek)

Fig. 53.5 Modeling an ancient dragonfly (with permission from: SMNS; photo: G. Bechly)

53.2 History

The history of the museum has been reviewed several times before (e.g., Lampert 1896; Rauther 1940; Schüz 1964; Adam 1991a, b; Fricke 1991; König 1991; Möhn 1991; Renner 1991; Ziegler 1991a, b; Warth and Ziegler 1991; Woog et al. 2003).

Fig. 53.6 Award-winning display of a white-faced saki (*Pithecia pithecia*) by SMNS taxidermist Jan Panniger, European Taxidermy Championship 2014 (with permission from: SMNS; photo: U. Stübler)

Much of the information given below—focusing on the zoological collections—is drawn from these contributions. For details on the museum's paleontological, mineralogical, and botanical collections, we refer to these sources.

The oldest parts of the collection date back to the sixteenth century and were originally incorporated in the private "Cabinet of Arts and Natural Curiosities" of the Dukes of Württemberg (Fig. 53.8). Like many European aristocrats at that time, they had a strong interest in collecting art and natural curiosities. Natural objects comprised minerals, fossils, marine, and terrestrial animals. Up to now, the skull remains of a Pleistocene giant elk collected in 1600 count as the oldest preserved object of the museum collection.

From 1762 to 1791, Johann Friedrich Vischer (1726–1811) was responsible for the natural history collections amounting to 10,000 objects at the end of his career. He arranged its zoological part in systematic categories like mollusk shells, insects (including all other arthropods), fish, amphibians (also including reptiles), birds, and tetrapods.

In 1783, the private "Cabinet of Arts and Natural Curiosities" was made accessible to the scholars of the local "Hohe Carls-Schule", a military academy and elite school for sons of distinguished families. Among them was Georges

Fig. 53.7 SEM and μ-CT laboratory (with permission from: SMNS; photo: A. Staniczek)

Fig. 53.8 The origins of the SMNS: The "Cabinet of Arts and Natural Curiosities" before 1746. Etching by Ludwig Som (Heimatmuseum der Stadt Ludwigsburg) (Photo: © with SMNS)

Cuvier (1769–1832), who studied in Stuttgart from 1784 to 1788. It was here that he received the stimulus for his groundbreaking studies on geology and paleontology.

The rapid growth of the collection soon made it necessary to separate natural objects from the remaining collectables. Through a decree in 1791, Carl Eugen, Duke of Württemberg, appointed four curators with respective responsibilities for art, minerals, plants, and the animal kingdom. This decree thus marked the starting point of what later would become the Stuttgart State Museum of Natural History. In 1792, Carl Friedrich Kielmeyer (1765–1844) became the first curator of zoology. His successors developed the rapidly growing collections according to the swift increase of knowledge in systematic zoology. In the 1820's, the museum moved to a new, more spacious building within the city center of Stuttgart (Fig. 53.9). The volume and species diversity of arthropods in the collection then resulted in the appointment of lepidopterist Ernst Hofmann (1837–1892) as the first curator of entomology in 1869. He would later be known as the author of his magnum opus "Die Gross-Schmetterlinge Europas".

During the nineteenth century, numerous explorers—many of them with Swabian roots—enlarged the zoological collections by donating or selling their collected specimens to the museum. The first significant contribution of material from overseas was made by Baron Carl von Ludwig (1784–1847), who traveled and worked in South Africa as a pharmacist and banker. In his spare time, he collected insects, plants, and vertebrates, which he regularly sent back to Stuttgart. Of considerable importance is his collection of smoky moths (Zygaenidae) from the Cape region. His South African collection was later complemented with two Cape lions donated by Stuttgart-born theologist Christian Barth (1799–1862). As only

Fig. 53.9 The historical museum building at Neckarstrasse around 1840 (Photo: © with SMNS)

seven specimens of this extinct lion form are preserved in the world's natural history museums, the two mounted skins in Stuttgart are a prominent part of the zoological collection.

The first major contribution to the insect collection was made by lawyer Karl von Roser (1787–1861), whose interest in natural history was sparked by lectures of Kielmeyer, the museum's first curator of zoology. Aside from his career in administration—he eventually became the minister of foreign affairs in Württemberg—von Roser managed to assemble a large number of insects, both local fauna and material from overseas, today still forming the historical basis of the museum's entomological collection. Duke Paul Wilhelm von Württemberg (1797–1860) was a globe-trotter with a particular focus on North America. After his death, considerable parts of his collection were acquired by the museum, among these many bird skins and drawings. The butterfly collection of bird wings (*Ornithoptera*), donated in 1865 by King Carl and King Wilhelm II, both royals of Württemberg, most probably also had been collected by Duke Paul Wilhelm von Württemberg.

The South American fauna, especially Surinam, was extensively sampled by August Kappler (1815–1887). His material now constitutes a major part of the museum's collection of South American butterflies, birds, and mammals. German naturalist Baron Ferdinand von Mueller (1825–1896) immigrated to Australia where he soon became director of the Melbourne Zoological and Botanical Gardens. Apart from plants, he collected numerous insects and also 2269 vertebrates of 837 species that were all sent to Stuttgart. This material is still preserved to a large extent and includes birds of paradise, kangaroos, and also a skillfully prepared specimen of the extinct Tasmanian tiger. Theodor von Heuglin (1824–1876) extensively traveled East and Northeast Africa. The many animals and skins he brought back to Stuttgart now form a substantial portion of the SMNS zoological collections. A key role in the history of the museum was played by Ferdinand Krauss (1812–1890). Not only did he travel to Africa himself to collect animals, but later on he was also appointed as curator at the Stuttgart museum. It was his merit to sort and arrange the comprehensive material collected by him and other naturalists thereby making it available for scientific studies. One of the most important collections secured by the museum in this time was a donation by Dutch physician Pieter Bleeker (1819–1878). During his duty as a military doctor in the Dutch East Indies (now Indonesia), he was able to collect an enormous variety of fish and other animals. He published 432 articles on the ichthyofauna of the region—among these a high number of first species descriptions—and sent his collections to natural history museums worldwide. His repeated gifts to the Stuttgart museum now make up a considerable part of its fish collection. Carl Benjamin Klunzinger (1834–1914), a Swabian physician, worked on the marine fauna of the Red Sea. The many specimens he sent to Stuttgart now form a prominent part of the SMNS inventory.

In 1913, Erwin Lindner (1888–1988) joined the museum as a curator of Diptera and was head of the Department of Entomology until 1953. He would work and collect for more than 70 years for the museum as a distinguished expert on soldier flies (Stratiomyidae). As the editor of some 15,000 pages of the "Die Fliegen der

Paläarktischen Region", he was known among entomologists all around the world. Important for the museum's collections were not only his frequent field trips to the European Alps but also his excursions to the South American Gran Chaco (1925/1926). In addition to several journeys to Africa (1951/1952 East Africa, 1958/1959 South Africa and Tanzania), these expeditions considerably enlarged the Diptera collection.

A major incision in the museum's history was its destruction during World War II. On the 21st of February 1944, allied bombs hit and destroyed the main museum building at Neckarstrasse. The adjacent collection buildings at Archivstrasse were hit on the 12th of September and also burned down. Although in the first years of war, the most valuable parts of the inventory had been transferred to numerous scattered locations in the countryside, considerable portions of the collections, catalogues, and libraries were lost. Previously outsourced collections were also affected, not only by destruction during the war but also due to postwar turmoil. Among the biggest losses were parts of the mollusk collection, the entire spider collection of Wilhelm Bösenberg (1841–1902) including 169 holotypes described by Strand, parts of the vertebrate collections, many invertebrate fossils, and also historically important mammoth tusks that had been excavated in 1815 under King Friedrich I. The geognostic collection of Oscar Fraas (1824–1897) was also destroyed.

After the war, the local government decided not to rebuild the museum at its previous location. Instead, it was decided to reerect "Schloss Rosenstein", built in 1822–1829 by Giovanni Salucci for King Wilhelm I of Württemberg and destroyed during wartime. This manor palace was chosen to house the zoological exhibition and large parts of the scattered collections. The museum—in 1950 renamed "Staatliches Museum für Naturkunde Stuttgart"—moved into the rebuilt palace in 1954; the public exhibition was opened 2 years later.

However, as there was by far not enough space to accommodate the entire museum inventory at this new location, from 1962 on, the paleontological and entomological collections along with associated staff were temporarily quartered in an arsenal in Ludwigsburg near Stuttgart.

In 1963, the Department of Phylogenetic Research ("Abteilung für Stammesgeschichte") was established for Willi Hennig (1913–1976), the founder of phylogenetic systematics. He worked at the SMNS outpost in Ludwigsburg until his untimely death in 1976. In 1998, this department was partly merged into the Department of Paleontology and into the Department of Entomology.

In 1969, Prof. Bernard Ziegler (1929–2013) was appointed as director of the SMNS. Owing to his persistent efforts, the 23-year interim at Ludwigsburg finally came to an end. In 1985, a new, modern building complex—the "Museum am Löwentor"—was opened in Stuttgart in immediate vicinity to "Schloss Rosenstein". It gave room for a new paleontological exhibition, modern labs, offices, paleontological, mineralogical, entomological, malacological, and botanical collections, and all other parts of the museum previously outsourced to Ludwigsburg.

53.3 Collections

As in other big natural history museums, the collections represent a nucleus for systematic research. In total, the museum harbors more than 12 million items, many of them of great historical and scientific importance. Apart from fossils (4.1 million), minerals (40,000), fungi, and plants (one million), the museum staff curate zoological collections of about seven million objects. Among these, arthropods number up to more than 5.5 million specimens, mollusks to one million, and vertebrates to 500,000. Increasingly, the collection information is made available digitally and in online databases to serve the needs of scientists all over the world. To this date, all type series and specimens are searchable online through the museum's type database.

Apart from its classic collections of specimens and parts thereof, the museum has been building up libraries of DNA, either obtained from museum specimens or during fieldwork of museum staff.

53.3.1 Arthropoda

Today, the entomological collections include more than 5.5 million specimens from all over the world, thus forming the largest part of the museum's inventory. Its major part holds about 4.8 million dried insects stored in more than 22,000 insect drawers. Additionally, a considerable amount of insects and other arthropods, mainly Crustacea and Arachnida, are stored in ethanol. In recent years, specimens are increasingly preserved in 100 % ethanol and stored in deep freezers at −80 °C for genetic studies.

Coleoptera (Beetles) With about two million specimens of almost 60,000 species, the beetle collection is the largest insect collection at the SMNS (Fig. 53.10). Its major part was acquired during the last 70 years. Geographical emphasis lies on the Old World, notably Europe, southern Africa, Central Asia, the Himalayan region, and tropical Asia including New Guinea. Its systematic focus encompasses Carabidae, Silphidae, Leiodidae, Prostomidae, Tenebrionidae, Chrysomelidae, and Scarabaeidae. The collection holds about 5000 type series.

Local collections of numerous contributors (Böhme, Buck, Gladitsch, Köstlin, Ulbrich, von der Trappen) together form a faunistic reference collection on South German beetles.

The beetle collection also includes numerous special collections with a large amount of type material: the collection Puthz on the staphylinid genus *Stenus* contains 80 holotypes and numerous paratypes. The collections Daffner and Angelini on Leiodidae and the collection Heinz on Carabidae also hold a large amount of type series. Due to the work of Wolfgang Schawaller, a beetle curator at the SMNS from 1983 to 2015 and from 1995 to 2015 also the head of the Entomology Department, the outstanding collection of Tenebrionidae includes

Fig. 53.10 Beetle collection (with permission from: SMNS; photo: J. Reibnitz)

type series of 210 species of the genera *Laena* and *Platydema*. It also encompasses type material of other specialists on Tenebrionidae, namely, Ando, Bremer, Grimm, and Masumoto.

Lepidoptera (Moths and Butterflies) The global collection of butterflies (Figs. 53.11 and 53.12) numbers up to about one million specimens with geographical emphasis on the Neotropical and Ethiopian region including a collection of 40,000 specimens from Kenya. Of particular importance are the worldwide collections of Hoppe on Noctuidae (80,000 specimens) and of Reisinger on Pieridae (70,000 specimens). The special collection of Sesiidae includes about 450 species. Further taxonomic emphasis lies on Erebidae and Geometridae. The latter also include 30,000 specimens from Ecuador (collections Brehm and Hilt). The Lepidoptera collection holds about 400 type series—among these types of the recently described family Pseudobistonidae (Geometroidea).

Hymenoptera (Bees, Wasps, and Ants) The Hymenoptera collection accounts for 350,000 pinned specimens including about 200 type series. Its oldest stock dates back to the nineteenth century and is based on the historic collection of von Roser. Also of historic importance is a reference collection of Ichneumonidae by an eminent German hymenopterist Otto Schmiedeknecht. Furthermore, the collection Gauss features numerous species relevant for forest entomology.

The recent focus of collection especially lies on the Aculeata of the Neotropics and on Scoliidae, Ichneumonidae, and Chalcidoidea on a worldwide scale. Also worth mentioning is a collection of bee and wasp nests (Apidae, Vespidae) from the Neotropical region.

Fig. 53.11 Lepidoptera collection in compact shelving (with permission from: SMNS; photo: A. Staniczek)

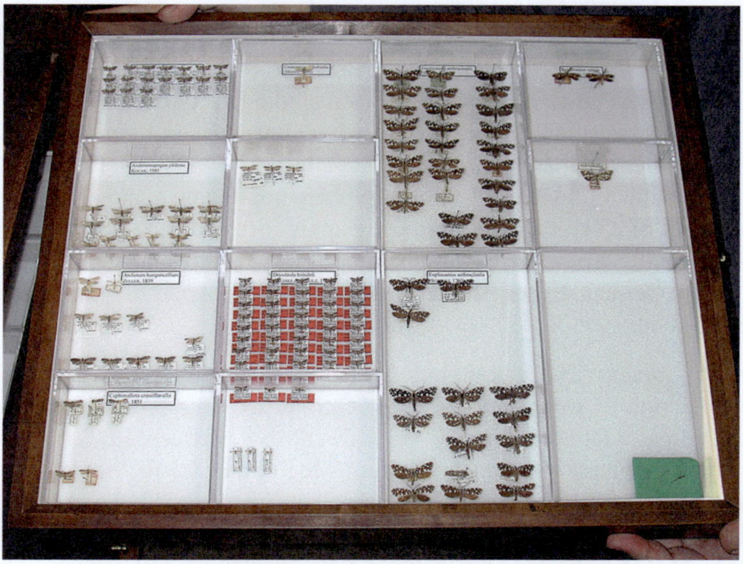

Fig. 53.12 Compartmentalized insect boxes allow for variable storage (with permission from: SMNS; photo: J. Holstein)

As part of the project German Barcode of Life, about one million specimens of Hymenoptera: Parasitoidea were collected throughout Germany by malaise traps. From these samples, a molecular reference collection of presently 2500 species has been compiled.

Diptera (Flies and Midges) About 450,000 specimens of about 12,000 valid species form the Diptera collection including about 1150 type series. Two-thirds of the specimens are stored in dry condition, one-third in ethanol.

The oldest specimens were collected by Karl von Roser (1822–1861) in the first half of the nineteenth century. When Erwin Lindner became a curator of the SMNS early in the twentieth century, his emphasis was the enlargement of the collection serving as a scientific source for the edition of his monumental series "Die Fliegen der Paläarktischen Region". Lindner collected many species himself but also motivated colleagues to donate or sell their material, e.g., Plaumann, Aharoni, Engel, Richter, and many others. Most of the flies and midges in the collection are from Europe, but Lindner's expeditions in the 1930's and 1950's also yielded much material from South America and Africa.

The collection's main focus lies on Stratiomyidae, Cecidomyiidae, Muscidae, Anthomyiidae, and Tachinidae, thus reflecting the main working areas of curators Erwin Lindner, Edwin Möhn, Willi Hennig, Benno Herting, and Hans-Peter Tschorsnig. Especially Tachinidae, which is of importance also in applied entomology, is studied in the SMNS since more than 50 years.

Other Insect Groups Of considerable importance is the Hemiptera collection comprising 160,000 specimens of cicadas (Auchenorrhyncha) and 80,000 species of true bugs (Heteroptera). Among these, there are 80 type series. The Orthoptera collection numbers up to 60,000 specimens including 50 types.

Recent efforts resulted in the buildup of an aquatic insect collection of about 150,000 specimens that are mainly stored in ethanol (Fig. 53.13). Emphasis is laid on a phylogenetic reference collection of Ephemeroptera on a global scale, including type series of 25 species. Its main geographic focus lies on Central Europe, the Caucasus, South Africa, and New Zealand. Additionally, a faunistic reference collection of German Ephemeroptera, Odonata, Plecoptera, and Trichoptera has been established.

Crustacea: Isopoda The entomological department also curates the worldwide largest collection of wood lice (Isopoda: Oniscidea). This is due to the ongoing work of Helmut Schmalfuss, a curator at the SMNS between 1977 and 2007. Its geographical emphasis lies on the Eastern Mediterranean; its current stock was accumulated through numerous field trips to Greece and adjacent countries, rounded off with bycatches of numerous colleagues. It comprises about 83,000 specimens of 1462 species and includes type species of 350 species. The type collection also includes important type material by Italian isopodologists Ferrara and Taiti (University of Florence, Italy). The entire collection is completely determined, catalogued, and available online.

Fig. 53.13 Aquatic insects, wet collection (with permission from: SMNS; photo: A. Staniczek)

53.3.2 Mollusca

The mollusk collection at the SMNS (Fig. 53.14) comprises more than 130,000 lots (>1 million specimens). Exotic marine shells represent the oldest objects already listed in the catalogue of the physician Pasquay (Frankfurt a. M.) in 1777. The collection Clessin (German, Austrian, and Hungarian mollusks, bought in 1903), the collection Zwiesele (freshwater bivalves of southern Germany and Switzerland, bought in 1925), and the collection Geyer (mollusks of the state Württemberg, bought in 1932) are especially worth mentioning among the historical parts. During World War II, some parts of the collection were destroyed, but the main portion survived undamaged.

More recent additions among the marine mollusks include the collections of tropical cone shells (Conidae, mainly collections Röckel and Da Motta, about 20,000 lots), olives (Olividae, coll. Greifeneder, nearly 3000 lots), West African volutes (genus *Cymbium*, coll. Stürmer), and Mediterranean mollusks including a substantial part of wet specimens and with special consideration of the "Lessepsian" migration (coll. Enzenross)—all of which are of worldwide importance. Continental mollusks of special significance encompass collections of door snails of the Western Palearctic (Clausiliidae, coll. H. Nordsieck, about 10,000 lots), European island faunas (coll. Rähle, Beckmann), European slugs, and freshwater mollusks (coll. Meier-Brook).

Fig. 53.14 Malacological collection, *Conus* shells (with permission from: SMNS; photo: A. Staniczek)

53.3.3 Vertebrata

Based on the work of past collectors and curators, the global vertebrate collections at the SMNS include a large amount of historic material (collected prior to 1900).

Fishes The fish collection (Fig. 53.15) now comprises 25,000 lots (ca. 150,000 specimens) of fishes from all over the world. Its main focus lies on European freshwater fishes, Mediterranean marine fishes, Indo-Pacific marine fishes, and Central and South American freshwater fishes. The collection includes over 1000 type specimens of about 450 species, mainly described by Bleeker, Klunzinger, and Fricke. Of particular interest is a large collection of dragonets (Callionymidae).

Amphibians and Reptiles The herpetological collection housed at the SMNS contains some 22,000 specimens spanning over all major zoogeographical regions. It includes 118 type specimens of amphibians and reptiles. Parts of the collection including valuable type material and the catalogue were destroyed during the bombing raids in World War II. After the war, herpetology curators Heinz Wermuth and Andreas Schlüter reconstructed most of the prewar condition. Schlüter much enhanced the collection and extended the zoogeographical focus to the Neotropics (Peru, Brazil, and Guyana shield).

Birds The bird collection (Fig. 53.16) comprises about 134,000 specimens of more than 5400 species (50,000 skins, 20,000 mounts, 8000 feather sets, 50,000 eggs, and

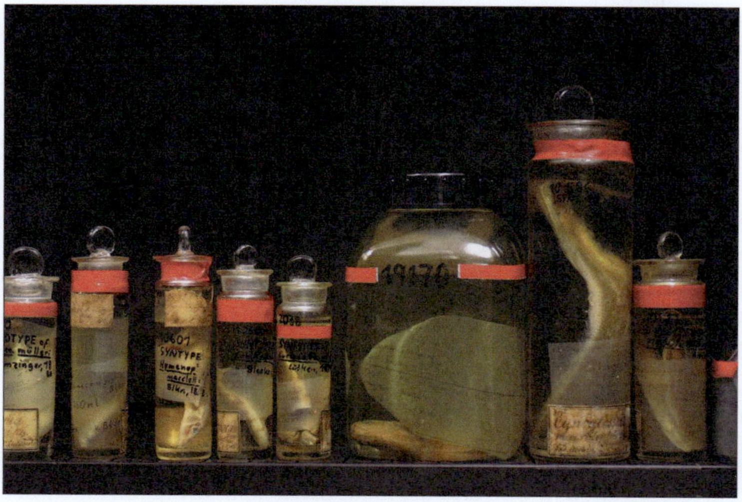

Fig. 53.15 Historical types of the fish collection (with permission from: SMNS; photo: J. Bergener)

Fig. 53.16 Bird collection (with permission from: SMNS; photo: U. Schmid)

6000 partial or complete skeletons) from all over the world. It includes 119 type specimens. The first entry in the comprehensive, handwritten catalogue dates back to 1837, and owing to the continuous and immense interest of explorers and collectors in colorful avifauna, the bird collection and its development played a major role in the course of the museum's history. Already in 1886, a total of 11,175 specimens of 4800 species were counted (Lampert 1896). This historical material is still well maintained; it was complemented by more recent additions including those of A. Fischer, K. Fischer, and Issel (Germany), Kipp (Europe), Hoy (Argentina), Gatter (Liberia), Nikolaus (Africa), and a substantial egg collection from Duve (Europe and Southern Africa).

Mammals The mammalogical repositories (Fig. 53.17) now hold more than 54,000 specimens of more than 1500 species, thus harboring one of the 25 largest mammalogical collections worldwide. Of special importance are the c. 240 type specimens including almost 80 primary types. The main focus of the collection lies on rodents, bats, and insectivores. Complementing the historical objects from Africa, former curator Fritz Dieterlen assembled a high number and diverse collection of Central African small mammals.

Osteology In the late twentieth century, osteologist Doris Mörike established a comparative collection of bird and mammal bones to assist quick determination of bone findings. Over the past years, this collection has proven its usefulness for answering numerous inquiries by the public, by the police needing assistance in identifying potential human remains, and by archeologists and paleontologists in order to assess fossil or subfossil discoveries.

Fig. 53.17 Mammalogical collection (with permission from: SMNS; photo: R. Baumann)

53.4 Collaborations and Research

As taxonomy and systematics on a global scale traditionally constitute core research areas of natural history museums, the SMNS is part of a multitude of collaborative research efforts. Preserving millions of biological specimens and associated information, it is also part of major institutional networks. Among these are CETAF (Consortium of European Taxonomic Facilities), GBIF (Global Biodiversity Information Facility), DNFS (Consortium "Deutsche Naturwissenschaftliche Forschungssammlungen"), Humboldt-Ring (Consortium of German Research Museums), and GFBio (German Federation for the Curation of Biological Data).

Major projects that the SMNS participates in include SYNTHESYS (Synthesis of Systematic Resources: an EC-funded project creating an integrated European infrastructure for natural history collections), GBOL (German Barcode of Life), and BiNHum (Biodiversity Network of the Humboldt-Ring).

SMNS collections can be partly accessed via data portals like GBIF, FishBase, and HerpNET. All type specimens and series are searchable in the type database on the museum's website.

The SMNS annually hosts between 150 and 200 guest researchers from all over the world. On average, visiting scientists publish between 50 and 90 articles per year including data obtained in the SMNS collections.

Scientists of the SMNS regularly engage in teaching activities at universities. Since 2013, the museum maintains a formalized cooperation with the University of Hohenheim, Stuttgart. This cooperation involves teaching duties of SMNS staff at graduate and undergraduate level and supervision of Ph.D., M.Sc., and B.Sc. theses. As an advantage to the museum, junior scientists are attracted to collection-based systematic research, and museum staff get the opportunity to qualify for a professorship.

To introduce young researchers to working at a natural history museum, the SMNS provides the opportunity for 14 scientific trainees to learn about collection maintenance, exhibition design, and public outreach and to conduct their own research projects under the supervision of an experienced staff. Trainees are usually contracted for 2 years.

Two peer-reviewed scientific journals—"Palaeodiversity" and "Stuttgart State Museum Contributions to Natural History A (Biology)"—are published by the SMNS. One popular science journal—"Stuttgart State Museum Contributions to Natural History C (Popular Science)"—addresses a wider German readership. The SMNS also serves the scientific community by hosting and maintaining the international website on mayflies, Ephemeroptera Galactica.

Museum staff frequently visits international and national cooperation partners, conducts fieldwork, and regularly presents research results at scientific conferences. Museum scientists contribute about 100 peer-reviewed publications on a yearly basis—with the large majority listed on the Science Citation Index (SCI)—and about 40 articles to popular science journals.

One major strength of the SMNS is the tight collaboration between its paleontologists and researchers working on extant fauna. Joint investigations on the evolution of amphibians (Schoch et al. 2015) and the early diversification of winged insects (Sroka et al. 2015; Staniczek et al. 2014) are examples for this fruitful interdisciplinary approach. Insects preserved in amber also spark the interest and collaboration of both entomologists and paleontologists at the SMNS (Huang et al. 2016). Of considerable importance to understanding the evolution of extant fauna are conclusions drawn from studying the significant paleontological collections of the SMNS (Schoch and Sues 2015).

As in other natural history museums, taxonomy is a traditional domain of the SMNS. Projects center around the research collections and strive to fill the gaps of documentation and classification of the natural world. These meticulous efforts unite many scientists working at the SMNS irrespective of their group of interest. They result in taxonomic descriptions and revisions, monographs on selected groups or regions, species catalogues, faunistic checklists, or field guides. Prominent examples are the world catalogue of wood lice (Schmalfuss 2003) and the description of Oriental and Eastern Palearctic Tenebrionidae of the genus *Laena* (Schawaller 2015). Other systematic studies include the description of whole extinct radiations in tropical land snails documenting the devastating effects of human settlement on small Pacific islands (Richling and Bouchet 2013).

Following the tradition of Willi Hennig, who established phylogenetic methodology at the museum in the 1960's, current phylogenetic studies constitute another focus at the SMNS. Following an integrated approach, most of these studies are nowadays not only based on comparative morphology but are complemented by comprehensive molecular analyses. Examples of recent work include studies on the phylogeny of the order Ephemeroptera (Ogden et al. 2009) and the phylogeny of chalcid wasps (Heraty et al. 2013). The discovery of Pseudobistonidae, a new family of Macrolepidoptera, is another recent highlight of phylogenetic research at the SMNS (Rajaei et al. 2015).

In a number of research projects, biogeographical questions are asked. These studies apply multiple methods, looking at molecular, morphological, or bioacoustic characters, and span a wide range of taxa. For example, a series of projects aims at reconstructing evolutionary processes leading to the diversification and present distribution of tarsiers, small nocturnal primates of Southeast Asia (Merker et al. 2009). Other studies focus on the phylogeography of African elephants, Irrawaddy dolphins, or European beavers.

Ecological and conservation-related research is also an imminent part of the museum's scientific work. Overseas, the SMNS looks on blood parasitic infections, site fidelity, and habitat use of Malagasy rainforest birds (Jonsson et al. 2012). A regional focus lies on the ecology of introduced Amazon parrots (Martens et al. 2013) and the ecology and behavior of graylag geese in urban habitats in order to assess their potential relevance for spreading avian influenza (Schwarz et al. 2012). These projects as well as studies on the diversity and life history of parasitic fly and wasp groups—important for biological pest control—spark significant public interest.

Life history evolution is also the topic of a series of studies on caecilians: among the highly variable parental care strategies in amphibians, skin feeding in some Caecilian species has been shown to be a particularly interesting mode of parental care (Kupfer et al. 2006).

With its roots deep in the history of Baden-Württemberg, the SMNS also takes a leading role in documenting and monitoring regional fauna. A statewide survey of amphibians and reptiles is coordinated by the museum. Beetles, butterflies, and bees in particular are monitored by members of the associated Stuttgart Entomological Society (established 1869). The SMNS also organizes regular meetings of German coleopterologists, hymenopterologists, and malacologists.

53.5 Exhibitions, Education, and Public Outreach

Transfer of knowledge into society is one of the museum's key competencies that largely account for its public perception. On average, about 200,000 visitors come to see the exhibitions throughout the year. The permanent exhibitions at the Museum Schloss Rosenstein (Biology, Fig. 53.18) and the Museum am Löwentor (paleontology) are tightly interwoven in form and content. New research findings of SMNS curators are constantly incorporated into the permanent and temporary exhibitions.

Fig. 53.18 Permanent exhibition at the Museum Schloss Rosenstein (with permission from: SMNS; photo: U. Schmid)

Inventive temporary exhibitions on different scales are conceived, designed, and implemented by interdisciplinary teams under the scientific guidance of the museum's researchers. Temporary exhibitions on a smaller scale are displayed at the Museum am Löwentor and often feature special parts of collections that are usually hidden in the magazines. As a recent example, the special exhibition "Antlers" illustrated the evolution of deer and cranial appendages (Fig. 53.19). At the Museum Schloss Rosenstein, temporary exhibitions on a larger scale are shown that mainly deal with general themes of biology. Recent examples include exhibitions on "150 Years of Evolutionary Theory", "Sex—Motor of Evolution" (Fig. 53.20), "Climate Change: A Tightrope Walk", and "Forever Young: Masterpieces of Taxidermy".

All exhibitions are accompanied by various educational offers, notably optional guided tours to impart widened background knowledge on the theme on display. Temporary exhibitions are also deepened by a series of expert talks on the topic. Additionally, special museum events (museum night, open house presentation, summer fete) foster public interest.

On a regular basis, the museum educational service offers programs for different target groups—from primary school and grammar school pupils up to families and adults. Numerous special projects and workshops for preschools and schools are conducted. In the project "Bioforum", teenagers actively explore domestic fauna and flora under qualified guidance by museum staff. The SMNS educational service

Fig. 53.19 Temporary exhibition "Museum Special—Antlers" at the Museum am Löwentor in 2014 (with permission from: SMNS; photo: R. Baumann)

Fig. 53.20 Temporary exhibition “Sex—Motor of Evolution” at the Museum Schloss Rosenstein in 2012 (with permission from: SMNS; photo: A. Staniczek)

also organizes various training courses for teachers to optimize their educative efforts before they visit the museum with their classes.

Besides traditional education, the museum educational service also strikes new paths by offering new forms of knowledge transfer, such as the “science pub”, where people meet in a pub to listen to scientific talks in a relaxed atmosphere.

Finally, the SMNS is also present on the World Wide Web in many ways: the museum’s home page is the main entrance gate for a digital visit. The SMNS maintains its own Facebook site and also uses Twitter to spread news in the social media. In the museum’s science blog “Understanding Nature”, SMNS scientists write online about their latest discoveries and present their scientific results intelligible to all.

References

Adam KD (1991a) Aus der 200jährigen Geschichte des Stuttgarter Naturkundemuseums. Stuttg Beitr Naturkunde (Ser A) 457:1–23

Adam KD (1991b) Die Württembergische Naturaliensammlung zu Stuttgart im Zweiten Weltkrieg. In: Aus der Geschichte des Stuttgarter Naturkundemuseums. Stuttg Beitr Naturkunde (Ser C) 30:81–97

Eder J (2015) Jahresbericht des Staatlichen Museums für Naturkunde 2014/Dokumentarische Ausgabe. SMNS, Stuttgart, p 67

Fricke R (1991) Bleeker und Klunzinger—zwei bedeutende Sammler im 19. Jahrhundert. In: Aus der Geschichte des Stuttgarter Naturkundemuseums. Stuttg Beitr Naturkunde (Ser C) 30:38–42

Heraty JM, Burks RA, Cruaud A, Gibson GAP, Liljeblad J, Munro J, Rasplus J-Y, Delvare G, Jansta P, Gumovsky A, Huber J, Woolley JB, Krogmann L, Heydon S, Polaszek A, Schmidt S, Darling DC, Gates MW, Mottern J, Murray E, DalMolin A, Triapitsyn S, Baur H, Pinto JD, van Noort S, George J, Yoder M (2013) A phylogenetic analysis of the megadiverse Chalcidoidea (Hymenoptera). Cladistics 5:466–542

Huang D-Y, Bechly G, Nel P, Engel MS, Prokop J, Azar D, Cai C-Y, van de Kamp T, Staniczek AH, Garrouste R, Krogmann L, dos Santos Rolo T, Baumbach T, Ohlhoff R, Shmakov AS, Bourgoin T, Nel A (2016) New insect order unravels the evolution of suction feeding in hemimetabolous insects. Sci Rep 6(23004):1–9

Jonsson KA, Fabre PH, Fritz SA, Etienne RS, Haegeman B, Ricklefs RE, Jorgensen TB, Fjeldsa J, Rahbek C, Ericson PGP, Woog F, Pasquet E, Irestedt M (2012) Ecological and evolutionary determinants for the adaptive radiation of the *Madagascan vangas*. Proc Natl Acad Sci USA 109(17):6620–6625

König C (1991) Forschungsreisende und ihre Verdienste um den Aufbau der Zoologischen Sammlung. Stuttg Beitr Naturkunde (Ser C) 30:21–37

Kupfer A, Müller H, Antoniazzi MM, Jared C, Greven H, Nussbaum RA, Wilkinson M (2006) Parental investment by skin feeding in a caecilian amphibian. Nature 440:926–929

Lampert K (1896) Zur Geschichte des K. Naturalienkabinets in Stuttgart nebst Bericht für die Jahre 1894 und 1895. Jahresh Ver Vaterl Naturkd Württemb 52:363–398

Martens J, Hoppe D, Woog F (2013) Diet and feeding behaviour of naturalised Amazon Parrots in a European city. Ardea 101(1):71–76

Merker S, Driller C, Perwitasari-Farajallah D, Pamungkas J, Zischler H (2009) Elucidating geological and biological processes underlying the diversification of *Sulawesi tarsiers*. Proc Natl Acad Sci USA 106(21):8459–8464

Möhn E (1991) Wegbereiter der Entomologie. In: Aus der Geschichte des Stuttgarter Naturkundemuseums. Stuttg Beitr Naturkunde (Ser C) 30:47–51

Ogden TH, Gattolliat J-L, Sartori M, Staniczek AH, Soldan T, Whiting MF (2009) Towards a new paradigm in mayfly phylogeny (Ephemeroptera): combined analysis of morphological and molecular data. Syst Entomol 34(4):616–634

Rajaei H, Greve C, Letsch H, Stüning D, Wahlberg N, Minet J, Misof B (2015) Advances in Geometroidea phylogeny, with characterization of a new family based on *Pseudobiston* (Lepidoptera: Glossata). Zool Scr 44(4):418–436

Rauther M (1940) Rückblick auf das Werden der Württ. Naturaliensammlung, vornehmlich auf die jüngst vergangenen 50 Jahre. 1. Teil. Aus: Zum 150-jährigen selbständigen Bestehen der Württ. Naturaliensammlung in Stuttgart, eine Gedenkschrift. Jahresh Ver Vaterl Naturkd Württemb 96(4):7–46

Renner F (1991) Die Spinnensammlung von Wilhelm Bösenberg. In: Aus der Geschichte des Stuttgarter Naturkundemuseums. Stuttg Beitr Naturkunde (Ser C) 30:43–46

Richling I, Bouchet P (2013) Extinct even before scientific recognition: a remarkable radiation of helicinid snails (Helicinidae) on the Gambier Islands, French Polynesia. Biodivers Conserv 22 (11):2433–2468

Schawaller W (2015) The genus *Laena* Dejean, 1821, in Nepal (part 2), with descriptions of twelve new species and new records of known species (Insecta: Coleoptera: Tenebrionidae). In Hartmann M, Weipert J (eds.) Biodiversität und Naturausstattung im Himalaya, Band V. Verein der Freunde und Förderer des Naturkundemuseums Erfurt e.V., Erfurt, pp 457–467

Schmalfuss H (2003) World catalog of terrestrial isopods (Crustacea: Isopoda). Stuttg Beitr Naturkunde (Ser A) 654:1–341

Schoch RR, Sues H-D (2015) A Middle Triassic stem-turtle and the evolution of the turtle body plan. Nature 523:584–587

Schoch R, Poschmann M, Kupfer A (2015) The salamandrid *Chelotriton paradoxus* from Enspel and Randeck Maar (Oligocene-Miocene), Germany. Palaeobiodivers Palaeoenviron 95:77–86

Schüz E (1964) Zur Geschichte der Heimatsammlung für Naturkunde. Jahresh Ver Vaterl Naturkd Württemb 118(119):359–369

Schwarz K, Hulme M, Woog F (2012) All you can eat—do peck rates of Greylag Geese (*Anser anser*) vary with age and social status? J Ornithol 153(4):1025–1029

Sroka P, Staniczek AH, Bechly G (2015) Revision of the giant pterygote insect *Bojophlebia prokopi* Kukalová-Peck, 1985 (Hydropalaeoptera: Bojophlebiidae) from the Carboniferous of the Czech Republic, with the first cladistic analysis of fossil palaeopterous insects. J Syst Palaeontol 13(11):963–982

Staniczek AH, Sroka P, Bechly G (2014) Neither silverfish nor fowl: the enigmatic Carboniferous *Carbotriplura kukalovae* Kluge, 1996 (Insecta: Carbotriplurida) is the putative fossil sister group of winged insects (Insecta: Pterygota). Syst Entomol 39(4):619–632

Warth M, Ziegler B (1991) Aus der Frühzeit des Naturalienkabinetts. In: Aus der Geschichte des Stuttgarter Naturkundemuseums. Stuttg Beitr Naturkunde (Ser C) 30:5–20

Woog F, Haas H, Heynen I (2003) Die Vogelsammlung des Staatlichen Museums für Naturkunde Stuttgart (SMNS)—Funde aus Deutschland mit Schwerpunkt Baden-Württemberg. Jahreshefte Ges Naturkunde Württemb 159:205–263

Ziegler B (1991a) Chronologie. In: Aus der Geschichte des Stuttgarter Naturkundemuseums. Stuttg Beitr Naturkunde (Ser C) 30:21–37

Ziegler B (1991b) Bilder aus der Geschichte des Stuttgarter Naturkundemuseums. Stuttg Beitr Naturkunde (Ser C) 31:1–64

Chapter 54
WAREN: Müritzeum with the Natural History Collections for the State of Mecklenburg-Vorpommern

Renate Seemann

Abstract The Mecklenburg-Vorpommern Natural History Collections (*Naturhistorische Landessammlungen für Mecklenburg-Vorpommern*) are part of the Müritzeum gGmbH (a non-profit limited company), which was founded in Waren (Müritz) on January 1, 2007. Located in the Mecklenburg Lake District, the Müritzeum is both a museum and, at the same time, a large nature experience centre. It covers an area of about 2,300 m^2 where exhibitions on nature and environment as well as on the collection's history and on the state of Mecklenburg-Vorpommern are on display. A freshwater aquarium showing fish species native to the area and the museum's garden complete the experience.

Dating back to a museum foundation by naturalist Freiherr Hermann von Maltzan (1843–1891) in 1866, the collections have documented the flora, fauna and geology of the Mecklenburg-Vorpommern inland for more than 150 years. Today, they comprise more than 285,000 exhibits. Amongst the zoological collections, particular emphasis is on the bird collection, on the collection of native molluscs including original exhibits from the museum's founder and on the insect collection. Both the natural science library and the extensive archives containing catalogues, collector diaries, correspondence and manuscripts from collector's estates and endowments are available for scientific research.

Keywords Mecklenburg-Vorpommern • Waren (Müritz) • Müritzeum • Freiherr Hermann von Maltzan • Natural History Collections

The "Natural History Collections for the State of Mecklenburg-Vorpommern" form part of the Müritzeum gGmbH, established in Waren on January 1, 2007. The Müritzeum is a museum and at the same time the biggest nature adventure centre in the Mecklenburg Lake District. Items covering nature and the environment, and the history of the state and its collections, are exhibited on 2,300 m^2 (Fig. 54.1). The aquarium for indigenous freshwater fish contains 26 tanks with a capacity of around

R. Seemann (✉)
Müritzeum, Zur Steinmole 1, 17192 Waren (Müritz), Germany
e-mail: r.seemann@mueritzeum.de

L.A. Beck (ed.), *Zoological Collections of Germany*, Natural History Collections,
DOI 10.1007/978-3-319-44321-8_54

Fig. 54.1 Aerophoto of the Müritzeum

200,000 l. The museum garden, having an area of around 2 ha and containing a natural lake, completes the nature experience for more than 160,000 visitors annually.

The beginnings of the collections are based on a museum foundation dating from 1866. They provide documentation of the flora, fauna and geology of the inland areas of Mecklenburg-Vorpommern and today consist of more than 285,000 exhibits.

54.1 History and Development

In 1866, Freiherr Hermann von Maltzan (1843–1891), an owner of a manorial estate, a natural scientist and a collector, founded the "von Maltzan'sche Naturhistorische Museum für Mecklenburg" in Waren (the von Maltzan natural history museum for Mecklenburg). The first official natural history museum in what was then the Grand Duchy of Mecklenburg-Schwerin was created. The founder's wish was "to show the friend of nature in an understandable way the results of research done, in order to thus move him to undertake more intensive studies of nature locally, but to give the newer researcher a desired overall picture which the systematic listing of the discoveries since made can only give in one direction". (Fig. 54.2) (1866) (Hecht 1994; Struck 1869)

He was able to gain the Waren schoolteacher Carl Struck as the honorary curator and director of the museum. Like Freiherr Hermann von Maltzan, Carl Struck was also a member of the "Verein der Freunde der Naturgeschichte in Mecklenburg" (association of friends of natural history in Mecklenburg). Here they established connections with the Mecklenburg collectors and natural scientists. In the following decades, the museum received numerous collections and valuable individual exhibits from them. Thus as long ago as the nineteenth century, a notable natural

Fig. 54.2 Freiherr Hermann von Maltzan (1843–1891)—estate owner, collector, natural scientist and museum founder

science collection was created. In the "Archiv des Vereins der Freunde der Naturgeschichte in Mecklenburg" (the archives of the association of friends of natural history in Mecklenburg), which appeared annually, there were regular reports about the progress of the museum, and further support was requested (Seemann 1997).

In 1884, with the support of the town of Waren, a museum building was erected. The sale of the building in 1919 and the provisional rehousing of the collections in various places interrupted the continuous work. As a result of the great commitment on the part of the "Vereinigung für Heimatschutz" (an organization for regional conservation) in Waren, the Maltzaneum was allocated rooms in the building of the former civic school in 1929 the cling (Fig. 54.3). For many years, some members of the organization looked after the collections in an honorary capacity.

The Second World War and the period immediately following left their mark on the collections. While the war was still in its last days, several museums removed valuable collections and stored them in Sophienhof castle near Waren. The Maltzaneum, too, had to deposit specimens from the bird collection there. Amongst these were rare vagrants in Mecklenburg, such as the short-toed eagle, the steppe eagle, the lesser kestrel and the Pallas's sandgrouse. The castle and nearly all collections were destroyed by arson. Some losses after the war were above all due to missing sheets of glass in the showcases and the lack of heating material.

The entomologist Carl Hainmüller (1875–1956) was in charge of the museum from 1927 to 1929, in an honorary capacity, and from 1946 to 1948 on a full-time basis. The extensive enlargement and reorganization of the insect collection and the creation of a specialist library are thanked to him. Karl Bartels (1884–1957), an ornithologist and from 1923 a nature officer and curator for the past history of the administrative district of Waren, looked after the bird collection, initially in an honorary capacity. From 1948 to 1956, he was the curator of the museum.

Fig. 54.3 "Haus der Sammlungen" ("House of the Collections")

In 1957, the natural history museum (Maltzaneum) was united with the local history museum, which had also been housed in the same building since 1929, under the name of "Müritz-Museum". In the years following, work focussed on the reorganization of the exhibition, educational tasks and nature conservancy commitment (Schönrock 1966; Schröder 1966).

After 1973, the museum was given the status of a special museum for state culture, nature conservation and environment. A renovation of the museum building which started in 1982 lasted until 1991 and involved great restrictions in the work concerning the collections. After the reopening of the house, the creation of an image which had been started was finalized with the exclusion of the historic collections of the former local history museum and the collections covering prehistory and early history. The Müritz-Museum was now rightly able to call itself the "Natural History Museum for the State of Mecklenburg-Vorpommern" and thus again refer to the tradition and significance of the collections. Enlarged storage areas and a new taxidermal workshop created new possibilities for the continuation of the collection work. A start was made on setting up computer-aided databases for the collections. The collections held increased considerably through specimens collected by the museum itself, through new specimen preparations and through gifts (Seemann 2001).

The historic museum building then again underwent renovation and alteration work which started in 2005, to become the "House of the Collections", the first stage of the "Müritzeum" project. Work on a second large exhibition building followed from 2006 onwards. With the founding of the Müritzeum gGmbH on January 1, 2007, the state natural history collections, for which however the

Fig. 54.4 The new exhibition building at the Müritzeum—"House of the 1000 Lakes"

administrative district continued to be responsible, became part of the Müritzeum. The project was successfully completed in August 2007 with the opening of the exhibitions in the "House of the 1000 Lakes" Fig. 54.4 (Seemann & Seemann 2011).

54.2 Specific Points Relating to the Collection: Its Importance and Contents

More than 285,000 items concerning nature in the state of Mecklenburg-Vorpommern have been collected in almost 150 years in the natural history museum in Waren, to form the botanical, zoological and geological collections. The collecting in this respect was always focussed on the inland areas. However, the acceptance of collectors' bequests brought about a number of fairly small collections or individual items from all over the world for the museum. Besides the natural history collections, an extensive stock of photos, maps, documents and a variety of archive items developed. A specialist library focussing on natural history and Mecklenburg items was created through gifts, purchases and collectors' bequests. More than 100 periodicals and a collection of special prints are available for scientific work.

54.3 Bird Collection

Prepared specimens and eggs which had been owned by Freiherr Hermann von Maltzan formed the beginning of the bird collection. By purchasing the extensive collection of the deceased Bützow forestry commissioner L.v. Grävenitz, Hermann Freiherr von Maltzan secured a valuable collection stock for his museum in 1867. Of the nineteenth-century collectors, the doctor Franz Schmidt (Wismar) and the pastor Heinrich David Friedrich Zander (Barkow) should be given special mention; valuable prepared specimens came from their collections. Particular items were also added to the collection with the help of the taxidermists Sievert Nicolai Steenbock (Rostock) and Carl Knuth (Schwerin), who worked for the museum for many years [e.g., glossy ibis (Warnemünde, 1842) and little bustard (Niekrenz near Sanitz, 1828)] (Jesse 1902).

A few rare collection items were lost during the Second World War.

Amongst the numerous additions in the twentieth century, three fairly large collections in particular must be emphasized: (1) the collection of the pharmacist Otto Held (Neukloster) comprised more than 100 prepared bird specimens; he donated this collection, along with other prepared specimens and valuable books, to the museum. (2) Due to the special efforts of Karl Bartels, the museum received the bird collection of the grand duchy court taxidermist Carl Knuth in 1951. (3) When the Neubrandenburg administration's nature conservancy station in Serrahn was wound up in 1991, 1,004 prepared bird bodies were given to the museum. The series, in part very extensive, are invaluable for comparative scientific examinations. The collections also included the prepared bird body of a red-flanked bluetail, an initial find for the area covered by the former GDR.

The oldest prepared specimen is an Alpine swift shot in Mecklenburg which became part of the collection of the Schwerin court counsellor Georg Lembcke between 1795 and 1804 (Fig. 54.5).

The state natural history collections have for many years been integrated in the state project "Research into causes of death". Here, the corpses examined, in particular those of white-tailed eagles and ospreys, are documented and entered in the archives (Seemann & Seemann 2011).

The collection grows continuously through the delivery of corpses found and their preparation and through the acceptance of collection bequests and donations. It comprises at present more than 3,030 mounted prepared birds and bird bodies, 4,115 clutches with more than 11,100 eggs, 696 plucked birds and around 300 skeletons (Fig. 54.6). The collection covers nearly all breeding bird species and also those species which are shown to be birds of passage or vagrants.

The work done in the taxidermal workshop is imperative for maintaining the collection and preparing new specimens.

Fig. 54.5 The oldest specimen—an Alpine swift—(collected between 1795 and 1804), one of the last remaining specimens from the large bird collection of the Schwerin court counsellor Georg Lembcke

Fig. 54.6 View of the bird collection

54.4 Insect Collection

The most extensive collection is that of the insects, with roughly 175,000 specimens. Beetles and butterflies/moths form the greater part of the collection. The first specimens in the nineteenth century were those donated, for instance, dragonflies/damselflies from Moritz Füldner (Neustrelitz), beetles from C.A.F. Peter (Neustrelitz), bees from Heinrich Friese (Schwerin) or cicada and bugs from Friedrich Wilhelm Konow (Schönberg). The oldest butterfly and moth collection (before 1835) is from the administrative district counsellor Karl von Oertzen at Brunn. It comprises approximately 1,800 specimens in 960 small veneered wooden boxes.

The most important contribution to extending the collection was that of the versatile entomologist Carl Hainmüller from Waren (1875–1956). There is evidence of his collection in nearly all insect groups. Hainmüller was above all out and about in the area around Waren and in the Müritz region. By exchanging specimens, he also built up a collection of around 20,000 beetles from all over the world, which is at present being reconstructed. As after Carl Hainmüller's death there was no longer a full-time entomologist working at the museum, it worked in cooperation with external entomologists for the revision and reorganization of the various collections.

In 1991, it was possible to buy the collection, comprising around 3,000 butterflies and moths, from the estate of H. Radtke (Eggesin). A valuable addition was also the donation of around 23,000 butterflies and moths by the teacher Jürgen Wulf (Bälau) in 2003. The collection contains specimens from Fritz Dörries jun. and sen. (Hamburg), Fritz Diehl (Hamburg) and Horst Weich (Elmshorn). The material comes mainly from northern Germany, but it also includes a Fritz Dörries lycaenid collection from Siberia.

The museum owns the most extensive small butterfly and moth collection for Mecklenburg-Vorpommern. The material in Waren stems mainly from Carl Hainmüller, Werner Eichbaum (Schwerin) and Uwe Deutschmann (Dobin am See). It comprises around 7,500 butterflies and moths from more than 520 species. The collection has been revised and digitalized.

The coleopteran collection covers around 42,000 beetles. In the past few years, there was a significant addition, mainly under the Cerambycidae, from the collections of Hans Dieter Bringmann (Rostock), Lothar Schemschat (Waren) and Wolfgang Döring (Stralsund). This makes the cerambycid collection in the Müritzeum, with its 6,317 specimens from 505 species and 61 subspecies, the most important collection for Mecklenburg-Vorpommern, and it also contains 90 % of all European species known at present.

In the bug collection, there are above all specimens from Carl Hainmüller, E. Wagner (Waren) and Friedrich Wilhelm Konow (Schönberg). Containing 1,100 prepared bugs, it is the most important collection in Mecklenburg-Vorpommern besides the collection at Rostock University.

The revision and reorganization of some hymenopteran groups (approximately 2,500 specimens) have already taken place.

54.5 Mollusc Collection

The founder of the natural history museum in Waren, Freiherr Hermann von Maltzan, was interested in natural history back in his childhood and soon discovered his particular interest in molluscs (Fig. 54.7). His finds from Mecklenburg formed the basis of the museum collection in 1866. In 1873, he published the "Systematic List of Mecklenburg Inland Molluscs".

The mollusc collection, too, contains items from many members of the "Association of Friends of Natural History in Mecklenburg", such as Carl Arndt, senior primary school teacher (Neubrandenburg); Friedrich Eduard Koch, state building supervisor (Güstrow); Carl Michael Wiechmann, estate owner (Kadow); and Carl Struck, teacher and museum curator (Waren).

In the twentieth century, there were significant additions to the collection from Nikolai von Nifontoff (Schwerin), Walter Karbe (Neustrelitz) and G. Stöckel (Neustrelitz). As a result of the close cooperation with the malacology working group, founded in 1987, of the three northern administrative districts of the GDR, new items were regularly added to the collection. Extensive material was given by Uwe Jueg (Ludwigslust), Michael Ludwig Zettler (Rostock), Renate Seeman (Waren) and Uwe Göllnitz (Rostock).

Fig. 54.7 Specimens from the mollusc collection with their original labels from H. v. Maltzan

In accordance with the museum's collection profile, mainly items from Mecklenburg-Vorpommern are collected. At present an inventory has been made of more than 7,000 exhibits. In addition to this, there are around 1,500 tropical land molluscs and marine snails and mussels. They come from gifts, bequests and former school collections.

54.6 Mammal Collection

The mammal collection is one of the smaller collections which have so far not undergone much revision. It comprises around 1,300 mounted prepared mammals, prepared bodies, skins and fells, skeletons and antlers. The oldest specimens date from the nineteenth century. Particular mention must be given to some bat specimens, such as the pond bat (Waren, before 1876, the first specimen for Mecklenburg), the lesser horseshoe bat (Neubrandenburg, 1875, the only specimen for Mecklenburg-Vorpommern) and the northern bat (Waren, 1907, today regarded as extinct in Mecklenburg-Vorpommern). The rarities also include the garden dormouse and the common dormouse, of which there are specimens dating from the nineteenth century. The collection was enlarged in the twentieth century above all by Helmut Richter (Waren), Otto Held (Neukloster), Carl Knuth (Schwerin) and Horst Schröder (Waren). In the past few years, additions were made to the collection above all in the form of small mammal specimens.

54.7 Other Collections

For some zoological taxa, there are only a few specimens or small collection items. Thus, there are around 400 specimens of amphibians, reptiles, fish, crabs, leaches and spiders. Special mention must be given to 36 specimens of Mecklenburg fish fauna stemming from 1870 to 1903 and an agile frog (1890, Schwerin).

54.8 Present Situation and Perspectives

Through the "Müritzeum" project, it was also possible to renovate the historic museum building. It became the "House of the collections" and on one floor shows the history of the museum and its collections. The storage area on the upper floors of the building was significantly increased and some rooms were newly furnished.

The collections are looked after by two curators, a taxidermist and a librarian. The inventory and cataloguing work is supported by voluntary helpers from the interest group "State Natural History Collections". In all collection areas, work has been started on setting up digital databases. In cooperation with external specialists, the revision and reorganization of the various collection sections are being continued. The collections also comprise extensive archives containing catalogues, collection diaries, correspondence and manuscripts; their reorganization is an important task. The contents of the valuable documents provide a variety of information concerning the collector and his/her activity.

Those undertaking scientific work in the collections also have access to an extensive specialist library (around 16,000 books and more than 100 series of specialist journals). A well-equipped taxidermal workshop and a specialist taxidermist provide the necessary requirements for looking after the conservation side of the collections. Regular additions to the collections come from the new specimens prepared from corpses found and through specimens collected in Mecklenburg-Vorpommern by the museum itself. The material examined as part of the project "Research into the causes of death of birds of prey in Mecklenburg-Vorpommern", once it has undergone the necessary conservation, is kept in the bird collection for further evaluations.

In the past few years, it has also been possible to add to the collections numerous valuable items stemming from bequests, gifts or purchases. Upon request, the collections are available to both amateurs undertaking nature research and to scientists for research work. The spectrum of questions posed has so far covered taxonomy to fauna.

The tasks for the coming years include the continuation of research in respect of the collections, the scientific processing of new additions and the continuation of the digital work in all areas of collection.

References

Hecht H (1994) Hermann Freiherr von Maltzan—eine Lebensbeschreibung—Neubrandenburger Mosaik No. 18

Jesse R (1902) Maltzaneum, naturhistor. Museum für Mecklenburg, 1. Teil—Verzeichnis der Säugetiere, Vögel, Eier und Nester, 2. Teil: Verzeichnis der Reptilien und Amphibien. Ratsbuchdruckerei C. Michaal, Güstrow

Schönrock H (1966) Zur Geschichte des Warener Museums. Heimatkundliches Jb Bez Neubrandenburg 1:21–43

Schröder H (1966) Bildungs-, Erziehungs- und Forschungstätigkeit der naturkundlichen Abteilung des Müritz-Museums. Heimatkundliches Jb Bez Neubrandenburg 1:44–58

Seemann R (1997) Der Verein und das "von Maltzan'sche Naturhistorische Museum für Mecklenburg" in Waren. Arch d Freunde d Naturgesch Mecklenburg XXXVI:53–67

Seemann R (2001) Die Vogelsammlung des Müritz-Museums, Teil 1: Standpräparate und Bälge. Veröff Müritz-Mus Waren 17:1–156

Seemann F, Seemann R (2011) Katalog der Vogelsammlung, Präparate, Eier, Skelette, Rupfungen. Veröff Naturhistorischen Landessammlungen 18/2011

Struck C (1869) Das Meklenburgische naturhistorische v. Maltzan'sche Museum zu Waren. Arch d Freunde d Naturgesch in Mecklenburg 22:114–121 (s.a. ANM 23 ff.)

Chapter 55
WESTLAUSITZ: The Zoological Collection of the Museum der Westlausitz Kamenz by Olaf Zinke

Olaf Zinke

Abstract The Museum der Westlausitz Kamenz is a regional museum in sponsorship of the administrative district Bautzen. The origin of its collections reaches back till the nineteenth century. From 1991 onwards, the zoological collection has been rearranged and redesigned. In the meantime, it has been actively expanded. Since 1991, the zoological collection is extended actively only with vertebrates. When the Sammelsurium—the public accessible depot of the "Museum der Westlausitz"—was built in 2000, an optimal working space and a modern depot have been created. The zoological collection grows annually by approximately 500 objects—it contains currently 14,500 objects with focus of skulls and skeletons. Scientific research of the material and objects focuses primarily on causes of death of endangered vertebrates and includes research in ecology of population, ecology of nourishment, parasitology, anatomy/morphology and faunistic developments. The collection is open to the public. The museum offers several educational guided tours leading visitors directly into the depots.

Keywords Regional museum • Eastern Saxony • Vertebrata • Depot accessible to the public

The Museum der Westlausitz Kamenz is a regional museum in sponsorship of the administrative district Bautzen. The origin of its collections reaches back till the nineteenth century. The foundation for nature studies in Kamenz was laid with the formation of the society "Gebirgsverein zu Kamenz" on 1 October 1887. The "constitution" of the "Gebirgsverein zu Kamenz" from 1912 states the goals of the society: "The distribution of scientific knowledge about the local landscape" was the guideline of the society which was to be achieved by acquisition of books, maps and antiquities concerning the homeland. The founding of natural historic

O. Zinke (✉)
Zoological Department, Museum der Westlausitz Kamenz, Macherstraße 140, 01917 Kamenz, Germany
e-mail: zoologie@museum-westlausitz.de; http://www.museum-westlausitz.de/sammlungen/fbzoologiebotanik/smzoologiebotanik

L.A. Beck (ed.), *Zoological Collections of Germany*, Natural History Collections,
DOI 10.1007/978-3-319-44321-8_55

collections and the effort to protect and maintain historical and scenic benchmarks were important tasks for the newfound society. Most of these tasks are still imperative today for the Museum der Westlausitz.

In the year 1931, the archivist Dr. Gerhard Stephan set out to find a town historical museum, which incorporated the collections of the "Gebirgsverein zu Kamenz". Since the 1930's, a branch division of the natural scientific society "ISIS Bautzen" carried out nature historical research in Kamenz.

World War II was a deep caesura into the collections and processing the local historical findings. The town historical museum has been shut down in 1943, and in December 1944, the last gathering of the natural scientific society "ISIS Bautzen" took place. The whole collection—including the nature historical objects—were packaged and relocated several times during the commotions of the war.

It took until 1954 to transfer the objects to the town hall, where they could be stored and exhibited to the public. Three years later, the town councilmen decided to reassign one of Kamenz' oldest middle-class houses—the Ponikauhaus—to the museum. Over the course of a decade, the departments of biology, geology and cultural history have been founded, and the museum became a landscape museum. To clarify this fact, the museum has been labelled as "Museum der Westlausitz" in 1968.

The "Museum der Westlausitz" established itself in the 1980's as an important pillar of natural scientific research and education in eastern Saxony alongside the natural scientific museums in Dresden and Görlitz. Its main focus was the western part of the upper Lusatia. In 1991, the museum became part of the administrative district Kamenz (today Bautzen). In the meantime, the museum realigned its policies, a process which influenced also the zoological collection.

The zoological collection included to this date approximately 3000 objects. The main focus of the collection was avian taxidermist specimens inherited from the "Gebirgsverein zu Kamenz" (till 1920), the Kamenz branch division of the natural scientific society "ISIS Bautzen" (till 1945), an extensive collection of scientific birds' hides (293 specimens of 71 species) and clutches (558 clutches including 2818 eggs). The collection has been purchased from the teacher and ornithologist Manfred Melde (1929–2012) from Biehla (Ulbricht 2009; Zinke and Cebulla 2009). Miscellaneously, the overall collection included two collections of insects—the extensive collection of Zscheische and the collection of Lepidoptera Lehmann. While the collection from Manfred Melde is exactly dated and scientifically significant, the old avian objects—apart from a few exceptions—remain without useable data. Approximately 150 well-dated habitus specimens have been created between the years 1985 and 1989—mostly birds from western Lusatia. In 1993, the avian collection of the teacher and ornithologist Paul Weißmantel (1893–1975) from Kamenz has been donated to the museum (see, e.g., Creutz 1993; Melde 1993; Neumann 2010). In parts the Weißmantel collection is well dated and originates from the time before 1914 when Weißmantel still lived in the western Saxonian town of Rochlitz. After 1914, he collected in western Lusatia.

From 1991 onwards, the zoological collection has been rearranged and redesigned. In the meantime, it has been actively expanded. Building upon the main emphasis of the museum, the goals of the collection were redefined:

- Long-term preservation of extensive series of eco-faunistic records about vertebrates in eastern Saxony (main focus is western Lusatia) for a species-related monitoring and research about biodiversity.
- Ecological research about causes of death by endangered vertebrates as a basis for consultations of authorities in terms of nature conservation.
- Documentation of faunistic peculiarities.
- Collections of comparable objects for matters of education and studies.
- Manufacturing of dermoplastics for exhibitions.
- Museum-related environmental education using original material and objects in a depot with public access.

It was almost impossible to achieve a satisfactory protection and expansion of the collection within the conditions the facility provided in 1990. When the Sammelsurium—the public accessible depot of the "Museum der Westlausitz"—was built in 2000, all these problems have been eradicated. The new facility provided full air-conditioned, windowless, depot with moveable cabinets and special racks: 200 m^2 of space and full public access. A state-of-the-art workshop has been built for taxidermist and preservation purposes. This new facility provides enough space and includes compartments for autopsies, maceration and refrigeration.

Since 1991, the zoological collection is extended actively only with vertebrates. Invertebrates are only added to the collection when they are delivered as chance findings.

Most of the new specimens are delivered by "gatherers" who secure carcasses and transfer them to the museum. Several district authorities support the museum in this task, namely, the lower nature protection agency, hunting agency and veterinarian agency. Further partners are supraregional authorities like the agencies for road maintenance and road construction. Additional support comes from local environmental protection stations, zoological gardens, colleagues in honorary appointment, huntsmen and dedicated citizens. The zoological collection grows annually by approximately 500 objects due to the help of all these institutions and people. While the collection included 3204 inventoried objects in 1991, it is currently comprised by 14,500 objects (see Fig. 55.1). Divided by the different vertebrate taxa, the zoological collection grew since 1991 by 10,200 mammals (Mammalia), 4.500 birds (Aves), 200 reptiles (Reptilia), 120 amphibians (Amphibia) and 130 fishes (Pisces) from eastern Saxony. Reptiles, amphibians and fishes are conserved in ethanol. Mammals and birds are preserved as skulls, skeletons, scientific hides and feather pickings or submerged in ethanol—dependent on their purpose of application.

Mammals are mostly conserved in form of skulls and skeletons, which you can see, for instance, with the family of Mustelidae in Table 55.1. Currently, the collection of mammals includes 1050 scientific hides or pelts, 8100 skulls (partly with skeletons or

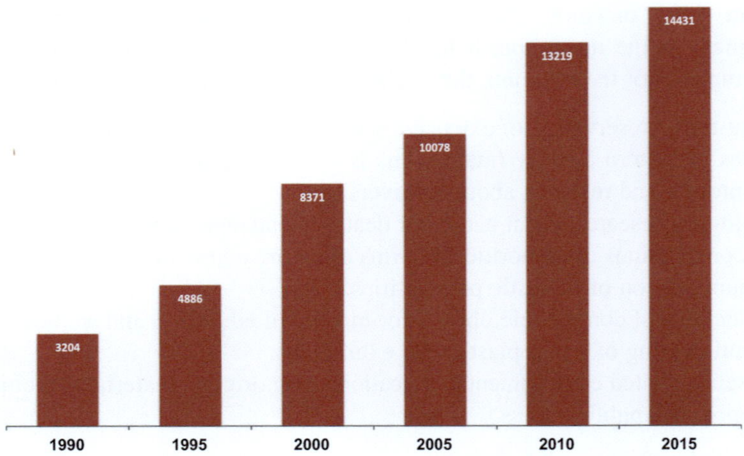

Fig. 55.1 Development of the zoological collection since 1990 by quantifying the inventoried objects (graphic: Olaf Zinke)

Table 55.1 Example for the current inventory: species of the family Mustelidae

Species	Skull with part of the skeleton	Scientific hide or pelt
Martes martes	51	16
Martes foina	234	18
Mustela erminea	71	44
Mustela nivalis	45	33
Mustela putorius	38	18
Neovison vison	43	11
Meles meles	114	1
Lutra lutra	577	111

part of skeletons) and 100 habitus specimens. 2600 mammals—including 1200 bats (Chiroptera)—are conserved in ethanol (see Fig. 55.2).

The collection of birds from eastern Saxony consists of 1000 scientific hides, 1600 feather pickings, 3500 skeletons, 500 habitus specimens and 1200 clutches. Sixty specimens are submerged in ethanol. Like the mammals, the birds are also mostly collected as skeletons and feather pickings (see Table 55.2 with the example of the family Fringillidae).

Habitus specimens are produced on a regular basis especially to be presented in exhibitions. Most prominent are two remarkable emulations of extinct species: a Tasmanian tiger (*Thylacinus cynocephalus*) and a dodo (*Raphus cucullatus*) (see Fig. 55.3).

Scientific research of the material and objects primarily focuses on causes of death of endangered mammals. Most prominent are the European otter (*Lutra lutra*)

Fig. 55.2 A glimpse into the depot. *Upper left*, collection of skeletons with the otter (*Lutra lutra*); *upper right*, collection of mammals with squirrel hides (*Sciurus vulgaris*) in the foreground; *lower left*, collection of birds with sparrow hawk hides (*Accipiter nisus*) in the foreground; *lower right*, collection of specimens conserved in ethanol showing parts of the collection of Chiroptera (photo: © with Olaf Zinke)

(Zinke 1991, 2000; Zinke et al. 2013a, b) and the Elbe beaver (Castor fibre albicus) (Zinke et al. 2013a, b). Causes of death aside, the research also grants insight into the ecology of population (see, for instance, Ansorge et al. 1996, 1997; Hauer et al. 2000, 2002), ecology of nourishment (Jauernig et al. 2008), parasitology (Mey and Zinke 1992, 1994; Christian 2012) and anatomy/morphology (Zinke 1993a, b; Zinke 1997; Zinke and Kümmel 1997; Zinke and Ansorge 1998). The data are also included into faunistic research (e.g., Krüger et al. 1998; Zöphel and Steffens 2002; Hauer et al. 2009). All inventoried objects are also recorded in a digitalised database and available for further special research.

It is important to mention the integration of the collection into the didactic concept of the museum. The museum offers several educational guided tours leading visitors directly into the depots. Approximately 25,000 people have visited the zoological collection as part of a guided tour since 2000. All these visitors were given an opportunity to directly take a peek backstage of the zoological department.

Table 55.2 Example for the current inventory: species of the family Fringillidae

Species	Skeleton	Scientific hide	Feather picking
Coccothraustes coccothraustes	52	13	35
Carduelis chloris	52	12	40
Carduelis carduelis	41	9	28
Carduelis spinus	77	10	45
Carduelis cannabina	24	6	15
Serinus serinus	5	1	4
Pyrrhula pyrrhula	27	12	14
Loxia curvirostra	34	31	1
Fringilla coelebs	75	24	44
Fringilla montifringilla	3	4	2

Fig. 55.3 Emulation of the dodo (*Raphus cucullatus*). The object was created together with an approximation of the Tasmanian tiger (*Thylacinus cynocephalus*) in the year 2006. Both objects were shown in the special exhibition "As dead as a dodo—About the fate of extinct species" (photo: © with Thomas Puttkammer)

References

Ansorge H, Schipke R, Zinke O (1996) Altersstruktur und Reproduktion in der Oberlausitz. In: Sächsisches Landesamt für Umwelt und Geologie (Hrsg.) Materialien zu Naturschutz und Landschaftspflege—Artenschutzprogramm Fischotter in Sachsen, pp 27–30

Ansorge H, Schipke R, Zinke O (1997) Population structure of the otter, *Lutra lutra*—parameters and model for a Central European region. Z Säugetierkd 62:143–151

Christian A (2012) Tick infestation (Ixodes) on the Eurasian Otter (*Lutra lutra*)—a longterm study. Soil Organ 84(2):481–487

Creutz G (1993) Paul Weißmantel zum Gedenken. Veröff Mus Westlausitz Kamenz Sonderh 5–8

Hauer S, Ansorge H, Zinke O (2000) A long-term analysis of the age structure of otters (*Lutra lutra*) from eastern Germany. Z Säugetierkd 65:360–368

Hauer S, Ansorge H, Zinke O (2002) Reproductive performance of otters *Lutra lutra* (Linnaeus, 1758) in Eastern Germany: low reproduction in a long-term strategy. Biol J Linn Soc 77:329–340

Hauer S, Ansorge H, Zöphel U (2009) Atlas der Säugetiere Sachsens

Jauernig O, Ansorge H, Zinke O (2008) Feeding ecology of the European otter Lutra lutra in eastern Saxony (Germany). Mamm Biol 73(suppl):18

Krüger S, Gliemann L, Melde M, Schrack M, Mädler E, Zinke O (1998) Die Vogelwelt des Landkreises Kamenz und der kreisfreien Stadt Hoyerswerda, Teil 1—Nonpasseres. Veröff Mus Westlausitz Kamenz Sonderh

Melde M (1993) Erinnerungen an Paul Weißmantel. Veröff Mus Westlausitz Kamenz Sonderh 9–10

Mey E, Zinke O (1992) Ein neuer Nachweis vom Fischotter-Haarling, *Lutridia exilis* (Insecta, Phthiraptera). Veröff Mus Westlausitz Kamenz 16:67–68

Mey E, Zinke O (1994) Eine Massenvermehrung und weitere Daten zur Biologie des Fischotter-Haarlings *Lutridia exilis* (Insecta, Phthiraptera). Veröff Mus Westlausitz Kamenz 17:22–26

Neumann J (2010) Paul Weißmantel. In: Neumann J, Holupirek H, Kneis P, Stamm HC, Weise W (eds) Lebensbilder sächsischer Ornitholgen. Mitt Ver Sächs Ornithol 10, Sonderheft 3:229–230

Ulbricht J (2009) Das ornithologische Schaffen von Manfred Melde im Spiegel seiner Publikationen. Veröff Mus Westlausitz Kamenz 29:115–120

Zinke O (1991) Die Todesursachen der im Museum der Westlausitz Kamenz von 1985–1991 eingelieferten Fischotter *Lutra lutra* (L., 1758). Veröff Mus Westlausitz Kamenz 15:57–63

Zinke O (1993a) Zur Präparation der Schlüsselbeine beim europäischen Fischotter *Lutra lutra* (L., 1758). Der Präparator 39(3):109–111

Zinke O (1993b) Erste Ergebnisse zu Untersuchungen am Schlüsselbein (Clavicula) des europäischen Fischotters *Lutra lutra* (L., 1758). Jahrestagung Fischottergruppe Österr Bad Radkersburg 29.04.–1.5.1993:85–89

Zinke O (1997) Ist der Penisknochen des Fischotters zur Altersbestimmung geeignet? Ber Naturforsch Ges Oberlausitz 6 Suppl., Abstracts 15. In: Internationales Marderkolloquium, 12–15 September 1996, Kollm/Oberlausitz 32

Zinke O (2000) Die Todesursachen der im Museum der Westlausitz Kamenz von 1992–1998 eingelieferten Fischotter *Lutra lutra* (L., 1758). Veröff Mus Westlausitz Kamenz 22:87–92

Zinke O, Ansorge H (1998) Remarks on age-determination criteria for otters (*Lutra lutra*). NERI Technical Report 262, Proceedings of the 16th Mustelid Colloquium, Arhus, Denmark, 9—12 October 1997, pp 26–27

Zinke O, Cebulla M (2009) Die ornithologische Sammlung von Manfred Melde—Beschreibung und Katalog. Veröff Mus Westlausitz Kamenz 29:87–114

Zinke O, Kümmel Th (1997) Bemerkungen zum Os clitoridis des Fischotters *Lutra lutra* (L., 1758). In: Proceedings 14. Mustelid Colloquium, Czech Republic, 14–17 September 1995, pp 27–30

Zinke O, Jeschke D, Ansorge H (2013a) Die Todesursachen ostsächsischer Fischotter aus dem Zeitraum 1990 bis 2010. Ber Naturforsch Ges Oberlausitz 21:73–81

Zinke O, Reuße P, Kramp T, Peters T, Ricklinkat A (2013b) Todesursachen von Elbebibern (Castor fiber albicus Matschie, 1907) aus Ostsachsen und Südbrandenburg im Zeitraum 1990 bis 2011. Säugetierkd Inf 9:153–158

Zöphel U, Steffens R (2002) Atlas der Amphibien Sachsens. Materalien zu Naturschutz und Landschaftspflege

Chapter 56
WIESBADEN: Museum Wiesbaden: Natural History Collections [MWNH]

Fritz Geller-Grimm

Abstract The natural history collections of the Museum Wiesbaden, State Museum for Art and Nature, was founded in 1829. The zoological collections encompass some 845,000 invertebrates, 20,000 vertebrates, and 3500 type specimens. The invertebrate collection by Johann Christian Gerning (1745–1802) is one of the oldest and biggest ones of the eighteenth century. Some animal preparations belong to Maria Sibylla Merian (1647–1717) and were the basis of more than 30 copperplates for her *Metamorphosis Insectorum Surinamensium*. In 2013, the original exhibition rooms of the natural history collections were refurbished with a new face, while parts of the storerooms were renovated and the collections restructured.

Keywords Museum Wiesbaden • Collection • Natural history • Zoology • Maria Sibylla Merian

56.1 Historical Background

The natural history collections were first made available to the public on the premises of the Palace of the Crown Prince as a museum initiated in 1829 by the Natural History Association of Nassau founded by the citizens of Wiesbaden. Ownership of the collections was transferred to the city of Wiesbaden from 1901 to 1973. With the growth and expansion of the collections, however, the museum required more space, and a new building was constructed at the close of the nineteenth century. The collections' final home is located today in the museum at Friedrich-Ebert-Allee. The structure, finished and furnished in 1915, was especially designed to house the three independent collections of antiquity, art, and natural history. In 1973, the State of Hesse united the collections to create a state museum. In 2009, ownership of the antiquity collection was transferred to the city of Wiesbaden. In 2013, the original exhibition rooms of the natural history collections

F. Geller-Grimm (✉)
Natural History Collections, Museum Wiesbaden, Friedrich-Ebert-Allee 2, 65185 Wiesbaden, Germany
e-mail: fritz.geller-grimm@museum-wiesbaden.de; http://www.museum-wiesbaden.de

L.A. Beck (ed.), *Zoological Collections of Germany*, Natural History Collections,
DOI 10.1007/978-3-319-44321-8_56

Fig. 56.1 A view of the crustacean display cabinet (with permission from: Museum Wiesbaden; photo: Bernd Fickert)

were refurbished with a new face, while parts of the storerooms were renovated and the collections restructured (Fig. 56.1) (Czysz 2004; Geisthardt 1985; Geller-Grimm 2004; Geller-Grimm et al. 2015).

56.2 Recent Collection

56.2.1 Structure

The natural history collections encompass basic geology, mineralogy, paleontology, botany, zoology, and ethnology. In the absence of a university in the Duchy of Nassau, the collection and documentation of the region's natural world became essential. Yet from the very beginning, the collections incorporated exemplars from other regions as well. For example, the most significant of the founding collections encompassed natural objects from almost all over the world. Johann Isaak von Gerning (1767–1837) contributed one of the largest arthropod collections of the eighteenth century to the museum in the year it was founded (Fig. 56.2).

The zoological collections today encompass some 845,000 invertebrates and 20,000 vertebrates. The approximately 120,000 mollusks date back to the nineteenth century, when numerous collectors sent their duplicate exemplars to Wiesbaden, while some 40,000 specimens have been acquired in the last 30 years. Our collection of nearly 720,000 insects includes abundant specimens of butterflies, beetles, and dipterans. Among the vertebrate collections, our collection of birds is the largest, encompassing some 13,500 specimens with 3500 display preparations.

Fig. 56.2 Sea heart and St. Thomas bean (with permission from: Museum Wiesbaden; photo: Bernd Fickert)

Fig. 56.3 Bobcat and saiga antelope (with permission from: Museum Wiesbaden; photo: Bernd Fickert)

Spiders, butterflies, cicadas, and dipterans make up the large part of our approximately 3500 type specimens (Fig. 56.3).

The collections are systematically organized and stored in one of the museum's four main archives, accessible through site and inventory catalogs. Some of the approximately 200,000 exemplars are recorded digitally, with the vertebrate collection making up the highest percentage of digital records at over 95 %.

Fig. 56.4 The movement room (with permission from: Museum Wiesbaden; photo: Bernd Fickert)

56.3 Recent Collection

56.3.1 Special Features

The exemplary collections of Johann Isaak von Gerning represent one of the most historically significant of its kind. The majority of the 40,000 preparations were collected originally by Gerning's father, Johann Christian Gerning (1745–1802), who acquired specimens from every region of the world, later, together with his son. Through circuitous routes, some animal preparations belonging to Maria Sibylla Merian (1647–1717) also made their way into this collection, some of which were the basis of more than 30 copperplates for her *Metamorphosis Insectorum Surinamensium*, though the precise correspondence in some cases cannot be established. The Gerning collection served as the basis for Eugen Johann Christoph Esper's (1742–1810) descriptions of numerous species of butterfly, as well as for the research of prominent entomologists of the time. While the preparations remain completely intact, many of those preserved in alcohol were destroyed in the war. There are, however, numerous type specimens by Embrik Strand (1876–1953) in the collection, used by arachnologists for their taxonomization. Like the Hymenoptera types of bee researcher Adolph Schenck (1803–1878), the cicada collection of Carl Ludwig Kirschbaum (1812–1880) was revised and newly structured. The same cannot yet be said of the Arnold Pagenstecher (1837–1913) collections, which are widely varied but, therefore, fragmentary. In recent years, our regional collections have grown, in particular the mollusks, through the generosity of Jens Hemmen (1944–2012) to whom we owe our gratitude for over 4200 taxa in more than 40,000 specimens and 10,400 series (Fig. 56.4).

Fig. 56.5 American lobster (with permission from: Museum Wiesbaden; photo: Ed Restle)

Our zoological collections undergo continuous revision and expansion. Our objectives place special emphasis on the region of the former Duchy of Nassau, today's State Museum of Hesse. The rapid digitalization of records and availability of data on the Internet immensely facilitate the incorporation of scientific private collections. Unlike many of the larger research collections, these private collections are preserved as individual collections. Shortages of both personnel and funding complicate the acquisition of general collections, though this is not always the case.

What is most significant for the collections remains to be the standing position of the storeroom administrator, who not only preserves order but ensures the safety of the collections (Fig. 56.5).

56.3.2 Exhibitions

The teaching and research collections are organized systematically in the storerooms and exhibited thematically. A new exhibition concept was developed in 2001

Fig. 56.6 Collection Gerning (with permission from: Museum Wiesbaden; photo: Fritz Geller Grimm)

and finally put on display in 2013. The museum's twofold focus on art and nature lends itself to the thematic presentation of phenomena, important to both natural scientists and artists—color, forms of nature, movement, and time.

The history and success of the natural sciences are due in large part to developments in illustrative techniques, a central educational medium. Maria Sibylla Merian serves as an example of the knowledge we can win through illustration. We took her example seriously for our new exhibitions, including superb specimens that not only provoke the viewers' interest and command their respect for nature but inspire them to put pencil to paper. Our younger guests, especially, enjoy trying their hand at sketching the specimens. Our exhibits follow the criteria of the natural sciences, though they are not explicitly devoted to general topics such as evolution, ecology, or biodiversity (Fig. 56.6).

In the coming years, the exhibition concept will be expanded to include displays devoted to metamorphosis in the natural world. Here, too, Merian will serve as a prominent example.

56.4 Research

The natural history collections of Museum Wiesbaden, in contrast to other state museums, have only two scientific staff members and one additional scientific volunteer position. Thankfully, our taxidermists have taken on some of this responsibility in the past and continue to do so at present. One of them is currently working on the bird and dragonfly collections. My work is currently devoted to the issue of

Fig. 56.7 Thyridia glasswing (Thyridia psidii) collection Merian (with permission from: Museum Wiesbaden; photo: Ed Restle)

taxonomy and phylogenetics concerning the robber fly (Asilidae), while our volunteer is busy with the genus *Gazella*. In the past, most curators have left their mark on their collections and in the form of scientific publications. The yearbooks of the Natural History Association of Nassau publish the results of research conducted by numerous volunteer workers, whose work is focused largely on the museum's collections. This publication has been in circulation since 1844 and has contributed significantly to natural scientific scholarship.

On occasion, the museum cooperates with universities in the region, though not in a leading role, as the natural history collections in most cases can only serve as a custodian to science due to the nature of its infrastructure. As a result, the museum hosts many international scientists and researchers who take recourse to our holdings or, sometimes, borrow specimens for their work. This contribution of the museum's natural history collection to the natural sciences as a whole, unfortunately, goes widely unnoticed outside the academy and is underfunded (Fig. 56.7).

Likewise underestimated is the power of our exhibits to inspire future generations of natural scientists. Our youngest "researchers", in particular, find their first home in the museum's natural history collections, and our preparations spur their interest in research.

56.5 Perspectives

The natural history collections of the Hesse State Museum of Art and Nature have a transregional and, in part, regional focus. As such, the natural history collections hold a special position in Germany between larger institutions with a primary focus on research and the far more plentiful, small-scale museums with a primarily local emphasis. The latter often do not employ specialist staff, while the former employs highly qualified expert staff. The curators on staff at Museum Wiesbaden might best be described as generalists whose fields of activity equally involve all of the classical disciplines of the museum: research, collection, preservation, and public relations.

In recent years, the natural history collection has been able to reclaim, restore, and newly design all of the exhibition space originally allocated for its use. Of the 2500 square meters of exhibition space, some 800 are reserved for special exhibitions for selected themes. For all of its exhibit space, the museum's workspaces and depots are cramped and in need of expansion. Not least, the number of positions must also be augmented to accommodate a collection of this size. On the whole, Wiesbaden is on the path of progress, thanks to its current directorship and the generosity of the State of Hesse, and largely able to fulfill its role as a public institution. Unfortunately, the city of Wiesbaden has never been a home to a university, which would, of course, in a variety of ways benefit a state museum. The hope remains, however, that the universities in neighboring cities will continue to maintain contact.

Acknowledgments I would like to express my gratitude to my colleagues at the natural history collections for their support and Dr. Staci von Boeckmann for the translation of our exhibit texts. My thanks go, as well, to Prof. Dr. Lothar A. Beck for initiating this project and helping it come to fruition. A current overview of the collections is an important tool for scientists and for all research collections.

References

Czysz W (2004) 175 Jahre Nassauischer Verein für Naturkunde und Naturwissenschaftliche Sammlung des Museums Wiesbaden. 1829–2004. Jahrb Nassauischer Ver Naturkunde 125: 1–372

Geisthardt M (1985) Die zoologisch-botanischen Sammlungen der Naturwissenschaftlichen Sammlung des Museums Wiesbaden. Ein Rückblick auf die Entwicklung seit 1829. Jahrb Nassauischer Ver Naturkunde 108:48–62

Geller-Grimm F (2004) Abbild der Natur—die Naturwissenschaftliche Sammlung im Hessischen Landesmuseum Wiesbaden. In: Streifzüge durch die Natur von Wiesbaden und Umgebung (Hrsg. Nassauischer Verein für Naturkunde). Jahrb Nassauischer Ver Naturkunde Sonderb 2:161–169

Geller-Grimm F, Kridlo S, Lerp H (2015) Ästhetik der Natur—aesthetics of nature. Museum Wiesbaden, Wiesbaden

Chapter 57
WÜRZBURG: The Zoological Study Collection at the Theodor-Boveri-Institute, Biocenter, Julius-Maximilians-University of Würzburg

Dieter Mahsberg and Gerhard Kneitz

Abstract The tradition of natural history museums in Würzburg is described. After World War II, a zoological study collection was reestablished. It comprises about 900 representatives of noninsect invertebrates, numerous boxes with insects, and more than 500 vertebrate species, with birds being a focal point. The collection also includes reference material and/or voucher specimens of insects from ecological field research conducted in Germany, Africa, and Asia. Examples of the application of the collection for the education of students are given. Current problems concerning the collection and potential perspectives for the future are mentioned.

Keywords University of Würzburg • Natural history museum • Zoological study collection • Biodiversity research • Educational function

57.1 Historical Roots of the Collection

The tradition of natural history museums in Würzburg dates back to the eighteenth century, when Bonavita Blank (1740–1827), a member of the Order of Friars Minor, established the "Blancksche Naturalienkabinett." Blank, a professor of science at the university, was the director of an enormous collection of 47,727 natural objects and 488 mosaics (Kneitz 1972). Only one of Blank's famous mosaic paintings made of feathers, hairs, and other biogenic material remained and may be admired even today in the university's Martin von Wagner Museum. After 1830, Blank's successors divided the collection of natural objects, which became part of the inventory of different institutes of the University of Würzburg. The smaller part

D. Mahsberg (✉)
Biocenter of the University, Theodor-Boveri-Institute, Zoology III, Am Hubland, 97074 Würzburg, Germany
e-mail: mahsberg@biozentrum.uni-wuerzburg.de

G. Kneitz
Hans-Gebhardt-Str. 40, 97280 Remlingen, Germany

L.A. Beck (ed.), *Zoological Collections of Germany*, Natural History Collections,
DOI 10.1007/978-3-319-44321-8_57

Fig. 57.1 Poster (partial view) of the "Fränkisches Museum für Naturkunde" (Franconian Museum for Natural History) (By courtesy of G. Kneitz)

of Blank's zoological collection was stored in the Zoological Institute near the city's downtown area.

It is very likely that many of Blank's items were integrated into the "Fränkisches Museum für Naturkunde" (Franconian Museum for Natural History), founded in 1919 by the "Naturwissenschaftlicher Verein Würzburg" (Natural History Society of Würzburg). The museum was located in the south wing of the Residence. Display collections with thousands of recent and fossil organisms and abiotic natural objects were presented in 40 rooms (Fig. 57.1) and served as a major and very popular source for communicating natural history and ecology to the Franconian public (Kneitz et al. 2000/2001). The museum also deeply impressed the senior author (GK), when he was still a schoolboy. Later on he became a professional zoologist at the university. To date, he is the strongest advocate for the idea of reviving the tradition of a natural history museum in Würzburg, with special attention to eco-sustainability.

This tradition initially ended in World War II, on March 16, 1945, when a firebombing almost completely destroyed Würzburg's historic town. The Zoological Institute was burned down, and parts of the famous Residence were seriously damaged, including the Natural History Museum. At the last moment, a few objects were evacuated. Their further fate is mostly unknown, but a recently discovered inventory register from the nineteenth century, covering the former collections of vertebrates, invertebrates, and anatomy, may shed light on the whereabouts of these objects.

57.2 The Collection from World War II to the Present Day

Some specimens of the old Natural History Museum might have been among the first objects of the study collection of the new Zoological Institute. In 1953, the reconstruction of the badly damaged building was completed at the former location in Würzburg at Roentgenring 10 (Autrum 1996).

In the following years, the collection grew constantly and became increasingly important for the education of students in biology, mainly for teacher trainees. The collection was supervised by a professor of the institute and/or by a scientific assistant. Strong technical support was provided by two employees, one of them (G. Belusa) being a capable taxidermist. After his retirement, no replacement was appointed.

In the early 1990's, most of the bioscientific institutes of the university were relocated in the extension area "Am Hubland." The zoological collection moved to the "Theodor-Boveri-Institute for Biological Research" in the Biocenter.

The present-day collection is not accessible to the public. It is located in a windowless room with a usable area of 70 m^2. In the present lack of space, not all objects can be adequately exhibited and are located in further rooms. Most specimens are kept in wooden glass-fronted vitrines of 120 × 220/2250 × 45 cm (width × height × depth) (Fig. 57.2). Insect boxes are stored in collection cabinets or stacked atop each other. Maintenance work is carried out in the collection room.

Table 57.1 gives an estimate of the number of species of the study collection, some of which are available in several items (anatomical specimens are not included here). Years ago, about 80 % of all items (insects excluded) were registered on index cards (scientific and common name, label, etc.; further collection data are often lacking). For some time now, most of the written information has been converted to a database management system.

The collection houses about 900 representatives of noninsect invertebrates and more than 500 vertebrate species, including skulls and skeletons from native and nonnative mammals (Fig. 57.3). European birds are a focal point of the collection. Additionally, a few old mounted skins of mainly Neotropical bird species from different families exist (e.g., cotingas, honeycreepers, hummingbirds, manakins, parrots, tanagers, trogons, turacos) (Fig. 57.4). They may point to the historical roots of the collection (see Sect. 57.1).

There are about 400 insect boxes with representatives of common insect orders for courses in basic and advanced insect taxonomy (mainly beetles, butterflies, and hymenopterans). As insects are in the scientific focus of the Biocenter, the collection also contains reference material and/or voucher specimens from ecological field research on German butterflies, wild bees and aculeate wasps, hover flies, and xylobiontic beetles as well as on tropical insects (sphingid moths and ants from the Comoé National Park, Côte d'Ivoire; ants from Sabah, Borneo; ants from Mt. Kilimanjaro, Tanzania; and ants from Southeast Asian myrmecophytes). For underlying publications and national and international networks, see the homepage

Fig. 57.2 Most objects of the collection are stored in wooden glass-fronted vitrines (Photo: © with D. Mahsberg)

Table 57.1 An estimate of the number of species of the zoological study collection at the Theodor-Boveri-Institute, Biocenter, Julius-Maximilians-University of Würzburg

Taxonomic unit	Wet collection	Dry collection	Mounted skin	Skull, skeleton	Artificial model
Porifera, Cnidaria	25				
Plathelminthes, Nemathelminthes	20				
Annelida	20				
Mollusca (shells)		>500			
Arthropoda (insects excluded)	100	10			
Insecta		>3000			
Echinodermata	50				
Other invertebrates	20				
"Pisces"	30				
Amphibia	20			5	20
"Reptilia"	60			5	15
Aves			270		
Mammalia	10		50	80	1
Total (rounded)	360	>3500	320	90	1

Fig. 57.3 Skeletons and skull from African mammals (Photo: © with D. Mahsberg)

Fig. 57.4 Among European bird species, the collection also contains some mounted skins of Neotropical birds (Photo: © with D. Mahsberg)

of the "Department of Animal Ecology and Tropical Biology (Zoology III)," address above.

57.3 Educational Work Integrating the Zoological Collection

Specimens of the collection are regularly used in bachelor and master courses on animal morphology and systematics, ecology, and faunistics also to better illustrate biological phenomena and concepts.

Six glass vitrines outside the room display selected objects to visitors of the Biocenter. Changing exhibitions are dedicated to particular biological topics [skulls (nutrition and dentition), bird skins and skeletons (avian flight), protected species (regulations on animal protection), etc.]. These exhibitions are planned and arranged by student teachers and can be used as a practical part of their written thesis.

Within the faculty, the division of "Teaching Methodology in Biology" (Gruppe Fachdidaktik Biologie) has enlarged this concept and recently opened a permanent exhibition called "Explore the Wild." Student teachers created dioramas in seven large vitrines, where special fauna and flora of different native natural environments are presented. The combination of classical exhibition (mounted skins, etc.) and modern media (interactive touchscreens) is part of the university's teaching–learning laboratory and primarily addresses prospective teachers in biology and their school classes.

Every summer semester, in the area of soft skills for bachelor students in biology, the first author (DM) offers a seminar entitled "Search, inform, present—Galleria zoologica," where participants are trained to present short talks with a focus on current research activities. Topics of most presentations are animal species selected from the zoological collection. So far, about 500 presentations were created, thereby also bringing the hidden value of a collection to the attention of students.

57.4 Problems and Perspectives

There are no officially nominated responsible and no permanent technical employees for the collection of the Biocenter. Two members of the academic staff look after the material, in addition to their duties. These academics and about half a dozen colleagues are the main users of scientific demonstration material for teaching purposes. The collection does not have its own budget. Over the last years, a one-off financing through tuition fees made it possible to replace badly conserved and unusable specimens and to buy SOMSO® life-size models of all native amphibian and reptile species, thus safeguarding the execution of an obligatory course in taxonomy.

The main problems facing the collection concern the lack of a permanent technical supervision and the storing of items in an unsuitable room, the temperature of which often rises to more than 25 °C. This is especially problematic for liquid stored material, of which the fluid level and alcohol concentration need to be checked regularly. One always comes across dried-out specimens that have to be disposed of. But also dry material requires much more care than currently possible.

It is our hope that the Faculty of Biology of the University of Würzburg will continue to appreciate the importance of organismic biology and biodiversity research, including the value of collections. The recent enlargement of the university's campus could provide opportunities for an improved storing of at least parts of the collection. In any case it is our concern to preserve the old tradition of natural history collections in Würzburg from falling into oblivion.

Acknowledgments We kindly thank Georg Krohne for comments on the manuscript and Emily Poppenborg Martin for improving our English.

References

Autrum H (1996) Mein Leben. Springer, Berlin, Heidelberg

Kneitz G (1972) Die Tradition der Würzburger Naturkundemuseen fortführen. Würzburg heute 14:58–61

Kneitz G, Kneitz H, Ullrich E (2000/2001) Ein Besichtigungsgang 1938 durch das Fränkische Museum für Naturkunde. Abh Naturwiss Ver Würzburg 41/42:65–94

Chapter 58
KARLSRUHE: The Zoological Collections of the State Museum of Natural History Karlsruhe

Hubert Höfer, Albrecht Manegold, Alexander Riedel, Robert Trusch, and Manfred Verhaagh

Abstract The State Museum of Natural History Karlsruhe is an institution of the state of Baden-Württemberg, with a staff of currently 57 permanent employees. It has emerged from the cabinet of natural history of Karoline Luise, Margravine of Baden and is open to the public since 1785. Permanent and temporary exhibitions extending to 5000 m^2 provide the basis for manifold educational activities. Living animals are integrated into the permanent exhibition, a 240,000 l seawater tank with a living coral reef and reef sharks being a major attraction for the public. Zoological research at the museum mainly deals with biodiversity, i.e., taxonomy, faunistics, biogeography, natural history, and ecology. It is strongly based on museum collections but also on field sampling in soil ecological studies. The zoological collections incorporate about 8.5 million specimens, mostly terrestrial, but also a mollusk collection (Conchylia) of historical importance and the Kiefer collection of fresh water Copepoda. Important collections are on Arachnida, Lepidoptera, Coleoptera, Hymenoptera, Mantodea, as well as small mammals from Baden-Württemberg and large series of mammal skulls from West Africa. Biogeographic emphases are on the fauna of Southwestern Germany, the Middle East (Iran, Afghanistan), the Indo-Australian region and the Neotropics.

Keywords Andrias • Arachnida • Max Auerbach • Karoline Luise von Baden • Baden-Württemberg • Curculionidae • Karl Christian Gmelin • Hans Himmelheber • Friedrich Kiefer • Mammal skulls • Microlepidoptera • Oribatida • Soil fauna • Southwestern Germany

H. Höfer (✉) • A. Manegold • A. Riedel • R. Trusch • M. Verhaagh
Staatliches Museum für Naturkunde Karlsruhe (SMNK), Erbprinzenstraße 13, 76133 Karlsruhe, Germany
e-mail: hubert.hoefer@smnk.de; albrecht.manegold@smnk.de; riedel@smnk.de; trusch@smnk.de; manfred.verhaagh@smnk.de

L.A. Beck (ed.), *Zoological Collections of Germany*, Natural History Collections,
DOI 10.1007/978-3-319-44321-8_58

58.1 History and General Facts

The State Museum of Natural History Karlsruhe emerged from the cabinet of natural history of Karoline Luise of Hesse-Darmstadt (1723–1783), Margravine of Baden-Durlach (1751–1771) and Baden (1771–1783) by marriage to Margrave Karl Friedrich (1728–1811) (Lauts 1990). Karoline Luise is renowned for her broad interest in both fine arts and sciences until today (Jacob-Friesen 2015). She was particularly interested in mineralogy and botany and eager to arrange her representative collections according to the latest classifications. Therefore, she corresponded extensively with collectors, dealers, and academics, most notably with Carl von Linné (1707–1778) (May 1906; Oberdorfer 1951; Lauts 1990). Probably, a small collection of shells of marine mollusks ("Conchylia") owned by Karoline Luise's predeceased father-in-law, Hereditary Prince Friedrich (1708–1732), formed the nucleus of her cabinet of natural history (Mayer 1978), which grew rapidly due to her activities between 1763 and 1783 (Trunko 1985; Lauts 1990: 216).

A lofty hall with a gallery at the ground floor of the court library at the east wing of Karlsruhe Palace was first designated to house the cabinet. Several rooms in the attic provided additional space for collections and equipment (Lauts 1990: 214). This new building was completed in 1768, and by 1776 "magnifiques pièces" of the collection were on display, but the cabinet was still a torso when Karoline Luise died in 1783 (Lauts 1990: 229).

Friedrich Samuel Schmidt was in charge of the Margravian library and all collections from 1764 to 1769 (Lauts 1990: 218). In the 1770s, Carl Friedrich Erhard was entrusted with the acquisition of new specimens and with the cataloging and arranging of the mineral collections (Erhard 1802). Thus, these two could be regarded as the first directors of Karoline Luise's collections, although Karl Christian Gmelin (1762–1837), a medical doctor and renowned botanist (Gmelin 1879), traditionally is regarded the first director of the cabinet. Gmelin was appointed in 1785 (Trunko 1985; Lauts 1990: 229), and the first entries in his catalogue of the Margravian bird collections date from the same year. In 1786, he started cataloging the "Conchylia." During Gmelin's directorship of more than 50 years, the collections grew markedly. In the course of the secularization, for example, the natural history collections of the prince-bishop of Meersburg and St. Blaise Abbey, Black Forest, became part of the Margravian cabinet of natural history (May 1906; Kretschmann 2006).

Gmelin's most influential successors as directors of the cabinet in the nineteenth and early twentieth century were the botanists Alexander Braun (1805–1877) and Moritz Seubert (1818–1878) and the zoologist Max Auerbach (1879–1968) (Trunko 1985).

The museum's current building erected at Friedrichsplatz between 1865 and 1872 was designed to house the cabinet of natural history, the court library, and the numismatic cabinet as well as the Margravian collections of antiques (lapidarium) and ethnographica (Schillinger 1973). The antiques were removed to Karlsruhe Palace during the German Revolution in 1918, and the ethnographica were transferred to the ethnological museum in Mannheim in 1935 (Auerbach 1937). Thus, additional exhibition halls became available, which allowed for a complete rearrangement of the permanent exhibition on natural history by 1937 (Auerbach 1937; Kretschmann 2006).

During air raids in 1942 and 1944, the museum building, the neighboring Erbprinzenschlösschen, housing laboratories, collections, and the taxidermist's workshop (Auerbach 1937), and also Karlsruhe Palace, to where major parts of the collections had been evacuated, were severely damaged or completely destroyed (Ritzi 1943/49; Schillinger 1973). The bird skin collection, a collection of marine invertebrates, the archive of photographic plates, as well as other items were lost (Ritzi 1943/49; Hauer 1951), but the mollusk shells, the complete entomological collections, a collection of mounted birds including the C. G. F. Riedel collection, and large mammals, including an okapi collected in 1914, as well as the spirit collection of vertebrates were rescued (Ritzi 1943/49). Further damages and losses occurred due to inadequate housing conditions during the following years.

The rebuilding process after the massive destruction of World War II was not completed before 1972 (Schillinger 1973). Since this time, the west wing of the museum building has been used as a depot of the Baden State Library, but between 2013 and 2015 it has been totally reconstructed for permanent and temporary exhibition purposes of the natural history museum.

58.2 Public Exhibitions

Karoline Luise regarded the cabinet of natural history her property; nevertheless, plans to make the collections accessible to the public date back to 1759 (Lauts 1990: 222f.). During the directorship of Gmelin, the cabinet of natural history apparently was accessible to visitors by prior arrangement. It became open to the public by decree of Margrave Leopold in 1831 (Trunko 1985), apparently in the course of the July revolution of 1830 in France. The separation of public exhibitions and research collections came into effect in 1902, once Max Auerbach was in charge of the natural history collections (Auerbach 1937; Trunko 1985). The systematic arrangement of specimens in the exhibition was gradually abandoned and replaced by a strong focus on biogeography and the local fauna (Auerbach 1904, 1937; Kretschmann 2006). Today, the permanent exhibitions extend to approx. 4000 m^2 of display area; another 1000 m^2 are available for temporary exhibitions and education activities. Local and exotic wildlife from different regions of the world are presented in realistic dioramas of their habitats and under several thematic arrangements, e.g., "*Climates and Habitats*" or "*World of Insects*." The concept of the exhibition "*World of Insects*" is to achieve the perspective of an insect living in a meadow: green pillars representing plant stalks contain numerous display cases and dioramas covering many entomological topics (Figs. 58.1 and 58.2).

Fig. 58.1 Exhibition "*World of Insects*"

The museum also presents living animals since 1948, at the beginning in a separate vivarium, meanwhile expanded and integrated into most of the permanent exhibitions in the form of aquaria, terraria, and paludaria. The giant salamander *Andrias* is part of the museum logo and is shown as living animal (*A. davidianus*) and 13 million-year-old fossil (*A. scheuchzeri*) from Öhningen in Baden-Württemberg, illustrating earth's biological and geological history.

Since 2015, the reconstructed west wing houses a large new permanent exhibition entitled "Form and Function—Role Model Nature." Prominent parts of this exhibition are a large seawater aquarium (240,000 l) with a living coral reef and reef sharks and a paludarium with Australian freshwater crocodiles and turtles.

58.3 Research and Publications

SMNK is part of the national networks German Natural History Research Collections (DNFS) and Humboldt-Ring (Association of Research Museums) and of the International Council of Museums (ICOM).

Zoological research at the museum mainly deals with biodiversity, i.e., taxonomy, faunistics, biogeography, natural history, and ecology. It is based on field sampling or museum collections. Current concern is modern collection and data management based on database systems, long-term data conservation, and supply of virtual research environments.

Fig. 58.2 Diversity of beetles in the exhibition "*World of Insects*"

Modern research at SMNK is illustrated by the following examples:

- In the cooperative project *Edaphobase*, soil zoological data are assembled and mobilized in a noncommercial data warehouse. The database comprises taxonomic thesauri, literature, geocoded collection and study data, and metadata.

Quality is maintained through a critical evaluation prior to data input. A client offers tools for data entry (e.g., GIS, semiautomatic literature analysis) and the portal (http://portal.edaphobase.org/) options for data exploration, from simple queries to sophisticated analyses of different data sources and prognostic tools to discern changes within soil biocoenotic assemblages as a reaction to land-use or climate change (Burkhardt et al. 2014).

- A *Species description pipeline for Curculionidae* was established and utilized to overcome the taxonomic impediment for tropical arthropods: a combination of molecular taxonomy and condensed morphological descriptions is used to name hundreds of new species of tropical weevils (Riedel et al. 2013a, b, 2014; Fig. 58.3).
- *Fossil avifaunas* from the Pliocene and Pleistocene of South Africa, Morocco, and Germany are described and analyzed for paleoecological reconstructions (Manegold 2014; Manegold et al. 2014).
- Studies on the *Phylogeny and evolution of birds* on the basis of morphological characters involve the application of CT scans and 3D models for the study of spirit specimens (Manegold and White 2014).

The museum publishes three series: *Carolinea* is published as a single annual volume presenting papers on the Natural History of Southwestern Germany and articles written by the staff of the museum. Volumes 1–39 were published as "Beiträge zur naturkundlichen Forschung in Südwestdeutschland" (Contributions to Natural History Research in Southwest Germany). *Andrias* comprises individual volumes on selected topics of natural history research. Since 2007, *Karlsruher Naturhefte* are published sporadically, explaining complex interactions between man and nature in a vivid and comprehensible style.

Examples of major publications edited by entomologists and zoologists of the natural history museum Karlsruhe are the series on Microlepidoptera Palaearctica, initiated by Hans-Georg Amsel in 1965, and several volumes of the outstanding series on the fundamentals of flora and fauna of Baden-Württemberg: [Mammalia, 2 vols: Braun and Dieterlen (2003, 2005); Coleoptera: Buprestidae, Lucanidae: Brechtel and Kostenbader (2002); Macrolepidoptera, 10 vols: Ebert (1991–2005)].

The museum presents itself with the homepage http://www.smnk.de, including an English version, and microsites on the Lepidoptera of Baden-Württemberg (www.schmetterlinge-bw.de), Amazonian wandering spiders (www.wandering-spiders.net), and butterflies (www.amazonian-butterflies.net). There is also a mobile version.

58.4 Infrastructure

Today, the museum is an institution of the state of Baden-Württemberg, with a staff of 57 permanent employees, with two scientific departments, i.e., Geosciences (two curators, two technicians) and Biosciences (seven curators, five technicians).

Fig. 58.3 New *Trigonopterus* weevil species

Zoological research and collection management is organized into the sections Entomology (three curators, two technicians) and Zoology (two curators, two technicians). Every section supervises one scientific trainee. The Botany section is represented by two curators and one technician.

Collection rooms, library, workshops for taxidermy and geological preparation, chemical lab, carpentry, and fitter's workshop are housed in the museum's main building. Parts of the zoological collections are currently in an external depot.

Offices for the scientific staff, a DNA lab, and parts of the collections are housed in the so-called Nymphengarten Pavilion, a separate building adjoining the museum's main building to the South. The Pavilion built in 1964 was originally designed to house catalogues, reading rooms, and offices of the Baden State Library (Syré 2015) but consigned to the museum in 2003.

58.5 Zoological Collections

The collections of today contain only a handful of the specimens dating back to the times of Margravine Karoline Luise and the director Karl Christian Gmelin (Fig. 58.4). Large parts of the zoological specimens were destroyed by fires during bombing of Karlsruhe in World War II. Further damages and losses occurred due to the rush evacuation and the disastrous conditions of storage during and after the war (Angst 1985).

58.5.1 Collections of the Section Zoology

58.5.1.1 Collections of Invertebrates

The limitation of staff allows only few collections to be managed and expanded actively. During the last 40 years, soil zoological and ecological research provided most of the vouchers of the arachnid collections, including valuable ecological and metadata. In these active collections, specimens are to a large part identified to the

Fig. 58.4 Lepidopteran vouchers dating back to the director Karl Christian Gmelin

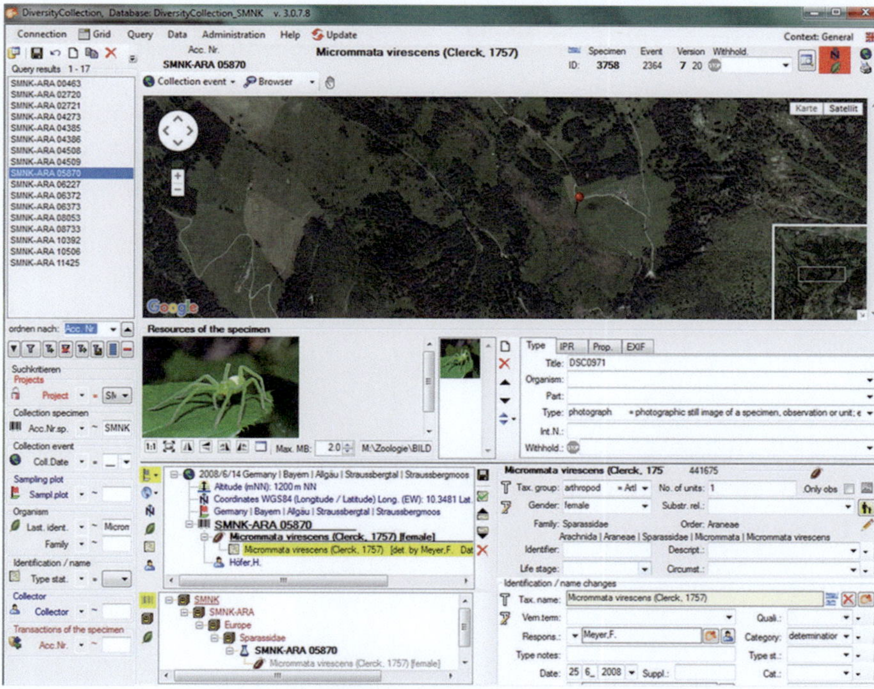

Fig. 58.5 Screenshot from the collection module (spider collection) of the database system Diversity Workbench

species level; nomenclature is based on modern thesauri and is kept up-to-date; samples are geocoded, and data digitized and managed in relational database systems (Fig. 58.5). The other collections mentioned herein are in varying conditions and of greater or lesser scientific significance and/or historical value.

58.5.1.2 Mollusca (Conchylia)

The collection had its origin in 14 drawers containing at least 714 specimens representing 134 "species" of predominantly marine mollusks collected by the Hereditary Prince Friedrich, son of the founder of Karlsruhe (Mayer 1978). It was further enlarged by Karoline Luise. However, soon after her death, the director of the cabinet of natural history K. C. Gmelin complained about the neglected condition of the collection (May 1906). He documented the inventory of Conchylia in a (still existing) catalogue of 384 pages, entitled *Testacea Musei Principio Badensis*, in the years 1786–1787 (Fig. 58.6). Unfortunately, in this catalogue individual specimens are not listed with collection numbers but only with numbers referring to species in the corresponding volumes of the 12th edition of Linné's

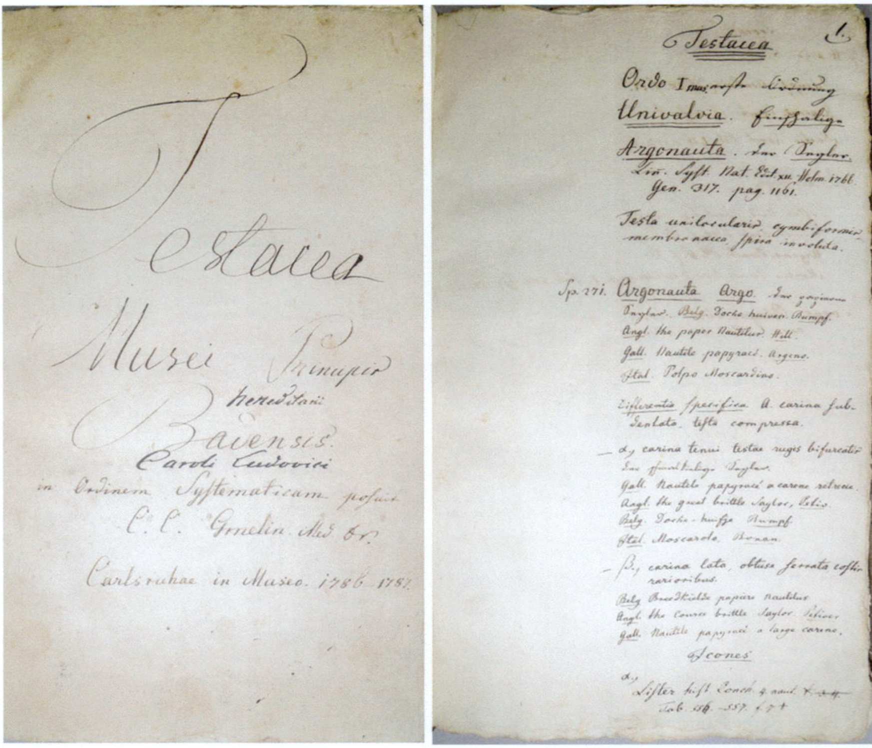

Fig. 58.6 Title and first page of the Conchylia catalogue of Karl Christian Gmelin

Systema Naturae. Presumably, shells from this and also from the Conchylia collection of the Meersburg Episcopalian cabinet of naturalia, acquired in 1803, still exist, but are not readily discernible within the current large mollusk shell collection (>20,000 specimens; see below).

Still under the direction of Gmelin, the mollusk collection grew strongly. Between 1830 and 1836, Wilhelm Schimper purchased shells from southern France and Colonel G.J. Peitsch donated 1243 shells of 600 exotic species, among them many rare and precious pieces of volutes, cowrie, and cone shells (May 1906).

Later, director Alexander Braun (1837–1845) sought to complete the collection, adding genera to date not included by purchases and donations, and not least by selling his own rich collection of terrestrial and freshwater mollusks to the cabinet (May 1906). This ambition was followed by his successor Moritz August Seubert (1846–1873), who itself donated Conchylia from Java and bought several species from Landauer (Frankfurt/Main) in 1866 and 1868 (May 1906). The last large additions came from donations of Carl Kreglinger in 1881 (>4000 specimens) and Adolf Schütt in 1889 (nearly 2000 specimens) under the direction of Adolph Knop (Angst 1985).

Between 1907 and 1911, Max Auerbach created a registry of entrances for the Conchylia in two volumes, containing 9091 entries, and notes on the collections of Kreglinger and Schütt. In this catalogue, the zoological curator Joseph Hauer noted in 1955 that most of the mollusk shells had been rescued from the fire in 1942, but got into disorder during the following rush. Only a part of the collection (from Kreglinger) was reordered; other parts (formerly registered as "Alte Sammlung," "Nat.-Cabinett," and "Sammlung Schütt") were merged as "Allgemeine Sammlung" without changing the systematic order or identifying specimens. Starting in 1959, the zoological curator Helmut Knipper (1914–1974) rearranged the whole collection based on a new taxonomic system in about 20 large cabinets. He also increased the collection by sampling land and freshwater mollusks in Europe and East Africa. Although there were some attempts to enlarge and register parts of the collection, a transfer from the main building to an external depot and to other cabinets did not improve matters, and until today the collection has more or less remained in this condition and order, making it extremely difficult to recognize historically important specimens or to mobilize data on the collection by a digital compilation.

58.5.1.3 Copepoda (Kiefer Collection)

(18,000 vouchers, 658 species, 76 subspecies)

The internationally renowned collection of copepods from the estate of Friedrich Kiefer (1897–1985) is hosted together with his scientific legacy at SMNK since he worked here as an honorary curator from 1934 to 1939. Friedrich Kiefer was a teacher, but he was also one of the most important copepod researchers of the last century: He had worked at the Max Auerbach Institute in Konstanz-Staad and published about 250 papers on this group, describing 9 subfamilies, 45 genera, 7 subgenera, and 292 new species and revising the system of Cyclopoida and of the Old World Diaptomidae. He also worked on ecological questions, e.g., studied and estimated the plankton stock of Lake Constance accurately by long-term surveys, providing the first evidence of the change of the plankton populations caused by the eutrophication of the lake. The voucher material of this study is still stored at our museum.

In total, his heritage comprises about 6500 glass vials with specimens preserved in an ethanol/glycerol solution. Kiefer also left ca. 11,400 microscope permanent mounts from these samples and 4500 drawings, which are stored in card boxes and always refer to the species locality and to the original specimen identification in the wet or slide collection. The wet collection is accompanied by a laboratory diary, listing species from the single sampling dates and localities; a laboratory book for the slides focuses on the species. The collection is supplemented by ca. 10,000 separata on Copepoda. Data of the Kiefer collection were catalogued by Franke (1989) and recently digitized.

58.5.2 *Soil Zoological Collections*

Under the direction of the zoological department (1976–2001) by Ludwig Beck, the focus of research and collection work was on soil invertebrate taxa, e.g., Enchytraeidae, Nematoda, Myriapoda, Arachnida, and Collembola, sampled in large projects with a focus on ecology (Höfer 2001). Over one million specimens of soil macrofauna (beetles, arachnids, centipedes and millipedes, snails, isopods) and mesofauna (roundworms, potworms, mites, and springtails) have been identified to species level and were deposited in sample-based collections. Their data were digitized early but are still not completely migrated to modern database systems.

The soil zoological studies aimed at describing and analyzing animal assemblages or communities (biocenoses) of temperate and Neotropical forests, in an earlier phase as basic research (inventories), later with strong applied aspects. Consequently, the data of the largest collections (i.e., Collembola and Oribatida) were integrated in the soil zoological database and information system Edaphobase, publicly available under http://portal.edaphobase.org/. In this database, not only the data of the taxonomic reference collection are accessible, but they can also be combined with data of the sample-based ecological voucher collection and data from studies of other institutions to solve modern ecological questions, like assessing soil quality based on biological criteria (Römbke et al. 2012).

58.5.2.1 Nematoda

(4173 permanent mounts, 125 types)

The collection was organized by Herbert Zell, who worked at the museum between 1983 and 1985 and later at the University of Aachen. He primarily identified the nematods sampled in the main study site of the soil zoological working group, a beech forest at the northern edge of the Black Forest (Schluttenbach), and in this context revised several genera and described 27 new species, for most of which types were deposited at the museum. He also defined and deposited neo- and topotypes of three species. More recently, 21 types of parasitic nematods (*Pristionchus* spp.) have been deposited by Matthias Herrmann of the Max Planck Institute for Developmental Biology in Tübingen. The type collection is available as a searchable list at www.smnk.de/sammlungen/zoologie/fadenwuermer/.

58.5.2.2 Enchytraeidae

(17,000 permanent mounts, 13 types)

From the long-term study of the soil fauna in the beech forest site at Schluttenbach (1978–1985), a large sample-based collection of 13 enchytraeid species (mounted in polyvinyllactophenol) (Römbke 1989) was deposited together with study metadata by Jörg Römbke. Data are not digitized. The paratypes of *Enchytraeus luxuriosus* Schmelz & Collado, 1999, are deposited at SMNK as permanent mounts, complemented by a vial with bouin-fixed specimens in ethanol.

58.5.2.3 Myriapoda

(597 vouchers, approx. 200 taxa)

The myriapod collection was (re-)organized by Jörg Spelda during his trainee period at the museum. Vouchers of 71 species from southern Germany had been bought and deposited at SMNK by the myriapod expert Karl-Wilhelm Verhoeff. 225 vouchers are of specimens sampled by Helmut Knipper (zoological curator 1959–1974) in Tanzania. This old part of the collection does not contain any types. More recent material is from ecological projects in southern Germany.

58.5.2.4 Protura

Types (1 holotype, 5 paratypes) of *Eosentomon rusekianum* Stumpp & Szeptycki, 1989, are deposited at SMNK.

58.5.2.5 Collembola

(4534 vouchers, 658 species, 20 types)

The collection contains 3236 specimens of 169 taxa from Germany and 1250 specimens of 42 taxa from Central Amazonia, Brazil, sampled by Ulrich Gauer (1989–1991). Included are 20 paratypes of 6 *Sphaeridia* species described by Bretfeld and Gauer (1994). Most specimens of the collection are vouchers of the long-term study of the soil zoological working group at the Schluttenbach beech forest site.

58.5.2.6 Arachnida

Oribatida

(14,800 vouchers, 967 taxa, 687 type specimens of 60 species)

The collection is an important legacy of L. Beck and Steffen Woas (both retired but still active), who have been working on ecology and taxonomy of oribatid mites since 1976. The taxonomic collection contains 7845 vouchers of 945 taxa. Included are 1800 specimens from South and Central America, where L. Beck did pioneering work on tropical ecology and soil zoology since the 1960s (Beck 1963). Most specimens, however, were collected in his subsequent ecological projects. Starting with the long-term study of soil fauna of a beech forest near Karlsruhe (Beck 1978, 2000), inventories of oribatid mites were expanded to more than 60 forest sites in Baden-Württemberg (Römbke et al. 2012). This locality-based collection rich in data and metadata is currently the central data basis for an evaluation of the use of oribatid mites for a biological soil quality assessment in the project Edaphobase. Collection and study data are accessible through the project's portal at http://portal.edaphobase.org/.

58.5.2.7 Small Arachnid Taxa

(732 vouchers, 125 taxa, 18 types)

This collection includes specimens of some exotic and in the twentieth century still rather unfamiliar tropical groups, e.g., Amblypygi (55), Schizomida (46, 16 types), Thelyphonida (13), and Ricinulei (31), but also Opiliones (54, 1 type) collected by L. Beck during his studies in Amazonia. This small collection continues to contribute to new scientific results (Friedrich 2014; Réveillon et al. 2014). 125 Scorpiones and 10 Solifugae specimens from all over the world are mostly unidentified, while more recent vouchers of Opiliones (333) and Pseudoscorpiones (65, 1 type) are mostly from ecological studies from Germany and identified to species level.

58.5.2.8 Araneae

(11,600 vouchers, approx. 1750 taxa, 127 types)

Starting in 1990 as a trainee at SMNK, and later working in several ecological projects mainly on Neotropical and German spider assemblages, Hubert Höfer became zoological curator in 2001. He expanded the originally small spider collection with thousands of specimens from his ecological studies in southern Germany and the Brazilian Amazon. Nearly all vouchers are geocoded. 2738 vouchers are from the Neotropical region, predominantly from Brazil; 7000 are from Germany representing 660 of approx. 1000 species registered for Germany, including vouchers of recently barcoded specimens in the German Barcoding of Life project (GBOL).

The collection includes types of wandering spider species from Brazil and Peru, collected by L. Beck, H. Höfer, or Manfred Verhaagh and of Salticidae from Greece, described by Metzner (1999).

The database in Diversity Workbench serves not only to manage the collection, but also as virtual research environment, containing about 45,000 study data (with vouchers) and a first set of 1300 literature data (no vouchers). Taxon names are currently linked to a complete and up-to-date taxon list for German spiders compiled by the curator and to Catalogue of Life for non-European spiders. Data are delivered to the Global Biodiversity Information Facility (GBIF).

58.5.3 Vertebrate Collections

58.5.3.1 Mammals

Apparently, no specimens dating from the eighteenth century or from the time of Gmelin's directorship can be traced in the collection. Several specimens dating from the first half of the nineteenth century were received from the zoological collections of Heidelberg University in 1989. A small number of skulls from Java and Abyssinia were donated by G. J. Peitsch in 1837 and by Wilhelm Schimper (1804–1878) in 1840 (Angst 1985). In 1845, the cabinet received a significant collection from Dr. Keller, Bahia, Brazil (May 1906). The mammal collection increased considerably between 1890 and 1914. Most notable are the donations of Bernhard Hagen (1853–1919), including specimens from Sumatra and New Guinea, among them the type series of the greater forest wallaby, *Dorcopsis hageni* Heller, 1897 (Manegold 2016). Most specimens, however, originated from South Cameroon and South-West and East Africa and were donated by employees of the German colonial administration such as Karl Grass and Karl Albert Haberer (1864–1941) or by adventurers and early environmental activists such as Carl Georg Schillings (1865–1921). Additional contributions were made by E. Rudel, who donated specimens from Deli, Sumatra, and Alfred Voeltzkow (1860–1947), who sent specimens from Madagascar.

This era of numerous new accessions came to an abrupt end in 1914, but is highlighted by a remarkable specimen. In a letter dated 4th August 1914, Max Auerbach announced to the Baden ministry of culture and education that the cabinet had received the skin of a female okapi, *Okapia johnstoni*, collected by a C. Müller in Belgian Congo. Actually, it was advertised as "one of the most beautiful skins of this species ever sent to Europe." Auerbach further informed the ministry that Gustav Krupp von Bohlen und Halbach (1870–1950) agreed to donate the sum of M 4000 requested by Müller and kindly asked for the ministry's authorization for accepting this offer. The authorization was given the following day, only a week after the outbreak of World War I and presumably about the same time, when the Fried. Krupp Company became Germany's leading armament factory.

No noteworthy accessions were made between 1914 and 1948. After World War II, the main focus was on the reconstruction of the permanent exhibition with a strong emphasis on the local fauna, but also on biogeography. Between 1956 and 1958, a small collection of mammals was purchased from J. Unger (1894–1959), Paraguay. Most of the skins were mounted for display by taxidermist Kurt Silber (1906–1979). About 60 study skins and skulls of small mammals were purchased from Thorkild Andersen, Tanganyika, in 1961. Helmut Knipper collected about 250 specimens of more than 38 species during his East Africa expedition between 1961 and 1963 (Knipper 1964).

Between 1973 and 1983, the Karlsruhe museum purchased a collection of approx. 6400 skulls of West African mammals gathered by Hans Himmelheber (1908–2003) in Liberia and Sierra Leone. Included were large series of monkeys such as *Cercocebus atys*, *Cercopithecus campbelli*, C. *petaurista buettikoferi*, *Colobus polykomos*, *Piliocolobus badius,* and *Procolobus verus*, duikers such as *Cephalophus dorsalis*, *Cephalophus niger*, and *Philantomba maxwellii* as well as rodents such as *Atherurus africanus*, *Cricetomys gambianus*, *Funisciurus pyrropus leonis*, *Thryonomys swinderianus*, and *Xerus erythropus*.

In 1985, the scientific trainee and later curator Manfred Verhaagh brought about 100 mammal specimens from his 2-years stay at the field station Panguana in Amazonian Peru, including several first records for the region (Hutterer et al. 1995).

In 1987, the museum acquired approx. 600 specimens of the bequest of Hubert H. A. Sternberg (1897–1987). This collection comprised hunting trophies from Tyrol, but also from Africa, Canada, and India. Sternberg's full mounts of a bongo *Tragelaphus eurycerus*, a lion *Panthera leo*, and a polar bear *Ursus maritimus* are currently on display.

The bat collection, set up and curated by Monika Braun and Ursel Häußler, comprises more than 8000 specimens of 23 species collected in Southwestern Germany. The collection of small mammals of Southwestern Germany comprises approx. 5900 specimens of 27 species. Both collections formed the basis for the most recent reference books on the mammal fauna of Baden Württemberg (Braun and Dieterlen 2003, 2005).

58.5.3.2 Birds

Karoline Luise was disappointed that bird specimens in her cabinet continued to falling apart shortly after their accession, but nevertheless bought specimens from Aernout Vosmaer (1720–1799), director of the cabinet of natural history in The Hague (Rookmaaker 1989), and continued spending money for additional specimens (Lauts 1990: 226). In 1810, Gmelin purchased 32 bird specimens from P.-A. Delalande and Bécouer in Paris (Mayer 1977), and similar acquisitions were made in 1825 and 1830 (May 1906). However, only three specimens of hummingbirds (Trochilidae) registered in 1811 and labeled by Gmelin have been located in the present collection among other hummingbirds acquired from Xaver Joseph Ackermann (1770–1836) before 1836. Franz Wilhelm von Kettner (1801–1874), a high-ranking forest officer in Karlsruhe, sold his entire bird collection in 1857, including a pair of passenger pigeons, *Ectopistes migratorius*, and several native passerines, while Johann Gerard Friedrich Riedel (1832–1911), Gorontalo, donated a collection of Indonesian birds in 1868 (Blasius 1883: 129). By the end of the nineteenth century, the bird collection was regarded to be quite representative by Oskar Heinroth (1898). Few additions were made in the first half of the twentieth century. For example, privy councilor W. Heinke donated an *Apteryx occidentalis* and a female *Heterolocha acutirostris* in 1931. Both specimens and few others still exist, but virtually the entire skin collection with specimens from north-east Africa (Baron Müller), Batavia (Casalotti), Brazil (Konsul Lämmert), and Peru (Wilhelm Reiß) (see May 1906) as well as from Central Asia (Julius Holderer; see Schalow 1901) was destroyed during World War II.

The complete destruction of the skin collection has never been redeemed since then. Between 1948 and 1958, the main focus was on the acquisition of native and exotic birds as replacement for the destroyed exhibitions, among them several specimens purchased from J. Unger, Paraguay, between 1956 and 1958. Helmut Knipper gathered a representative collection of 1200 specimens, collected during his field trip to Tanganyika and Kenya in 1961–1963 (Knipper 1964: 20), which was supplemented by 627 skins of East African birds purchased from Thorkild Andersen in 1961. Few bird skulls from West African species were purchased together with the Himmelheber collection between 1973 and 1983. In 1997, the Karlsruhe museum received the bird collection of the zoological institute of Freiburg university with 770 specimens, representing the local avifauna of the Upper Rhine valley and the northern part of the Black Forest.

58.5.3.3 Fishes, Amphibians, Reptiles

Compared to birds and mammals, the collections of the remaining vertebrate groups are very small, although the collections of wet-preserved specimens suffered less from perditions during World War II than the skin collections. Reptiles collected by Rudel in Sumatra and by Hagen in Sumatra and New Guinea have been identified by Boettiger. Amphibians and reptiles from Paraguay were purchased from J. Unger-Peters in the 1980s. There are also about 100 specimens of the herpetological collection M. Verhaagh gathered in Panguana, Peru, during 1983–1985, mainly frogs, lizards, and snakes. The entire spirit collection and especially the fishes have to be revised and catalogued.

58.5.4 Collections of the Section Entomology

The first entomological curator, Hermann Leininger, was hired in 1920 and since 1955 (curatorship of Hans-Georg Amsel) entomology is an independent section at the museum. The total number of mounted insects at SMNK attains approx. three millions, stored in more than 20,000 insect drawers. In 2012, the entomological magazine got a new compact shelving ensuring additional space for the following years. Additional material (e.g., Collembola, Formicidae, Coleoptera, Diptera), representing mainly voucher collections of ecological projects, is stored in a large wet (ethanol) collection. This includes the vouchers of the long-term monitoring of the Rhine by the Landesanstalt für Umwelt, Messungen und Naturschutz Baden-Württemberg (LUBW). The main focus of the (dry) collection is on butterflies and moths (Lepidoptera). Other important parts of the collection are ants, bees and wasps (Hymenoptera), beetles (Coleoptera), praying mantises (Mantodea), and true bugs (Heteroptera).

58.5.4.1 Lepidoptera

Butterflies and moths form the bulk of the insect collection at SMNK, comprising approx. 2.52 million specimens (Ebert 1964, 1977, 2004; unpublished data 2015), representing the third-largest Lepidoptera collection in Germany. The collection of Microlepidoptera is the largest one in Germany, with more than 650,000 specimens.

The collection achieved its current status after World War II, due to the field sampling of the lepidopterists H.-G. Amsel, Günter Ebert, R. Ulrich Roesler, Michael Falkenberg, and Robert Trusch, careful acquisition of specialized collections, and donations from private persons. The geographic focuses are on Europe (especially Southwestern Germany), the Middle East (Iran, Afghanistan), and the Indo-Australian region. The collection contains more than 20,000 species, more than 500 primary types, and thousands of paratypes. There are also approx. 40,000 microscopic slides of dissected genitalia.

The Lepidoptera collection of the SMNK is composed of more than 300 individual collections or sampling results from expeditions. Major contributions came from the following entomologists: H.-G. Amsel, G. Baisch, G. Belter, W. Bender, Ch. Boursin, K. Burmann, O. Engelhardt, M. Daub, G. Ebert, W. Eckweiler (Lycaenidae), M. Fiebiger (Geometridae), J. Froitzheim, A. Gremminger, W. Glaser, D. Heidelberger, Ph. and R. Henriot, G. Hesselbarth, O. Hoegh-Guldberg, G. Junge, G. Kabis, J. Kaisila, J. F. Klapperich, O. Kudrna, H. Lienig, A. Mees, J.-U. Meineke, H. Messmer, C. M. Naumann, H. Noack, G. Reich, H. Reisser, K. Reutti, R.-U. Roesler, M. Sälzl jun., A. Scheubel, K.-F. Schüller, M. Schlusche, L. Sieder, F. Smetacek, W. Staib, K. Strobel, R. Sutter, R. Trusch, H. Wagner, and C. Wyatt.

A few outstanding parts of the collection are discussed here in detail:

Of historical significance is the collection M. Daub (Karlsruhe), which was the most valuable individual donation to the Margravian collection (Auerbach 1917). It comprises 55,000 specimens of butterflies and moths, which survived World War II without damage. At the time, it was the second-largest private collection of Lepidoptera in Germany containing many historical syntypes, since M. Daub had a purchase option at Staudinger and Bang-Haas. Thus, it is an important source of type material regarding designation of lectotypes or neotypes.

SMNK keeps the largest collection of Lepidoptera from Afghanistan, continually enlarged until the recent past, comprising approx. 43,000 specimens.

Of equal significance are the collections from Iran (approx. 50,000 specimens) and from Turkey (120,000 specimens, collections of H. Noack, G. Junge, and A. Ö. Koçak).

Under taxonomic aspects, the 20,000 specimens of Psychidae from L. Sieder, K. Burmann, and A. Biebinger are especially noteworthy.

Finally, important recent additions to the Microlepidoptera are the collections of G. Baisch (Biberach, approx. 46,000 specimens) and K. Burmann (Innsbruck, approx. 48,000 specimens).

58.5.4.2 Coleoptera

The beetle collection of SMNK currently comprises approx. 330,000 specimens. In recent years, large parts of it were rearranged and databased on species level. Thus, the collection is well accessible and can be used for scientific identification and for faunistic research.

The majority of specimens were accessed after World War II. The value of the "old Baden collection" is limited because of the insufficient labeling and the absence of collecting data. A few drawers containing mainly Curculionidae and Scarabaeidae collected by Privy Council Methner in Tanzania arrived at the SMNK through donations of W. Holtz in 1921 and 1924.

An important focus is on the German fauna with important contributions by local collectors (H. Gräf, O.-K. Hebestreit, J. Hillger, P. Hozman, Hünsch, H. Nowotny, G. Riegelbauer, W. Rössler, H. Steude). Moreover, some beetles were collected by lepidopterist H.-G. Amsel on expeditions to Afghanistan and Iran more than 50 years ago. The Curculionidae (weevils) of this collection were identified by E. Voss and the Tenebrionidae (darkling beetles) by Z. Kaszab. Between 1993 and 2001, Fritz Brechtel enlarged the collection of xylophagous beetles, especially of Buprestidae (jewel beetles) and Scolytidae (bark beetles). Moreover, a major collection of Elateridae (click beetles) of P. Cechovsky reached the SMNK (>15,000 specimens). In 2010, more than 30,000 specimens of Palearctic longhorn beetles (Cerambycidae) were purchased from M. Sláma (Prague).

In August 2003, Alexander Riedel established a focus on weevils (Curculionoidea; Fig. 58.7). Meanwhile, the SMNK has a large collection of this

Fig. 58.7 Drawer with part of the weevil collection (*Eulophus* spp.)

group, mainly of Indo-Australian species, especially from New Guinea. Since 2005, a refrigerated collection stored in absolute ethanol suitable for DNA extraction was established. So far, it was mainly used for studies on the weevil genus *Trigonopterus* (Riedel et al. 2013b, 2014); specimens were extracted nondestructively and vouchers transferred to the dry collection. Many of these DNA vouchers became or are in the process of becoming type specimens.

58.5.4.3 Hymenoptera

The Hymenoptera collection comprises approx. 200,000 prepared specimens. Ants (Formicidae) constitute with about 110,000 prepared specimens the largest portion of the collection, with further hundreds of thousands of specimens preserved in ethanol. The foundation of the ant collection was laid by H. Leininger (1920–1936), but large parts of his personal collection were sold after his death to the natural history museum of Kassel. Important regional additions came from H. Nowotny. The global coverage of today's collection was provided by additions of Manfred Verhaagh, Krzysztof Rościszewski, Jochen Ketterl, Jochen Bihn, Andreas Schulz, Lubos Dembicky, and Stefanie Berghoff. Ant taxa from Europe, South America, and Southeast Asia are especially well represented. In 2014, the ant collection started to be newly arranged by W. Hohner and M. Verhaagh in small unit trays of various sizes and in systematic order, according to the system of Bolton (1995, 2003), with updates from http://www.antcat.org/. Genera and species as far as determined were registered in a spreadsheet.

For wasps and bees, the geographical focus of the collection is Southwestern Germany, but there are also types from Afghanistan (expedition of H.-G. Amsel). Important parts of the collection form the collections of Walter Stritt (mainly Tenthredinidae and Sphecidae), Hubert Hilpert (Ichneumonidae, Diapriidae), Fritz Brechtel, and Julian Fricke (Apidae).

58.5.4.4 Mantodea

The Mantodea collection currently comprises nearly 13,000 specimens in approx. 500 drawers with 1486 species/subspecies and 300 genera/subgenera, thus encompassing a major portion of the world fauna of praying mantises (approx. 453 genera/subgenera and 2452 species/subspecies). In addition to the adult insects, the collection also includes exuvia and oothecae (i.e., egg cocoons). Specimen data are completely digitized and managed in a collection database.

The Mantodea collection began with 185 specimens of H. Knipper's collection of Orthoptera (sensu lato), purchased in 1970. It grew into its present size largely by activities of Reinhard Ehrmann, collection manager in the entomological section from 1996 to 2008. He visited various countries (Dominican Republic, France, Italy, Yemen, Jordan, Kenya, Portugal, Spain, Sri Lanka, Turkey, and Venezuela) to collect Mantodea. In 1998 and 1999, SMNK purchased his collection of 5865

Fig. 58.8 Part of the Mantodea collection

specimens in 150 drawers. Subsequently, the Mantodea collection was constantly enlarged by acquisition and donations (Fig. 58.8).

58.5.4.5 Other Insects

Besides the large collections described above, the collections of some other orders have also grown to considerable significance:

The collection of Odonata (dragonflies and damselflies) comprises approx. 7000 specimens of approx. 800 species from all biogeographic regions. It is divided into a collection of pinned specimens in drawers and a collection of papered specimens arranged in an index-card file cabinet. Both parts were recently rearranged and databased on species level. The main part comes from the donations of K. Kormann and G. Jurzitza.

The Orthoptera (crickets, katydids, and grasshoppers) are represented by more than 10,000 specimens, comprising 259 species in 53 drawers. Among them are collections made by H. Knipper in the 1960s and 1970s in Europe and East Africa. Later, material was added from Yemen, Turkey, Costa Rica, Ecuador, and Malaysia. The collection of regional Orthoptera provided important data to the book of

Detzel (1998) on the Orthoptera of Baden-Württemberg. The identified material of the collection was digitally databased on species level.

The Heteroptera (true bugs) collection comprises approx. 25,000 specimens that were recently rearranged and digitally databased on species level. Almost half of the material comes from the collection of E. E. Wolfram. More than 8500 specimens from the state of Baden-Württemberg were revised and their data published in a catalogue (Heckmann 1996).

Diptera (flies) are strongly underrepresented in the insect collection relative to the diversity and ecological relevance of the group, and a major part of the collection is not yet sorted (approx. 30,000 specimens). A large proportion is formed by the collection of Ceratopogonidae of Peter Havelka. Within the other groups, the collection of hoverflies (Syrphidae), recently revised by Dieter Doczkal, is a remarkable exception.

The collection of Neuropteroidea, rearranged in 2007 by R. Güsten, consists of approx. 600 specimens of 205 species. Type material originates from a revision of species from the Near East (Hölzel 1972).

References

Angst R (1985) Kurze Geschichte der Zoologischen Abteilung. In: Rietschel S (ed) Vom Naturalienkabinett zum Naturkundemuseum 1785–1985. Landessammlungen für Naturkunde, Karlsruhe, pp 49–64

Auerbach M (1904) Führer durch das Großherzoglich Badische Naturalien-Kabinett zu Karlsruhe. Zoologische Abteilung. Braun'sche Hofbuchdruckerei, Karlsruhe, pp 1–100

Auerbach M (1917) Große Stiftung für das Großherzogliche Naturalienkabinett zu Karlsruhe. Deutsche Entomol Z "Iris" 31:58–60

Auerbach M (1937) Die badischen Landessammlungen für Naturkunde zu Karlsruhe. Beitr Naturk Forsch Südwdtl 2:16–21

Beck L (1963) Zur Ökologie und Taxonomie der neotropischen Bodentiere I. Zur Oribatiden-Fauna Perus. Zool Jb Syst 90:299–392

Beck L (1978) Zur Biologie eines Buchenwaldbodens 1. Einleitender Überblick und Forschungsprogramm. Beitr Naturk Forsch Südwdtl Band 37:93–101

Beck L (2000) Streuabbau und Bodenfauna in Wäldern gemäßigter und tropischer Breiten. Carolinea 58:243–256

Blasius W (1883) Ueber neue und zweifelhafte Vögel von Celebes. J Ornithol 31:8–162

Bolton B (1995) A new general catalogue of the ants of the world. Harvard University Press, Cambridge, MA, pp 1–504

Bolton B (2003) Synopsis and classification of Formicidae. Mem Am Entomol Inst 71:1–370

Braun M, Dieterlen F (eds) (2003) Die Säugetiere Baden-Württembergs, vol 1. Ulmer, Stuttgart, pp 1–687

Braun M, Dieterlen F (eds) (2005) Die Säugetiere Baden-Württembergs, vol 2. Ulmer, Stuttgart, pp 1–704

Brechtel F, Kostenbader H (2002) Die Pracht- und Hirschkäfer Baden-Württembergs. Ulmer, Stuttgart, pp 1–632

Bretfeld G, Gauer U (1994) Diagnostic description of the males of new *Sphaeridia* species (Insecta, Collembola) from South America. Andrias 13:113–136

Burkhardt U, Russell DJ, Buryn R, Decker P, Döhler M, Höfer H, Lesch S, Rick S, Römbke J, Trog C, Vorwald J, Wurst E, Xylander WER (2014) The Edaphobase Project of GBIF-Germany—a new online soil-zoological data warehouse. Appl Soil Ecol 83:3–12

Detzel P (1998) Heuschrecken Baden-Württembergs. Ulmer, Stuttgart, pp 1–580

Ebert G (1964) Die Macrolepidopteren-Sammlungen der Landessammlungen für Naturkunde in Karlsruhe und ihre Neugestaltung. Beitr Naturk Forsch Südwdtl 23:87–106

Ebert G (1977) Die Macrolepidopteren-Sammlungen der Landessammlungen für Naturkunde in Karlsruhe und ihre Neugestaltung (2. Teil). Beitr Naturk Forsch Südwdtl 36:247–260

Ebert G (ed) (1991–2005) Die Schmetterlinge Baden-Württembergs Band 1–10. Ulmer, Stuttgart

Ebert G (2004) Die Macrolepidopteren-Sammlungen des Staatlichen Museums für Naturkunde Karlsruhe und ihre Neugestaltung (3. Teil). Carolinea 62:129–144

Erhard CF (1802) Badisches Mineralreich, Erster Bericht. Magazin von und für Baden I/1

Franke U (1989) Katalog zur Sammlung limnischer Copepoden von Prof. Dr. Friedrich Kiefer. Carolinea Beiheft 5:1–433

Friedrich S (2014) The Peruvian short-tailed whip scorpion *Surazomus chavin*, rediscovery after three decades and first photographic documentation. Spixiana 37:131–134

Gmelin M (1879) Gmelin, Karl Christian. Allg Deutsche Biographie 9:271–272

Hauer J (1951) Die Bad. Landessammlungen für Naturkunde im Wiederaufbau. Beitr Naturk Forsch Südwdtl 10:90–95

Heckmann R (1996) Katalog der Wanzen aus Baden-Württemberg in der Sammlung des Staatlichen Museums für Naturkunde Karlsruhe (Insecta, Heteroptera). Carolinea Beiheft 10:1–146

Heinroth O (1898) Über den Verlauf der Schwingen- und Schwanzmauser der Vögel. Sitzber Ges Natforsch Freunde Berlin 1898:95–118

Höfer H (ed) (2001) Festband zu Ehren von Prof. Dr. Ludwig Beck mit Beiträgen zur Taxonomie und Faunistik von Arthropoden, Biologie und Ökologie südamerikanischer Bodenarthropoden und Ökotoxikologie der Bodenfauna. SMNK, Karlsruhe, pp 1–218

Hölzel H (1972) Die Neuropteren Vorderasiens. IV. Myrmeleonidae. Carolinea Beiheft 1:1–103

Hutterer R, Verhaagh M, Diller J, Podloucky R (1995) An inventory of mammals observed at Panguana biological station, Amazonian Peru. Ecotropica 1:3–20

Jacob-Friesen H (2015) Karoline Luise—Hessische Minerva und Vielwisserin von Baden. In: Jacob-Friesen H, Müller-Tamm P (eds) Die Meister Sammlerin Karoline Luise von Baden. Deutscher Kunstverlag, München, Berlin, pp 23–51

Knipper H (1964) Zoologische Reise 1961/63 nach Ostafrika der Landessammlungen für Naturkunde Karlsruhe. Beitr Naturk Forsch Südwdtl 23:7–30

Kretschmann C (2006) Räume öffnen sich—Naturhistorische Museen im Deutschland des 19. Jahrhunderts. Akademie Verlag, Berlin, p 365

Lauts J (1990) Karoline Luise von Baden—Ein Lebensbild aus der Zeit der Aufklärung, 2nd revised edn. C.F. Müller, Karlsruhe, p 481

Manegold A (2014) Two new parrot species (Psittaciformes) from the early Pliocene of Langebaanweg (South Africa) and their paleoecological implications. Ibis 155:127–139

Manegold A, White J (2014) Morphologische Anpassungen an das Klettern und Hacken beim Hüpfspecht *Nesoctites micromegas* (Picidae, Piciformes) dargestellt mit Hilfe der Computertomografie. Vogelwarte 52:324–325

Manegold A, Pavia M, Haarhoff P (2014) A new species of *Aegypius* vulture (Aegypiinae, Accipitridae) from the early Pliocene of South Africa. J Vertebr Palaeontol 34:1394–1407

Manegold A (2016) Rediscovery of the complete type series of the White-striped Forest Wallaby Dorcopsis hageni Heller, 1897 in the collections of the State Museum of Natural History Karlsruhe. Vertebr Zool 66:217–219

May W (1906) Zur Geschichte des Großh. Bad. Naturalienkabinetts in Karlsruhe (1751–1878). Verhandl Naturwiss Verein 19:2–19

Mayer G (1977) Beiträge zur Geschichte der Badischen Landessammlungen für Naturkunde in Karlsruhe. VII. Eine Paris-Reise Carl Christian Gmelins im Jahre 1810 und dessen

Beziehungen zu Gelehrten, Gärtnern, Naturalienhändlern und Besuchern dieser Stadt. Beitr Naturk Forsch Südwdtl 36:45–50

Mayer G (1978) Das Naturalienkabinett des Erbprinzen Friedrich (1708–1732). Beitr Naturk Forsch Südwdtl 37:29–34

Metzner H (1999) Die Springspinnen (Araneae, Salticidae) Griechenlands. Andrias 14:1–279

Oberdorfer E (1951) Zweihundert Jahre Landessammlungen für Naturkunde. Beitr Naturk Forsch Südwdtl 10:69–74

Réveillon F, Maquart PO, Höfer H (2014) Nota científica: first record of *Heterophrynus elaphus* (Pocock, 1903) (Amblypygi: Phrynidae) for Bolivia. Revista Ibér Aracnol 25:102

Riedel A, Sagata K, Suhardjono YR, Tänzler R, Balke M (2013a) Integrative taxonomy on the fast track—towards more sustainability in biodiversity research. Front Zool 10:15

Riedel A, Sagata K, Surbakti S, Tänzler R, Balke M (2013b) One hundred and one new species of *Trigonopterus* weevils from New Guinea. ZooKeys 280:1–150

Riedel A, Tänzler R, Balke M, Rahmadi C, Suhardjono YR (2014) Ninety-eight new species of *Trigonopterus* weevils from Sundaland and the Lesser Sunda Islands. ZooKeys 467:1–162

Ritzi M (1943/49) Die Badischen Landessammlungen für Naturkunde in Karlsruhe während und nach dem Kriege. Beitr Naturk Forsch Südwdtl 8:1–4

Römbke J (1989) Zur Biologie eines Buchenwaldbodens 12. Die Enchytraeen. Carolinea 47:55–92

Römbke J, Jänsch S, Roß-Nickoll M, Toschki A, Höfer H, Horak F, Russell D, Burkhardt U, Schmitt H (2012) Erfassung und Analyse des Bodenzustands im Hinblick auf die Umsetzung und Weiterentwicklung der Nationalen Biodiversitätsstrategie. UBA Texte 33(2012):1–386

Rookmaaker LC (1989) The zoological exploration of southern Africa: 1650–1790. Balkema, Rotterdam, p 368

Schalow H (1901) Beiträge zur Vogelfauna Centralasiens. Übersicht über die von Herrn Oberamtmann Dr. Holderer während der Durchquerung Asiens gesammelten Vögel. J Orn 49:393–456

Schillinger F (1973) Das Gebäude der Landessammlungen für Naturkunde in Karlsruhe. Beitr Naturk Forsch Südwdtl 32:21–29

Schmelz RM, Collado R (1999) *Enchytraeus luxuriosus* sp.nov., a new terrestrial Oligochaete species (Enchytraeidae, Clitellata, Annelida). Carolinea 57:93–100

Syré L (2015) Ein Glaspavillon im Nymphengarten. Badische Heimat 1(2):217–229

Trunko L (1985) Vom fürstlichen Naturalienkabinett zum modernen Naturkundemuseum. Kurze Chronik der Landessammlungen für Naturkunde. In: Rietschel S (ed) Vom Naturalienkabinett zum Naturkundemuseum 1785–1985. Landessammlungen für Naturkunde, Karlsruhe, pp 7–34

Index

L.A. Beck (ed.), *Zoological Collections of Germany*, Natural History Collections,
DOI 10.1007/978-3-319-44321-8

E

H

I

J

K

N

O

Y

Z

MIX
Papier aus verantwortungsvollen Quellen
Paper from responsible sources
FSC® C105338

If you have any concerns about our products,
you can contact us on
ProductSafety@springernature.com

In case Publisher is established outside the EU,
the EU authorized representative is:
Springer Nature Customer Service Center GmbH
Europaplatz 3, 69115 Heidelberg, Germany

Printed by Libri Plureos GmbH
in Hamburg, Germany